CONTINENTS AND SUPERCONTINENTS

# Continents and Supercontinents

JOHN J.W. ROGERS AND M. SANTOSH

OXFORD
UNIVERSITY PRESS

2004

Oxford  New York
Auckland  Bangkok  Buenos Aires  Cape Town  Chennai
Dar es Salaam  Delhi  Hong Kong  Istanbul  Karachi  Kolkata
Kuala Lumpur  Madrid  Melbourne  Mexico City  Mumbai  Nairobi
São Paulo  Shanghai  Taipei  Tokyo  Toronto

Copyright © 2004 by Oxford University Press, Inc.

Published by Oxford University Press, Inc.
198 Madison Avenue, New York, New York 10016

www.oup.com

Oxford is a registered trademark of Oxford University Press

All rights reserved. No part of this publication may be reproduced,
stored in a retrieval system, or transmitted, in any form or by any means,
electronic, mechanical, photocopying, recording, or otherwise,
without the prior permission of Oxford University Press.

Library of Congress Cataloging-in-Publication Data
Rogers, John J. W. (John James William), 1930–
Continents and supercontinents / John J. W. Rogers and M. Santosh.
p. cm.
Includes bibliographical references and index.
ISBN 978-0-19-516589-0
1. Earth—Crust. 2. Continents. 3. Continental drift.
4. Plate tectonics. I. Santosh, M. II.Title.
QE511.R59 2004
551.1′36—dc22      2003022676

Printed in the United States of America
on acid-free paper

# Preface

This book deals with continents, where more than 95% of the earth's population lives. We discuss the present state of knowledge on the origin and evolution of continents. We pose a number of questions and attempt their answers. Many of these are controversial, and potential clues have to be sought from future research.

The book begins with two chapters on the history of geologic thought about continents and the development of basic concepts of plate tectonics. We follow with discussions of the evolution of the rocks that make up continental crust and the development of cratons, which we regard as the building blocks of continents mortared together by orogenic activity. Two chapters describe the accretion of continents and the further aggregation of most of the world's continental blocks together to form supercontinents. The next two chapters discuss the five supercontinents that were proposed to have existed during earth history, ranging from a very hypothetical one in the Late Archean to the much better known Pangea at the end of the Paleozoic. Two chapters then describe the history of ocean basins and continents that developed after the breakup of Pangea, and we conclude the book with discussions of the effects of continents and supercontinents on climate and organic evolution.

This book was written for two purposes. Primarily it is intended as a text for upper-level undergraduate and beginning graduate courses. We assume that readers have a basic background in mineralogy, petrology, and structural geology, but we understand that many will not have had geochemistry or geophysics. For this reason, we include short general appendices (A to D) that explain how to use information obtained from studies in seismology, heat flow, paleomagnetism, and heavy isotopes.

We also hope the book is sufficiently informative and provocative that it can be used for general reading by professional geologists and perhaps as a basis for graduate seminars. Partly for this purpose, the book contains five special appendices (E to I) that provide detailed information that is not generally needed in an undergraduate text. These special appendices include details about several cratons, anorogenic magmatic suites, orogenic belts of Grenville age, belts with ages ranging from 2.1 to 1.3 Ga, and orogens commonly referred to as Pan-African or Brasiliano.

We have tried to keep the book current to the year 2002, and to include some references which were published as recently as early 2003. However, new information on every subject mentioned here continually becomes available through the publication of recent books and issues of journals. In order to provide more timely information after the publication of the book, we have established a website at http://www.gondwanaresearch.com/csbook.

Although the book contains over 700 references, many important papers had to be omitted. In order to keep the bibliography as short as possible, we have generally listed only recent publications on each topic, particularly choosing those that have extensive references to earlier publications.

The book has been greatly improved by discussions with several people, some of whom kindly reviewed different chapters. We are particularly grateful for the assistance given by Tovah Bayer, Joe Carter, Drew Coleman, Allen Glazner, Tony Hallam, Joe Meert, Brent Miller, Calvin Miller, Lisa Sloan, and Debbie Thomas. All of the line drawings and many of the photographs in the book were prepared by us, and we acknowledge people who provided other photographs in the captions. The University of North Carolina at Chapel Hill and the Kochi University are acknowledged for extending facilities. Miriam Kennard and the staff of the UNC Geology Library provided exceptional help with references. Rachel Cottone, Darrell Sandiford, and Yvette Thompson helped in preparation of the manuscript. Finally, we also thank our wives for understanding the amount of time that preparation of this book has taken from our family lives.

# Contents

1. Continental Drift—The Road to Plate Tectonics, 3
2. Plate Tectonics Now and in the Past, 13
3. Creation, Destruction, and Changes in Volume of Continental Crust Through Time, 31
4. Growth of Cratons and their Post-Stabilization Histories, 50
5. Assembly of Continents and Establishment of Lower Crust and Upper Mantle, 66
6. Assembly and Dispersal of Supercontinents, 85
7. Supercontinents Older than Gondwana, 100
8. Gondwana and Pangea, 114
9. Rifting of Pangea and Formation of Present Ocean Basins, 131
10. History of Continents after Rifting from Pangea, 147
11. Effects of Continents and Supercontinents on Climate, 159
12. Effects of Continents and Supercontinents on Organic Evolution, 176

Appendix A. Seismic Methods, 190

Appendix B. Heat Flow and Thermal Gradients, 194

Appendix C. Paleomagnetism, 199

Appendix D. Isotopic Systems, 204

Appendix E. Cratons, 211

Appendix F. Anorogenic Magmatic Suites, 217

Appendix G. Orogenic Belts of Grenville Age, 221

Appendix H. Orogenic Belts of 2.1–1.3-Ga Age, 227

Appendix I. Orogenic Belts of Pan-African–Brasiliano Age, 234

References, 243

Author Index, 273

Subject Index, 281

# CONTINENTS AND SUPERCONTINENTS

# 1

# Continental Drift—The Road to Plate Tectonics

Alfred Wegener never set out to be a geologist. With an education in meteorology and astronomy, his career seemed clear when he was appointed Lecturer in those subjects at the University of Marburg, Germany. It wasn't until 1912, when Wegener was 32, that he published a paper titled "Die Entstehung der Kontinente" (The origin of the continents) in a recently founded journal called *Geologische Rundschau*. This meteorologist had just fired the opening shot in a revolution that would change the way that geologists thought about the earth.

In a series of publications and talks both before and after World War I, Wegener pressed the idea that continents moved around the earth independently of each other and that the present continents resulted from the splitting of a large landmass (we now call it a "supercontinent") that previously contained all of the world's continents. After splitting, they moved to their current positions, closing oceans in front of them and opening new oceans behind them. Wegener and his supporters referred to this process as "continental drift."

The proposal that continents moved around the earth led to a series of investigations and ideas that occupied much of the 20th century. They are now grouped as a set of concepts known as "plate tectonics." We begin this chapter with an investigation of the history of this development, starting with ideas that preceded Wegener's proposal. This is followed by a section that describes the reactions of different geologists to the idea of continental drift, including some comments that demonstrate the rancorous nature of the debate. The next section discusses developments between Wegener's proposal and 1960, when Harry Hess suggested that the history of modern ocean basins is consistent with the concept of drifting continents. We finish the chapter with a brief description of seafloor spreading and leave a survey of plate tectonics to chapter 2.

## Prevailing Tectonic Ideas in 1912

Although Wegener is credited with first proposing continental drift, some tenuous suggestions had already been made. We summarize some of this early history from LeGrand (1988).

As soon as Europeans began to colonize the Americas, and geographers started to draw maps of the Atlantic Ocean, some scholars noticed similarities between the western and eastern borders of the ocean and suggested reasons for the similarity. Perhaps the most bizarre was in a 1858 book by Antonio Snider (Snider-Pellegrini) titled "La Creation et ces Mysteres devoiles" (The creation and its mysteries unveiled). Snider thought that the present form of the earth developed between the biblical creation and the biblical (Noachian) flood, with one of the last events being a massive catastrophe that opened the Atlantic Ocean, flooded the Old World, and left Adam's descendants alive only in the Americas.

Geologic reality is slightly more evident in other suggestions. One was that catastrophic separation of the moon from the earth would have left a void that the Americas rushed in to fill, leaving Europe and Africa behind. Origin of the moon from the earth in the early days of formation of the solar system is still a viable hypothesis, but it clearly occurred before the evolution of continents and had no effect on drift. In 1881 Eduard Suess proposed the name "Gondwana Land" (chapter 8) for the southern continents (further publication in English, 1904–1909). By 1910 the American geologist F.B. Taylor suggested that the world's mountain belts developed because of drift of continents toward the equator from both the north and the south.

Wegener's proposal was very different, and much better documented, than all of these earlier ideas. It also brought Wegener and his supporters into direct conflict with two prevailing, and somewhat contrasting, views of the evolution of the earth: "contractionism" and "permanentism" ("fixism"). Both of these views have now disappeared from geological thinking, but they need to be summarized for their historical importance.

Contractionism held that the earth was shrinking as it cooled down; it was supported by many geologists, including Eduard Suess, who found it compatible with his concept of Gondwana Land. Contractionism's supporters thought that contraction exerted sufficient pressure on the crust to generate mountain ranges as the earth's circumference became smaller. In order for the earth to remain a sphere, it would have to contract equally along all great circles, and proponents of the theory pointed out that this could be approximately accomplished by compression along two great circles perpendicular to each other. In the modern earth, compression is mostly along the rim of the Pacific Ocean and along the Alpine–Himalayan belt, which are nearly perpendicular, and some contractionists felt this proved their claim.

Contractionism suffered a major blow in the 1890s when radioactivity was discovered. Radioactive decay generated heat, and an earth containing radioactive elements was not cooling down as rapidly as one in which the only heat source was its initial accretion. Contractionism did not disappear immediately, however, because no one knew how much radioactive material the earth contained, how old the earth was, or how effectively heat was transferred from the interior to space. Therefore, some geologists still maintained that contraction caused by cooling of the earth could explain earth history much better than the movement of continents.

The second major concept attacked by drifters was "permanentism" ("fixism"), which held that the earth's continents and ocean basins had always occupied the position that they do now. It was also supported by numerous geologists, including James D. Dana, who, when he was not thinking about tectonics, established the basic classification of minerals that is still used today. Some continental margins had clearly undergone orogeny and possible enlargement, but it was assumed that this was simply minor growth that displaced small areas of ocean basins. Identical fossil species found on continents now separated by ocean basins may have crossed when temporary "land bridges" rose in the oceans. Sediments in mountain belts that appeared to have been derived by erosion of areas in adjacent ocean basins were likewise derived from temporary uplifts in the oceans. Some geologists still espoused permanentism even after seafloor spreading was proposed in 1960 (see below). They included Maurice Ewing, a pioneer in geophysical exploration of the oceans and founder of the Lamont Geophysical Laboratory. A.A. Meyerhoff and H.A. Meyerhoff, two of the world's leading petroleum geologists, found flaws in the concepts of drifting and plate tectonics into the 1970s (e.g., Meyerhoff and Meyerhoff, 1974).

The ideas discussed above are now discredited, but much that geologists had learned by the early 20th century remains an important part of the discipline. One of the most significant concepts is isostasy, which was developed in the 1800s when surveyors found that they could not produce an exact triangulated base map of India. When the mathematician J.H. Pratt, who was also the Anglican Archdeacon of Calcutta, worked through the data, he decided that the surveyors had not corrected accurately for the pull of the Himalayas on the plumb bob that they used to set their survey instruments horizontal. The surveyors had known that the mountains would attract the plumb bob and had calculated the effect by assuming that the mountains were simply an additional mass added to the surface of the Indian plains. What they did not know originally was that the Himalayas actually are not an additional mass but the same mass as the plains areas because the mountains extend farther down into the denser part of the earth (mantle) than the rocks underlying the plains.

Pratt suggested that the mass equivalence of mountains and plains could be explained by assuming that areas with different elevations all extended upward from a flat level in the subsurface, with higher elevations underlain by rocks of lower density than those in lower elevations (fig. 1.1). An alternative model proposed by G.B. Airy, the Astronomer Royal of the United Kingdom, suggested that mass equivalence could also be explained if most continental rocks had the same density and those in higher elevations extended farther down into the earth than those at lower elevations.

Ultimately mass equivalence became known as the "principle of isostasy" (fig. 1.2). The principle broadly states that most areas of the earth's surface are underlain by equal masses of rock, with elevation difference caused either by variation in density from one part of the earth to another or variation in thicknesses of rocks of equal density. For the Himalayas and other mountain belts, the accepted explanation soon became that all mountain ranges are high because of their roots (an Airy model). The difference between the elevations of continents and oceans is now known to be the result of both the

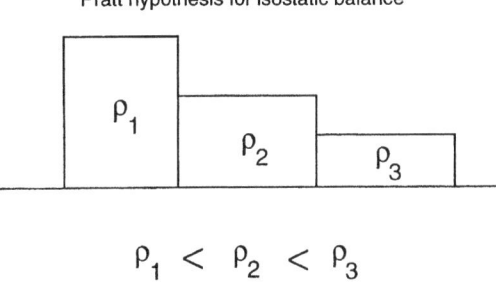

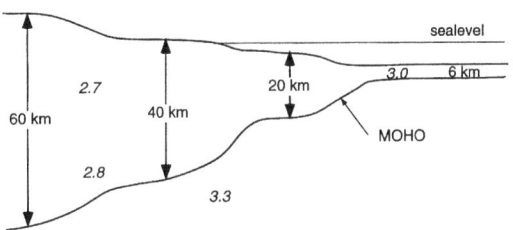

Figure 1.2. Cross section of mountains, continental plains, continental shelf, and ocean basin. Values in italics are typical densities in g/cc. The different depths to the Moho and the higher density of the thin oceanic crust than the thicker continental crust maintain isostatic balance throughout the section.

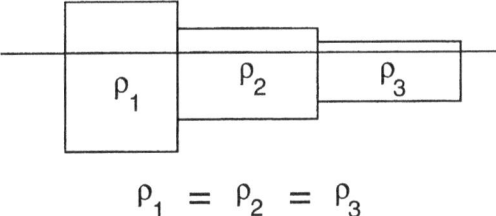

Figure 1.1. Comparison of Pratt and Airy models of isostasy.

higher density and the lesser thickness of oceanic crust (a combined Airy and Pratt model). The base of the crust is now referred to as the Moho, which we discuss below and in chapter 3.

Isostasy not only requires variations in density and thickness of rocks but also implies that surface rocks are "floating" on denser material within the earth. (A simple demonstration of this is to note that large ice cubes float higher above water than smaller ones but also displace more water by extending farther below the surface.) A further inference is that surface rocks can float only if there is some region in the earth's interior that is mobile enough to be displaced both vertically and laterally. This evidence of mobility was extremely important to people who believed in continental drift because any large-scale movements along the earth's surface would be impossible if the interior of the earth is completely rigid.

## The Controversy about Wegener's Proposal of Continental Drift

The proposal of continental drift brought immediate reactions both pro and con. Most geologists in the United States were opposed, while those in Europe and other parts of the world were more inclined to support the idea. We use this section to summarize a few of the principal arguments and refer readers who want a more complete survey to books by Marvin (1973), LeGrand (1988), and Oreskes (1999).

The concept of drift was supported by a considerable body of evidence. The easiest to recognize is the fit of North and South America to Europe and Africa around the Atlantic Ocean. Further support came from the concept of Gondwana, which originated when paleontologists realized that Paleozoic and Early Mesozoic fossils in South America, Africa, India, and Australia were all very similar although the continents are now separated by large ocean basins. Various explanations had been proposed for this similarity, but after continental drift was proposed it became clear that the simplest explanation for Gondwana was that the southern continents had once been joined together in a "supercontinent" and later drifted apart.

Once the opening of the Atlantic was proposed, it was possible to add Eurasia and North America to Gondwana and place all continental masses into one supercontinent called "Pangea" (chapter 8). The proposed configuration of Pangea (fig. 1.3) was supported not only by the fit of continents but also by more geological information. Paleontologists soon confirmed that animals that lived before Pangea rifted are similar not only in the Gondwana continents but also among most of the world's modern continents. Animals that evolved after rifting, however, are quite distinct from one modern continent to another. In similar fashion, mountain (orogenic) belts such as the Appalachians and Caledonides that were older than the time of drift could be correlated from

Figure 1.3. Configuration of Pangea.

one modern continent to another, but younger ones could not.

People opposed to continental drift found no merit in these arguments. Before we discuss some of the general objections we set the tone of the debate with quotes from publications of three of the leading geologists in the United States (in the interest of civility we omit several others):

> Can we call geology a science when there exists such difference of opinion on fundamental matters as to make it possible for such a theory as this to run wild?
> R.T. Chamberlain (1928)

> [Wegener's method] is not scientific, but takes the familiar course of an initial idea, a selective search through the literature for corroborative evidence, ignoring most of the facts that are opposed to the idea, and ending in a state of auto-intoxication.
> E.W. Berry (1928)

> the theory of continental drift is a fairy tale ...
> B. Willis (1944)

Many of the contrary arguments expressed in these and other papers were based on detail. Drift couldn't explain the location of some particular mountain range, the distribution of some fossil genus, the inferred climate of deposition of some deposits. The reply of Wegener and his supporters was that a sweeping new theory couldn't possibly include all observations that might ultimately prove to be exceptions to the concept. They also suggested that people who were concerned about these details were so interested in a few "trees" that they could not see the "forest" of geologic thought.

That reply led to more problems. People accused the "drifters" of getting an idea and then looking around for evidence to support it (a "deductive" approach to science). Science, so some people argued, is an "inductive" discipline, where grand theories are slowly established by the process of accumulating facts until the conclusion is obvious to everyone.

One attack on continental drift was by people who were concerned by the lack of mechanism by which the continents moved or a force to move them. Before information about the low-velocity zone and other mantle structures became available (see below), geologists assumed that continental drift meant continents moving across the top of the mantle, where no zone of weakness or mobility was apparent. It also meant that continents would have to plow into oceanic floor, either shoving it aside or somehow destroying it. Either mechanism required a very powerful force, and some people would not accept its existence although Joly (1925), Holmes (1928), and other geologists pointed out very early in the debate that the convection currents already presumed to exist in the mantle could move continents.

This debate illustrates a problem that earth scientists face all the time. The earth is large. It has a long and complicated history that no one was around to observe. Much of the evidence of past events has been destroyed by younger ones. With these difficulties, it becomes impossible to draw conclusions that can be "proved," and geologists are left with the necessity of doing "the best they can" with the limited information available to them.

Proponents of continental drift, therefore, resorted to the argument that the preponderance of evidence showed that drift had occurred, even if they were not able to provide all of the details. Furthermore, they said, if the evidence showed that the continents moved, then they did, even if the force to move them was unknown. It would be nice to figure out what the mechanism and force were, but their inability to do so did not mean that the continents hadn't moved.

As information about the earth accumulated during the 50 years after Wegener's first publications, the evidence for continental drift became stronger (see next section). Recognition that the mobile asthenosphere was overlain by a rigid lithosphere that contained both upper mantle and crust showed that continental drift probably meant drift of the lithosphere. More information about ocean basins showed that land bridges and other uplifts of oceanic lithosphere were impossible. Paleomagnetic information demonstrated the necessity of different continents moving separately. Recognition that the mantle must undergo convection showed that powerful forces for lateral movement existed in the upper mantle.

Even as recently as the 1970s, however, some earth scientists, particularly in the United States, did not accept the concept of continental drift (see above). This opposition began to disappear quickly after seafloor spreading was recognized and had almost completely vanished by the 1980s. In the next two sections we discuss the amount of information that had accumulated for geologists to use in 1960 and then the proposal of seafloor spreading.

## Information Available in 1960

Information available to earth scientists in the early 1900s rapidly expanded by the continuation of traditional geologic studies and also by the development of new techniques for studying the earth. One of the most productive was interpretation of seismic waves that had passed through the earth's interior (appendix A). The first advance came in a series of papers in the early 1900s, when A. and S. Mohorovicic used seismological data to measure the depth of the crust and the differences between crust and mantle (figs. 1.2, 1.4). They found that the basic distinction between crust and mantle is a rapid increase in the velocities of seismic waves downward across a "discontinuity" at the base of the crust. Velocities of P waves are in the range 6.5–7 km/sec in the lower part of the continental crust and are about 8 km/sec in the mantle just below the discontinuity. The depth to the discontinuity in continents varies from about 30 to 40 km in most regions but is commonly greater than 50 km under mountain belts. In ocean basins the discontinuity is much shallower, commonly about 10 km, with a change of P-wave velocities from about 7 km/sec in the crust to 8 km/sec just below the discontinuity.

This "Mohorovicic discontinuity," which we fortunately shorten to the "Moho," firmly established the distinction between crust and mantle and clarified numerous relationships that had only been vaguely known in the past. It verified the presence of mountain roots and the relative thinness of oceanic crust, both of which had been surmised from the concept of isostasy. When it was demonstrated that P-wave velocities are roughly proportional to rock density, the higher density of oceanic crust than that of the upper part of continental crust could also be confirmed. It also became possible to demonstrate that material with the seismic velocities of continental crust extended outward from the shoreline to approximately the edge of continental shelves, requiring that the pre-drift fit of continents be determined at shelf edges rather than the continental margins seen on ordinary geographic maps.

In the early 1900s seismic studies were beginning to outline the deep interior of the earth. The existence of a heavy region somewhere within the earth was already known from gravity studies, which revealed that the earth had an average specific gravity of 5.5. Because the specific gravities of almost all surface rocks are less than 3, this meant that some part of the earth's interior had specific gravities much higher than 5.5, and most earth scientists had concluded that this was probably caused by a "core" of iron and nickel. The question was exactly where this region is.

Beno Gutenberg and his colleagues began working on this problem in the early 1900s, but it wasn't until 1934 to 1936 that a series of papers provided the first definitive information (Gutenberg and Richter, 1934–1936). The basic observation was that arrival time of P waves at different distances from the location (focus) of an earthquake increases progressively up to a distance of $103.5°$ from the earthquake and then jumps to a much longer time beyond that distance (see appendix A for a more complete explanation). This established that a discontinuity ("Gutenberg discontinuity") at a depth of 2900 km separated a core with lower P-wave velocities from an overlying mantle with higher velocities (fig. 1.4). This decrease in velocities into denser material also indicated that at least the outer part of the core is liquid.

Further seismic investigations of the mantle soon revealed more structural details. The one most clearly related to continental drift is the presence

8  Continents and Supercontinents

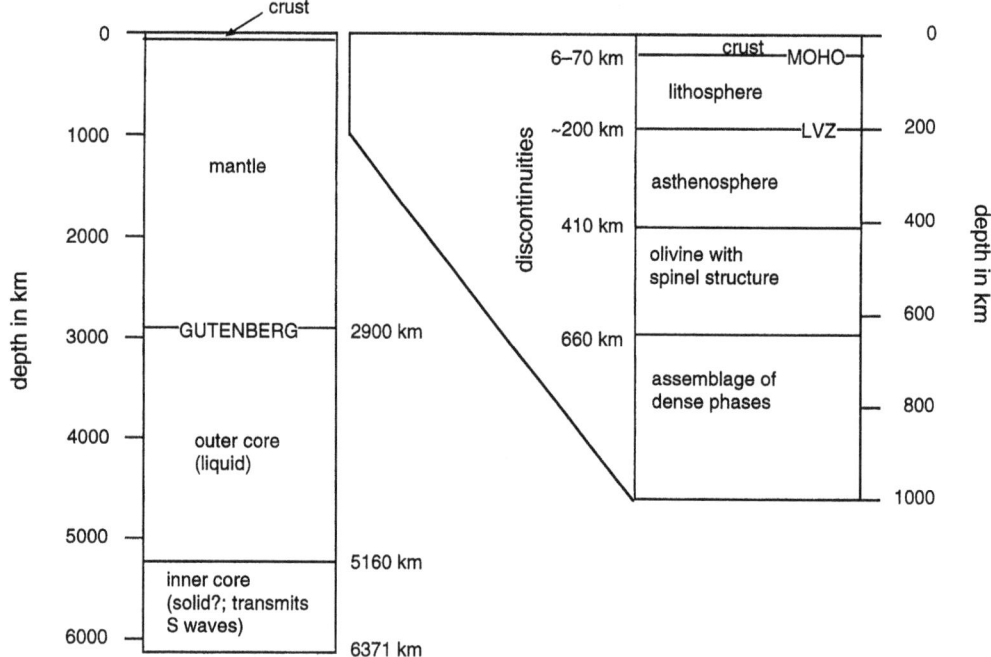

Figure 1.4. Cross section of earth from crust to core.

of a "low-velocity zone" (LVZ) where velocity either decreases slightly downward or, at least, does not increase as rapidly as it does above or below the LVZ (fig. 1.4). This lower velocity may indicate the presence of a partial liquid or simply an increased plasticity, but either interpretation leads to a designation of a mantle with a mobile asthenosphere overlain by a rigid lithosphere consisting of both upper mantle and crust. The LVZ occurs at depths of about 150–200 km in ocean basins, and a zone of mobility presumably also exists under continents, although its depth and nature are still controversial. Regardless of its exact nature, however, the presence of an LVZ strengthens the idea that rigid lithosphere is underlain by a zone mobile enough to permit lithospheric movement.

Other seismic discontinuities in the mantle were found to occur at depths of about 410 km and 660 km, where there are small and abrupt increases in P-wave velocity downward (fig. 1.4). At first, they may not seem to be closely related to continental drift, but recently it has been proposed that the 660-km discontinuity is a zone along which subducted lithosphere accumulates, and its episodic collapse into the lower mantle is related to superplumes. Because both of these processes may be associated with the formation of supercontinents, we return to this issue in chapter 6.

Seismic studies of the earth yielded information on not only its internal structure but also an increasingly accurate view of the locations of earthquakes. More information on the surface locations (epicenters) of earthquakes steadily reinforced the idea that large, roughly equidimensional, areas of the earth's crust are stable, and deformation occurs only in linear belts between these stable blocks. Whereas this had been known for continents since the beginning of the 20th century, the new information expanded the concept of stability and linear mobile belts to the ocean basins. These areas were later referred to as plates and plate margins (chapter 2).

Information on the depths of earthquakes led to an unexpected result first described by K. Wadati in 1929 (see Wadati, 1940). He showed that some seismically active areas contain earthquakes that extend from the surface to depths of several hundred kilometers along a plane that dips from the ocean margin under a continental margin or island arc. The idea was expanded by Hugo Benioff in 1954, who demonstrated that the dips of these planes vary from margin to margin. These zones have commonly been referred to as "Benioff zones," although more accurately they should be "Wadati–Benioff zones." After the terminology of plate tectonics was refined, they have been known as "subduction zones" (chapter 2).

The identification of the earth's major layers and their properties was combined with newly emerging information about the earth's heat budget to identify another major earth process—mantle convection currents (appendix B). Geologists have always known that the interior of the earth is hot, partly because temperatures in deep mines are high and also because there needs to be a heat source to melt the lavas erupted by volcanoes. Before the discovery of radioactivity in the 1890s, however, geologists had assumed that the internal heat of the earth was merely a remnant of the heat generated when the earth accumulated. This implied that the earth was rapidly cooling down and could not be very old if the interior was still hot.

The presence of radioactive elements in the earth, however, meant that the earth had a continual heat source and was cooling down much more slowly than previously thought (the rate of cooling is highly controversial and well beyond the scope of this book). Realization that the earth's interior is very hot led to early proposals of convection currents (see above), and more evidence gave additional credibility to this idea. As soon as seismic information confirmed that the mantle consists mostly of material with the composition of olivine, as petrologists had long presumed, it was easy to calculate the maximum temperature gradient the mantle could have without becoming unstable and undergoing convection (the thermodynamic reason for this instability is discussed in appendix B). Calculations yield some uncertainty because the properties of the deep mantle are not well known, but the equilibrium gradient cannot be higher than 0.3°C/km and may be as low as 0.1°C/km.

A gradient of 0.3°C/km for a mantle 2900 km thick means that the base of the mantle could be only ~900°C hotter than the top in order to remain stable. Studies of metamorphic rocks and inclusions in lavas had already shown petrologists that the base of the continental crust had a temperature on the order of several hundred degrees, for a temperature gradient of a few tens of degrees per kilometer. And if the top of the mantle has a temperature of a few hundred degrees, then the bottom of the mantle could be only slightly hotter than about 1000°C.

For several reasons this temperature is impossible. The outer part of the core could not be liquid at this temperature. Volcanoes erupting lavas at temperatures of 1000°C and higher could not originate at the bottom of the mantle. Even a small amount of radioactive material in the mantle would generate higher temperatures. Estimates of actual temperature gradients in the mantle were on the order of 1°C/km, yielding temperatures at the base of the mantle of several thousand degrees (consistent with modern estimates).

These temperature estimates indicate that the mantle could be stable only if it is rigid and unable to flow and/or if its composition varies so much with depth that the equilibrium gradient is much higher than has been calculated. Both conditions seemed unlikely, and geologists concluded that the mantle is unstable and is continually undergoing convective overturn (discussion of work by Joly and Holmes in preceding section). When this was first realized, the next problem was locating the convection currents. Placing the upwelling limbs at mid-ocean ridges was relatively easy. Mountain belts might be places where two colliding currents generated the horizontal compression that caused orogeny. But where are the downgoing limbs of the convective cells? They could be confidently located under Wadati–Benioff zones where those zones exist, but many ridges, such as the mid-Atlantic ridge, are in oceans whose margins show no deformation. In these areas, the location of the downgoing limbs was unclear. This problem is still unsolved, and most geologists simply assume that downgoing limbs are somewhere under continents or their margins, although firm evidence is lacking.

About the middle of the 20th century, geologists discovered that rocks containing significant concentrations of iron might be used to show the position of the magnetic poles at the time the rocks formed (appendix C). Combined with the ages of the rocks, these magnetically determined pole positions yielded results that were originally very hard to explain because the magnetic orientations in the rocks usually did not point to the present magnetic poles.

The first conclusion was that the magnetic poles must have moved since the rocks formed, a process called "polar wandering" ("true polar wandering"). One explanation was that the magnetic axis of the earth is independent of the rotational axis, and it is only an accident that the present magnetic poles are near the geographic poles. This possibility was quickly rejected when earth scientists realized that the magnetic field is almost certainly generated by rotation of the earth's core, and although there is continual movement of the field, it is unlikely ever to have produced magnetic poles far from the geographic ones.

A better explanation became clear when paleomagneticists working in different continents found different pole positions for rocks of the same age. The only explanations for this observation were either that the earth's magnetic field was not a dipole, which seemed impossible, or that the rocks had moved since their magnetic orientations were imposed on them. This led to the realization that the magnetic orientations measured in rocks do not show the real magnetic poles but only "apparent" ones, and "polar wandering curves" were then called "apparent polar wandering curves" (APW curves; further explanation in Appendix C).

One of the earliest observations by S.K. Runcorn in 1956 showed that APW curves for North America and Europe during the Mesozoic and Cenozoic started about 5000 km apart and gradually approached the present North Pole as the measured rocks became younger (fig. 1.5). Because the North Atlantic Ocean is about 5000 km wide, the best explanation for these curves was that the two continents were joined early in the Mesozoic, when the ocean started to open, and continually moved farther apart since then. This movement caused all apparent poles older than Early Mesozoic to be displaced by 5000 km, whereas younger apparent poles were displaced only by the amount of opening since the age that they represent.

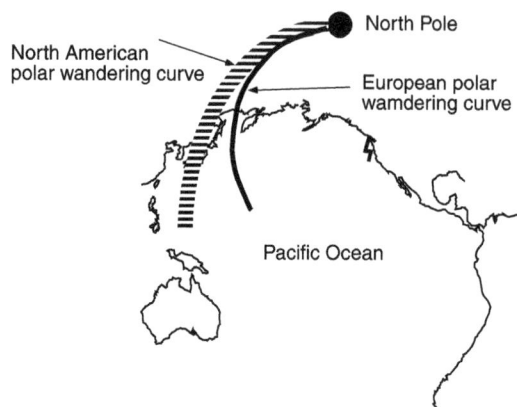

Figure 1.5. Apparent polar wandering (APW) curves for North America and Europe for the past 500 million years. Cambrian pole positions for both continents are in the Pacific Ocean. The curves become closer toward the present and meet at the current North Pole. For more detailed information see Runcorn (1956, 1962).

## Seafloor Spreading Proposed in 1960

In 1959 Harry Hess began to talk about a new idea for the evolution of ocean basins. He described it in a research report submitted that same year, and then Robert Dietz amplified the idea in a paper published in 1961. In 1962 Hess published one of the most famous papers in the history of the earth sciences. The title was "History of ocean basins," and Hess referred to it as "an essay in geopoetry" because, although his idea answered many questions then bothering geologists, at that time there was no specific proof that seafloor spreading really occurred.

The concept of seafloor spreading required a fundamental change in the way that earth scientists regarded the earth (fig. 1.6; Dietz, 1961). Midocean ridges had been considered merely to be the places where convection currents rose and then moved away beneath immobile ocean crust. The new idea stated that ridges are not just the sites of rising limbs of convection currents but are also the place where new oceanic lithosphere is created. This new lithosphere then moves progressively outward to make room for younger material, thus continually enlarging ocean basins. Ocean basins that expand in this way are places where continents drift away from each other.

In an earth with a constant radius, oceanic lithosphere must be destroyed at the same rate at which it forms. The only way for this to happen is for lithosphere to reenter the mantle, and the obvious place for this to occur is the Wadati–Benioff zones (soon named "subduction" zones) that dip downward to depths of several hundred kilometers. Seafloor spreading requires these zones to be places not just of minor fault slippage but places where large volumes of lithosphere descend to the depth of the deepest earthquakes or possibly beyond.

Simultaneous construction and destruction of oceanic lithosphere meant that some oceans, or portions of them, are expanding while other oceans are becoming smaller. The Atlantic Ocean is almost entirely an opening ocean because the only subduction zone is along the Lesser Antilles, whereas the Pacific Ocean is completely surrounded by subduction zones and is contracting. Geologists had long known that "Atlantic-type" margins without earthquakes were very different from "Pacific-type" margins, with earthquakes and volcanism, but the reason for this difference was not clear until seafloor spreading was understood.

Creation and destruction of oceanic lithosphere explained another problem that had bothered geologists. Early exploration of the oceans had retrieved some rocks from the ocean floor, but none of them geologically very old. Furthermore, early seismic measurements of the thickness of oceanic sediments had shown that the sediment load of rivers draining continents could easily fill the basins in a few tens of millions of years. These

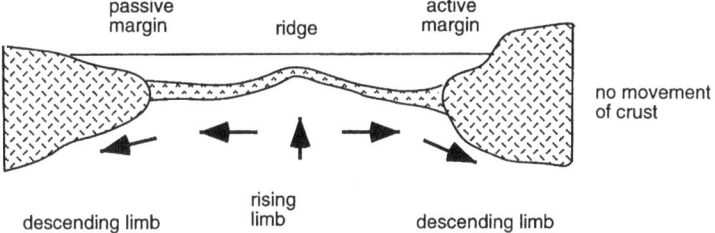

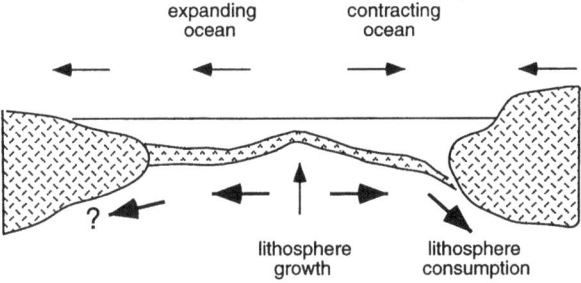

Figure 1.6. Convection beneath ocean basins as understood before and after seafloor spreading.

observations seemed impossible if ocean basins were as old as continents, and several explanations had been proposed. One was that old rocks at depth below the seafloor had never been observed either by dredging or seismic studies. Another possibility was that past sedimentation rates were much slower than modern ones. Both the problem and the explanations disappeared when measurement of spreading rates and further ocean dredging and drilling showed that the oldest oceanic lithosphere was no more than about 200 million years old. The absence of older rocks in the oceans simply results from their being drawn down in subduction zones and recycled into the mantle.

The concept of seafloor spreading needed to be tested, and the simplest method was to demonstrate that the ocean floor becomes progressively older away from ridge crests. This was accomplished by dredging or drilling rocks from exposed ridges, measuring their ages radiometrically, and observing young crust at the ridge crest and older crust farther away. Conclusions based on ages were soon augmented by showing that magnetic stripes are symmetrical on either side of ocean ridges and have patterns (finger prints) that can be correlated from one ridge to another (further explanation in caption for fig. 1.7; Vine and Matthews, 1963; Morley and Larochelle, 1964; appendix C).

A second demonstration of the validity of seafloor spreading came from studies of the offsets of oceanic ridges along spreading centers. All ridges are cut by perpendicular faults spaced an average of ~100 km apart. They are referred to as "transforms" between ridge crests and fracture zones beyond the ridge crests (further explanation in chapter 2). The transforms offset ridge crests by up to several hundred kilometers and appear to be ordinary strike-slip faults. If ridges are spreading, however, then blocks moving past each other along the transform between the offset ridge crests must actually be moving in the opposite direction of the offset of the crests (fig. 1.8). This direction of movement was demonstrated by Sykes (1967) from studies of first motions in earthquakes along these inter-ridge sections.

Almost all earth scientists accepted seafloor spreading within 10 to 20 years after it was proposed. With both continental drift and seafloor spreading firmly established, the next step was to provide a conceptual framework and clarify terminology to describe major earth processes. We call it "plate tectonics" and discuss it in the next chapter.

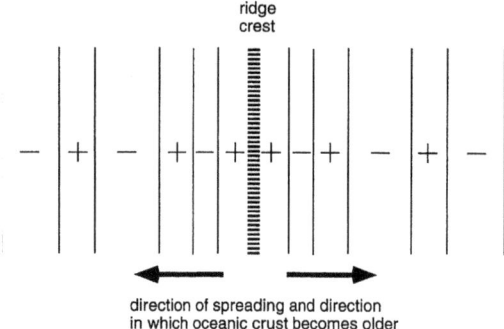

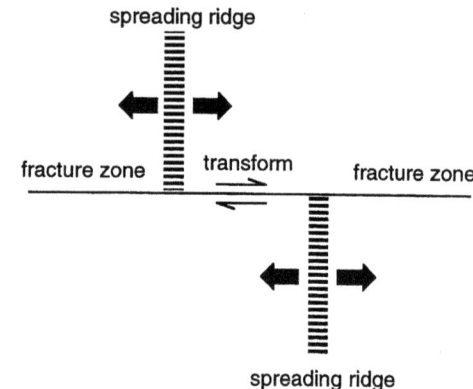

Figure 1.7. Magnetic stripes around ridges. Plus signs indicate enhancement of total magnetic intensity because magnetic north in underlying rock is oriented in the same direction as present magnetic north. Minus signs indicate reduction of total magnetic intensity because magnetic north in underlying rock is oriented in the opposite direction of present magnetic north. The alternation of plus and minus results from episodic reversal of magnetic north and south, and the symmetric pattern around ridges results from seafloor spreading (further explanation in text and appendix C).

Figure 1.8. Movement on transform faults. Spreading on ridges on opposite sides of the transform causes the movement shown by the arrows along the transform as the two ridges move apart (opposite to the movement on the transform itself). Continued movement on either side of the fault zone beyond the two ridges forms a fault ("fracture zone") with normal strike-slip motion.

## Summary—The Problem of New Ideas

In less than one century, continental drift, seafloor spreading, and the broad concepts of plate tectonics have come to dominate the thinking of geologists in place of the old ideas of contractionism and permanentism. Progress toward this understanding has been slow, partly because new factual information was needed to support new ideas and partly because of the inherent difficulty of introducing new ideas to an established profession. Some scholars are understandably cautious because many new ideas have evaporated under scrutiny by someone other than the original author. Other scholars, however, resist new ideas because their egos are tied to old ones, and they do not want to see a lifetime of research proven wrong. Regardless of the reasons, all of the argumentation pro and con is usually good for the profession because it strengthens the ultimate product of research.

# 2

# Plate Tectonics Now and in the Past

The concepts known as plate tectonics that began to develop in the 1960s built on a foundation of information that included:

- The earth's mantle is rigid enough to transmit seismic P and S waves, but it is mobile to long-term stresses.
- The earth's temperature gradient is so high that convective overturn must occur in the mantle.
- The top of the mobile part of the mantle is a zone of relatively low velocity at depths of about 100 to 200 km. This zone separates an underlying asthenosphere from a rigid lithosphere, which includes rigid upper mantle and crust.
- Seismic activity, commonly accompanied by volcanism, occurs along narrow, relatively linear, zones in oceans and along some continental margins.
- The zones of instability surround large areas of comparative stability.
- Ocean lithosphere is continually generated along mid-ocean ridges and destroyed where it descends under the margins of continents and island arcs. This causes oceans to become larger, but shrinkage of oceans can occur where lithosphere is destroyed around ocean margins faster than it is formed within the basin.
- Some of the belts of instability are faults with lateral offsets of hundreds of kilometers.
- Some continental margins are unstable (Pacific type), but others are attached to oceanic lithosphere without any apparent tectonic contact (Atlantic type).
- Different areas containing continents and attached oceanic lithosphere move around the earth independently of each other.

Most of this chapter consists of a summary of plate tectonics in the present earth, including processes along plate margins and the types of rocks formed there (readers who want more detailed information are referred to Rogers, 1993a; Kearey, 1996; and Condie, 1999). We also briefly discuss plumes and then finish with a word of caution about interpreting the history of the ancient and hotter earth with the principles of modern plate tectonics.

## Plates and Plate Margins

Starting from the body of continually expanding information summarized above, numerous earth scientists in the 1960s and 1970s began to establish a conceptual framework that would organize scientific thinking about the earth's tectonic processes. This required a new terminology, and it arrived rapidly (Oreskes, 2002). Geologists decided to call the stable areas "plates" and the unstable zones around them "plate margins." Thus, the concept became known as "plate tectonics."

Plates are essentially broad regions of lithosphere, although the failure to detect low-velocity zones under many continents leaves unresolved questions. Because lithosphere is known to be relatively cool and rigid, geologists realized that it was capable of transmitting stresses over long distances without internal deformation. This presumably explains the restriction of tectonic activity to the margins of plates.

Two terms became widely used for the two different types of continental margins. Atlantic-type margins were referred to as "passive," with the continental margin consisting of thin continental crust formed during continental rifting and joined to the oldest seafloor in the adjacent ocean. Pacific-type margins became known as "active" margins because of their tectonic activity. The join between continental and oceanic lithosphere in active margins is almost everywhere a trench where oceanic lithosphere is drawn downward into the mantle.

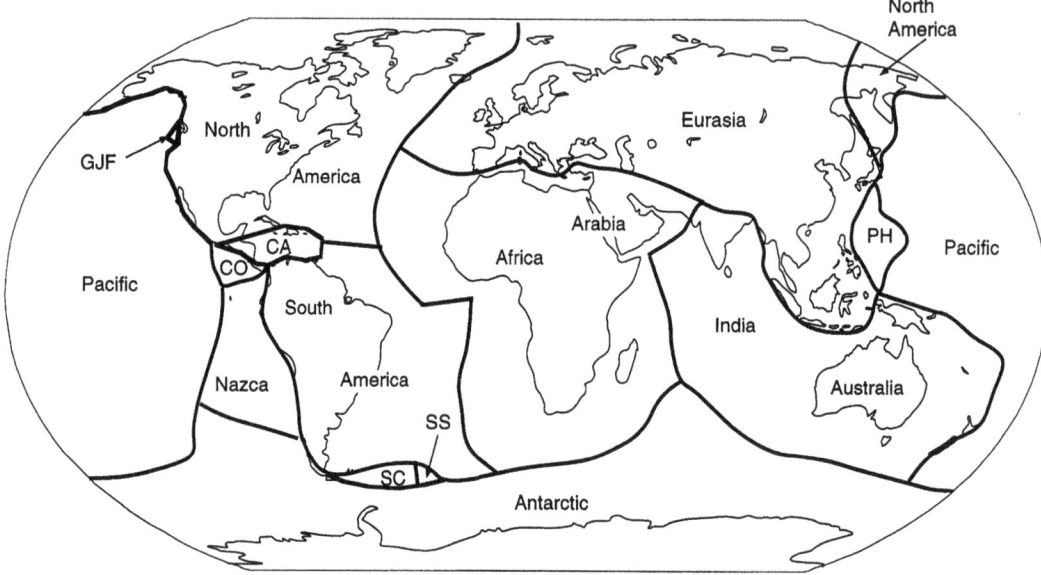

Figure 2.1. Major plates on the present earth. The North American plate extends into eastern Asia and is separated from the Eurasian plate by a boundary through the Arctic Ocean (chapter 9). Abbreviations are: GJF, Gorda and Juan de Fuca; CA, Caribbean; CO, Cocos; SC, Scotia; SS, South Sandwich; PH, Philippines.

With this distinction established, most plates that contain continents also contain some oceanic lithosphere attached to a passive margin. For example, the South American plate extends west from the mid-Atlantic ridge to the active margin with the Pacific Ocean, thus including both continent and the western South Atlantic Ocean. The Pacific plate, however, includes much of the Pacific Ocean and consists solely of oceanic lithosphere. Figure 2.1 shows the present distribution of plates on the earth's surface.

Plate margins are classified into three types. Accreting (rifting) margins include mid-ocean spreading centers and areas of rifting within continental areas, each of which is discussed separately below. Destructive (subducting) margins also occur in two different situations, and we include separate sections for subduction that forms island arcs within ocean basins and subduction along continental margins. The third type of margin is transform (strike-slip) and includes most of the world's longest faults. We conclude this section with a very important feature where three margins intersect to form a "triple junction."

### Accreting Margins

"Accreting" margins occur along mid-ocean ridges, where new oceanic lithosphere is formed. "Rifted" margins, where two plates separate, are primarily oceanic spreading ridges, and whether zones of rifting within continents, such as the East African rift system, should be considered plate margins is simply a matter of definition.

The top of all mid-ocean ridges is at a depth of ~2.5 km (fig. 2.2). This is the isostatically balanced elevation of exposed asthenosphere as new lithosphere forms and starts to move away from the ridge crest. As this movement continues away from the source of heat at the ridge crest, rocks become progressively cooler and the elevation of the ocean floor decreases to the ~5-km depth of abyssal plains (fig. 2.2). This decrease in elevation

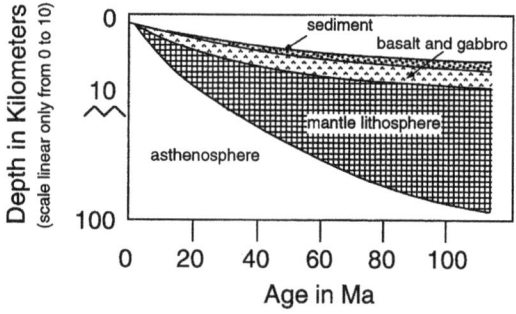

Figure 2.2. Features of spreading centers.

Figure 2.3. Extension of the Reykjanes spreading center from the North Atlantic into Iceland, where it forms a valley. The photograph is taken from the western shoulder of the ridge, where some of the basalt has pulled away from the shoulder and fallen down into the valley toward the right (photo by J.J.W.R.).

is almost entirely the result of cooling, which causes asthenosphere with a relatively low density to convert to lithosphere with higher densities, leading to a general "shrinkage" of the rock column.

Because cooling causes the elevation of ridges to be proportional to age of the lithosphere, ridges with different spreading rates have different topographic profiles. Fast spreading centers such as the East Pacific Rise, with spreading rates of ~7 mm/yr on each side of the ridge ("half spreading rates") remain at high elevations for hundreds of kilometers beyond the crest, but ridges with low spreading rates, such as the mid-Atlantic ridge with a half rate of ~1 mm/yr, show rapid decreases in elevation away from the ridge crest and have sharp topographic profiles. Ridges with slow spreading rates also have a better defined valley at the top of the ridge, where the valley is surrounded by rift "shoulders" uplifted as the new lithosphere pulls away from the ridge crest (fig. 2.3).

The transform faults that separate segments of ridge crests (chapter 1; fig. 1.8) extend beyond the crests as fracture zones that cause strike-slip offsets between sections of the ridges. Differences in ages between rocks on either side of the fracture zones cause elevation differences that may be as high as one kilometer on very large fractures. All of the known fracture zones are perpendicular to the ridge crests, which distinguishes them from the faults that offset segments of linear rift systems within continents (see below).

The types of rocks that form along ridge crests are very similar to those exposed in suites of rocks called "ophiolites" where they have been emplaced on land by tectonic processes. We discuss them here, starting with the history of their recognition and also pointing out that the exact relationship between exposed ophiolites and mid-ocean ridges is controversial.

Ophiolites   One of the most influential papers in the history of geology was published by G. Steinmann in the Proceedings of the 14th International Geological Congress in Madrid (1927). Although he had discussed ophiolites and deep-sea sediments for more than 20 years, his 1927 paper was the first one to convince geologists of the tectonic significance of a sequence of rocks that came to be known as the "Steinmann trinity."

The ophiolites studied by Steinmann are in the Alps, and they became a standard for recognition of ophiolites worldwide. A complete ophiolite of this type consists of a basal peridotite (harzburgite), a series of gabbroic rocks and overlying basalts/spilites that are commonly pillowed, and a top suite of various seafloor sediments, including red clays and cherts (fig. 2.4). In many places, the

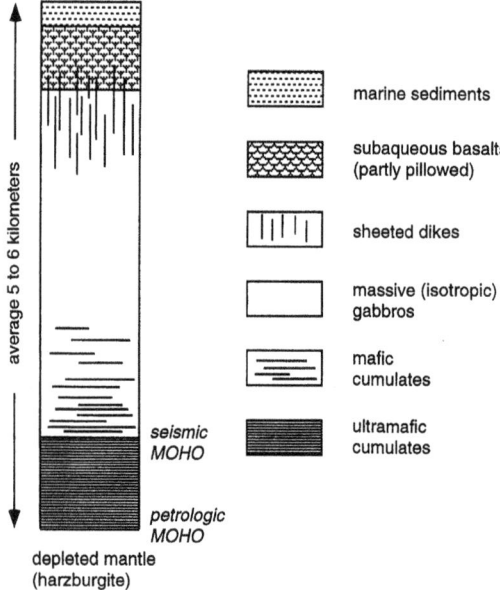

Figure 2.4. Cross section of ophiolite.

gabbros and the lower part of the basalt section are intensely intruded by vertical basalt dikes that are so closely spaced that they are referred to as "sheeted" (fig. 2.5). Both in the Alps and elsewhere, many ophiolites are highly metamorphosed, with peridotite converted to serpentine and all rocks intensely deformed. Many of the suites also have been tectonically fragmented ("dismembered") and contain only one or a few rock types.

Because all of the rock types in ophiolites also occur along oceanic spreading centers, the standard interpretation is that ophiolites are fragments of oceanic crust that have been thrust ("obducted") onto continents during orogeny (R.G. Coleman and Irwin, 1974). This simple explanation runs into problems involving the trace-element compositions of the different types of basalts. All basalts sampled from oceanic spreading centers have comparatively high concentrations of Zr, Ti, and other "high-field-strength" (HFS) elements. Basalts in ophiolites, however, have lower concentrations of HFS elements, similar to those concentrations found in

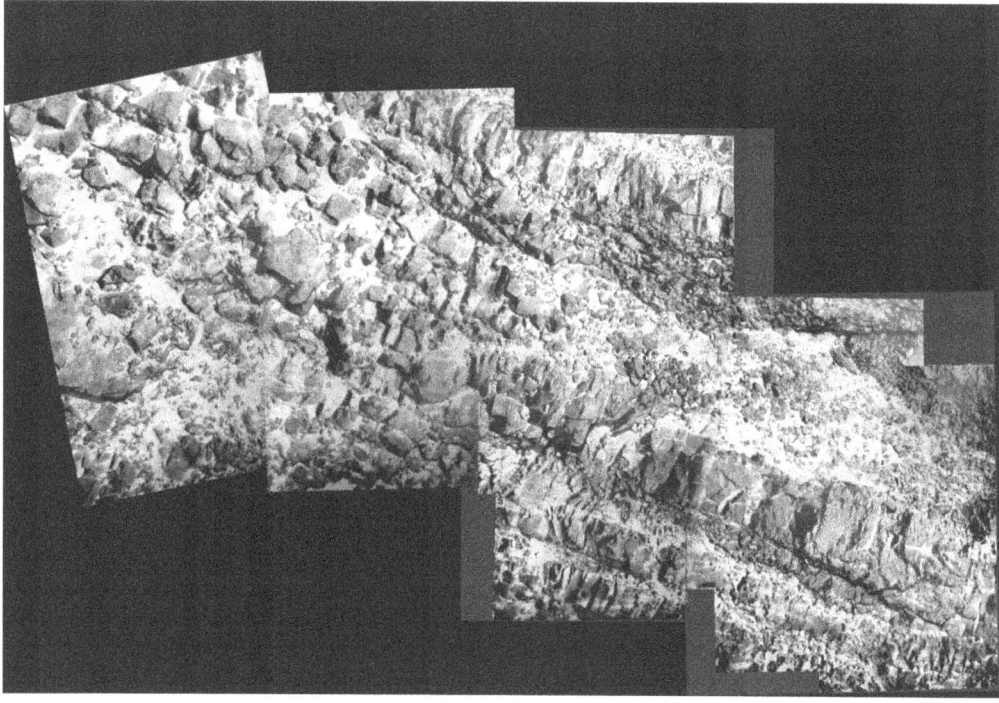

Figure 2.5. Basalt magmatism at oceanic spreading center. The picture looks north along the North Wall of the Hess Deep Rift along the East Pacific Rise and was taken ~2500 m below sealevel and ~1000 m below the seafloor at the top of the rift. The full spreading rate here is ~12 cm/yr, and the rocks are ~1 million years old. Gabbroic rocks toward the bottom of the picture are overlain by pillowed basalts. Dikes ~1–1.5 m wide appear as continuous tabular bodies cut by cross-joints. (Photo courtesy of Jeff Karson; Karson, 1999.)

subduction-zone basalts. The problem of ophiolites with rock types clearly formed along spreading centers and basalts with the compositions of rocks formed in subduction zones has been referred to as the "ophiolite conundrum" (Moores et al., 2000). Moores et al. explain this problem by pointing out that some spreading centers, particularly those that were probably the source of ophiolites in the Alps and other Tethyan orogenic belts (chapter 10), may have evolved in oceanic crust that had recently been involved in subduction along older continental margins and was already depleted in HFS elements when spreading centers were established.

## Intracontinental Rifts

Rifts within continents can be subdivided into three basic types. One is a single rift valley or a series of rift valleys oriented to form a linear rift system, and a second is a broad basin consisting of multiple rifts. The third type of intracontinental rift now occurs as passive margins, and their evolution from the other types of continental rifts is unclear.

Until the past few decades it was assumed that linear systems of rifts were simple grabens dropped down between two blocks that now form rift shoulders similar to those on oceanic spreading centers (fig. 2.3). Recent work, however, demonstrates that the linear systems consist of a series of individual rifts that are highly asymmetrical (they have a "polarity"), with one side uplifted to form a fault-bounded shoulder and the other side pulled away to form the rift basin (fig. 2.6). The individual rifts commonly alternate polarity along the rift system, causing the uplifted (mountainous) edge of the rift to alternate from side to side of the system. Where individual rift blocks overlap each other, they form a compressional region of uplifted rocks, and where individual rift blocks move past each other, they form "transfer" faults at angles to the linear rift system. The best developed example of a long linear rift system is the East African rift valleys, which we discuss briefly in chapter 10.

Extension in some continental areas develops broad basins that contain numerous individual rifts. Sections across the basins pass through several parallel half grabens, with the total extension of basin roughly equal to the sum of the extension on each of the individual rifts. A type example is the Basin and Range of the western United States and northern Mexico (fig. 2.7), and we describe similar examples in our discussion of the extension of northern Eurasia during the Mesozoic and Cenozoic.

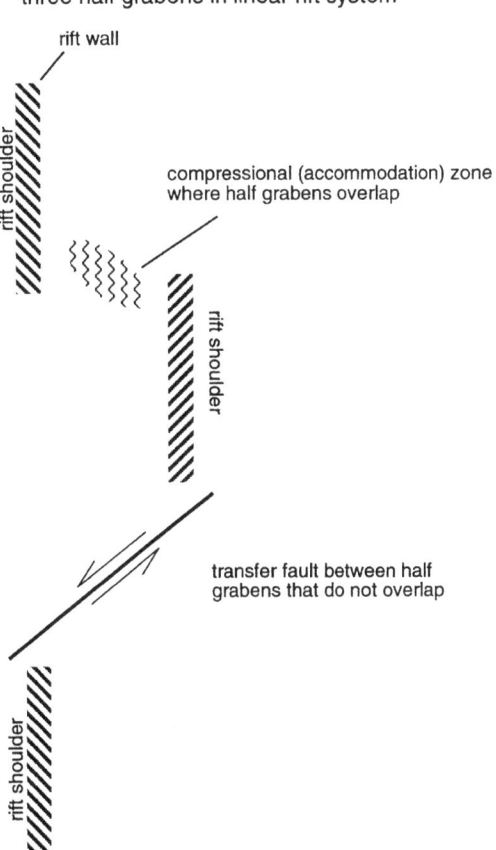

Figure 2.6. Features of linear rift systems.

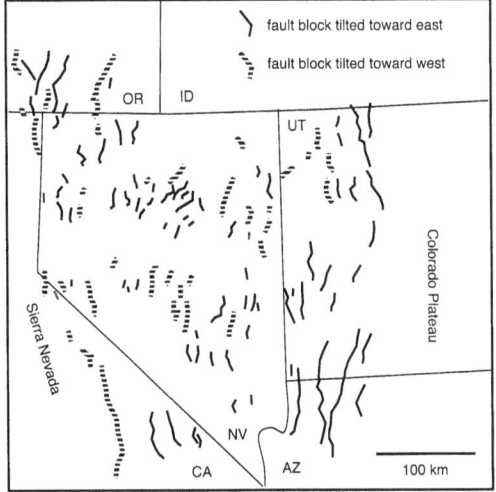

Figure 2.7. Map of Basin and Range in western United States (adapted from J. Stewart, 1978).

18  Continents and Supercontinents

Figure 2.8. Longonot volcano in eastern branch of East African rift system (photo by J.J.W.R.).

The upwelling asthenosphere reaches the surface in oceanic ridges, but not in intracontinental settings. In their early stages of development, however, both linear rifts and broad basins develop in areas of uplift that indicate significant asthenospheric upwelling at depth. Presumably this occurs because mantle heating causes dense lithosphere to convert to lower density asthenosphere (the opposite of the process that causes ocean ridges to subside on their flanks). The mantle heating not only causes uplift but also produces an enormous variety of volcanic rocks in extensional environments. The eastern branch of the East African rift system is mostly filled by high-Na basalt and lavas apparently fractionated from it (figs. 2.8 and 2.9). The western branch has only a few volcanic cones that consist

Figure 2.9. Eastern branch of East African rift valley looking west. The far wall is ~30 km away. (Photo by J.J.W.R.).

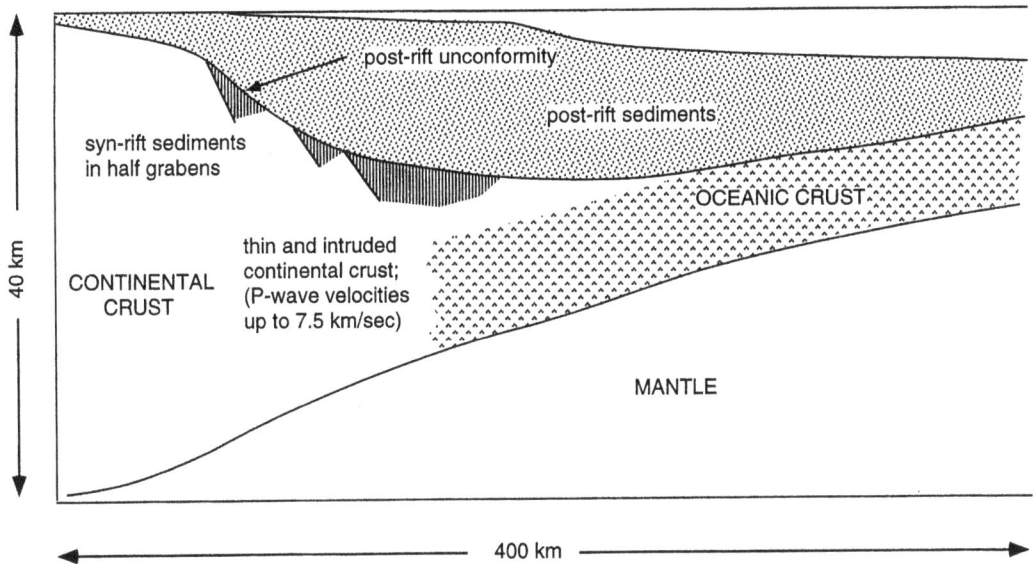

Figure 2.10. Features of passive continental margins. This section is adapted from Grow and Sheridan (1988) and illustrates an area of the east coast of North America where sediment accumulation is thicker than normal.

of extremely high-K lavas. The Basin and Range contains rock types ranging from rhyolite to alkali-olivine basalt. Thus far, there has been no successful effort to relate these different rock types to the different extensional environments in which they were erupted.

Passive continental margins formed by separation of continents have a mixture of characteristics that do not fit either linear or basinal intracontinental rift zones. They are clearly places where continental crust becomes thinner toward the oceans and then disappears at approximately the outer edge of the continental shelf (fig. 2.10). The outermost edge of continental crust is difficult to investigate because it is under water and buried by sediments, but apparently it is highly intruded by basaltic dikes that were probably injected during the fragmentation that pulled the continents apart. Syn-rift sediments formed in half grabens that developed as the two continents began to separate. The post-rift unconformity developed on the rifted surface, and post-rift sediments accumulated on the subsiding margin as the continents drifted away from each other.

Some continental margins that were attached to each other before rifting apparently separated along low-angle "detachment" faults (fig. 2.11). This mechanism creates an upper-plate margin and a lower-plate margin. The upper plate commonly displays a steep fault bounding the rift and one or more thick half grabens filled by sediments. Conversely, the lower plate shows little uplift and is covered only by thin sediments in shallow half grabens.

The traditional view of continental rifting is that ocean basins develop either from linear rift systems or from broad basins, but the evidence for this sequence is unclear. No present continents contain broad extensional basins near their rifted margins, suggesting that regions such as the Basin and Range were not precursors to continental separation. Evolution of oceans from linear zones such as East Africa seems more plausible, but no characteristic sets of alternating half grabens and transfer faults have been found along modern margins.

## Intraoceanic Subduction

A term for zones of lithosphere destruction was not formally defined until a paper by D.A. White et al. (1970) suggested that they be called "subduction" zones. The term had originally been used to describe very large thrusts that brought one part of a mountain belt on top of another (Ampferer and Hammer, 1911). With the expansion of the term to all areas where lithosphere descends to significant depths, some geologists use the term "A-type" subduction (Ampferer) for subduction of continental crust and "B-type" subduction (Benioff) for subduction of

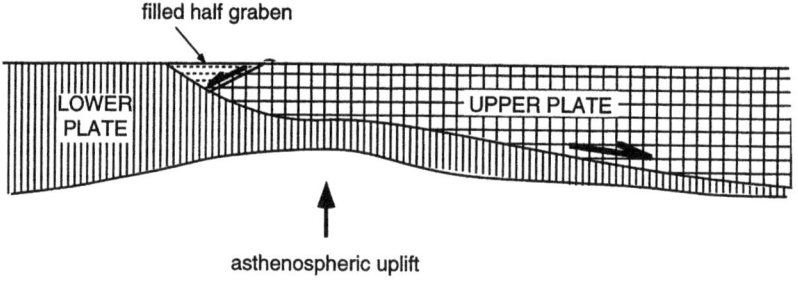

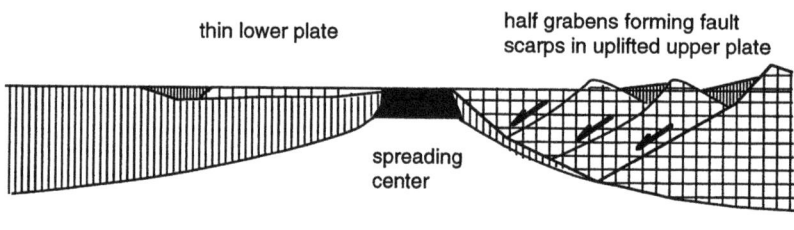

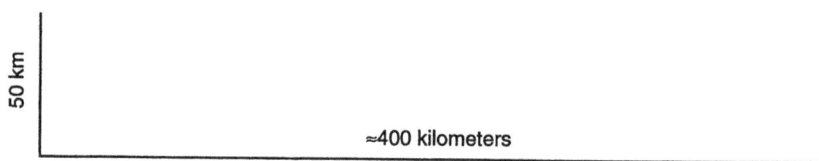

Figure 2.11. Development of rift along detachment fault.

oceanic crust. The importance of subduction in earth history was emphasized in numerous papers during the development of plate tectonics, including Dewey and Bird (1970), who showed its relationship to mountain building, and W. Ernst (1970), who studied continental-margin accretion in California and also showed how subduction could cause the development of blueschist-facies rocks.

Subduction of oceanic lithosphere beneath other oceanic lithosphere produces chains of islands referred to as "island arcs" (figs. 2.12 and 2.13). The arcuate shape as seen on the earth's surface results from the intersection of a flat plane (the slab) with a sphere (the earth), with the curvature of the arc ranging from nearly a straight line (great circle) where the slab is steep to a highly curved arc where the angle of subduction is shallow. The frontal part of the arc consists of a foredeep trench between the descending slab and the overlying plate and commonly a forearc wedge of sediments and volcanic rocks derived from the emerging arc. As the slab descends it becomes isolated from the overlying lithosphere by a mantle wedge. Intersection of the slab with the low-velocity zone generates melting of some mixture of slab, overlying mantle wedge, and older parts of the arc to produce a volcanic suite

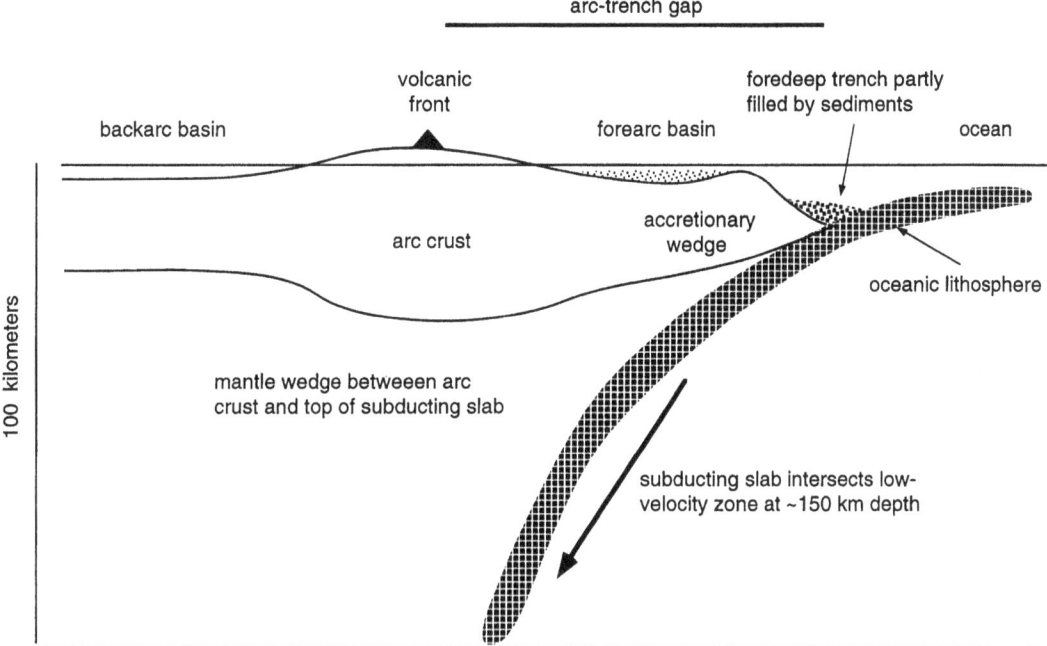

Figure 2.12. Features of intraoceanic island arc.

Figure 2.13. The volcanic island of Statia (St. Eustatius) as seen from the island of St. Kitts in the Lesser Antilles (photo by J.J.W.R.).

22   Continents and Supercontinents

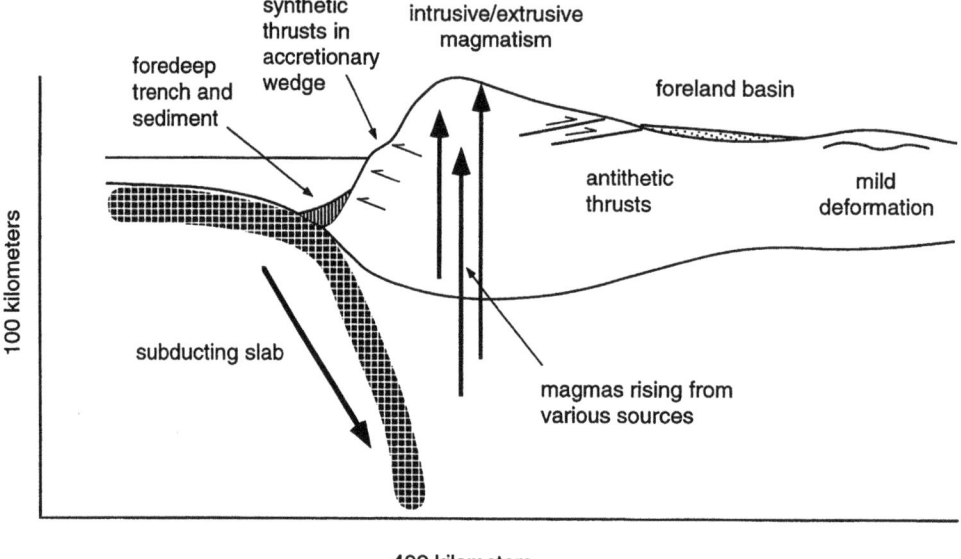

Figure 2.14. Major features produced by subduction of oceanic lithosphere beneath a continental margin.

generally referred to as "calcalkaline" (chapters 3 and 5) but of highly controversial origin.

Further descent of the slab behind the arc commonly places the slab in a zone of extension as the lower part (toe) tends to break away and descend as an isolated body of lithosphere. Some combination of this breaking and mantle upwelling generates a "backarc" basin, where volcanism very similar to mid-ocean ridge basalt (MORB) results from melting of this rising mantle.

*Subduction Beneath Continental Margins*

Descent of lithosphere beneath continental margins generates a more complicated series of events than subduction beneath oceanic lithosphere (fig. 2.14). The foredeep trench, forearc deposits, and mantle wedge are similar to those of intraoceanic arcs, but the overlying continental lithosphere causes differences in both structures and magmatic products. Structures in compressional continental margins tend to be both "synthetic" and "antithetic." Synthetic faults are approximately parallel to the descending slab and form thrusts that verge toward the ocean. Antithetic thrusts, conversely, develop on the landward side of the orogen and may carry overlying rocks far into the continental interior.

Many continental-margin arcs contain rock suites that were metamorphosed under conditions of high pressure and low temperature. Where subduction is particularly rapid, sediment and basalt in the slab enter the blueschist and/or eclogite stability field (fig. B3; discussion of Japan in chapter 5). Typical rock types include metamorphosed varieties of basalt, peridotite, oceanic mud and chert, and sandstones and shales formed in the subduction-zone trench. In some places these rocks were returned to the surface as large blocks along thrust or normal faults, but many suites are referred to as "melanges" because they contain a chaotic assemblage of small fragments of all of these rock types. The origin of the melanges has been very controversial, but they have now been explained as the result of upward ("return") flow of subducted materials along one or more wedges within the subduction complex (fig. 2.15; Cloos, 1982; V. Hansen, 1992).

Magmatic products in continental margins are more complicated than in island arcs because they can be produced by partial melting of subducted sediments and oceanic lithosphere, by the mantle wedge, and also by a complex sequence of continental rocks already in place above the subduction zone. The major products of this magmatism are batholiths and related volcanic rocks that are generally classified as calcalkaline (chapter 3). The proportion of these magmas derived from remelting of older lithosphere and the proportion newly derived from asthenosphere are closely tied to the problem of changes in the volume of continental crust through time. We discuss this question in chapter 4 and also in a description of the Andes in chapter 5.

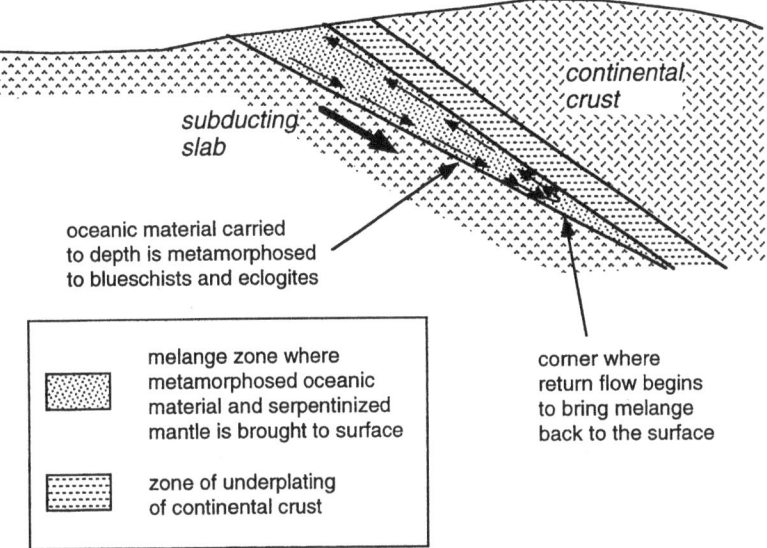

Figure 2.15. Development of melange along subduction zones. Arrows show downward and return flow in wedge between subducting slab and continental crust.

Subduction beneath some continental margins brings another continent, continental fragment, or island arc into collision with the continental margin. This collision causes intense deformation and creates some of the world's largest mountain ranges. The principal development of the ranges is on the lower plate as it descends beneath one of the colliding blocks. Much of the terminology related to the collisional process evolved from studies in the Alps, and we illustrate this in fig. 2.16. The major features of the Alps include an uplifted zone of thrusting, formation of "nappes," and high-grade metamorphism, which is locally intense enough to generate granite by melting ("anatexis") of local rocks. Sediments deposited in the evolving orogen are called "flysch," and the deposits in the relatively undeformed foreland basin are referred to as "molasse," which was deposited in a foreland basin ("molasse basin") that subsided because of the weight of the developing orogen.

In addition to rocks from the descending slab, the Alpine orogen contains some suites that had originally been deposited in the Tethyan ocean between Europe and Africa and also a few suites from Africa. The oceanic suites consist partly of ophiolites (see above) and partly of thick sequences of sediments deposited in the foredeep above the descending European slab before collision with Africa closed the ocean. Occurrence of the oceanic suites within the deformed Alpine orogen demonstrates that the Alpine collision was so intense that these members of the suture zone were thrust upward and northward instead of remaining as a linear suture between the colliding blocks (see discussion of Himalayas in chapter 5). The intensity of the collision was also sufficient to thrust some parts of the African continent across both the European block and the Tethyan suites.

The upper plate of a collision zone is commonly much less deformed than the lower plate and has a lower current elevation (fig. 2.17). In some collision zones the upper plate is the site of extensive calcalkaline magmatism, shown by the development of batholiths and accompanying volcanic rocks. This magmatism develops during subduction of oceanic lithosphere but ceases when the two continental blocks collide, probably because of the difficulty of subducting continental lithosphere (see discussion of Himalayas in chapter 5).

## Transform Margins

It is geometrically impossible for rigid plates to be bounded only by accreting and subducting margins as they move around a sphere. Some margins must be places where the plates simply move past each other. This necessity led J. Tuzo Wilson to designate a "new class" of faults in 1965. He named them "transforms," and unfortunately this term has been used both to describe long faults (like the San Andreas; fig. 2.18) that separate some plate margins

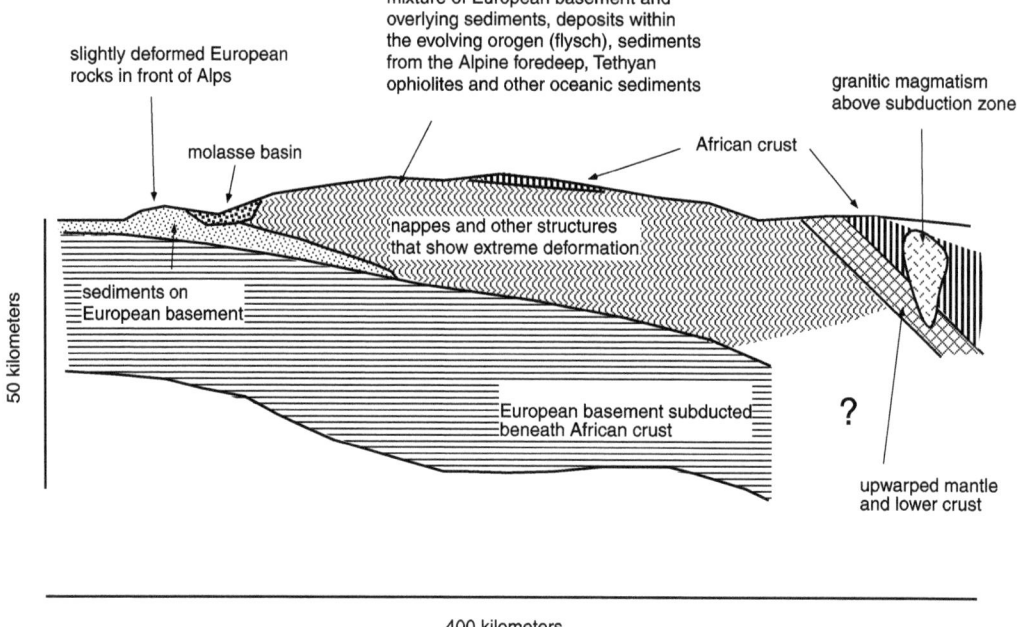

Figure 2.16. Diagrammatic relationships of major units of the Alps. Collission and post-collision movements have warped the mantle of the African plate to the surface just south of the collision zone. More complete information is in Schmid and Kissling (2000).

Figure 2.17. Mountains on the right are part of an orogenic belt thrust onto the Tanzanian craton to the left (photo by J.J.W.R.).

Figure 2.18. San Andreas fault in California looking north, with the right side moving south (the fault is right lateral). Rocks in the fault zone are easily eroded and form a series of shallow valleys. (Photo from U.S. Geological Survey Photo Library.)

and also to describe much shorter faults that offset oceanic spreading centers (see above).

Transform zones in continents are essentially large strike-slip faults that are not ordinarily referred to as plate margins (fig. 2.19). They are commonly generated where plates collide obliquely (transpression) or where a colliding continental block penetrates into another continent and pushes crust out of its way by forming thrust faults in front and strike-slip faults to the side. One of the largest areas of this type of activity results from the collision of India and Asia, which we discuss in chapter 10.

Large strike-slip faults are seldom exactly straight and form complex structures at bends. Where a bend in a fault causes blocks on opposite

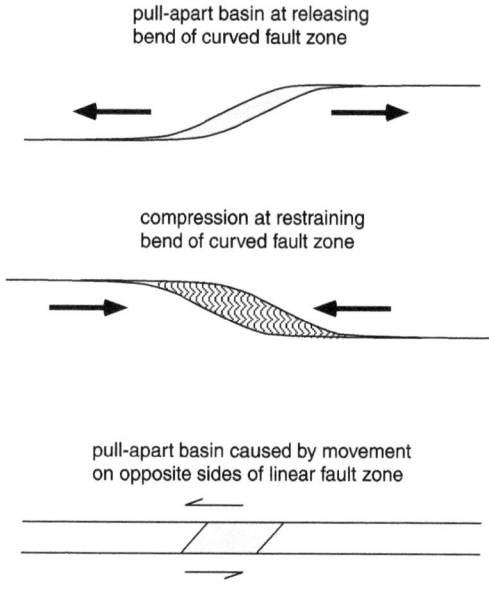

Figure 2.19. Features along transform faults.

sides to move toward each other, the area is called a "restraining bend," and thrusts and related structures develop in the compressional zone. Where the opposite sides of a fault move away from each other, the area is a "releasing bend" dominated by extension. "Pull-apart" basins form in these bends and also along straight sections of strike-slip faults, forming some of the world's lowest continental areas, including the Dead Sea (nearly 400 m below sealevel) and the Turfan depression of western China, which is more than 100 m below sealevel (chapter 10).

Large transforms that develop as strike-slip faults during continental rifting may also become plate margins. The principal example is the separation of Africa, South America, and North America during the rifting of Pangea; fig. 2.20 shows a diagrammatic version of the process (see chapter 9 for more information). It began with the development of a strike-slip fault within Pangea and continued with rifting and formation of spreading centers, first between North America and Africa

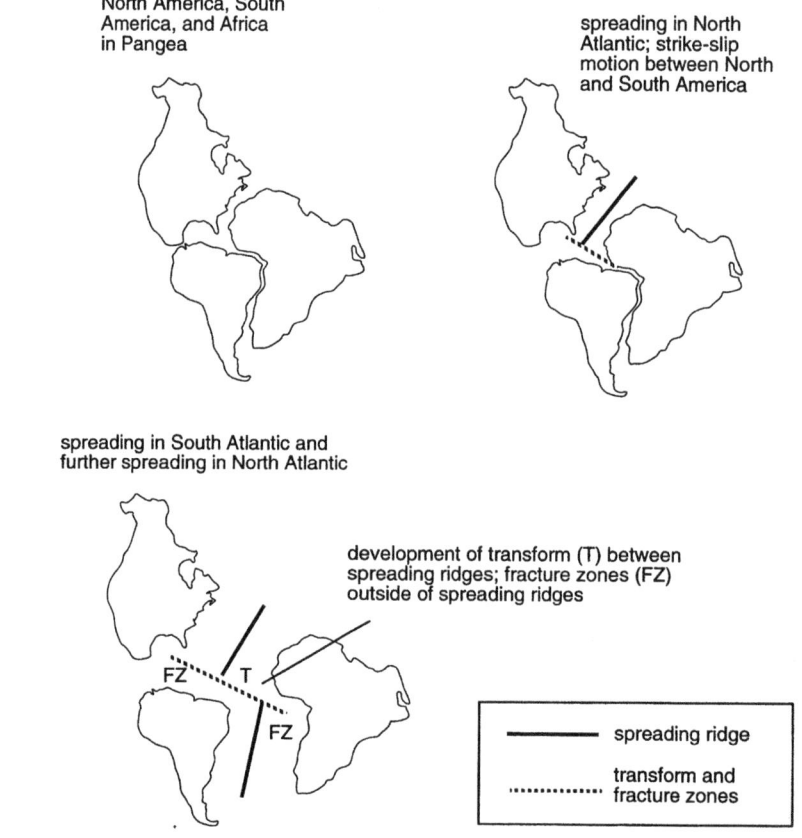

Figure 2.20. Development of transform fault as South America, North America, and West Africa split apart.

and later between South America and Africa. After spreading had continued long enough for the northern and southern Atlantic Oceans to join, the former intracontinental fault continued within the ocean basin as a transform between the two spreading ridges and as a fracture zone with strike-slip movement outside of the area between the ridges.

## Triple Junctions

The three types of plate margins commonly intersect, and in a few places they form "triple junctions" where three margins come together. Places where three rift margins intersect ("r–r–r triple junctions") may be particularly important in the development and movement of plates. A type example is the junction of the Red Sea, Gulf of Aden, and the East African rift system (fig. 2.21). Initial activity is recorded by volcanism on both the African and Arabian margins, apparently generated above a plume (see below) that became active at ~30 Ma. Following this initial volcanism, both the Gulf of Aden and the Red Sea began to split open, although most of the opening in the Red Sea has been caused by stretching of the continental crust and mantle lithosphere rather than by creation of new oceanic crust. The third arm of the triple junction is the East African rift valleys, which are regarded as a "failed arm" ("aulacogen") because no ocean basin has opened within it.

## Plumes

One important process that does not fit neatly into the concepts of plate tectonics is "plumes," which were identified by W.J. Morgan (1972). Plumes are generated somewhere in the mantle below the litho-

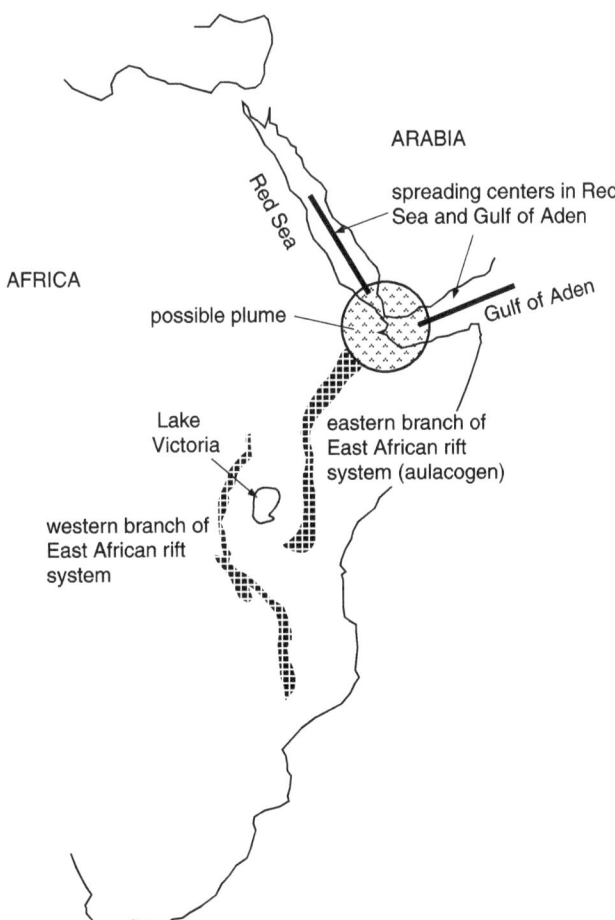

Figure 2.21. Triple junction between Africa and Arabia.

sphere (the exact depth is still unknown) and rise rapidly to the earth's surface. Most of them form volcanic edifices, such as the Hawaiian Islands and Iceland (fig. 2.22), and remain active for tens to a few hundred million years. The standard interpretation of plumes is that they maintain relatively stationary positions in the mantle as lithospheric plates move above them, but recent evidence suggests that at least some plumes move rapidly (see below).

Magmatism caused by plumes follows a very different pattern from the volcanism at oceanic spreading centers. Ocean ridges form basalts at a nearly constant rate during the life of the ridge, which is almost 200 million years for the oldest ridges. Plumes, however, seem to begin with an enormous burst of volcanism that gradually dwindles to a much slower rate. This change is commonly described as the difference between the "head" of a plume, which contains a large amount of pent-up magma, and the "toe" of the plume, where a small amount of magma is produced as asthenosphere moves up along the conduit vacated when the head erupted.

Eruption of the heads of plumes may have been responsible for some continental fragmentation. The r–r–r triple junction at the southern end of the Red Sea (see above) was the site of such extensive eruption of basalts when it formed that it almost certainly developed above the head of a plume. Large basalt plateaus such as the Deccan (India) and Parana (South America) also developed during the first stages of rifting to form the Indian and South Atlantic Oceans respectively (chapter 9). Several oceanic plateaus have also been regarded as plume heads. An unresolved question is whether initial plume activity caused continental separations or whether it simply resulted from the opportunity for plumes to rise into lithosphere that had already been weakened by the early stages of rifting.

If plumes maintain stationary positions in the mantle, geologists have an opportunity to measure the absolute movements of individual lithospheric plates instead of only the relative movements between two or more of them. A principal method is to plot the course of "hotspot" tracks as remnants of progressively older volcanism move with the plate away from the present center of activity (fig. 2.23). A prime example is the track made by the plume now centered under Hawaii, which suggests that the Pacific plate turned from a northerly direction to a more northwesterly one at about 47 Ma (fig. 2.23).

This interpretation of the Hawaii–Emperor hotspot track has been challenged in several ways.

Figure 2.22. Strokkur, a geyser in the Mid-Atlantic Rift Valley where it crosses the Iceland plume. Strokkur is in the same geyser field as Geysir, which is now inactive and from which the term "geyser" was originally derived. (Photo by J.J.W.R.).

Tarduno and Cottrell (1997) used paleomagnetic data from rocks along the track to show that the plume itself has moved at speeds of up to 30 mm/yr. Paleomagnetic data also led Torsvik et al. (2002) to suggest that Hawaii and other hotspots moved southward during the Early Tertiary. Other doubts about the fixed position of Hawaii arise from the lack of any other evidence within the ocean basin that the Pacific plate changed direction and from the absence of any changes in the geologic evolution of western North America at ~47 Ma, which should have been affected by major changes in the Pacific.

## Plate Tectonics in a Hotter Earth

The earth's present thermal gradients are reasonably well known, but past gradients were higher and must be inferred from limited evidence available in preserved rock suites.

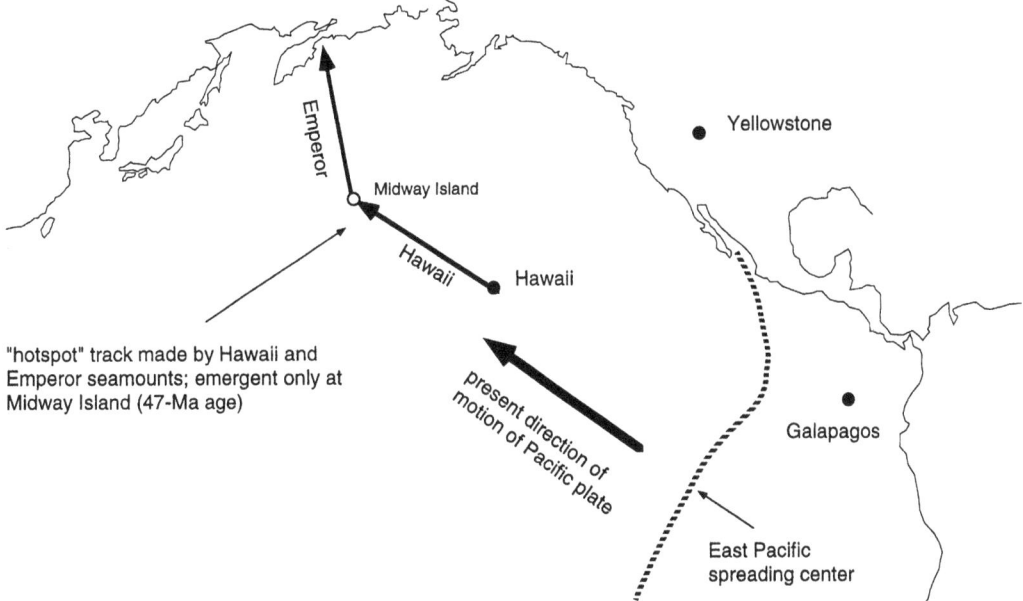

Figure 2.23. Outline of Hawaiian plume track.

Figure 2.24 summarizes the probable ranges (high and low) of thermal gradients at present and in the Archean. Much of the information comes from the presence or absence of komatiites and blueschists. Komatiites are volcanic or shallow intrusive rocks that have high concentrations of olivine, presumably indicating derivation from the mantle by a higher degree of partial melting of peridotite than the melting that produces typical basalt. No komatiites formed in the Phanerozoic, showing that all present gradients must be low enough that they intersect the mantle solidus below the temperatures needed to form komatiites. The presence of komatiites in the Archean, however, shows that some gradients were high enough to intersect the mantle solidus at a shallow depth.

Blueschists are metamorphic rocks that contain minerals that are stable only under conditions of high pressure and low temperature (fig. B.3). They are formed where subduction of oceanic lithosphere cools the surrounding mantle as it descends. The abundance of blueschists in Phanerozoic orogenic complexes clearly shows that some Phanerozoic gradients are low enough to be within the blueschist stability field. By contrast, Archean rock suites contain no blueschist, demonstrating that the lowest Archean gradients were too high to intersect the blueschist stability field.

Further inferences about Archean thermal gradients are provided by diamonds and the thickness of old continental crust. Archean diamonds could have developed only if some Archean gradients were low enough to intersect the graphite–diamond boundary. Preservation of Archean continental crust with thicknesses of a few tens of kilometers shows that at least some Archean gradients must have been below the melting curve of continental crust at those depths. Many gradients, however, may have been so high that most Archean continental crust was destroyed shortly after it formed (chapter 3).

The earth may have been too hot during the Archean for the creation of the type of organized convection that the mantle now undergoes, and some earth scientists have speculated that movement in this very hot mantle is better described as "turbulent." Oceanic areas with many, frequently shifting, convection cells would not have had simple spreading ridges, and their history would be almost impossible to determine. The earth may also have been hot enough not only to cause mantle turbulence but also to prevent the development of rigid lithospheres over broad areas. In that situation, modern types of subduction would have been impossible, and both oceanic and continental lithosphere would have been circulated back into the mantle frequently and in local areas.

One consequence of higher thermal gradients in the past is development of oceanic crusts that were thicker than those at present. Because thermal gra-

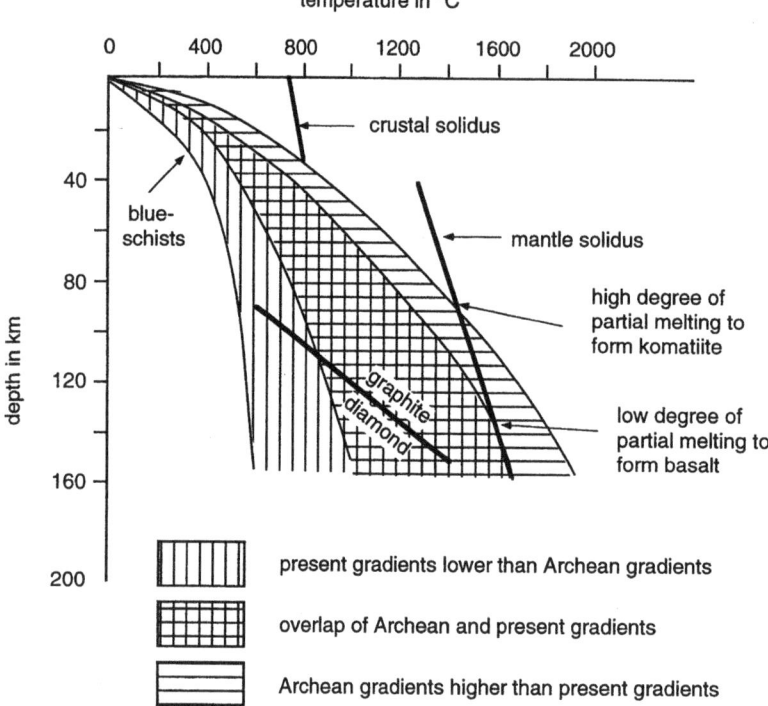

Figure 2.24. Evidence of change in thermal gradients through time.

dients would have intersected the mantle solidus at deep levels, a higher proportion and greater thickness of suboceanic mantle would melt in the past than at present. Sleep and Windley (1982) proposed that this may have developed ocean crusts at least 20 km thick. A thicker oceanic crust would have affected continental evolution in two ways. Because the crust may have been too thick to subduct, it could have contributed to the development of greenstone belts by accreting to continental margins (chapter 4). Thick crust also would have been isostatically compensated by floating higher on the mantle, thus creating shallower ocean basins. This would have pushed seawater out of the ocean basins and onto continents, thus causing much of the evolution of ancient continental crust to take place under water.

## Summary

The concepts of plate tectonics developed over a century of conflict and are now well established. The locations of plates and various types of plate margins clearly control much of the present activity on the earth and can explain a large part of its past history. Whether these concepts can be applied throughout earth history, however, is not clear. In the early stages of earth history, the interior of the earth was much hotter because it contained much higher contents of radioactive elements and had not yet lost much of its original heat of accretion.

An understanding of the concepts of plate tectonics is crucial to earth scientists who wish to decipher the history of the earth. Consequently, we use it throughout this book but frequently refer to processes that may not be explained adequately by the concept of rigid plates.

# 3

# Creation, Destruction, and Changes in Volume of Continental Crust Through Time

The low density of continental crust creates areas of land that cover about 25% of the world's surface. The ultimate source of this crust must have been the mantle, but the mechanisms and times of extraction are not well known. Because the rate at which continental crust has been destroyed and incorporated back into the mantle is also uncertain, the volume of continental crust that existed in the earth throughout its history is highly controversial. We investigate these problems in this chapter, beginning with a discussion of how the wide variety of rocks that constitute the crust may have evolved from the mantle. The next section summarizes information on the rates of destruction of crust, which leads to the issue of changes in the volume of continental crust through time.

We leave to chapter 4 the discussion of "cratons," which are large areas of continental crust that became stable at different times in the past 3 billion years and then moved around the earth as relatively undeformed continental blocks. Cratons are commonly referred to as "shields" where their crystalline basement rocks are exposed. The discussion in this chapter and several later chapters requires some knowledge of the significance of isotopic information, and we provide a brief summary in appendix D.

In order to describe continental crust and its relationships to other parts of the earth, we must start with several definitions. "Sialic" (Si + Al) refers to rocks, such as granite, that are particularly rich in $SiO_2$, with $Al_2O_3$ as the second most abundant oxide; they are commonly referred to as "felsic." "Simatic" (Si + Mg) similarly designates basalts and other low-silica "mafic" rocks. "Ultrasimatic" and "ultramafic" refer to rocks extremely rich in olivine, mostly different varieties of peridotite. Some sialic (felsic) rocks have high concentrations of large-ion-lithophile (LIL) elements, which are ions that are "incompatible" with (do not fit into) typical minerals in the mantle, including olivine, pyroxene, and calcic plagioclase. LIL elements include K, Rb, Ba, and U, which is coordinated with oxygen to form the low-density uranyl complex ($UO_2^{+2}$). Because of their incompatibility, LIL elements tend to concentrate in fluids that move them upward in the crust as part of the density stratification of the earth (see below).

This chapter makes extensive use of isotopic information summarized in appendix D.

## Origin of Continental Crust

The mantle and crust of the earth represent the "stony," silicate-bearing, components of the solar cloud that accreted to form the inner, "terrestrial," planets of the solar system. The 15% of the earth that is in the core came from the "metallic," mostly iron and nickel, part of the solar cloud. During and shortly after planetary accretion, the Fe/Ni formed a liquid that was "immiscible" with (did not mix with) the silicate fraction of the earth. In the earth's gravity field, this dense Fe/Ni liquid separated downward into the core, and the silicates fractionated upward. Throughout its history, the earth underwent further gravitational segregation that developed a mantle, oceanic crust, and continental crust (figs. 1.4 and 3.1).

The earth's crust extends from the surface down to the Mohorovicic discontinuity (Moho; chapter 1). As originally defined by Mohorovicic, the boundary between crust and mantle is a seismic discontinuity that separates much higher velocities in the mantle (typically ~8 km/sec for P waves) from those in the overlying crust. Normal values of P-wave velocities in the crust range from very low in near-surface sediments to 6.8–7.2 km/sec just above the Moho. In many areas a minor discontinuity (Conrad) at a depth of ~20 km separates upper crust with velocities of 6–6.7 km/sec from the region of higher velocities in the lower crust.

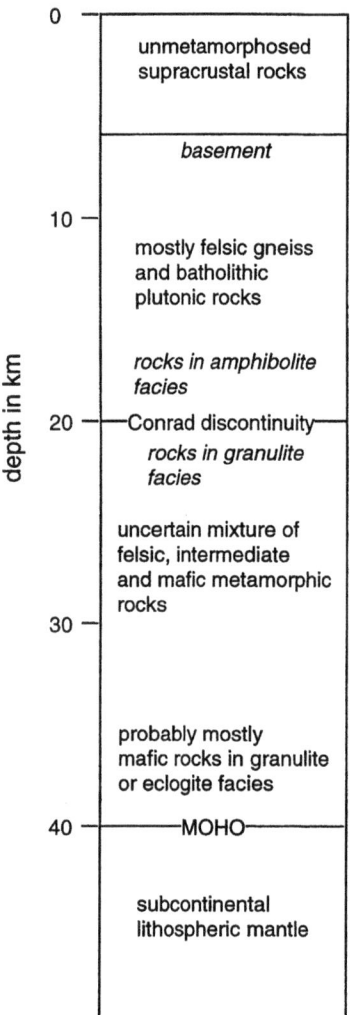

Figure 3.1. Cross section of typical continental crust. The mantle underlying the continent is designated as subcontinental lithospheric mantle.

The different P-wave velocities correspond to different rock densities and rock types. Velocities in the upper crust indicate densities of ~2.7 g/cc, consistent with a crust consisting of gneisses, granitic intrusions, and volcanosedimentary suites metamorphosed to amphibolite facies. The densities of 2.8–3.0 g/cc in the lower crust are consistent with mineral and rock stability ranges that require all of the crust to be in granulite facies below approximately 20 km (6 kbar, 600 MPa; fig. B.3). Whether this granulite is felsic or mafic, however, cannot be determined uniquely from seismic data. In areas where the crust is thickened by orogenic or other processes, the mafic lower crust may become denser than 3.3 g/cc and founder ("delaminate") into the mantle (further discussion in chapter 5).

Upper-mantle densities of 3.3 g/cc are typical of mantle peridotites but could also represent gabbros metamorphosed to eclogites. This uncertainty raises the possibility of a difference between a "seismic Moho," which is defined as the top of a layer with a P-wave velocity of 8 km/sec, and a "petrologic Moho," which is commonly regarded as the top of a layer of peridotite. The problem is similar to the one posed by ophiolites that contain thick sections of mafic cumulates (chapter 2).

## Source and Generation of Mafic Rocks in the Crust

The rocks that constitute the earth's lower continental crust are poorly known because of lack of exposure (summary edited by Fountain et al., 1992). Information about them is obtained from a few places where uplifts expose rocks that have equilibrated at pressures higher than ~6 kbar and also from studies of granulite-facies xenoliths in volcanic rocks. Two of these uplifts are in shield areas where the transition between the upper and lower crust is exposed, and we discuss them briefly here.

Uplift of the crust along east-vergent thrusts exposes an oblique cross section of both upper and lower Archean crust from amphibolite-facies rocks to depths as much as 30 km (~10 kbar) in the Kapuskasing zone of the Superior province of Canada (fig. 4.3 in chapter 4; D. Shaw et al., 1994). Rocks in granulite facies (lower crust) have a large range of lithology and composition from felsic to mafic. Their average composition shown in table 3.1 is approximately the composition of a typical andesite (intermediate). Estimated compositions of lower and upper crust at Kapuskasing differ primarily in the lower abundance of LIL elements in the lower crust, but the similarity of major-element abundances does not support the concept that the lower crust above a depth of 30 km has a very different composition from the more felsic upper crust. The section does not reach the Moho, however, and it is very possible that crustal rocks below the level of exposure are almost entirely mafic.

A well exposed transition from amphibolite- to granulite-facies in southern India reaches pressures as high as 8 kbar. Rocks equilibrated at pressures higher than 7–8 kbar, however, occur only in structurally complex areas south of the transition zone. They are predominantly felsic (table 3.2), but whether they represent typical lower crust or upper crust structurally transported to great depth is unclear.

Neither of the uplifts discussed above expose the Moho, and Bohlen and Mezger (1989) pointed out that exposed granulite massifs generally represent only rocks that equilibrated at intermediate crustal depths. Consequently, we must rely on xenoliths in volcanic rocks to provide information about the deepest parts of the crust.

Alkalic basalts commonly contain xenoliths, which range from mantle peridotites to granulites from the lower crust (figs. 3.2 and 3.3). Granulite-facies xenoliths are predominantly mafic, amounting to more than 90% of the xenolith suite in some areas. They have a variable mineralogy of calcic plagioclase, pyroxenes, and garnet +/− amphibole, biotite, and quartz. Equilibration temperatures range from ~600°C to ~800°C, and pressures from ~7–12 kbar. A few areas contain eclogite xenoliths, with a mineralogy of garnet and omphacite that shows equilibration at pressures higher than ~11 kbar.

Compositions of the mafic xenoliths provide some information on the origin of the lower crust, and we compare them with mafic rocks from other environments in table 3.1. The crustal xenoliths contain higher concentrations of LIL elements, such as K, than MORB contains and clearly were not formed by partial melting of depleted asthenosphere by the same process that produces modern MORB. Derivation from depleted asthenosphere is even less likely if the xenoliths are the residue from fractionation of felsic rocks of the upper crust, implying that they originally had higher concentrations of LIL elements than they do now.

If the lower crust could not have formed by fractionation of the depleted asthenosphere that forms MORB, then it must have originated from a mantle that was enriched in elements that characterize continental crust. The question is how and when that enrichment occurred, and we discuss two of the numerous possibilities. One is simply that continental crust evolves only in areas of the earth where the mantle has undergone widespread metasomatism before crustal segregation begins. This proposal is not well documented, but it is supported by a growing body of evidence for compositional alteration of the subcontinental mantle throughout its history (further discussion in chapter 4).

A second explanation for origin of an enriched mantle is based on the compositional similarity between mafic crustal xenoliths and the typical rocks of mantle plumes, including oceanic islands and both oceanic and continental plateaus (examples in table 3.1; Condie, 1999, 2001). All of the rocks generated by plumes contain higher concentrations of LIL elements than MORB contains, presumably because they originate from magmas generated in deep and undepleted regions of the mantle. Some plume sources, such as Iceland, are also rich enough in silica to generate rhyolite, whereas felsic rocks occur only as sparse plagiogranites in oceanic crust formed along spreading ridges (table 3.2). For this reason, many geologists believe that mantle plumes are highly involved in the development of continents, including mafic and dacitic rocks of greenstone belts that may have been derived partly from plumes (Abbott, 1996; Puchtel et al., 1998; Naqvi et al., 2002a; further discussion in chapter 4). Plumes could form lower crust by congealing before they reached the surface, and it is also possible that oceanic plateaus formed by plumes could be subducted under continental margins and incorporated into the lower crust.

Age relationships between upper crust and apparent samples of underlying lower crust are variable. Similar ages of ~2.6–2.5 Ga for upper and lower crust in the Kapuskasing zone are reported by Krogh and Moser (1994), but studies of lower-crustal xenoliths to the southeast of Kapuskasing show ages a few hundred million years younger than the stabilization age of the overlying Archean upper crust (Moser and Heaman, 1997). Davis (1997) proposed that large parts of the mafic lower crust of the Slave craton of the Canadian shield were formed by underplating of plumes related to mafic dike swarms intruded more than 1 billion years after the 2.5-Ga stabilization of the craton.

Summary  The preponderance of evidence from studies of xenoliths and geophysical properties suggests a mafic composition of the lowermost crust. Compositions of the xenoliths also indicate that they were derived from a mantle that had been enriched in LIL elements before the crust separated from it. Significant volumes of lower crust are not mafic but felsic rocks that equilibrated at pressures not more than ~7 kbar (20–25 km depth). Some felsic granulites with equilibration pressures greater than 7 kbar may have reached these depths by younger tectonic processes instead of during the evolution of the overlying upper crust. Ages of rocks in the lower crust may be younger than those of the upper crust, either because of thermal resetting of old rocks or because new mafic material was added by younger processes.

Tables 3.1 and 3.2 show the compositions of rock suites that illustrate discussions in the text of chapter 3. Because of the enormous number of rocks that have been analyzed, we cannot claim that the values shown in these tables are comprehensive samples of the different environments in which the rocks formed, but we believe that they closely represent rocks generated in each environment. Some of the suites shown here were selected in preference to others, including those that are more widely known, because we wanted all of our suites to have values for LIL and other trace elements that are regarded as diagnostic of environments of formation.

The tables contain both averages and ranges. We use averages if the investigators who did the work regarded a rock suite as sufficiently homogeneous to calculate an average in their published paper. Where averages were not reported, we show ranges of data for each element. For clarity, we have rounded off many values and eliminated outliers. Concentrations of major elements are shown as weight percent and of trace elements as part per million (ppm) by weight. Values for iron are shown as total Fe calculated as FeO. Initial ($^{87}Sr/^{86}Sr)_I$ are normal, and all $\varepsilon_{Nd}$ are relative to CHUR except for suite 9 in table 3.2.

Table 3.1. Compositions of mafic rocks

| Component | 1 | 2 | 3 | 4 | 5 | 6 | 7 | 8 | 9 |
|---|---|---|---|---|---|---|---|---|---|
| $SiO_2$ | 49–50.5 | 50–52 | 47–47.5 | 48–52 | 49–54 | 50.0 | 57.9 | 43–48 | 40–48 |
| $TiO_2$ | 1–2 | 1.5–3 | 2–2.1 | 1.7–2.4 | 0.6–1.6 | 1.2 | 0.65 | 0.8–3 | 0.7–3 |
| $Al_2O_3$ | 14–15 | 13–16 | 14.4–15.2 | 12–14 | 14.5–21 | 14.9 | 17.3 | 13–18 | 11–16 |
| $FeO_T$ | 10–14 | 11–14 | 12–13 | 11–13 | 8–11 | 9.4 | 7.1 | 11–16 | 11–18 |
| MgO | 6–8 | 5–6.5 | 4.3–6.4 | 8–15 | 5–9 | 8.1 | 4.3 | 5–10 | 7–9 |
| CaO | 10.5–13 | 8.5–10.5 | 7.2–10.8 | 8–10 | 8–13 | 10.5 | 7.6 | 12–15 | 6–12 |
| $Na_2O$ | 2–3.5 | 2.8–3.3 | 2.5–3.0 | 1.2–2.3 | 1.3–3.5 | 2.8 | 2.7 | 1.2–2.5 | 1.5–2.8 |
| $K_2O$ | 0.03–0.2 | 0.5–1.2 | 0.5–1.7 | 0.04–0.5 | 0.14–0.6 | 0.48 | 1.5 | 0.03–0.4 | 0.3–1.1 |
| Ba | 4–10 | 200–400 | 100–300 | 30–100 | 50–200 | 201 | 523 | 30–190 | 90–800 |
| Rb | 0.6–4 | 10–35 | 9–62 | 0.1–6 | 5–20 | 10 | 41 | 1–5 | 4 or 40 |
| Sr | 70–100 | 250–400 | 175–340 | 250–380 | 150–300 | 311 | 447 | 50–250 | 200–400 |
| Zr | 75–200 | 125–225 | 140–170 | 90–12.5 | 20–50 | 124 | 114 | 70–300 | 50–150 |
| Y | — | 30–50 | 30–50 | 20–30 | 12–20 | 28 | 16 | 20–45 | 20–100 |
| $(^{87}Sr/^{86}Sr)_I$ | 0.702–0.703 | 0.703–0.705 | 0.704–0.706 | 0.703–0.703 | 0.7045–0.705 | 0.7035–0.704 | — | 0.701–0.705 | 0.703–0.704 (or variable) |
| $\varepsilon_{Nd}$ | +10 | +3 to 0 | +2 to −3.5 | +3 to 2 +3 to −3 | | | — | +2 to −5 | negative |

34

1. Mid-ocean ridge basalt (MORB) from a segment of the North Atlantic ridge as far from plume influence as possible. Zr and Y were not reported by the authors and are added by us as typical values for MORB. (Schilling et al., 1983.) Sr and Nd values are typical of hundreds of analyses.
2. Imnaha basalts of Columbia River basalt plateau. They represent a continental flood basalt, probably developed from the head of a plume. Major and trace elements are from Hooper and Hawkesworth (1993). Sr and Nd values are from Hooper (1997).
3. Kerguelen plateau in the Indian Ocean, a typical intraoceanic plateau possibly related to plume activity. All values from Davies et al. (1989).
4. Tholeiitic basalt of Lanai, Hawaii. It represents the Hawaiian plume in an island that is no longer volcanically active. (West et al., 1992.) Sr and Nd values are typical of hundreds of analyses of Hawaiian basalts.
5. Primitive subduction-zone magmas from intraoceanic island arcs in the southwest Pacific. Major and trace elements are from I. Smith et al. (1997). Sr and Nd values are from Gamble et al. (1997). More information is provided by other papers in volume 35 of the *Canadian Mineralogist* (edited by Nixon et al., 1997).
6. Basalts of the Clarno Formation, central Oregon. The Clarno is a predominantly andesitic suite that ranges from basalt to rhyolite and was proposed to have been produced by subduction under a thin continental margin. Samples chosen for averaging are the lowest-$SiO_2$ members of the suite, and values are normalized to 50.0% $SiO_2$. Major and trace elements are from Rogers and Ragland (1980). Sr values are from Suayah and Rogers (1991).
7. Weighted average of the granulite-facies rocks of the Kapuskasing zone of the Superior craton (D. Shaw et al., 1994).
8. Xenoliths typical of mafic granulites that have very low LIL element concentrations. Ranges of major elements, trace elements, Sr ratios, and Nd values are from Rudnick (1990) and Rudnick and Taylor (1991) in eastern Australia and from Kempton et al. (1997) in the Pannonian basin of Hungary.
9. Range of major and trace element concentrations in xenoliths of mafic granulites reported by Holtta et al. (2000) in Finland and by Markwick and Downes (2000) in northwestern Russia. They have higher concentrations of LIL elements than the xenoliths shown in column 8 except for Rb in the suite reported by Markwick and Downes; this averages about 4 ppm Rb, whereas the suite studied by Holtta et al. averages about 40 ppm Rb. Initial Sr values are reported as 0.703–0.704 by Markwick and Downes and highly variable by Holtta et al. Nd values are reported as negative and variable by Markwick and Downes.

Table 3.2. Compositions of felsic rocks

| Component | 1 | 2 | 3 | 4 | 5 | 6 | 7 | 8 | 9 | 10 |
|---|---|---|---|---|---|---|---|---|---|---|
| $SiO_2$ | 69–74 | 74.3 | 62–63 | 77.7 | 67–76 | 73.0 | 73.0 | 72–73 | 68.8 | 72.5 |
| $TiO_2$ | 0.1–1.1 | 0.31 | 0.5–0.7 | 0.12 | 0.2–1.0 | 0.22 | 0.26 | 0.31–0.33 | 0.79 | 0.45 |
| $Al_2O_3$ | 10.4–16 | 11.8 | 17.0–17.3 | 13.3 | 14.1–18.7 | 14.2 | 13.5 | 13.7–14.4 | 14.7 | 12.0 |
| $FeO_T$ | 0.8–4 | 2.6 | 4.2–4.5 | 1.0 | 1.2–5.7 | 1.7 | 2.2 | 1.6–2.3 | 4.4 | 4.0 |
| MgO | 0.2–4.2 | 1.2 | 1.9–4 | 0.5 | 0.3–1.3 | 0.42 | 0.37 | 0.64–0.75 | 1.0 | 0.35 |
| CaO | 0.8–3.6 | 1.4 | 4.3–4.8 | 1.0 | 1.7–6.6 | 1.4 | 1.0 | 2.0–2.5 | 2.5 | 2.2 |
| $Na_2O$ | 2.7–6.3 | 4.9 | 3.9–4.1 | 6.1 | 2.1–6.4 | 3.6 | 3.6 | 3.4–3.9 | 3.2 | 3–5 |
| $K_2O$ | 0.2–4 | 0.37 | 2.5–4.4 | 0.4 | 0.32–1.9 | 4.2 | 4.8 | 3.6–3.8 | 5.0 | 2.8 |
| Ba | 50–630 | 126 | 1250–1350 | 100 | — | 740 | 410 | 720–780 | 2060 | 600 |
| Rb | 1–90 | 7 | 50–110 | 9 | — | 220 | 230 | 130–150 | 150 | 6.0 |
| Sr | 70–590 | 109 | 490–790 | 155 | — | 200 | 71 | 220–260 | 340 | 120 |
| Zr | 105–557 | 107 | 110–150 | 94 | — | 140 | 250 | 200–500 | 400 | 520 |
| Y | 13–130 | 31 | 17–28 | 12 | — | 27 | 61 | — | 51 | 120 |
| $(^{87}Sr/^{86}Sr)_I$ | 0.700–0.701 | 0.704 | | variable | — | 0.705–0.707 | 0.702–0.704 | 0.705 | — | 0.703–0.704 |
| $\varepsilon_{Nd}$ | −1 to +2 | +6 to +10 | +6 to +10 | +6 to +10 | — | negative variable | +5 to 0 | 0 | −13 to −22 | +6 to +10 |

1. Tonalite–trondhjemite–granodiorite (TTG) gneisses from the Western Dharwar craton of southern India. The wide range of values is from 18 different suites analyzed by various workers and includes typical ranges for TTG suites elsewhere. Major and trace elements from Rogers et al., (1986). Sr and Nd values from Meen et al. (1992) and Peucat et al. (1993).
2. Keratophyre (Na rhyolite) produced in the early stages of subduction in the Virgin Islands, northeastern Caribbean (Donnelly and Rogers, 1980). The same Sr and Nd values are listed for columns 2 and 3 and are based on general studies of Mesozoic igneous activity in the northeastern Caribbean by Frost et al. (1998).
3. Batholithic rocks with greater than 60% $SiO_2$ produced by subduction in the northeastern Caribbean. (A. Smith et al., 1998; additional information is in other papers in Geological Society of America Special Paper 322, edited by Lidiak and Larue, 1998.) The same Sr and Nd values are listed for columns 2 and 3 and are based on general studies of Mesozoic igneous activity in the northeastern Caribbean by Frost et al. (1998).
4. High-$SiO_2$ plagiogranite (Na granite) from an oceanic shear zone (Flagler and Spray, 1991). Sr and Nd values are from the study of a plagiogranite in the Semail ophiolite (Amri et al., 1996).
5. Range of compositions produced by experimental melting of basaltic compositions under different conditions of temperature, total pressure, and water pressure (Winther, 1996).
6. Compositions of granites produced during the later stages of continental-margin subduction (e.g., granites of Sierra Nevada). One of four suites normalized to 73.0% $SiO_2$ from average $SiO_2$ contents ranging from 72.9% to 74.2% (LO suite of Rogers and Greenberg, 1990). Sr and Nd values are typical of hundreds of analyses of rocks from the Sierra Nevadas, Andes, and other continental-margin batholiths.
7. Compositions of post-orogenic granites (identified in text). One of four suites normalized to 73.0% $SiO_2$ from average $SiO_2$ contents ranging from 72.9% to 74.2% (PO suite of Rogers and Greenberg, 1990). Sr and Nd values are typical of numerous analyses of post-orogenic granites worldwide.
8. Cantarito rhyolites produced by subduction in central Andes. Zr and Y were not reported by the authors and are added by us as typical values for subduction-zone rhyolites. (All data from Kay et al., 1991; more information is in other papers in Geological Society of America Special Paper 265, edited Harmon and Rapela, 1991.)
9. Felsic granulites in Cardamom massif of southern India. Major and trace elements are from Chacko et al. (1992). Nd values (relative to depleted mantle) are from Brandon and Meen (1995).
10. Rhyolites from Iceland. Icelandic rhyolites are referred to locally as liparites and constitute ~15% of the volcanism in Iceland. Major elements, trace elements, and Nd values are from Kempton et al. (2000). Sr values are typical of Icelandic volcanic rocks (from R. Taylor et al., 1997).

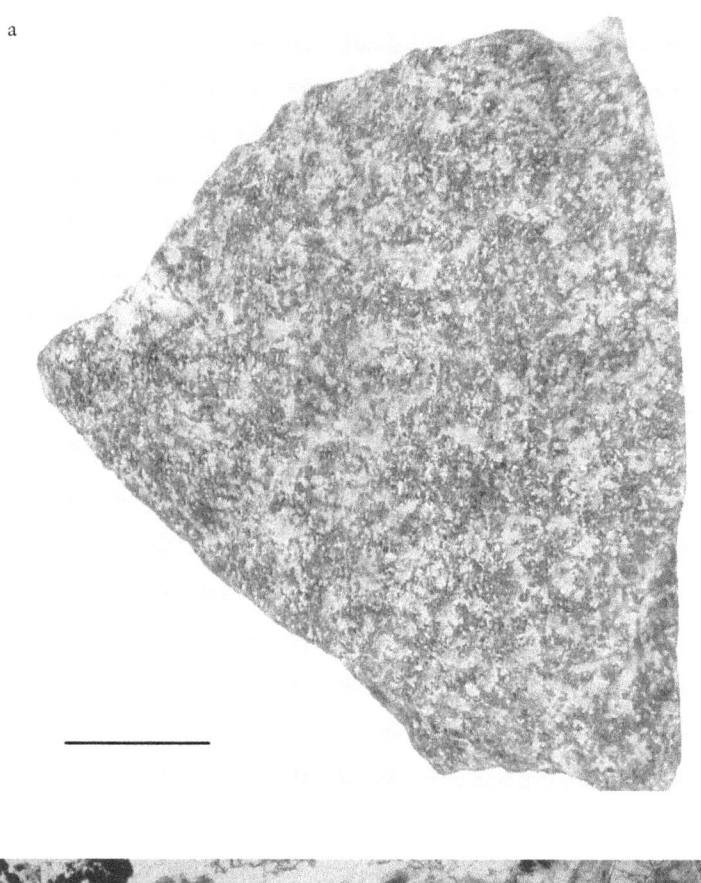

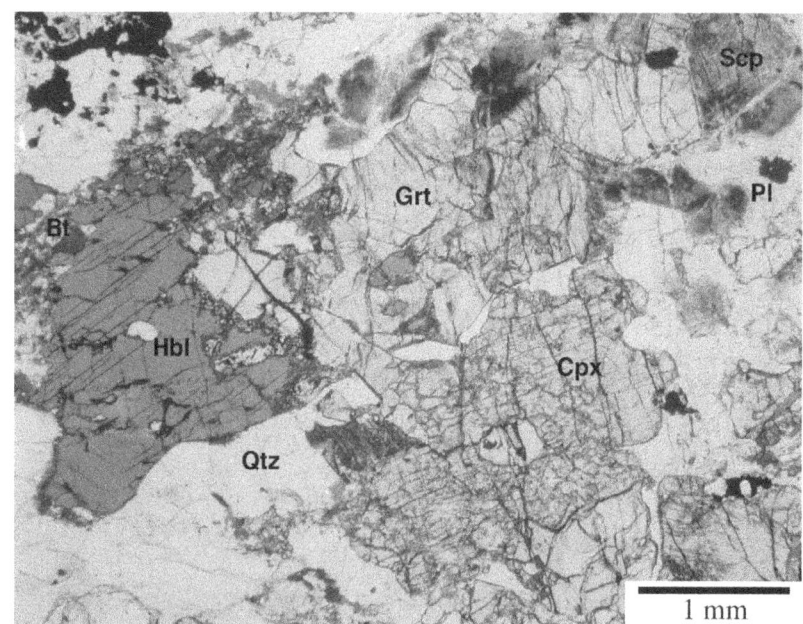

Figure 3.2. Mafic granulite from Salem, South India. a. Hand specimen; scale bar is 5 cm. b. Thin section. Scp, scapolite; Grt, garnet; Pl, plagioclase; Bt, biotite; Hbl, hornblende; Cpx, clinopyroxene; Qtz, quartz. (Photos by M.S.)

Figure 3.3. Peridotite xenoliths in basalt at Dish Hill, California (courtesy of Allen Glazner).

## Source and Generation of Felsic Rocks in the Crust

The upper continental crust is mostly felsic, and the first task in explaining its origin is to find a method of generating highly deformed rocks that consist of a simple mineralogy of quartz + oligoclase/andesine + minor K feldspar + biotite, +/− hornblende. Compositionally they are classified as tonalite–trondhjemite–granodiorite (TTG; fig. 3.4). Rocks with this mineralogy, composition, and structure have ages ranging from older than 3.5 Ga to about 2 Ga, and suites from numerous parts of the world yield remarkably similar compositions shown in table 3.2. Rocks of the TTG suite constitute most of the earth's exposed shields and apparently form the basement of most continental areas that are covered by younger rocks. They are commonly referred to as "Archean gray gneiss" even though many of them are Early Proterozoic (Martin, 1994; discussion of Archean and Proterozoic in chapter 4).

Because all of the earth's crust segregated from the mantle at some time, the TTG and other felsic rocks that constitute the upper continental crust must have begun their evolution by partial melting of ultramafic rocks. Melting of peridotite in the mantle can form small amounts of quartz-normative liquid, partly by the incongruent melting of

Figure 3.4. Outcrop of deformed TTG in Brazil (photo by J.J.W.R.).

pyroxene to form olivine and a silica-rich basalt. Large volumes of silicic magmas, however, cannot be produced by direct partial melting of mantle peridotite, and most geologists propose that high-silica magmas must be generated by a two-stage process in which the mantle produces basalt magmas and the resulting basalts/gabbros are later remelted to generate siliceous melts. This can be accomplished by a variety of processes of partial melting, and Winther (1996) experimentally generated tonalites and trondhjemites similar to those of table 3.2 from basalt/gabbro under a large range of temperatures, total pressures, and water-vapor pressures.

Although the felsic rocks of both oceanic and continental areas are ultimately derived from the mantle, they show significant differences in composition (table 3.2). At similar percentages of $SiO_2$, TTG suites and other continental rocks are much richer in LIL elements than the rocks of oceanic islands (such as Iceland) and intraoceanic arcs (such as the Caribbean) are. These differences in composition indicate that continental rocks have been derived from a more LIL-element-rich ("fertile") upper mantle than the mantle underlying oceanic crust. This is the same conclusion that we reached for the evolution of mafic rocks in the lower crust (see above) and further supports the likelihood that all continental crust evolved only from mantle enriched in LIL elements.

The time at which mantle enrichment occurred is clarified by isotopic data, which invariably show that the TTG have high positive $\varepsilon_{Nd}$ values, high negative $\varepsilon_{Sr}$ values, and $T_{DM}$ ages only a few hundred million years older than the measured ages of gneiss emplacement. These data show that enrichment must have occurred only shortly before separation of the basalts that melted to form the TTG. If enrichment had occurred much earlier than basalt separation, high Rb/Sr and Sm/Nd ratios would have generated lower (less positive) $\varepsilon_{Nd}$ values, and higher (less negative) $\varepsilon_{Sr}$ values, than those of asthenospheric mantle. The $T_{DM}$ ages also indicate extraction of TTG magmas from the mantle after only a brief residence in some basalt/gabbro intermediate.

Some gneisses in the TTG suite are classified as granodiorites because they have slightly higher concentrations of K and other LIL elements than tonalites and trondhjemites. They have similar isotopic properties to the tonalites and trondhjemites, however, and presumably they were also derived from a similar mantle source (Stern and Hanson, 1991). The higher concentrations of LIL elements may have resulted from fractional crystallization of

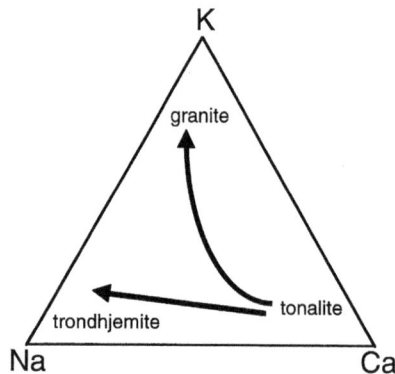

Figure 3.5. Diagram showing common sodic fractionation trend in Archean plutonic suites and potassic trend in younger suites.

tonalite/trondhjemite magmas before the gneiss precursors were emplaced, but it is also possible that some of these rocks were metasomatized after they originally crystallized (see below). In parts of some shield areas, the TTG are not as intensely deformed as gneisses, and they have locally been referred to as batholiths (Nutman et al., 1999).

The TTG suites differ in three ways from typical subduction-zone ("Sierran-type") batholiths developed along Phanerozoic continental margins. One is that they contain almost no diorite and exhibit a range of $SiO_2$ contents that is smaller than the gradational series of rocks from diorite to granite that characterizes modern batholiths. A second difference is that, within their limited range of $SiO_2$ contents, the TTG show increasing $Na_2O$ with increasing $SiO_2$, whereas the dominant variation in modern batholiths is strong increase in $K_2O$ (fig. 3.5). The third distinction is that Phanerozoic batholiths have $\varepsilon_{Sr}$ values that are positive, instead of negative, and $\varepsilon_{Nd} \sim 0$, instead of positive. All of these differences suggest that Phanerozoic suites were generated from subcontinental mantle that had been enriched in LIL elements long before the batholiths were produced (further discussion in chapter 5).

Summary The world's upper continental crust consists mostly of TTG suites formed during the Archean and Early Proterozoic. They clearly developed by a two-stage process of partial melting of mantle peridotite to form basalt/gabbro and then melting of the basalt/gabbro to form felsic rocks. Apparently the mantle had been enriched in LIL elements shortly before initial partial melting occurred. The original mantle source shows some

Figure 3.6. Enchanted Rock pluton, a post-orogenic granite in the Llano uplift, Texas, United States (photo by J.J.W.R.).

similarities to the source of magmas in intraoceanic subduction zones and plumes but contains lower concentrations of LIL elements than the source of subduction-zone magmas along Phanerozoic continental margins.

*Source and Generation of LIL-Element-Rich Rocks*

Some continental rocks contain significantly higher concentrations of LIL elements than any of the TTG gneisses and accompanying batholiths. Different processes that cause high LIL element enrichment produce different types of rocks, including granite plutons that are comparatively undeformed, pegmatite dikes, and areas of pegmatization. In addition to these recognizable areas of enrichment, many TTG suites seem to have been subjected to metasomatism that altered their concentrations of LIL elements without changing the megascopic or microscopic appearance of the rocks.

Massive granite plutons with high concentrations of K feldspar are present in most cratons (fig. 3.6; table 3.2). They are among a suite of rocks classified as "post-orogenic" granites (definition by Rogers and Greenberg, 1990). Most of these plutons formed during the last stages of stabilization of continental crust (chapter 4; Anderson and Morrison, 1992), and their positive $\varepsilon_{Nd}$ and negative $\varepsilon_{Sr}$ show that they were derived largely from asthenospheric mantle or from crustal rocks recently separated from the mantle. Some workers have suggested that intrusion of basalt magmas into the lower crust provided the heat for crustal melting, with the granite magmas formed by this process incorporating components from both the continental crust and the intrusive basalts. This is an example of "intracrustal melting," which S. Taylor and McLennan (1985) proposed operates on a craton-wide scale to move LIL elements upward and leave a depleted lower crust. Generation of these magmas by partial melting under conditions ranging from dry to water-saturated was described by W. Collins (1993) and Landenberger and Collins (1996).

Rocks of granitic composition occur not only as granite plutons but also as widespread pegmatitic dikes and locally as metasomatic permeations into the gray gneisses. Pegmatite dikes and permeations may have formed from late-stage fluids released from crystallizing granitic magmas, but LIL-element-rich fluids were active in many areas that do not contain granite outcrops or show other evidence that magma intrusion occurred. Presumably they were generated metasomatically by fluids moving upward through the gray gneisses, and most well-studied cratons show evidence of widespread fluid movement both during and after their major periods of magmatic and tectonic activity (chapter 4).

Metasomatism that changes rock compositions can occur without changing their lithology or general appearance. Two examples are the addition of

K to graywackes of supracrustal suites in Canada and southern India at uncertain times after the original deposition of the sediments (Fedo et al., 1997; Naqvi et al., 2002b). This addition increased the $K_2O$ content of the clay matrix, resulting in a higher sericite/chlorite ratio, without affecting the proportions of other minerals. These observations raise the possibility that the process of LIL element enrichment is widespread throughout the world's upper continental crust but has not been widely recognized because of the lack of change in the visible appearance of the rocks. The extreme restriction of U to the uppermost continental crust (Zartman and Doe, 1981) may also indicate widespread metasomatism, but the mechanism remains unknown.

## Destruction of Continental Crust

Crust can be destroyed by two methods: erosion and direct reincorporation into the mantle by some process similar to modern subduction. The relative significance of these two processes has almost certainly changed from the early stages of earth history to the present time, with erosion dominant now and direct reincorporation more significant in the past.

### Present Destruction

The most important method of crust destruction at the present time is clearly erosion. The erosion rate is particularly intense from mountains along active continental margins, and the debris is almost immediately deposited in the adjacent trench and carried down the underlying subduction zone. Even sediments deposited in Atlantic-type oceans must ultimately be carried back into the mantle as all oceanic crust older than ~200 million years is destroyed (chapter 1). Judging by present erosion rates, von Huene and Scholl (1991) suggested that sediment being eroded from present continents is subducted and destroyed almost as rapidly as new rock is added to continents by magmatic activity. Despite this high erosion and recycling rate, large regions of the earth's continents still expose very ancient rocks, and we examine this problem in chapter 4.

The most intense destruction of continental crust today is taking place in the Himalayas, where the Indian subcontinent descends under Asia (fig. 3.7). The complex layering of mantle and crust under the Himalayas, including some seismic sections that reveal molten material in the middle crust (K.

Figure 3.7. Himalayan river carrying eroded debris (photo by A.K. Jain, copyright Gondwana Research Group, Japan).

Nelson et al., 1996), suggests that some of the subducted crust is being melted or directly incorporated into the mantle by some other process. There is no compelling evidence, however, that subduction without erosion is responsible for any significant destruction of Himalayan crust. Conversely, uplift rates in the Himalayas (locally much greater than 1 km/million years) cause erosion and produce an enormous amount of debris that is ultimately washed into the Indian Ocean. This debris represents a minimum 25 km of erosion in the Himalayas in the past 20 million years, and a combination of erosion and normal faulting maintains a constant elevation of the Himalayas as they are tectonically uplifted (M. Johnson, 1994; Einsele et al., 1996; Fielding, 1996; Zeitler et al., 2001; further discussion of Himalayas in chapter 5). These data show that more than 90% of the destruction of crust in the Himalayas is accomplished by surface erosion.

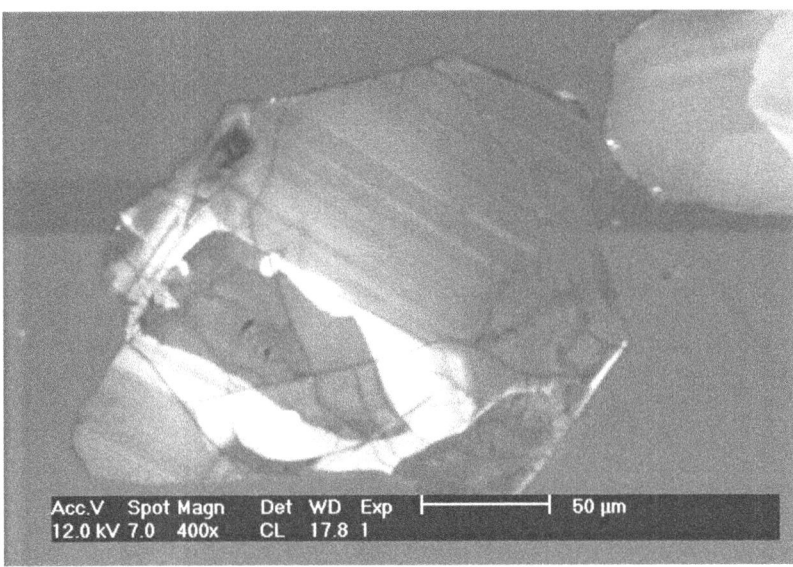

Figure 3.8. Cathodoluminescence image of zircon crystal W74/2–36, the oldest terrestrial fragment known on earth. It was extracted from the Jack Hills Conglomerate in the Narryer Gneiss of western Australia. The zircon is a broken fragment of a large grain composed of a dark central core surrounded by a bright zone that grew in a granite magma undergoing low-temperature interaction with liquid water. The oldest part of the grain is the triangular tip in the extreme lower left corner, with a sensitive high-resolution ion microprobe (SHRIMP) $^{207}Pb/^{206}Pb$ age of 4404 ± 8 Ma (2 sigma). (Photo courtesy of Simon Wilde; see Wilde et al., 2001.)

## Ancient Destruction

(Further information on creation and destruction of ancient continental crust is provided in appendix E by our discussion of the Barberton Mountain region of the Kaapvaal craton of southern Africa.)

Evidence for generation and later destruction of continental crust is present in all ancient cratons. No entire cratons were stabilized at an age older than ~3 Ga (chapter 4), but almost all well-studied cratons contain small enclaves of highly deformed gneiss with ages of ~3.8 Ga that are now incorporated in younger gray gneiss terranes. Rocks with ages between 4 Ga and 3 Ga, however, occupy less than 5% of the current outcrop area of ancient cratons, suggesting either a very low growth rate of continental crust earlier than 3 Ga or a very effective method of destroying the newly formed crust.

Further indication of early generation and destruction of continental crust comes from zircons. The oldest zircons known have an age of 4.4 Ga and occur in fluvial metasediments in western Australia (fig. 3.8; Wilde et al., 2001). Zircons with ages of 3.5 Ga or older have also been found in metasediments in most of the world's shields, indicating significant erosion beginning in the Early Archean. In almost all locations, the sialic continental rocks in which the zircons originally crystallized have not been found, suggesting nearly complete destruction of a significant volume of continental crust.

Much of the discussion about destruction of old continental crust centers around the significance of metasedimentary suites scattered through TTG gneisses. These metasediments locally form coherent suites several kilometers long, but most of them occur in outcrops only tens of meters long or smaller. All of the suites pose questions about the environment in which they were deposited and the source of the debris in the sediment.

The most complete evidence of 3800-Ma crust that was later destroyed is in the Isua region of northwestern Greenland. The Isua supracrustal sequence is exposed along strike for more than 30 km in gneisses on the central west coast of Greenland (fig. 3.9; Rosing et al., 1996; Fedo, 2000). The lower part of the Isua section is dominated by basaltic amphibolites and intruded by other metabasaltic rocks and meta-ultramafic rocks. The lower and middle parts of the section contain chert and ironstone (silica–magnetite rock). The upper part of the Isua sequence con-

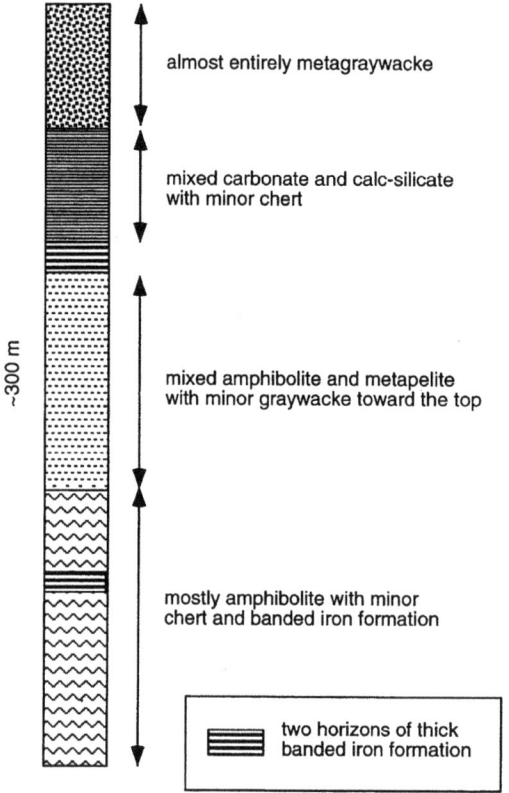

Figure 3.9. Generalized stratigraphic section at Isua section, Greenland. The thicknesses shown in the section have been greatly reduced from those of the original sediments by subsequent deformation. For more detailed information see Nutman and Collerson (1989).

tains metamorphosed pelites and graywackes plus calcsilicates (metamorphosed sandy/shaly carbonates). The lithologies indicate that deposition occurred on a platform covered by shallow water. The basement of this platform is unknown because it was destroyed when the supracrustal suite was engulfed by younger gneisses, but the presence of rounded zircons in some Isua rocks clearly signifies erosion of continental gneissic/granitic precursors.

Early Archean supracrustal suites smaller than Isua are commonly scattered as remnants through gray gneiss terranes. They do not exhibit a coherent stratigraphy but consist of lithologies similar to those at Isua, including quartzite, quartz-pebble conglomerate, ironstone (fig. 3.10), calcsilicates, and mica schists (presumably metamorphosed siltstones and shales). As at Isua, all of these rocks appear to have been deposited in shallow water, but their basements have not been preserved, and their composition is unknown.

The origin of quartz in the sandstones and conglomerates is highly controversial. Much of it is probably clastic, derived by erosion of pre-existing sialic rocks. This source is supported by the presence of zircons in some siliceous metasediments, but whether they are rounded (and thus transported) or oddly shaped because of metamorphic growth is generally unclear. If the quartz and zircon are detrital, then presumably some of the surrounding gneiss is older and formed a basement and/or a source area for detritus. The abundance of quartz in the sandstones and conglomerates now may be much higher than when they were originally deposited because of diagenetic destruction of less stable minerals (Cox et al., 2002; fig. 3.11).

Although erosion of continental crust clearly contributed debris to some Archean sediment, large volumes of sediments older than 3 Ga do not contain quartz or other debris from a continental source. Examples include the oldest greenstone belts in the Western Dharwar craton of southern India (Naqvi et al., 1983), the Warrawoona sequence of the Pilbara craton of western Australia (Barley, 1993), and parts of the Isua sequence described above. All of these suites contain cherts deposited by precipitation and debris from mafic source rocks despite the presence of adjoining gneissic terranes at the same time as the sediments were deposited.

Because large amounts of sialic crust clearly existed much earlier than 3 Ga, the absence of quartz and resistant "granitic" minerals such as zircon in many Archean sediments requires some kind of plausible explanation. One possibility is that the sialic crust that existed before 3 Ga was too thin to establish a subaerial landmass that could be eroded. A second possibility is that oceanic crust was so thick in the early earth that continental areas had very little "freeboard" and seawater covered most of them (chapter 2). This explanation is consistent with observations that show evolution of greenstone belts of the same age as surrounding gneisses but without any sedimentary debris from the gneisses (chapter 4). Furthermore, if most crust was submerged, then its destruction must have been caused by direct reincorporation into the mantle rather than by surface erosion.

Summary  Destruction of continental crust at present is largely caused by surface erosion, but direct reincorporation into the mantle was probably more important than erosion in the past. Because

Figure 3.10. Banded iron formation, showing alternating bands of silica and iron oxide with widths of a few centimeters. The outcrop is partly covered by leaves. Sudan Formation, northern Minnesota (courtesy of John Goodge).

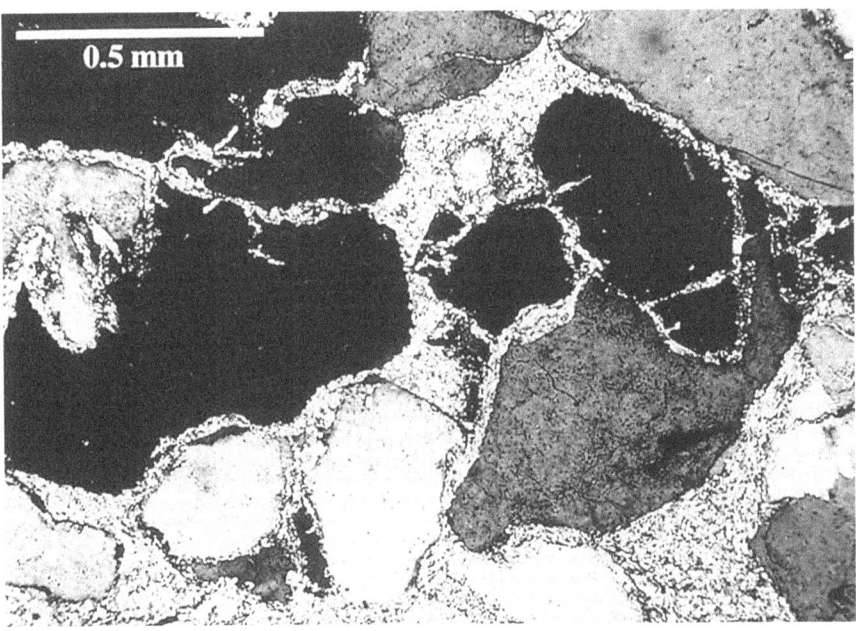

Figure 3.11. Photomicrograph of quartz sandstone showing diagenetic destruction of less stable minerals and replacement by $SiO_2$ (courtesy of Ronadh Cox; copyright by Geological Society of America, 2002).

of the scarcity of rocks older than 3 Ga, the total rate of destruction at that time must have been only slightly less than the rate of production. After 3 Ga, however, production clearly exceeded destruction, although continental crust is now being destroyed at a rate approximately equal to the rate at which new crust is forming.

## Changes in Volume of Continental Crust Through Time

Because both production of juvenile crust and crustal destruction occurred simultaneously throughout the history of the earth, and neither of the rates are well known, many investigators now rely almost completely on isotopic information to calculate growth rates, destruction rates, and changes in the volume of continental crust though time. We discuss below the inferences that can be drawn from the distribution of zircon ages, $T_{DM}$ values, and limited information from Os isotopes (see appendix D for explanation of isotopic systems).

### Isotopic Information

Zircons crystallize from magmas that are rich enough in silica to be quartz normative. Small zircons form in some tholeiitic (silica-saturated) basalts, and they occur in almost all quartz-bearing igneous rocks with compositions ranging from quartz diorite to granite. Therefore, initial ages of crystallization of zircons generally show the times when granites and other silicic igneous rocks were emplaced, and the distribution of these ages through time can be used to estimate rates of formation of continental crust. Detailed study of individual zircons may also reveal whether they crystallized from magmas ultimately derived from a mantle source or from magmas produced by partial melting of older crustal rocks. Remelting of zircon-bearing rocks generally does not consume all of the zircons, and the small remnants commonly form tiny cores around which the newer zircons develop. These rocks may have been derived wholly by remelting of older crust or by mixing of magmas with both juvenile and crustal sources. Conversely, silicic magmas produced by partial melting of mafic rocks from the mantle may not contain zircon cores, thus implying that the entire granite is a juvenile addition to the crust.

Plotting the frequency of zircon crystallization ages, commonly the upper intercepts of discordia, within individual cratons usually reveals one or

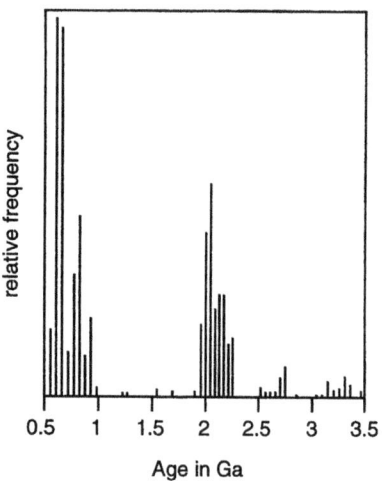

Figure 3.12. Histogram of ages of concordant zircons in South America (based on data in Hartmann, 2002).

more very narrow peaks. Figure 3.12 (Hartmann, 2002) shows the distribution for the Amazonian shield (Brazilian and Guiana cratons; appendix E). Several peaks suggest crystallization of silicic rocks at various times, but the peaks centered around 2.1 Ga and 0.7 Ga are clearly dominant. The 2.1-Ga age is referred to as the Transamazonian event and represents the time at which the Amazonian craton became a stable block (chapter 4). We use $T_{DM}$ ages below to show that 2.1 Ga was a time of major crustal growth, but granites produced during the event at 0.7 Ga consisted almost entirely of remelted older crust.

Values for $T_{CHUR}$ and $T_{DM}$ based on Nd isotopes have been calculated for numerous areas, such as individual cratons and orogenic belts. As an example, we compare $T_{DM}$ data for the Amazon shield with zircon ages discussed above (fig. 3.13; Sato and Siga, 2002). With the exception of a few ages between 0.5 and 1 Ga, mantle separation ages range broadly from ~2 Ga back to ~3.5 Ga and show a significant peak at 2.1 Ga. This corresponds to the 2.1-Ga zircon peak and confirms segregation of large volumes of crust from the mantle at that time. The scarcity of $T_{DM}$ ages at 0.7 Ga, however, demonstrates that the 0.7-Ga event shown by zircon data represents melting of older crust rather than generation of juvenile crust.

The frequency of zircon crystallization and $T_{DM}$ ages throughout the entire earth suggests that the earth's preserved continental crust was developed at a relatively constant rate, with two peaks ~100 million years long at ~2.7 Ga and ~1.9 Ga (fig.

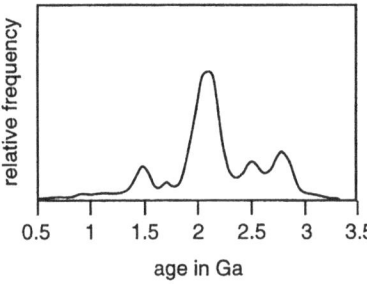

Figure 3.13. Frequency distribution of $T_{DM}$ ages in South America (based on data in Sato and Siga, 2002).

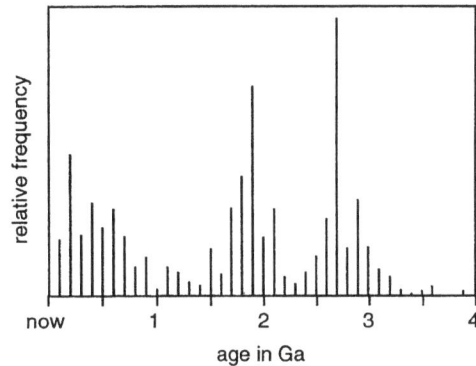

Figure 3.14. Episodic crustal growth based on zircon ages and partly modified by $T_{DM}$ data (based on data in Condie, 2001).

3.14; Condie, 2001). Condie proposed that times of high rates of crustal production were related to periods when numerous slabs of oceanic lithosphere descended below the 660-km discontinuity ("slab avalanches"; chapter 6). This signifies rapid rates of mantle overturn and the possibility of generating large volumes of enriched mantle that can be remelted to form continental sial.

Because numerous measurements are available for some areas and no information for other places, it has not yet been possible to construct a comprehensive frequency distribution of $T_{DM}$ values for the entire earth. The most complete survey available in 2003 was one prepared by Burrett and Berry for 100 separate continental terranes in Australia and western North America (web address in Burrett and Berry, 2001). About half of the values in this survey center around 1.9–1.8 Ga, with the remainder scattered broadly throughout the rest of the Precambrian. This age corresponds to one of the peaks of crust production proposed by Condie (2001), and we discuss its significance further in chapter 7.

The oldest preserved whole rocks known at the time this book was written (2003) formed in several places at 3.8–3.9 Ga or earlier. One of these suites is the Acasta Gneisses of northwestern Canada, and Bowring and Housh (1995) show that $\varepsilon_{Nd}$ values for 10 rocks of this suite range from as high as +3.6 (mantle source) to as low as −4.8 (remelted crustal rocks). This variation in $\varepsilon_{Nd}$ values requires that the earth had already undergone considerable separation into crust and mantle earlier than 4 Ga. Similar data from other investigators summarized by Bowring and Housh (1995) suggest that much of the earth's continental crust evolved earlier than 4 Ga, and younger crust was produced largely from old crust that had been destroyed and recycled back into the mantle. This conclusion has been rejected by other investigators largely on the basis that Sm and Nd may have fractionated from each other during the high-grade metamorphism to which all rocks of this age have been subjected (Gruau et al., 1996; Vervoort et al., 1996; Moorbath et al., 1997).

Progressive depletion of Re from the mantle during various melting episodes has caused the $^{187}Os/^{186}Os$ ratio in the mantle to increase more slowly than in the whole earth (Pearson et al., 1995a, 1995b; Nagler et al., 1997; Meisel et al., 2001). Consequently, $^{187}Os/^{186}Os$ ratios in mantle-derived rocks provide an approximate estimate of the times when various magmas were extracted from the mantle. All conclusions are still tentative, but the limited information available on Os isotopes generally supports conclusions that crustal segregation from the mantle began very early in earth history.

### Rates of Increase in Volume of Continental Crust Through Time

Using some combination of the information that we have discussed above, numerous geoscientists have attempted to quantify changes in the volume of continental crust through time. Early estimates that were based on the ages and areas of exposed terranes used maps that show the areas of continental terranes produced at different times and by different processes. Assuming relatively uniform thicknesses of crust (surface to Moho) in these continents, then the areas of rocks of different age are proportional to the volume of crust generated in these various episodes, and we apparently can use this information to show the rate of production of continental crust.

Unfortunately, these simple calculations do not work very well. One reason is the problem of distinguishing "juvenile" crust that has come directly from the mantle from crust that contains pre-existing sial. Only juvenile crust represents true crustal growth, and suites such as granite batholiths that formed by melting of older basement of the terrane cannot be included in the volume of new crust. Distinction of juvenile and recycled components is extremely difficult even in modern batholithic suites such as the Andes, an issue that we discuss further in chapter 5.

Data obtained from all of the isotopic systems discussed above indicate that the crust and upper mantle of the earth followed different paths of isotopic evolution since early in the history of the earth. Selective movement of LIL elements upward began depleting the mantle at an early age, leaving a mantle with higher ratios of Sm/Nd and lower ratios of Rb/Sr, Lu/Hf, Re/Os, and U/Pb than in the crust. The exact rates of crustal evolution are highly controversial, but all estimates show such rapid production that there must also have been a high rate of crustal destruction throughout most of earth history.

All of the difficulties of determining the volume of continental crust at any one time in earth history have led geologists to propose an extraordinary variety of curves to show changes in volume with time. They all start with the assumption that the initial earth accreted with a relatively homogeneous composition and that all sial separated at a later time. We show four generalized examples in fig. 3.15.

- At least 100 years ago geologists began to realize that the sizes of continents had increased through geologic time. The antiquity of continental "basements" was recognized even before radiometric dating became possible, and in the past half century these ages were quantified. Because of these ages, most geologists assumed that the volume of continental crust increased more rapidly in the past than in more recent times, and some geologists suggested that almost all growth took place in the Archean. An approximate growth curve representing these beliefs is shown as "traditional" in fig. 3.15.
- Suggestions of episodic growth of continental crust have been based on both geologic observations and isotopic data. The possibility that much of the growth of continents occurred at about 1.9–1.8 Ga is suggested by the large number of orogenic belts of this

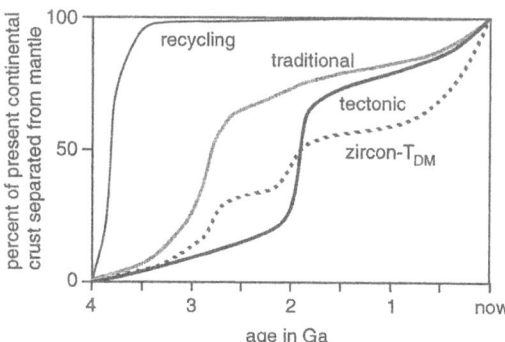

Figure 3.15. Proposed rates of growth of continental crust.

age (chapter 5), and recent summaries of $T_{DM}$ ages tend to support this conclusion (see above). We designate a curve based on these data as "tectonic" in fig. 3.15 (further discussion in chapter 4).
- An episodic growth curve using zircon and $T_{DM}$ ages (from Condie, 2001) is similar to the tectonic curve in showing one peak at 1.9–1.8 Ga, but it also shows another major peak at ~2.7 Ga. We label it as a "zircon–$T_{DM}$" curve in fig. 3.15.
- Some isotopic studies (see above) have been interpreted to show that nearly all of the continental crust fractionated from the mantle very early in earth history and has undergone recycling since then. This possibility was suggested by Hargraves (1976) based on the possibility that the earth was totally covered by oceans very early in its history, which would not have been possible unless distinction between continents and ocean basins did not exist because continental sial of uniform thickness covered the entire earth. We label this curve "recycling" in fig. 3.15.

## Summary

Continental crust evolved from the mantle as part of the process of gravitative segregation of the earth. Separation of crust from the mantle began within a few hundred million years after the earth accreted, but we have only indirect (mostly isotopic) and controversial evidence of the quantity of crust produced at different times in earth history. Generation of large volumes of siliceous magmas required at least a two-stage process in which mantle ultramafic rocks partially melted to form basaltic rocks and the basalts melted again to form magmas that crystallized to quartz-bearing rocks. All preserved continental crust appears to have evolved from regions of the upper mantle that had been enriched in LIL elements shortly before segregation began. Further generation of LIL-element-rich rocks may have been the result of intracrustal melting, metasomatism, and other uncertain processes.

Continual recycling of crust back into the mantle has taken place throughout earth history, presumably starting at about the same time that crustal generation began. This recycling was caused both by erosion of continental crust above sealevel and probably also by direct reincorporation of thin crust into the mantle by a process similar to modern subduction. Direct reincorporation was the principal method of recycling during the early history of the earth, but erosion is the dominant process now. Recycling has destroyed almost all of the rocks older than 3 Ga, but many cratonic areas escaped significant destruction since they were formed.

# 4

# Growth of Cratons and their Post-Stabilization Histories

As we have seen in chapter 3, continental crust evolved from regions of the mantle that contained higher concentrations of LIL elements than regions that underlie typical ocean basins. The most complete record of this evolutionary process is in cratons, which passed through periods of rapid crust production to times of comparative stability over intervals of several hundred million years. After the cratons became stable enough to accumulate sequences of undeformed platform sediments, they moved about the earth without being subjected to further compressive tectonic activity. Because many of the cratons are also partly covered by sediments that are unmetamorphosed or only slightly metamorphosed, they appear to have undergone very little erosion since the sediments were deposited. Thus, a craton may be considered as a large block of continental crust that has been permanently removed from the crustal recycling process.

This chapter starts with a discussion of the history of cratons as interpreted from studies of the upper part of the crust. We describe the Superior craton of the Canadian shield and the Western Dharwar craton of southern India within the chapter and use appendix E for brief summaries of other typical cratons. These cratons and numerous others elsewhere developed at different times during earth history, and we look for similarities and differences that may have been caused by progressive cooling of the earth (chapter 2). This section concludes with a summary of the general evolution of cratons and the meaning of the terms "Archean" and "Proterozoic."

The following section is an investigation of processes that occurred following stabilization, all of which take place in the presence of fluids that permeate the crust. We include a summary of these fluids and their effects on anorogenic magmatism and separation of the lower and upper crust.

The final section discusses the relationship between cratons and their underlying subcontinental lithospheric mantle (SCLM). Continual metasomatism and metamorphism of the SCLM after cratons develop above it apparently has not destroyed the relationship between the ages of the cratons and the concentrations of major elements in the SCLM. This provides us with an opportunity to determine whether cratons evolved from the mantle beneath them or by depletion of much larger volumes of mantle.

The discussions in this chapter are based partly on information summarized in appendices B (heat flow) and D (isotopes).

## History of Cratons

This section describes the evolution of the upper crust of cratons that were stabilized at widely different times. Within the chapter we include summaries of the Western Dharwar craton of southern India (3.0 Ga) and the Superior craton of Canada (2.7 Ga). Appendix E contains summaries of the Barberton Mountain region of the Kaapvaal craton of southern Africa (3.1 Ga), the Guiana craton of South America (2.1–2.0 Ga), and the Nubian–Arabian craton of Africa–Arabia (0.6–0.5 Ga).

### Western Dharwar Craton (Stabilized at 3.0 Ga)

The Western Dharwar craton (WDC) of southern India began to evolve about 3.5 billion years ago and possibly earlier (fig. 4.1; summaries by Rogers, 1986, and Rogers and Mauldin, 1994). The western margin is the Indian Ocean, where the WDC rifted away from Madagascar. The northern margin is overlain by Neoproterozoic/Early Phanerozoic sedimentary basins and the Cretaceous/Tertiary Deccan basalts. The southern margin is a transition to a 2.5-Ga Granulite terrane (see below). The

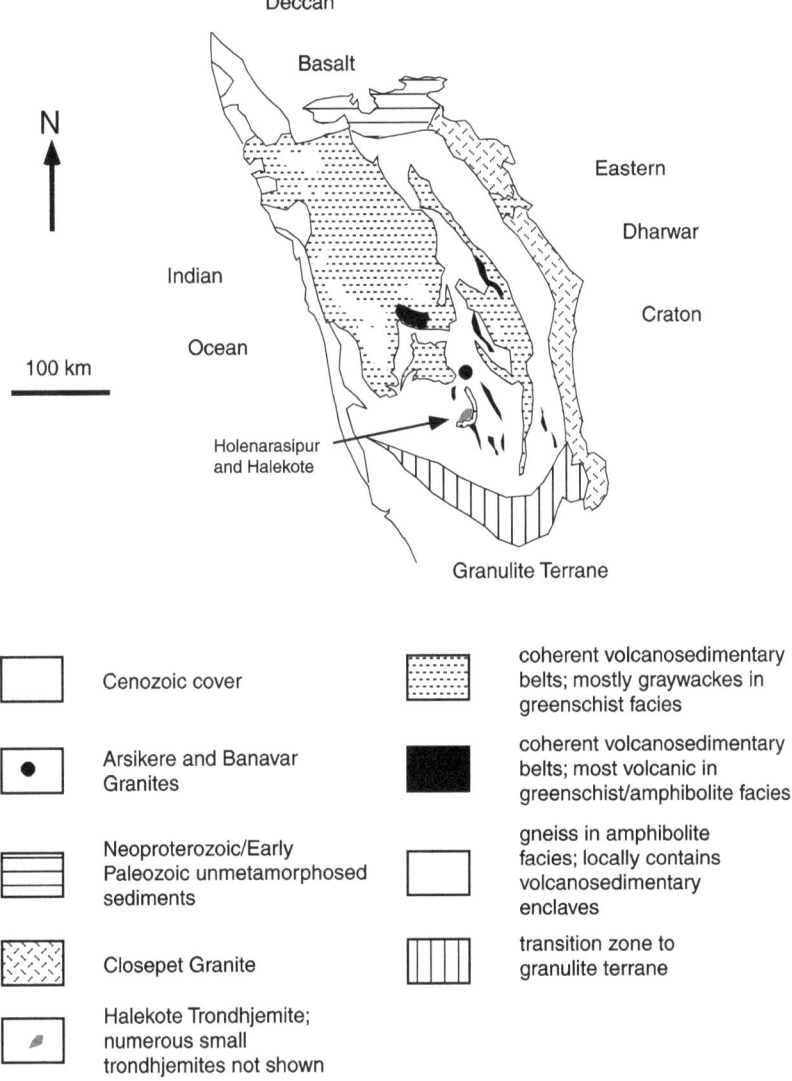

Figure 4.1. Map of Western Dharwar craton. Halekote trondhjemite and Holenarasipur greenstone belt are specifically mentioned in the text.

border between the WDC and the eastern Dharwar craton is commonly regarded as the Closepet granite, although shear zones farther west have also been proposed. The Closepet granite consists of several plutons along a nearly straight belt that extends from the Granulite terrane northward to Deccan cover (Moyen et al., 2001). It formed at 2.5 Ga, contemporaneous with metamorphism in the transition zone.

Most of the WDC basement consists of tonalite–trondhjemite–granodiorite (TTG; "Archean gray") gneisses that are highly deformed. Distinction is difficult between orthogneisses of igneous origin and paragneisses composed of metasediments. Although some small suites near the transition zone are clearly metaquartzites and/or metavolcanics, most of the gneisses are probably metamorphosed TTG plutons. Derivation from the mantle is shown by $(^{87}Sr/^{86}Sr)_I$ values of 0.700–0.701 and $\varepsilon_{Nd}$ of $-1$ to $+2$ (table 3.2). Measurement of $T_{DM}$ ages shows that segregation of sial to form the gneisses began earlier than 3.5 Ga and probably continued until about 3 Ga (Peucat et al., 1993). Zircons show complex overgrowths, with the older parts having ages of 3.2 Ga and older (Peucat et al., 1993).

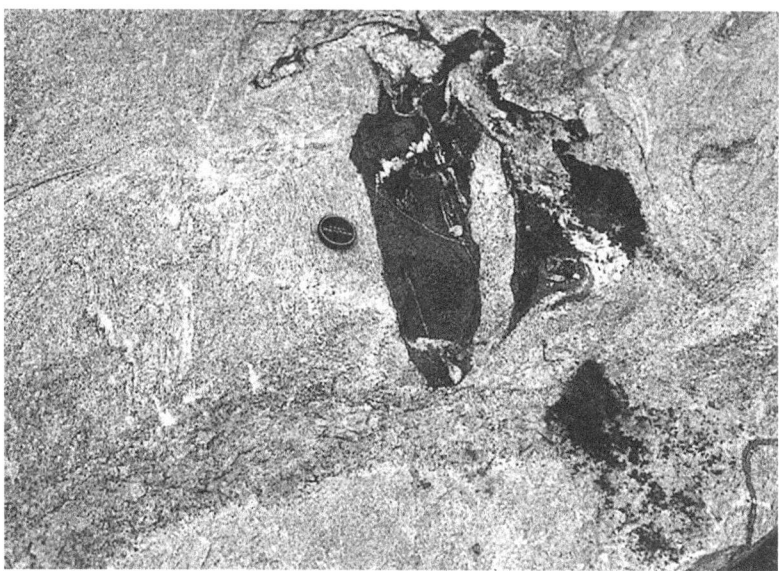

Figure 4.2. Halekote trondhjemite of Western Dharwar craton, southern India. The intrusion contains inclusions of older greenstones and is cut by a younger pegmatite (photo by J.J.W.R.).

Greenstone belts are interspersed through the gneisses (Chadwick et al., 2000). Rocks in them are highly deformed and metamorphosed from greenschist to amphibolite facies, and although ages are difficult to obtain, they appear to be approximately coeval with the gneisses. A typical lithologic assemblage is shown by the Holenarasipur belt, one of the largest greenstone belts in the WDC (fig. 4.1). It contains metamorphic varieties of a large number of rock types, including: claystones and siltstones, some of which contain quartz and some of which are quartz-free; basalts and andesites; komatiites and other ultramafic rocks ranging from extrusive to shallow intrusive; and very minor lenses of quartzite and quartz conglomerate. Limited isotopic studies of greenstone belts show $T_{DM}$ values that indicate derivation from the mantle at approximately the same time as segregation of the gneisses (Peucat et al., 1995).

The intensely deformed gneiss/greenstone terrane contains several small bodies of diapiric, massive trondhjemites, which are megascopically identical to the gneisses except that deformation is absent or minor in the diapiric rocks (fig. 4.2). The principal minerals of both suites are also similar, consisting mostly of a simple assemblage of quartz, oligoclase, minor K feldspar, and small amounts of hornblende and biotite. Accessory minerals in both suites consist of zircon, titanite, and various opaques. The only significant mineralogical distinction between the gneisses and trondhjemites is the morphology of the zircons. Gneiss zircons consist mostly of irregular grains that show no identifiable crystal form, but zircons in the trondhjemites consist mostly of simply or doubly terminated crystals, many of which contain small inclusions with rounded to irregular shapes. They crystallized in the trondhjemite magmas and have not undergone growth or deformation since that time.

Abundant isotopic evidence shows a major event in the WDC at 3.0 Ga. In the gneisses, it is a time of closure of whole-rock Pb–Pb and Rb–Sr systems, and it is also the age of emplacement of diapiric trondhjemites (Meen et al., 1992). The 3.0-Ga age of the trondhjemites is shown as an upper-intercept zircon age and by whole-rock Pb–Pb and Rb–Sr (Meen et al., 1992). Because both gneisses and diapiric trondhjemites show isotopic closure at 3.0 Ga, this age is regarded as the age of stabilization of the WDC.

The 3.0-Ga rocks form a basement for several basins of supracrustal rocks known as the Dharwars, which apparently began to form shortly after 3.0 Ga (Chadwick et al., 2000). They consist of mafic to felsic volcanic rocks plus large volumes of quartz-bearing graywackes. Some of the older suites have been slightly deformed and metamorphosed, but most of them are relatively undisturbed and are regarded as some of the world's oldest platform assemblages.

The 2.5-Ga granulite event in the transition zone is only sparsely recorded in most of the WDC. The

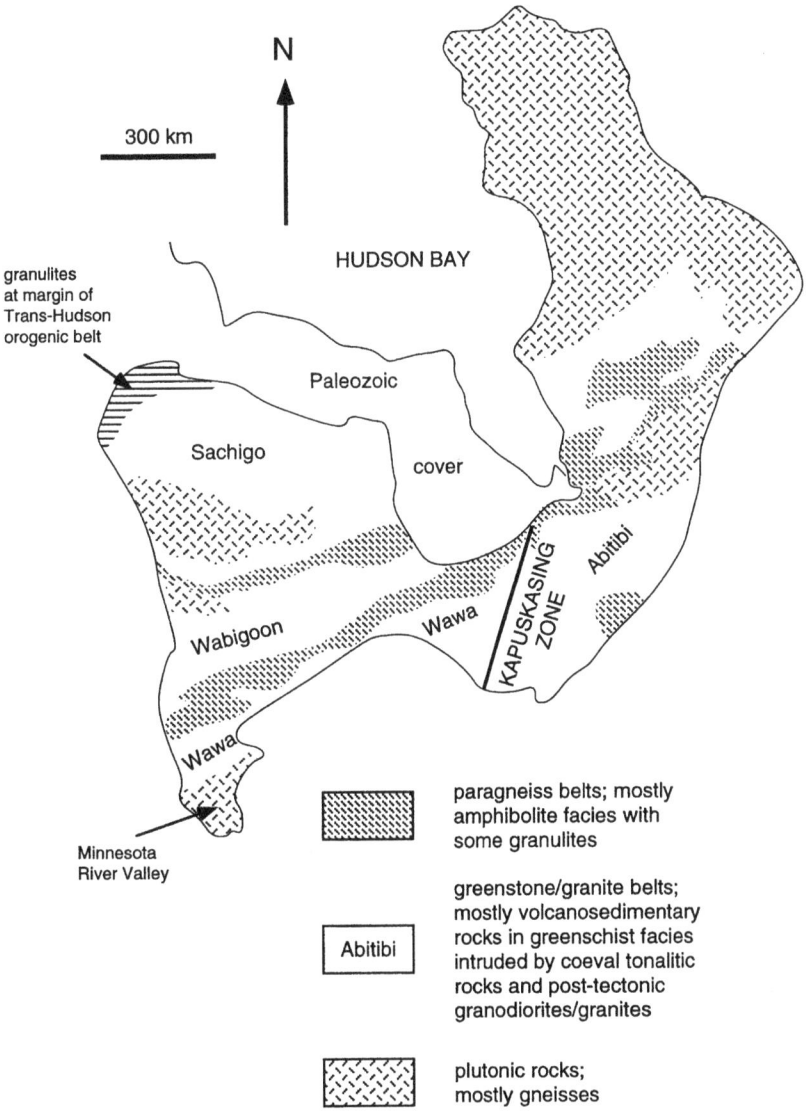

Figure 4.3. Map of Superior craton (generalized from Card, 1990, and Stott, 1997). Greenstone/granite terranes show complex relationships between supracrustal suites and approximately synchronous intrusive rocks; names of the major belts are Abitibi, Wawa (two parts), Wabigoon, and Sachigo. The Kapuskasing zone is discussed in chapter 3.

only igneous activity consists of the Closepet Granite along the eastern margin, the Arsikere Granite (appendix F), some pegmatite dikes, and areas of pervasive pegmatization by K-rich fluids near the Closepet outcrop. Except for the possible resetting of a few mineral isochrons, there is no evidence of metamorphism, deformation, or sufficient heating to affect isotopic systems at this time. The Arsikere Granite is discussed below in our section on post-stabilization processes.

## Superior Craton/Province (Stabilized at 2.7 Ga)

The Superior craton occupies much of the eastern Canadian shield (summaries by Card, 1990, and Stott, 1997; fig. 4.3). Most of the eastern margin is the Grenville granulite belt of approximately 1-Ga age, and much of the belt may be reworked Superior crust (chapter 7). The southern margin is covered by Early Proterozoic metasediments and, farther south, by Phanerozoic platform sediments of midcontinent North America. The western and northern margins are terminated by the Trans-Hudson orogenic belt, and part of the contact is covered by Phanerozoic sediments. The northeastern part of the Superior craton is mostly surrounded by various Proterozoic orogens.

Five lithologic assemblages are recognizable in the Superior craton. The oldest, ~3.5 Ga, is the Minnesota River gneiss terrane, which is apparently exotic to the craton and attached at some time in the Proterozoic. The four other suites were produced almost completely between 3.2 Ga and 2.7 Ga, when the craton was stabilized. They include TTG gneisses, greenstone belts, graywacke suites, and late-stage diapiric intrusions. The gneisses, greenstone assemblages, and graywacke suites are arranged in approximately ENE–WSW belts that were compressed and welded in the Late Archean by subduction down to the NW beneath the southward-growing margin of the emerging craton. Sparse outcrops of quartz conglomerates and sandstones indicate local exposure of gneisses to erosion (Donaldson and de Kemp, 1998), but most of the area was apparently submerged prior to 2.7 Ga.

The TTG gneisses began to form at approximately 3.2 Ga and continued to evolve for about 500 million years. Their $T_{DM}$ ages range from 3.4–2.7 Ga, and $\varepsilon_{Nd}$ and $\varepsilon_{Hf}$ range from small positive values to small negative values (Henry et al., 2000). Precursors of the oldest gneisses apparently separated from the mantle, but younger magmas incorporated some melts from slightly older gneisses and components of greenstone belts. Deformation of the gneisses caused some fragments to split off the major bodies and become incorporated as exotic blocks in the greenstone belts.

Greenstone belts occupy much of the Superior craton, but only part of them consists of greenstone (Corfu et al., 1998). The remainder consists largely of syntectonic and post-tectonic plutons ranging from diorite to granodiorite. Although approximately 35 belts have been identified, almost all of the area of greenstone outcrop is in the few large belts named in fig. 4.3. The Abitibi belt (fig. 4.4) contains the following supracrustal assemblages (summarized from Jackson et al., 1994, and Wyman et al., 2002):

- tholeiitic metabasalt and metakomatiite (fig. 4.5) with variable amounts of intermediate to silicic metavolcanics; probably formed in areas of extension;
- calcalkaline intermediate to silicic metavolcanic flows and pyroclastic rocks, probably

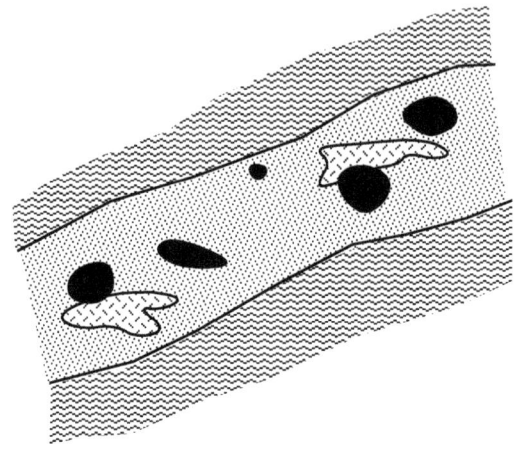

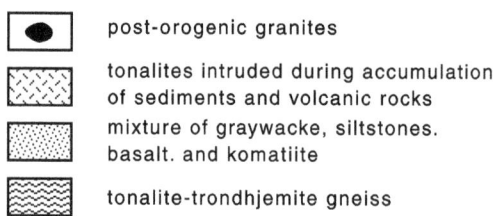

Figure 4.4. General relationships among lithologic suites in greenstone belts. Detailed discussion of Abitibi belt is in Wyman et al. (2002).

formed in island arcs; the volume of andesite is minor, with most of the suite being high-$SiO_2$ dacite;
- minor boninite (basalt with high $SiO_2$ and MgO and low LIL element concentrations);
- minor adakite (andesite with high $Al_2O_3$ and MgO);
- iron formation deposited on older gneisses and possibly forming part of the basement for younger supracrustal rocks;
- turbiditic metasediments that are younger than most of the volcanic activity; most of the debris is from metavolcanic rocks within the greenstone belt;
- fluvio-deltaic metasediments that are younger than all of the other suites and are locally associated with alkalic metavolcanics; they contain debris from older rocks within the greenstone belts and also from exposed TTG gneisses.

The volcanic rocks were produced in an oceanic setting and derived from a large variety of sources (Polat and Kerrich, 2001; Wyman and Kerrich, 2002). These sources apparently include subducted

Figure 4.5. Thin section showing spiniflex texture in komatiite from Munro Township, Ontario, Canada. Field of view approximately 4 mm wide. Large olivine crystals are bladed in three dimensions. Area between olivine crystals consists partly of altered glass and partly of clinopyroxene spherulites formed during quenching of the melt. (Photo courtesy of Tony Fowler; for further information about spiniflex texture see Shore and Fowler, 1999.)

oceanic crust, above subduction zones, and probably a high contribution from mantle plumes. The plume sources could include both plumes directly beneath the Abitibi belt and oceanic plateaus formed above plumes elsewhere and transported into the Abitibi belt.

All of the supracrustal suites in the Abitibi and other belts were intruded by plutonic rocks with compositions ranging from tonalite to granite. Figure 4.4 shows general locations of plutons in part of the Abitibi belt, but the pattern of small plutons and intervening supracrustal rocks is far more complex. The plutonic rocks formed during a range of a few tens of million years centered around 2.7 Ga, commonly referred to as "Kenoran," and their emplacement coincides with the end of pervasive deformation throughout the craton. Work in the central Abitibi belt summarized by Davis et al. (2000) shows that the plutons have $(^{87}Sr/^{86}Sr)_I$ ranging from 0.701–0.703 and $\varepsilon_{Nd}$ of 0 to +3, both of which are consistent with derivation of most of the magmas from the mantle.

Metasedimentary (paragneiss) belts consist of amphibolite-facies assemblages that show the same ENE–WSW trend as the greenstone belts. Sources of the sedimentary debris include gneisses and all rock types of the greenstone belts. Current marks and other sedimentary structures show that the sediments were deposited on shallow-water platforms and along the margins of developing microcratons. Despite this appearance of deposition in environments of crustal stability, the metasedimentary suites were intensely deformed along with greenstones and gneisses before and during the stabilization of the craton.

Flat-lying Proterozoic sedimentary suites are preserved at various places in the Canadian shield, but not on the Superior craton. They are known at depth under surrounding Phanerozoic cover, however, and may have covered much of the Superior craton before being removed by erosion.

## General History of Cratons and the Meaning of "Archean" and "Proterozoic"

The cratons described above and in appendix E illustrate the general pattern of development of continental cratons (fig. 4.6). Cratonic basement develops during an early stage lasting a few hundred million years, when deformation and metamorphism of all rocks is intense. All cratons show roughly synchronous development of mantle-derived TTG magmatic rocks and greenstone belts that contain a wide variety of volcanic, volcaniclastic, and sedimentary rocks. Source rocks of the sediments are primarily within the greenstone belts until late in the development of the craton.

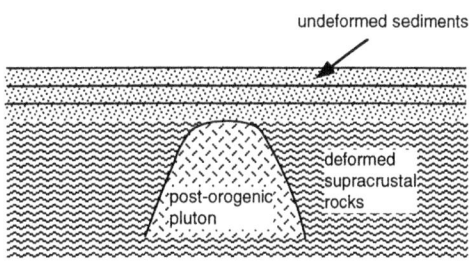

Figure 4.6. General history of cratons.

Crustal stability ("age of stabilization" or "age of cratonization") is attained at an age determined from four observations (1 to 4). The most important (1) is the beginning of deposition of undeformed platform sediments in shallow basins on the stabilized crust. These sediments commonly lie on top of undeformed granite plutons (2) that are only a few tens of million years older than the oldest sediments. The age of intrusion of the granites also represents the youngest age of compressive deformation of basement gneisses (3) and the youngest age of whole-rock isochrons (4).

All cratons seem to have followed the same pattern of development at all ages from 3 Ga to 0.5 Ga, and we can recognize only three differences between old and young cratons. One is that komatiites are restricted to old crust developed when the earth's thermal gradients were high. A second is that older volcanic suites tend to be bimodal basalt–rhyolite, whereas younger ones are complete basalt–andesite–rhyolite sequences. The third difference is that plutons intruded during stabilization of the Guiana and Nubian–Arabian cratons consist of granite dominated by K feldspar, whereas equivalent plutons in the older Superior and Western Dharwar cratons are granodiorites and tonalites/trondhjemites dominated by sodic plagioclase. All of these differences may be related to a mantle that was more "primitive" in the early earth than at more recent times.

Differences between cratons of different ages raise the issue of the meanings of the terms "Archean" and "Proterozoic." When it was first used, Archean referred to rocks that were presumed to be so different from younger rocks that they must have evolved under very different conditions. Further investigation, however, showed that they consist of sedimentary, volcanic, and intrusive rocks similar to recent rocks, with degrees of metamorphism also in the same ranges as young rocks in Phanerozoic orogenic belts. Comparison of Archean and younger environments is discussed in more detail by Nisbet (1987).

Rocks referred to as Archean, however, are different from younger ones in the amount of deformation that most of them have undergone, with the oldest sialic rocks in most areas creating a basement of highly deformed gneisses that are overlain by less-deformed suites. Thus, geologists commonly refer to the Archean as a time in earth history when the earth was hotter and more mobile than it was later. When geologists working in Canada discovered that this change took place at about 2.5 Ga, a standard time scale was developed in which the Archean–Proterozoic boundary was established at 2.5 Ga.

After an age of 2.5 Ga was designated for the boundary, work in other Precambrian areas began to show similar transitions from a very mobile environment to a more stable one at different times. For example, our discussion of cratons shows that the transition occurred in the Western Dharwar craton when it was stabilized at ~3.0 Ga. Similarly, the Guiana craton became stable at ~2.0 Ga and the Nubian–Arabian craton at ~0.5 Ga (appendix E). These ages have led some geologists to regard the Archean–Proterozoic boundary as a time of tectonic change, which occurred at different times in different areas (Master, 1990).

As we discuss above, we regard the transition from mobility to stability as a very important time in the history of each craton, and we recognize that it took place at different times in different cratons. For simplicity, however, we will continue to refer to events older than ~2.5 Ga as Archean and those between ~2.5 Ga and ~0.5 Ga as Proterozoic.

Using 2.5 Ga as a starting date, the Proterozoic extends to the base of the Cambrian, which is marked by the development of skeletal organisms at 540 Ma (chapter 12). Subdivisions of the Proterozoic are not standard, and we use a classification in which Paleoproterozoic extends from 2.5 Ga to about 1.8 Ga, Mesoproterozoic from 1.8 Ga to about 1.0 Ga, and Neoproterozoic from 1 Ga to the Cambrian at ~0.54 Ga.

## Processes During and Following Cratonic Stabilization

Crustal stabilization merely signifies the end of pervasive compression and the development of broad sedimentary platforms. It does not mean that a craton has become inert to further internal processes. All cratons are permeated by fluids throughout their post-stabilization history, locally causing anorogenic magmatism more than 1 billion years after the craton originally formed. We start this section with a general discussion of fluids and amplify it by discussions of four anorogenic igneous suites in appendix F. We finish with a discussion of a transition zone between upper and lower crust in southern India and show that it was developed largely by fluid movement.

### Fluids

Fluids play an important role in modifying the composition of the earth's crust by transporting and redistributing various elements. They are also

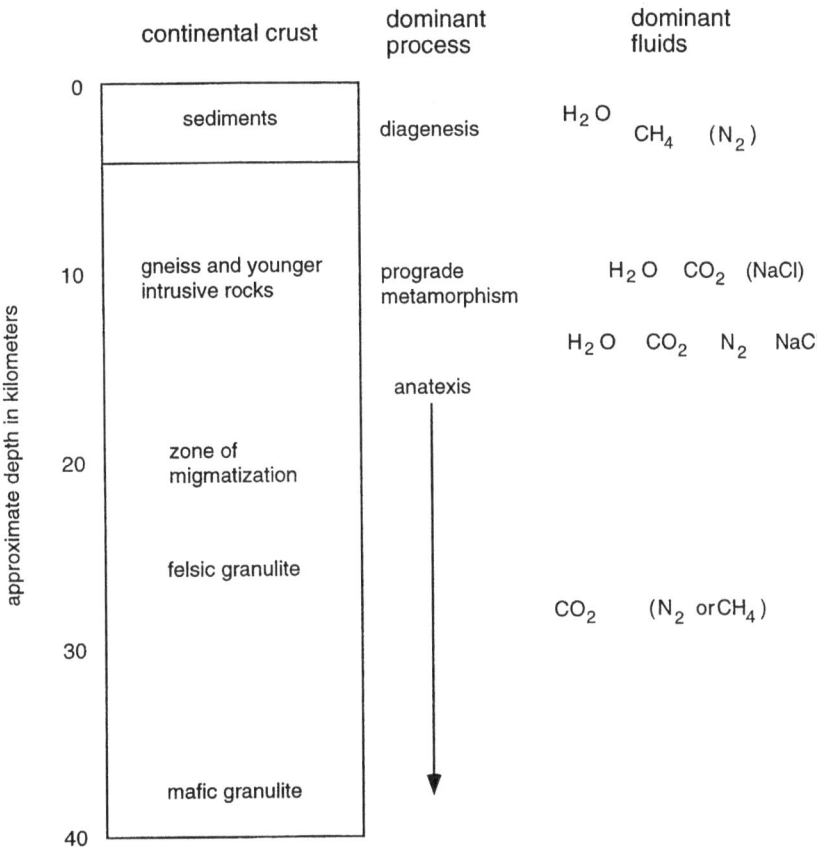

Figure 4.7. General distribution of fluids in continental crust (based on Touret and Dietvorst, 1983).

important scavengers of metals and have been instrumental in the formation of a variety of economic mineral deposits. Metal-enriched fluids focused along crustal pathways such as shear zones or fault zones give rise to rich veins or concentrations of metallic mineral deposits. Seepage of ore-bearing fluids over larger areas of the crust gives rise to disseminated mineral deposits.

Evidence for the nature and activity of fluids in the continental crust comes from field observations, petrographic examination, and geochemical studies. Their pervasive effect has been noted in most cratons, including the Superior province (Kerrich and Ludden, 2000). Distribution of major and minor elements in minerals and rocks, their isotopic patterns, and calculations of mineral–fluid equilibria based on various mineral assemblages also yield potential information on the activity of fluids.

Although a variety of fluids and their mixtures have acted upon the earth's crust, the dominant fluid species are $CO_2$ and $H_2O$ with variable concentration of salts (mainly chlorides and carbonates) and traces of other volatiles such as $CH_4$, $N_2$, and $SO_2$. A general stratification in the distribution of fluids shows that: the upper crust is dominated by $H_2O$ and brine solutions with various amounts of $CH_4$ and/or $N_2$, as trace admixtures derived from supracrustal sources; $CO_2$–$H_2O$ fluids dominate the deeper parts of the upper crust; and $CO_2$-rich fluids characterize the lower crust (fig. 4.7; Touret and Dietvorst, 1983).

Good outcrops, commonly in quarry sections in granulite terranes, show structurally controlled transformation of garnet- and biotite-bearing upper-amphibolite facies gneisses to orthopyroxene-bearing greasy green zones of dry granulite (incipient charnockite) on a mesoscopic scale (Santosh, 2000; Santosh and Yoshikura, 2001; Santosh et al., 2001). The arrested charnockite patches, lenses, and trees are reminiscent of fluid pathways along which copious amounts of $CO_2$ have influxed from deeper sources, leading to the desiccation of the gneisses and stabilization of the dry granulite assemblages which characterize the

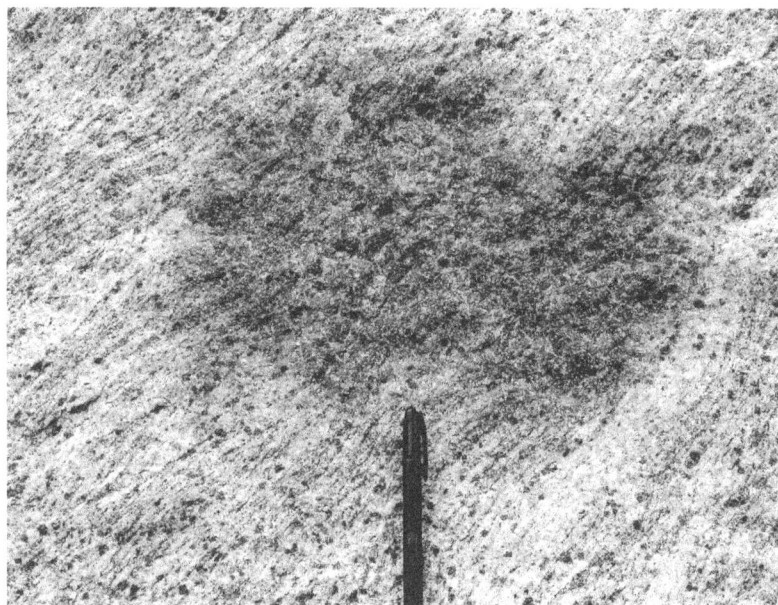

Figure 4.8. Gneiss–charnockite relationship at Kottavattom, southern India. Incipient charnockite (dark) with coarse orthopyroxene crystals developed within garnet–biotite gneiss as a result of influx of $CO_2$-rich fluids along structural pathways. (Photo by M.S.)

incipient charnockites. Spectacular examples of fluid pathways in these areas are displayed by the incipient charnockites of southern India and Sri Lanka (fig. 4.8). Some workers believe that these oriented features are the harbingers of regional granulites, although others consider them to be a local phenomenon.

Fluid-induced transformations similar to those in southern India and Sri Lanka are preserved by the eclogite veins in anorthosites of Bergen, southwest Norway (Austrheim, 1986). At Bergen, strongly channelized fluid influx occurred into the lower crust, and the dense networks of channels locally coalesced by permeation along foliation to convert tracts of many square meters to eclogite.

One of the potential tools to detect fluid-induced alterations in rocks and minerals is stable isotope geochemistry, particularly that of carbon, oxygen, and sulfur. Gigantic oxygen isotopic shifts of the order of 24 parts per mil brought about by meteoric waters have been traced from single crystals of calcite in carbonate rocks of Skallen region, East Antarctica (Satish-Kumar and Wada, 1997). Spectacular carbon isotopic zonation has been observed in some large crystals of graphite from southern India (Santosh and Wada, 1993a, 1993b). While the cores of these graphites are characterized by isotopic values resulting from biogenic origin, their rims show precipitation by reduction of $CO_2$-rich fluids derived from ultimate mantle sources.

Direct evidence for the nature, composition, and activity of fluids has come from studies on fluid inclusions trapped within minerals (Santosh et al., 1990). When minerals grow from a medium such as liquid, gas, or melt, portions of the mineral-forming medium are trapped within various growth zones to form different generations of fluid inclusions. These micron-sized sealed cavities offer important clues on the composition and density of the fluids, pressure–temperature conditions of formation of minerals, and exhumation history of the rocks. Our present knowledge on the nature of fluids in the deep crust relies heavily on observations of fluid inclusions in granulite-facies mineral assemblages and mantle xenoliths. Although a variety of fluids and their mixtures are reported from these rocks, the dominant fluid phase in most cases is high-density $CO_2$. The highest-density $CO_2$ so far observed in crustal rocks comes from ultrahigh density fluid inclusions trapped within garnet from a mafic granulite in southern India (fig. 4.9; Santosh and Tsunogae, 2003; $CO_2$ density: 1.164 g/cc, entrapment pressure 10 kbar at 800°C).

The abundance of $CO_2$-rich fluid inclusions has contributed substantially to the debate on the origin of the continental deep crust. The increase in amount of $CO_2$ from lower-grade to higher-grade

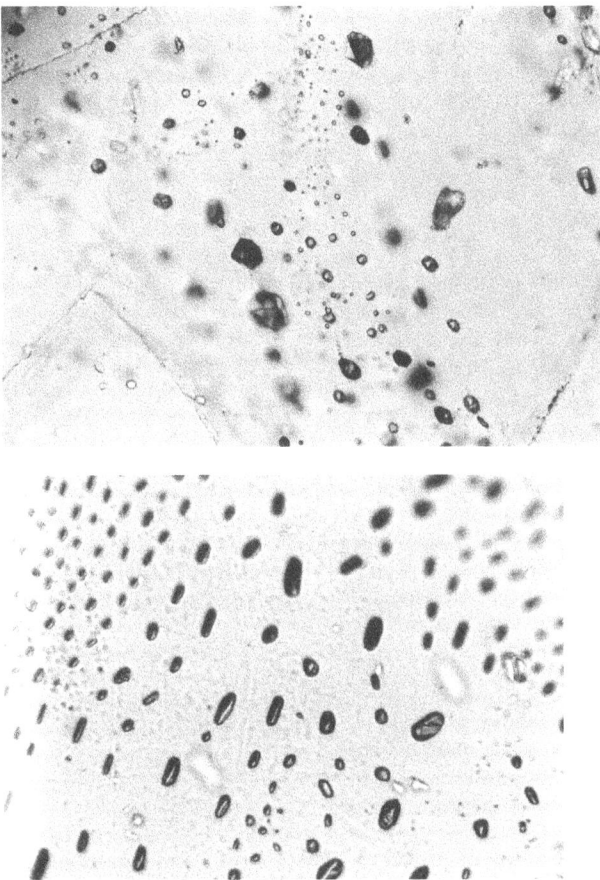

Figure 4.9. Photomicrographs of fluid inclusions containing very high-density $CO_2$ in dark ovoid and rounded cavities up to 30 microns long. The upper photograph from a garnet granulite from southern India contains $CO_2$-filled inclusions with densities up to 1.1 g/cc in quartz (Santosh and Tsunogae, 2003). The lower photograph shows high-density (1.0 g/cc) $CO_2$ trapped within fluid inclusions in garnet from a mafic granulilte in Sri Lanka. (Photos by M.S.)

rocks, and the dominantly anhydrous composition of the rocks found in most of the exposed continental deep crust, prompted some workers to propose pervasive flooding of the deep crust by $CO_2$ degassed from the mantle. However, some deep crustal terranes do not preserve evidence for $CO_2$-induced transformations. They are thought to have been derived either through vapor-absent metamorphism by the intrusion of hot magmas, or through the extraction of water in granitic melts, leaving an anhydrous residue.

Indeed, $CO_2$ and $H_2O$ could have played a major role in the genesis of the continental crust and some of the related mineral deposits. However, many questions remain unanswered. Where does the $CO_2$ originate? We know that $CO_2$ can be generated through a number of mechanisms such as decarbonation of carbonate rocks, or by oxidation of carbonaceous matter. The $CO_2$-rich fluids commonly associated with deep crustal rocks are thought to have been derived from ultimate mantle sources and transferred to higher crustal levels through either magmatic conduits or mantle-rooted shear zones. Since $CO_2$ has only limited solubility in normal silicic magmas, it is possible that underplated alkali basalts could have transported $CO_2$ in a dissolved state and released it upon freezing. Some $CO_2$ could also have been transported as suspended carbonate minerals in magmas rather than as a free fluid phase.

Many other questions also are unanswered. If a substantial amount of $CO_2$ is liberated from the mantle, in what form is it stored there? Does the weathering of $CO_2$-enriched rocks contribute significantly to the $CO_2$ budget of the atmosphere, with the continental crust acting as a temporary store of $CO_2$ on its passage from the mantle to the crust? Opinions are diverse, and the topic continues to be a challenge to geologists.

## Anorogenic Magmatism Related to Fluids

Anorogenic magmatism appears to occur in almost all of the world's cratons at times ranging from a few hundred million years to more than 1 billion years after stabilization. Similar rocks also intrude orogenic belts long after the final compressional event. Most of the rocks produced by anorogenic magmatism have high concentrations of alkalies and other LIL elements, and many of them contain abundant Sn, rare earth elements (REE), and other ore elements that are mobilized by fluids. Even rocks that are not ore-bearing were affected by fluids containing $H_2O$, $CO_2$, HF, and other components discussed above. Many of them are broadly regarded as "ring complexes," some are carbonatitic, and some are mantle-derived kimberlites.

The time of emplacement and types of rocks distinguish anorogenic suites from the post-orogenic granites discussed in chapter 3. Post-orogenic suites commonly consist solely of granite intruded within a few tens of millions of years following cratonic stabilization or the end of compressive orogeny, whereas anorogenic suites contain the highly variable lithologies mentioned above and follow crustal stabilization after a much longer time.

We describe four anorogenic suites in appendix F and briefly summarize what we learn from them in this section.

- The Arsikere Granite of southern India was intruded at 2.5 Ga, approximately 500 million years after the stabilization of the Western Dharwar craton (see above). The major minerals are typically granitic, but restriction of Ti to sphene instead of magnetite, high F contents in mica, and primary fluorite indicate that the granite magma was generated when HF invaded a relatively anhydrous lower crust.
- Alkaline rocks and carbonatite at Alno Island, Sweden, were intruded about 1.5 billion years after stabilization of the Baltic craton at ~2 Ga. Fluid phases included both $H_2O$ and $CO_2$, and some of the $H_2O$-rich phases contained as much as 40% NaCl. Fluids not only mobilized magmas but also penetrated wall rocks and caused widespread alteration.
- Alkaline and silica-undersaturated rocks formed a ring complex at Abu Khrug, Egypt at 90 Ma. They are part of a series of anorogenic complexes intruded into the Nubian–Arabian shield over a period of several hundred million years after it became stable (appendix E). Fluids at the various complexes included $H_2O$, $CO_2$, and HF, all of which caused generation of the magmas and also penetrated surrounding rocks.
- The volcano Ol Doinyo Lengai in the Eastern Branch of the East African rift valleys was constructed mostly by nepheline-rich flows and pyroclastic rocks and is now the only active volcano in the world to erupt natrocarbonatite (sodium carbonate). This eruption of carbonate lavas shows extreme concentration of $CO_2$ at depth in addition to $H_2O$ and other fluids.

*Summary* Anorogenic magmatism shows continued escape of fluids from the earth's interior more than 1 billion years after stabilization of the overlying crust. Although the fluids are dominated by $H_2O$ and $CO_2$, some of them also contain HF and dissolved NaCl.

## Separation of Upper and Lower Crust in Southern India

Exposure of the lower continental crust and the transition zone between the lower and upper crust is very rare. The best known region occurs in southern India, where the Indian subcontinent has been tilted northward as continuing spreading along the Carlsberg ridge in the northwestern Indian Ocean pushes the Indian plate farther north beneath the southern margin of Asia. The resultant northward tipping of India exposes deep crustal rocks over a larger area than anywhere else in the world, and the northern margin of this granulite terrane is a transition zone to amphibolite-facies gneisses. We describe it where it forms the southern margin of the Western Dharwar craton (fig. 4.1).

Along the southern boundary of the WDC the transition zone is a few tens of kilometers wide (Pichamuthu, 1960; Condie et al., 1982). Rocks to the south are typical felsic granulites with a primary mineralogy consisting of quartz, K feldspar, sodic plagioclase, and a small amount of orthopyroxene ± biotite and accessory minerals. Farther to the north, exposures show a mixture of gneisses and

Figure 4.10. Transition between amphibolite-facies gneiss and granulite-facies charnockite at Kabbaldurga, India. The charnockite is dark layers and lenses in the lighter colored gneiss. (Photo courtesy of M. Jayananda.)

granulites, with the granulites occurring mostly in veins or "stringers" surrounded by amphibolite-facies gneiss (fig. 4.10). This mixed rock suite passes farther northward to a homogeneous mass of gneiss with amphibole and biotite.

Because both amphibolitic gneiss and silicic granulites are closely intermingled in the transition zone, both rocks clearly came to metamorphic equilibrium at the same temperatures and pressures. The only reason the amphibolitic and granulitic rocks could coexist must have been variation in the compositions and pressures of fluids, which were extremely important in forming the entire granulite terrane (see above). This conclusion is reinforced by studies of fluid inclusions that show a predominance of $CO_2$ in the granulite stringers and water in the surrounding gneisses (E. Hansen et al., 1995). Apparently expulsion of $CO_2$ from the mantle and lower crust along zones of weakness in the transition zone caused dehydration and conversion of hydrous amphibole to anhydrous orthopyroxene.

Water driven upward by $CO_2$ encroachment carried LIL elements with it because of their solubility in hydrous media. This left a depleted lower crust, which has very low concentrations of Rb and Ba, and produced an upper crust enriched in U, K, and other LIL elements. Evidence of fluid movement, LIL element depletion, and production of silicic granulites has major implications for the nature of the lower continental crust worldwide (chapters 3 and 5).

## Relationship Between Cratons and their Underlying Upper Mantle

The mantle directly beneath continents is referred to as "subcontinental lithospheric mantle," commonly abbreviated as SCLM. The SCLM is now known to be significantly different from mantle underlying oceanic crust, and it also seems to have slightly different compositions from one cratonic area to another. This relationship remains intact despite continued thermal metamorphism and metasomatism of the SCLM from the formation of the overlying crust to the present. We begin with a discussion of this metasomatism and then continue with the significance of compositional patterns.

### Metasomatism of the Subcontinental Lithospheric Mantle

Studies of the concentrations of LIL elements in the SCLM produce remarkably inconsistent results. The differences probably result from the extreme variability of the metamorphic/metasomatic process from place to place and also with time in the same place. We illustrate this variability by discussing only a few of the numerous studies that have investigated the process.

Compositional alteration of the SCLM is well demonstrated by xenoliths in the 90-Ma Matsoku kimberlite pipe in Lesotho (Olive et al., 1997; appendix F). The peridotite was intruded into the Kaapvaal craton, which was stabilized at 3.1 Ga, but osmium isotope studies of peridotite xenoliths show a range of younger ages, with apparently unaltered peridotite yielding ages of 2.2 to 1.2 Ga, and ages of veins, dikes, and metasomatized peridotite ranging from ~300 Ma to ~150 Ma. These ages show that alteration of the mantle removed Re from different phases and established their Os isotopic systems at different times throughout the entire 3-billion-year history of the craton.

Similar young alteration is shown by peridotite (spinel lherzolite) xenoliths in Quaternary volcanic rocks erupted through Early Paleozoic to Mesozoic crust in the Yukon, Canada (Carignan et al., 1996). The xenoliths have similar ratios of the various Pb isotopes, which indicates that they should have had similar U/Pb ratios throughout their history. The $^{238}U/^{204}Pb$ ratios range from 12.2 to 46.5, however, suggesting that U and Pb metasomatism occurred as recently as 30 Ma. Earlier metasomatic alteration is also shown by $^{87}Sr/^{86}Sr$ ratios that range from 0.7033 to 0.7050 in different xenoliths.

Several complexes of carbonatite and alkaline rocks occur along the transition zone between upper and lower crust in southern India (see above; summary from Pandit et al., 2002). A complex intruded in the Paleoproterozoic has $(^{87}Sr/^{86}Sr)_I$ of ~0.702 and $\varepsilon_{Nd}$ of 0 to +1, indicating derivation from a depleted mantle. Neoproterozoic complexes formed approximately 1 billion years later in the same area, however, have $(^{87}Sr/^{86}Sr)_I$ ranging from 0.705 to 0.707 and $\varepsilon_{Nd}$ of −6 to −17, indicating enrichment of the mantle source in LIL elements at some time during the middle part of the Proterozoic.

Despite these demonstrations of mantle metasomatism, some studies show that the SCLM has been unaltered since the overlying craton was formed. Bell and Blenkinsop (1987) described more than 25 carbonatites with ages from 2.7 Ga to 120 Ma in the Superior craton of the Canadian shield. Most of the carbonatites are associated with Phanerozoic rifts or other major faults, but some suites are in locations that are not clearly associated with zones of structural weakness. The carbonatites of the Superior craton also occur as isolated bodies that are not comagmatic with silicate rocks of alkaline ring complexes, as they are in many other regions. Because carbonatites apparently are derived from

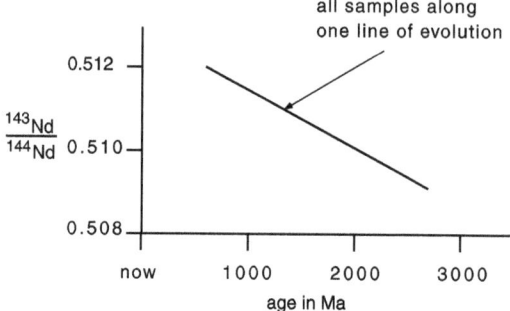

Figure 4.11. Nd isotope relationships in carbonatites of Superior craton, Canada (modified from Bell and Blenkinsop, 1987).

the mantle (presumably SCLM; see above), Bell and Blenkinsop (1987) used the isotopic composition of the Superior carbonatites as representative of mantle composition at the time they were formed. Variations in initial $^{143}Nd/^{144}Nd$ ratios (fig. 4.11) and in $^{87}Sr/^{86}Sr$ ratios with age show that Rb/Sr and Sm/Nd ratios in the mantle source regions did not change between 2.7 Ga and 120 Ma. Furthermore, the Rb/Sr = 0.020 is lower, and the Sm/Nd = 0.358 is higher, than the ratios for the unfractionated silicate portion of the earth. These ratios indicate that the upper mantle beneath the Superior craton had undergone significant depletion of Rb and Nd, and presumably other LIL elements, by the time of stabilization at 2.7 Ga and apparently has not been enriched since that time.

Summary Most studies of xenoliths from the upper mantle beneath continents show that fluids passing through the mantle have added or removed LIL elements that affect isotopic systems. This alteration apparently has continued throughout the lifetime of the overlying craton and presumably is still in operation at present. A few studies, however, suggest that the SCLM has remained inert since the cratonic crust was extracted from the mantle. Available data suggest that the metasomatic process is too variable to provide useful generalizations at the present time.

## Relationship Between Composition of SCLM and Age of Overlying Cratons

The most compelling information about the SCLM and its relationship to cratons comes from studies of the compositions and equilibration conditions of mantle xenoliths in volcanic rocks and kimberlites, and we discuss this before adding the information provided by studies of heat flow and isotopes.

Worldwide investigations of compositions, equilibration conditions, and ages of mantle xenoliths have been summarized in a series of papers by Griffin et al. (1998, 1999), O'Reilly et al. (2001), and Poudjom Domani et al. (2001). They show progressive changes in SCLM composition from Archean cratons through Proterozoic cratons to areas that became stabilized during Phanerozoic orogeny (see discussion of Archean and Proterozoic above). One observation is that SCLM under younger crust contains higher concentrations of CaO and $Al_2O_3$, which leads to the conclusion that younger mantle is less depleted than older mantle. This difference may be the consequence of decrease in thermal gradients through time, but it may also be partly the result of anorogenic magmatism and movement of fluids through cratons following their stabilization.

Another major finding is that SCLM beneath Archean cratons contains much higher Mg/Fe ratios than under Proterozoic cratons. The difference is almost certainly caused by the decrease in the earth's thermal gradients with time (chapter 2). Iron preferentially fractionates into a liquid when ultramafic rocks undergo partial melting, and higher temperatures of melting produce larger percentages of liquid. Thus, more iron is extracted from ultramafic rocks at high temperatures, leaving a mantle that is relatively depleted in iron and richer in magnesium. The high Mg/Fe ratio makes the SCLM of Archean cratons less dense than mantle with higher Fe concentrations, which contributes to the ability of these old rocks to remain buoyant.

Some confirmation of differences in thermal gradients and heat production in mantle and crust can be obtained from studies of surface heat flow.

Rudnick et al. (1998) proposed that mean surface heat flow increases progressively from Archean regions to areas of Phanerozoic orogeny (table 4.1). This relationship may signify progressive increase of mantle heat production toward younger areas, possibly because thickness of mantle lithosphere also decreases toward younger cratons (Nyblade and Pollack, 1993). This interpretation is controversial, however, because relationships between age and reduced heat flow suggest that at least some of the differences in surface heat flow are caused by low heat production in older crust (Jaupart and Mareschal, 1999).

Isotopic studies of the differences between the SCLM of cratons of different ages are highly controversial. The Rb–Sr isotopic system is clearly unreliable because of metasomatic alteration after cratonization, and many workers have suggested the Sm–Nd system is also subject to modification. At present, the most acceptable isotopic information on mantle age comes from the R–Os system, which unfortunately has been investigated in only a few areas.

Several Re–Os studies of ancient SCLM indicate very old ages of stabilization of $^{187}Os/^{188}Os$ in mantle peridotites. They include $^{187}Os/^{188}Os$ ratios as low as 0.106 in the Kaapvaal craton (Pearson et al., 1995a), 0.104 in the Tanzanian craton (Burton et al., 2000), and 0.108 in the Aldan craton (Pearson et al., 1995b). All of these values are below the range of bulk-earth $^{187}Os/^{188}Os$ in the Archean and indicate that the SCLM under these cratons was depleted before 2.5 Ga.

Although these and other studies of Os isotopes show that isotopic systems were established at about the same age as the overlying crust, the SCLM beneath individual cratons tends to blend into surrounding mantle. One demonstration of

Table 4.1. Relationship of heat flow to crustal age

|  | Surface heat flow | Contribution from crust | Contribution from mantle (reduced heat flow?) |
| --- | --- | --- | --- |
| Archean | 41 | 20–30 | 11–21 |
| Proterozoic near Archean | 47 | 25–35 | 12–22 |
| Proterozoic far from Archean | 55 | 33–43 | 12–22 |
| Phanerozoic | 49–55 | high | low |
| Continental average | 48 | 35–40 | 10–15 |

Heat flow is in milliwatts/m². Archean is older than 2.5 Ga and Proterozoic from 2.5 Ga to 0.5 Ga. Reduced heat flow is shown as the same for both Proterozoic terranes. Measured values of reduced heat flow may include lower crust as well as mantle. Based on data in Rudnick et al. (1998).

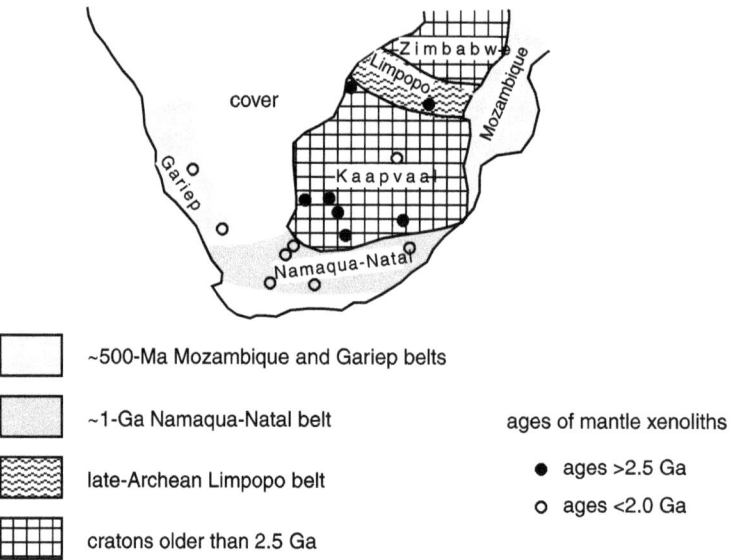

Figure 4.12. Ages of xenoliths in and around Kaapvaal craton (modified from Carlson et al., 2000). The Archean Limpopo belt separates the Kaapvaal craton from the slightly younger Zimbabwe craton (further discussion in chapter 5). Late Proterozoic mobile belts are Damaran, Mozambique, Gariep, and Namaqua–Natal.

this gradation is studies by Carlson et al. (2000) and Schmitz and Bowring (2001) that show that Re‑Os ages greater than 2.5 Ga in the Kaapvaal craton gradually decrease to younger ages in the Late Proterozoic and Phanerozoic orogenic belts around the southern margin of the craton (fig. 4.12). Similar studies of gradation in Re–Os ages are less available for younger cratons, such as those in South America and North Africa, but there is some indication that $^{187}Os/^{188}Os$ ratios in the SCLM of these areas evolved to higher values than in regions of Archean cratons. Another demonstration of modification of the SCLM is a study by Lee et al. (2001), who used $^{187}Os/^{188}Os$ ratios to show alteration of the SCLM during the Proterozoic in the southwestern United States.

As we discussed in chapter 3, continental crust was produced by some process that removed felsic material rich in LIL elements from the mantle. The source of this material is constrained by the relationship between ages of cratons and the composition of their underlying SCLM, which implies that cratons are still lying above the mantle that existed at the time they formed. This suggests that each craton was derived directly from its underlying SCLM, possibly because of depletion of stationary plumes or because the mantle source was continually resupplied with LIL elements by some other process. Conversely, if the continental crust was produced by depletion of large areas (or all) of the mantle, possibly by long-continued subduction of fresh mantle lithosphere under the craton, then the SCLM should show little variation from one craton to another. The issue is unclear at present and is closely related to the question of whether modern styles of subduction operated in the Archean (chapter 2)

## Summary

Rocks now exposed in cratons evolved from the mantle over periods of time that generally lasted several hundred million years. After this evolution, cratons became stable at times when penetrative compression stopped, undeformed post-orogenic granites were intruded, and platform sediments were deposited on the cratonic basement. Different cratons became stable at times both before and after the traditional Archean–Proterozoic boundary at 2.5 Ga, leading to the possibility that the boundary should be defined tectonically instead of as a specific time.

Preservation of shallow-water sediments indicates that cratons apparently were removed from the crustal recycling process shortly after they became stable. Following stabilization, however, all cratons have undergone extensive modification by fluids, which generate anorogenic magmatism more than 1 billion years after stabilization in some cratons. This modification seems to have had a significant effect on the concentrations of LIL elements in the underlying mantle, but general relationships between ages of cratons and the compositions of the mantle beneath them demonstrate that the sialic rocks of cratons probably evolved largely from the mantle beneath them.

# 5

# Assembly of Continents and Establishment of Lower Crust and Upper Mantle

Continents are very large areas of stable continental crust. After their initial accretion, they rift and move about the earth but undergo compressional deformation almost entirely on their margins. We start our discussion of continents by identifying the varieties of terranes that come together to create them. Because accretion of terranes requires closure of oceans between them, we continue our discussion by describing two different processes of closure. Then we recognize that assembly merely develops a group of terranes, and they must be "fused" or "welded" together before they can be a coherent continent. This process takes place partly during assembly, but most of it appears to be the result of post-collisional processes that continue for tens to hundreds of millions of years.

This fusion develops lower continental crust and subcontinental lithospheric mantle (SCLM), the part of the upper mantle directly underlying continental crust, that have similar, although slightly variable, properties across the entire continent. The lower crust and SCLM are separated by the seismic discontinuity known as the Moho (chapter 1), and we finish this chapter by describing the lower crust and SCLM and variations in the depth of the Moho through time.

## Types of Continental Terranes

Many of the blocks involved in continental accretion are "exotic" terranes that formed somewhere away from the continent and became "allochthonous" when they accreted to the continent. They include large continental blocks that collide with each other, small continental fragments that accrete to the margins of existing continents, intraoceanic island arcs, and small amounts of oceanic lithosphere. Terranes formed on the margin of a growing continent are regarded as "autochthonous" terranes. We recognize two of them: continental-margin magmatic arcs and sediments accumulated on passive margins.

### Large Blocks that Collide with Each Other

Collision of large continental blocks causes intense orogeny. We illustrate this process with the collision of the Russian platform and the Siberian plate to form the Urals in the Late Paleozoic (fig. 5.1; Fershtater et al., 1997; Puchkov, 1997; Friberg and Petrov, 1998; Brown and Spadea, 1999).

The East European (Russian) platform was formed by the fusion of the Baltic and Ukrainian cratons at ~2 Ga. It was the site of widespread intracontinental rifting from ~1.5 Ga to 0.5 Ga (chapter 6), and from the Late Proterozoic through the Paleozoic it formed a broad platform for the accumulation of shallow-water sediments. The eastern margin of the platform grew eastward (present orientation) by passive-margin sedimentation during this entire time.

The Kazakhstan block also developed during the Late Proterozoic and Paleozoic. A core of exposed Proterozoic rocks grew westward by accretion of island arcs as intervening oceanic lithosphere was subducted under the cratonic core. At some time the Kazakhstan block was sutured to the Archean Anabar–Angara craton, forming a broad Siberian plate.

At the start of the Paleozoic the Russian and Siberian blocks were separated by an unknown distance (fig. 5.1). By the Middle Paleozoic the two blocks were approaching each other, separated by several ocean basins that contained at least one small block of continental crust (microcontinent) and two or three island arcs. In the southern Urals eastward subduction of ocean lithosphere developed the large Magnitogorsk arc, beginning in the Devonian with bimodal basalt–rhyolite suites and progressing upward to andesitic assemblages.

Assembly of Continents, Establishment of Crust and Mantle 67

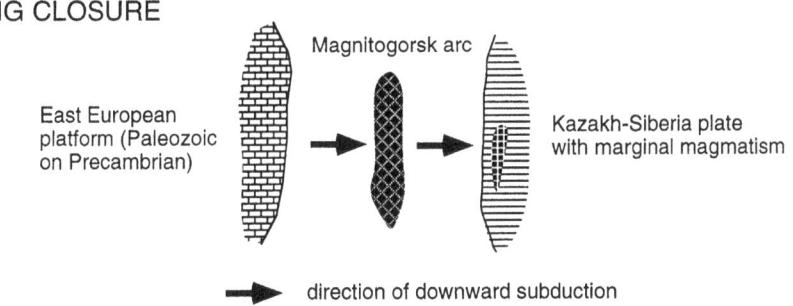

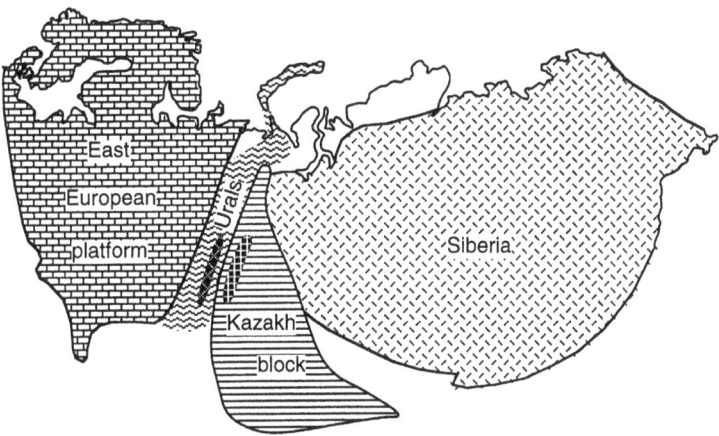

Figure 5.1. History of collision that formed the Urals (based on Puchkov, 1997).

Final collision of the Russian platform and Siberia caused the Uralian orogeny in the Permian.

During closure, the Siberian block, numerous arcs and oceanic fragments, and the large Magnitogorsk arc were thrust over the Russian platform on the "main Uralian fault" (fig. 5.2). The rising Urals formed a foreland basin to the west, and sediments deposited in it and on the European platform were deformed by thrusting and folding. The major orogeny, however, is exposed in uplifted rocks to the east. This area contains numerous ophiolites, the deeply exposed Magnitogorsk arc, and metamorphosed oceanic sediments and arcs collected by the Siberian plate as it moved westward.

Subduction down to the east caused intrusive magmatism throughout the development of the Urals, both within the orogen and in the Kazakh–Siberian plate to the east. Plutonic rocks began to form in the western part of the orogen in the Devonian, and suites become progressively younger toward the east. This eastward progression in age is

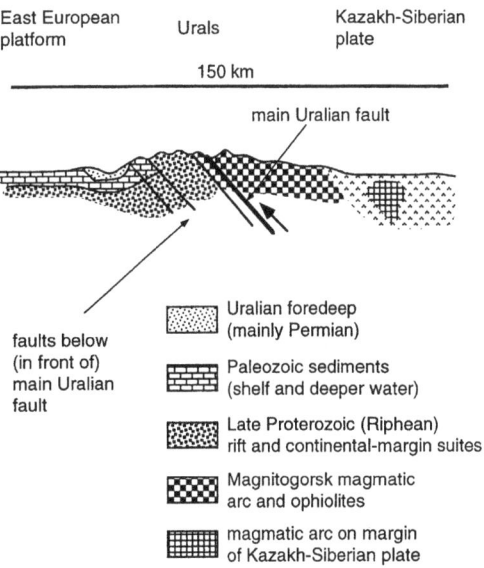

Figure 5.2. Cross section of Urals (based on Brown and Spadea, 1999).

accompanied by increase in LIL elements, which are particularly enriched in the anatectic granites that formed toward the end of the Paleozoic.

The orogeny that produced the Ural Mountains welded a large area of Asia to the European platform. It occurred at the same time as numerous other orogenies in Asia led to the formation of the Eurasian landmass at the end of the Paleozoic. All of this activity was a major part of the formation of the supercontinent Pangea, which we discuss more completely in chapter 8.

## Small Continental Blocks that Accrete to Continental Margins

Many orogenic belts are partly occupied by small crustal blocks that were originally parts of older continents, including continents on either side of the orogen. Some are fragments of cratons covered by platform sediments, and they are commonly referred to as "microcontinents." Some fragments are parts of continental-margin orogenic belts, locally containing magmatic suites only slightly older than the orogens in which they occur. Fragments of passive margins thinned by rifting are locally overlain by younger volcanic suites before or after incorporation in the orogen.

All of these types of blocks occur in a suite of terranes that are called Avalonian in eastern North America and Cadomian in southern Europe (fig. 5.3; papers in D'Lemos et al., 1990, and Nance and Thompson, 1996). Development of an understanding of these terranes is an important part of the history of geology and contributed greatly to modern concepts of the evolution of continents and supercontinents.

In the late 1800s, paleontologists working in Europe and North America recognized that Early Paleozoic fossils in parts of eastern North America and southern Europe were very different from those of comparable age elsewhere in the two continents. Gradually they realized that the fossils in these terranes were much more like fossils in Africa and South America than those in northern Europe and the interior of North America. These fossils were then called "Gondwanan" because Africa and South America had clearly been part of the supercontinent Gondwana (chapter 8), and this raised the question of how these fossils could have traveled to other continents.

Before the acceptance of continental drift, some geologists devised elaborate explanations for movement of fossils or their larvae around the earth. Then, the earliest concepts of continental drift showed how movement of blocks could occur, but they did not provide an explanation for this particular distribution of Early Paleozoic fossils. The first acceptable explanation was found only when geologists realized that much of the tectonic history of the Paleozoic could be explained by progressive closure of ocean basins between Gondwana and North America–northern Europe (fig. 5.3). The Ordovician–Silurian Taconic orogeny in North America and Caledonian in Europe were the earliest phases of deformation that continued with only minor interruption throughout the rest of the Paleozoic. Orogenies in the Devonian and Carboniferous are referred to as Acadian in North America and Variscan or Hercynian in Europe. By the end of the Paleozoic, complete closure of the Iapetus Ocean between North America and Africa and the Tornquist Ocean between Europe and Africa was a major part of the development of Pangea.

Exotic terranes are major components of the Acadian and Variscan/Hercynian belts. All of them contain Lower Paleozoic suites with Gondwana fossils, but their basement lithologies are remarkably diverse. The Carolina slate belt is a series of metavolcanic and volcaniclastic rocks that includes abundant rhyolites erupted on a thin continental crust. The type Avalonian terrane of Newfoundland is dominated by a thick suite of shallow-water shales and quartzose siltstones that have not been deformed and contain one of the world's best transitions from latest-Precambrian soft-bodied fauna to lowest-Cambrian skeletal forms. The Armorican massif of northwestern France consists largely of volcanosedimentary and subduction-related plutons with ages centered around 600 Ma emplaced into and against a 2-Ga basement (B. Miller et al., 2001).

Fossils show that the Avalonian/Cadomian terranes were still attached to Gondwana in the Ordovician before they rifted away and began traveling toward North America and Europe. Tens of millions of years later, the ocean between North America/Europe and the Avalonian/Cadomian terranes closed by subduction under the exotic terranes, which contain subduction-generated plutoni suites. All of the terranes were incorporated in volcanosedimentary assemblages that were developing along the European and American margins.

Several models have been proposed for the location of the Avalonian/Cadomian terranes in Gondwana and their method of travel to North America and Europe. Rast and Skehan (1983) proposed that Avalonia was a small continent that rifted away from Gondwana and collided with

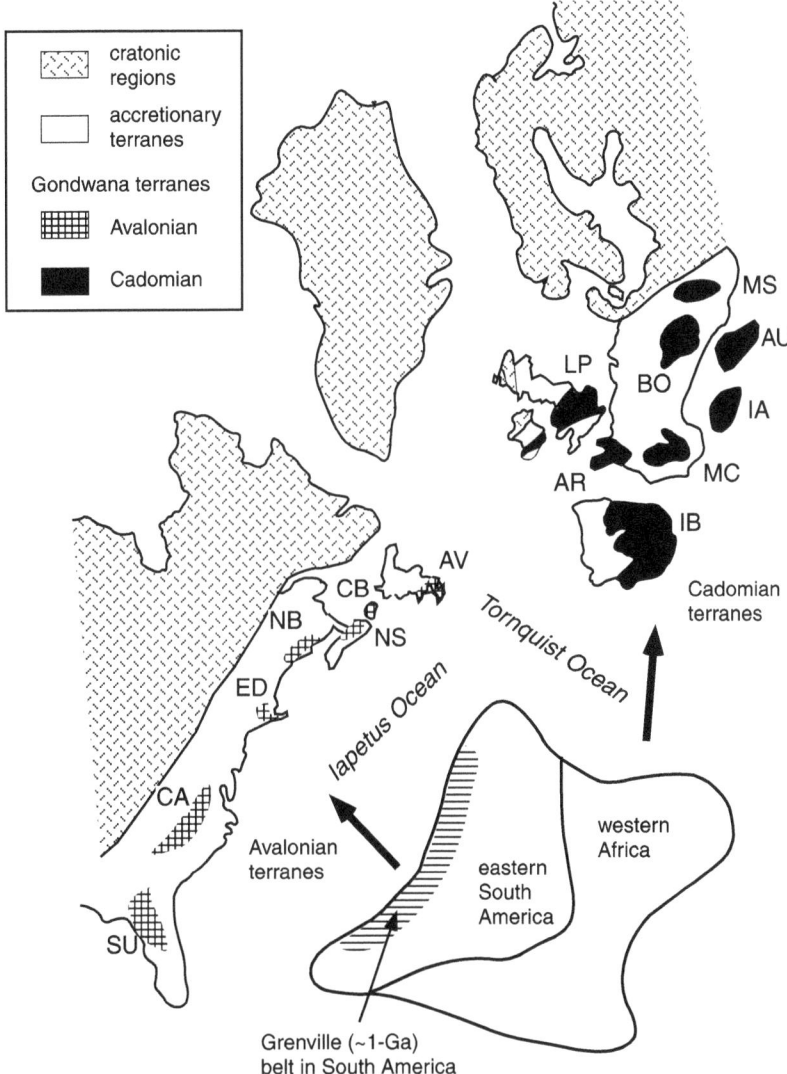

Figure 5.3. Avalonian–Cadomian blocks (from Mallard and Rogers, 1997). Avalonian blocks are: SU, Suwanee; CA, Carolina; ED, Esmond–Dedham; NB, New Brunswick; NS, Nova Scotia; CB, Cape Breton; AV, Avalonian. Cadomian blocks are: IB, Iberian; AR, Armorican; LP, London platform; MC, Massif Central; IA, Intra-Alpine; AU, Austro-Alpine; BO, Bohemian; MS, Malopolska. Source location and migration of Avalonian–Cadomian blocks are discussed in the text.

North America/Europe as a single block, with the individual terranes tectonically separated during and after collision (we discuss a similar proposal for the Cimmerian terranes of southern Asia in chapter 10). Other proposals describe rifting of individual fragments from Gondwana and accretion to North America/Europe as separate terranes.

The part of Gondwana where the terranes originated is only partly known. Rifting of North America from South America seems to require that the Avalonian terranes came from western South America, although some paleomagnetic data suggest an African source (McNamara et al., 2001). The source of the Cadomian terranes is also uncertain. Mallard and Rogers (1997) used $T_{DM}$ and upper-intercept zircon ages to show that the basements of all of the Avalonian terranes had ages of about 1 Ga (Grenville), but the basements of the Cadomian terranes are older. Because ~1-Ga belts occur along the western margin of Amazonia but

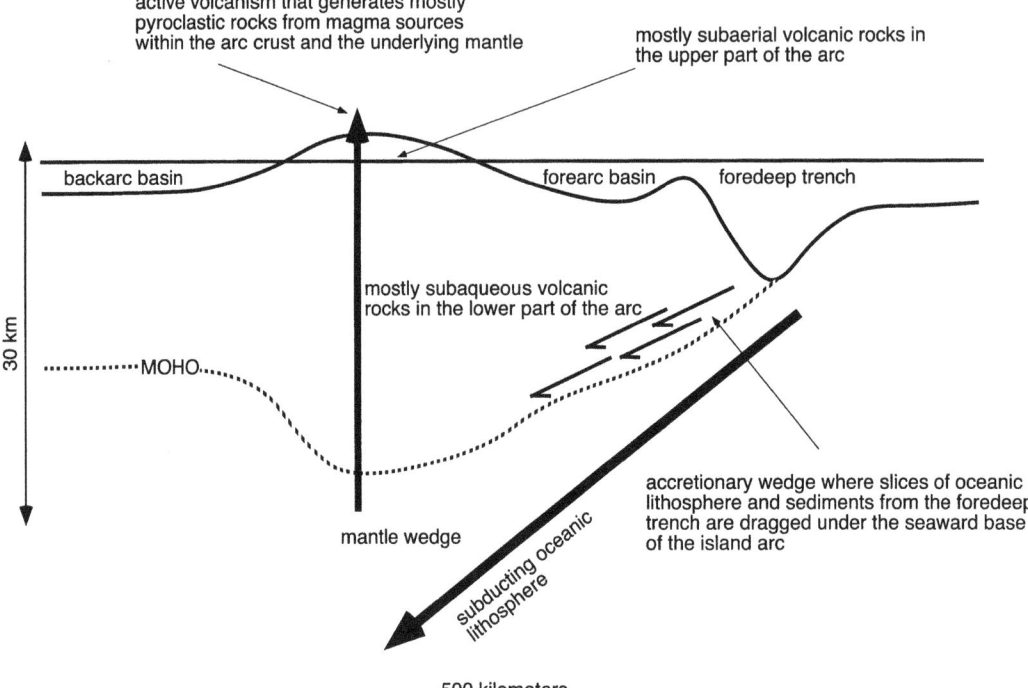

Figure 5.4 Typical features of island arc. The accretionary wedge (accretionary prism) separates the foredeep trench (subduction zone) from the elevated island arc. The forearc basin is shown with an accumulation of sediments in the collapse structure on the top of the wedge. The mantle wedge is mantle between the subducted slab and the crust of the arc (including accreted material). The back-arc basin commonly overlies extensional parts of the subducted slab.

are absent from northern Africa, this difference in basement ages suggests a model for rifting and accretion of the Avalonian/Cadomian terranes as shown in fig. 5.3.

*Intraoceanic Island Arcs*

Island arcs form in different tectonic settings. Some are purely intraoceanic, whereas narrow ocean basins separate others from continents. Some arcs pass laterally into continental-margin fold belts or are built against continental crust. A typical arc system (fig. 5.4) comprises: a trench filled by pelagic and turbidite sediments; a tectonic melange of thrusted oceanic and terrigenous sediments plus small layers of ophiolites; an active arc trench gap that includes a forearc basin with diverse sediments; and a wide magmatic arc composed of high-level volcanic rocks and deep-seated plutonic intrusions. The backarc or interarc basin (also called marginal sea) forms by backarc spreading and rests on thin oceanic crust.

The Japanese arc represents a typical mature intraoceanic arc system with a thick subarc crust of 30–35 km (fig. 5.5; Hashimoto, 1991; Kano, 2001). It began to form during subduction beneath Asia and continued as the opening of the Japan Sea separated the arc from mainland Asia. A large component of the crust includes intrusive rocks and possible detached segments of continental basement. The volcanic rocks are predominantly andesites but show a range of compositions that enables them to be classified into three series: tholeiitic on the seaward side of the arc, calcalkali farther inland, and alkali toward the northwest.

This magmatism occurred on and within accretionary complexes of Late Paleozoic to Cenozoic age. They show a distinct zonal arrangement as narrow linear belts several tens of kilometers wide and 100 to 1000 km long, all of them with E–W longitudinal axes. The complexes are older on the Japan Sea (continental) side and become younger toward the Pacific Ocean.

Major pre-Jurassic units comprise the ~300-Ma (Permian) accretionary complex of the Hida mar-

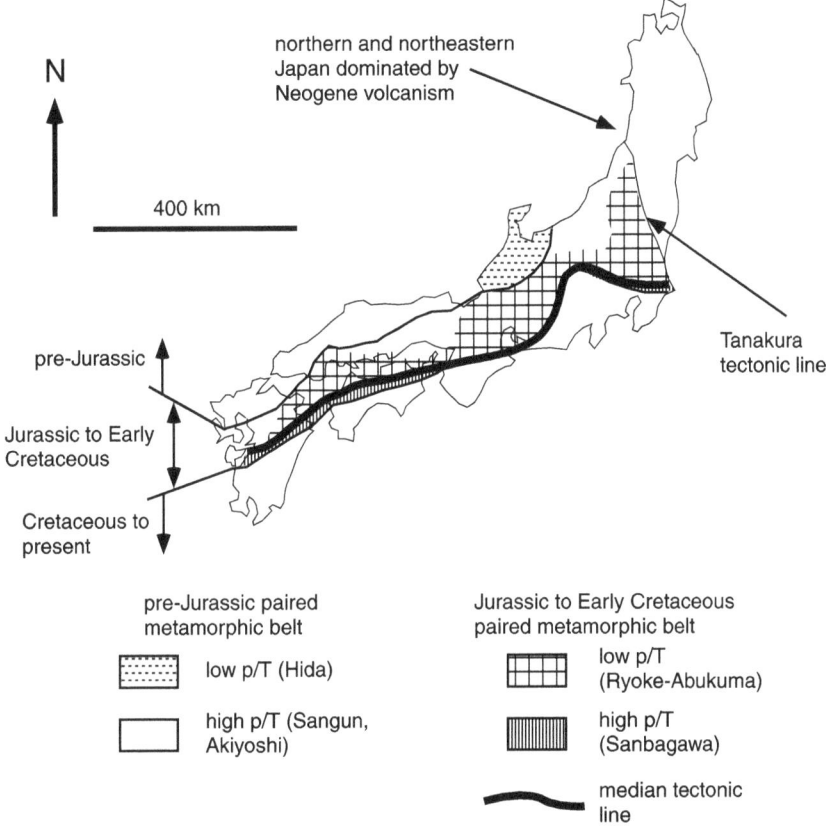

Figure 5.5. Geology of southern Japan (omitting Hokkaido and northern Honshu). The median tectonic line separates high $p/T$- and low $p/T$- facies in suites of Jurassic to Early Cretaceous age. A similar relationship may have occurred in pre-Jurassic rocks between the Sangun/Akiyoshi and Hida terranes, but no median tectonic line is drawn there because of uncertainty about its position. The Cretaceous Kurosegawa terrane consists of small outcrops in the Chichibu belt just seaward of the Sanbagawa terrane. The Tanakura line is a major tectonic boundary. Based partly on Ichikawa (1990).

ginal belt and high $p/T$ metamorphic rocks of 230–180-Ma age in western Japan. From northwest to southeast, Jurassic to Early Cretaceous accretionary complexes include metamorphic rocks of the Ryoke belt (including the Mino–Tanaba terrane), the Sanbagawa belt, and Chichibu belt, which contains outcrops of the Kurosegawa terrane. Belts farthest to the south and east, along the Pacific coast, are less metamorphosed and commonly grouped as the Shimanto complex.

The Ryoke and Sanbagawa belts together constitute one of the best-known examples of paired metamorphic belts (Miyashiro, 1961), extending for a length of more than 700 km in SW Japan. The Sanbagawa is a high $p/T$ metamorphic belt, while the Ryoke represents a low $p/T$ belt. The two belts are separated by a major right-lateral active fault known as the Median Tectonic Line.

In addition to magmatic and sedimentary rocks formed within the arc, island arc systems such as Japan also contain exotic blocks derived from continental fragments that have traveled far and been incorporated into a chaotic melange matrix. A typical example is the Kurosegawa terrane which crops out as a narrow (maximum width of 10 km), elongate, discontinuous E–W belt extending over 1000 km along southwestern Japan. The dominant units are igneous and high-grade metamorphic basement rocks, volcaniclastic sediments with carbonate horizons, and chaotic serpentinite melanges with cover sediments.

All of the Kurosegawa rock units are allochthonous with respect to the surrounding terranes. Paleomagnetic data and faunal and floral correlation equate the rocks in this belt to fragments of a paleocontinent or continental margin located far away from their present position. They probably formed in continental blocks close to a paleo-equatorial latitude at the northern margin of Gondwana and also close to the eastern margin of South China and/or Southeast Asia. The blocks rifted from Gondwana during the Late Devonian, drifted northward, and amalgamated with the proto-Asian continent during the Mesozoic.

## Oceanic Crust

Different types of oceanic crust can be incorporated into continents in different ways. Crust of normal thickness that consists of mid-ocean ridge basalts (MORB) and overlying sediments is locally obducted as ophiolites (chapter 2; fig. 2.5) or drawn down into subduction zones where it can return to the surface in melanges of sediment, basalt, serpentine, and eclogite (chapter 2; fig. 2.16). The amount of crust added to continents by this process is probably small, but oceanic plateaus may become significant parts of a growing continent.

Most oceanic plateaus form during the initial eruption of a plume onto the earth's surface (chapter 2). Many plumes form during continental breakup, with the heads now exposed as continental flood basalts and the tails as hotspot tracks on either continents or the adjacent oceans. Plumes initially erupted into ocean basins create thick plateaus before producing hotspot tracks. Present oceanic plateaus range in thickness from about twice that of oceanic crust up to 33 km (Ontong Java plateau of the southwest Pacific; Neal et al., 1997). Rock compositions are almost entirely basaltic, with layers of basalt and fractured basalt underlain by gabbros or gabbroic granulites. The mafic rocks have LIL element concentrations and isotopic ratios between those of MORB and continental rocks.

Incorporation of plateaus into continents is unclear. Because of their low density, thick plateaus may be impossible to subduct, but thin plateaus could be drawn down beneath continental margins and added to the lower continental crust. Their "intermediate" LIL element concentrations and isotopic characteristics are consistent with the possibility that large volumes of lower crust are composed of plateau or other plume-derived rocks (Condie, 1999).

## Continental-Margin Arcs

Continental-margin arcs form where oceanic lithosphere is subducted beneath continental lithosphere. The most extensive modern example is the Andes, and we briefly discuss it before examining the significance of the magmatic suites formed in this environment.

Subduction began along the western margin of South America as Pangea was developing in the Paleozoic (Davidson et al., 1991; Petford et al., 1996; Lamb et al., 1997), but orogenic activity was not intense until the Cenozoic, when rearrangement of plates in the Pacific and spreading in the South Atlantic caused a rapid increase in the rate of subduction. This increase caused the Andes to develop along the entire 6000-km length of the South American coast. In fig. 5.6 we show a generalized transect from the Pacific coast across the mountains into roughly the center of South America.

The Andes orogen is constructed mostly on Mesoproterozoic continental crust. The Arequipa massif crops out between the Andes and the coast and consists of gneisses that were metamorphosed from ~1.2 Ga to ~1.0 Ga (Wasteneys et al., 1995), and outcrops of similar age have recently been found within the Andean chain (Restrepo-Pace et al., 1997). These ages suggest that the Andean basement consists of fragments of eastern North America overprinted during the Grenville orogeny (chapter 7) and emplaced against the western edge of the Amazon shield at that time.

The westernmost part of the Andes is the active magmatic arc, with Miocene to Recent andesitic volcanoes erupted through older Cenozoic and Mesozoic suites that include ignimbrites, various lavas, and deformed sedimentary suites. It is also the site of the Late Cenozoic Coastal Batholith, mostly in Peru, that consists largely of granite. Some of the plutons were emplaced at such shallow levels that they intruded their own volcanic suites. East of the frontal arc is a broad region of Paleozoic to Cenozoic volcanosedimentary sequences lying on Proterozoic basement. The eastern edge of the Andes is dominated by a region of Cenozoic thrusts that verge to the east. They displace sedimentary suites as old as Paleozoic and as young as modern sediments now being eroded eastward from the Andes. The intense Cenozoic deformation associated with the Andean orogeny locally overprints earlier deformation caused by subduction throughout the Phanerozoic.

Subduction beneath intraoceanic island arcs and continental margins generates magmatic suites that are broadly similar but significantly different in detail. Intraoceanic subduction initially produces bimodal basalt–rhyolite volcanic suites that are followed by basalt–andesite–dacite (a typical "calcalkaline" trend) after the framework of the island

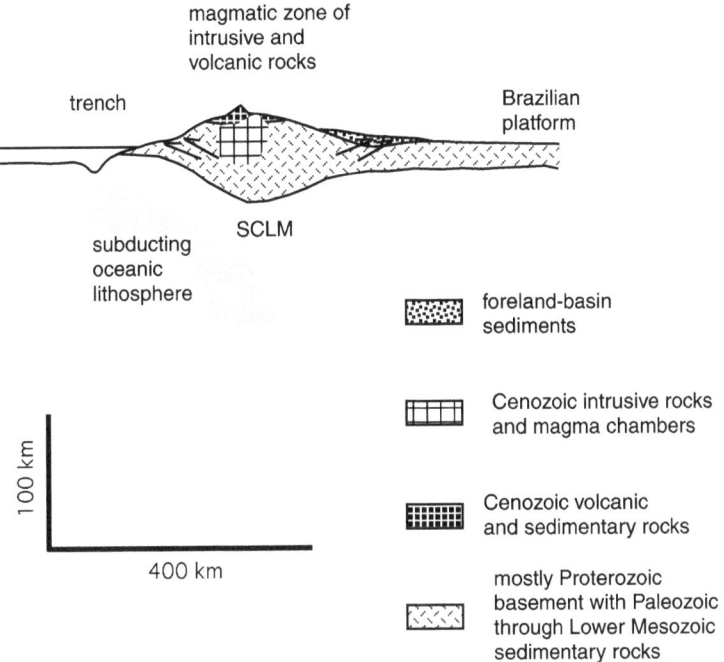

Figure 5.6. Diagrammatic cross-section of Andes Mountains (based mostly on Lamb et al., 1997).

arc is firmly established (chapter 2). The most siliceous volcanic suites and the rare granodioritic plutons in intraoceanic arcs generally have less than 70% SiO$_2$ and comparatively low concentrations of LIL elements. Derivation from oceanic lithosphere results in $\varepsilon_{Nd}$ values in the range of +4, and initial $^{87}Sr/^{86}Sr$ ratios of modern suites are less than 0.704.

Subduction beneath continental margins in the Andes and elsewhere produces magmatic suites in which both intrusive and extrusive rocks commonly have more than 70% SiO$_2$ and high concentrations of LIL elements. They include large volumes of granodioritic–granitic batholiths, which show increasing concentrations of K$_2$O at given SiO$_2$ concentrations inland from the continental margin. Isotopic data show $\varepsilon_{Nd}$ negative for almost all suites, with initial $^{87}Sr/^{86}Sr$ ratios of modern suites greater than 0.705. Rocks with these compositional properties could not have been derived from subducting oceanic lithosphere, and numerous proposals have been made for their source and method of formation. All of these proposals are closely related to the problem of change in the volume of continental crust with time (discussion in chapter 2), and we discuss three of them here (fig. 5.7).

- One early proposal for the origin of continental-margin magmatic suites was that they are juvenile (new) continental crust produced by partial melting of subducted oceanic lithosphere. The principal difficulty with this process is that the concentrations of K and other LIL elements are extremely low in oceanic lithosphere and even lower in the mantle. Therefore, continental crust could be produced only by complete extraction of components from a volume of lithosphere about 100 times the volume of crust produced (more for most incompatible elements). This volume of oceanic lithosphere would be available for partial melting only if new lithosphere were subducted over a very long period of time, perhaps longer than the period of evolution of the crustal rocks. This proposal had to be almost completely abandoned as increasing isotopic information showed that oceanic lithosphere could not produce the continental-margin suites.

- A second early proposal was that continental-margin magmas were produced by remelting of older continental crust or by mixing this old crust with juvenile magmas from oceanic lithosphere and overlying mantle. The volume of new continental crust produced by this process is variable, depending on the proportions of old crust and juvenile

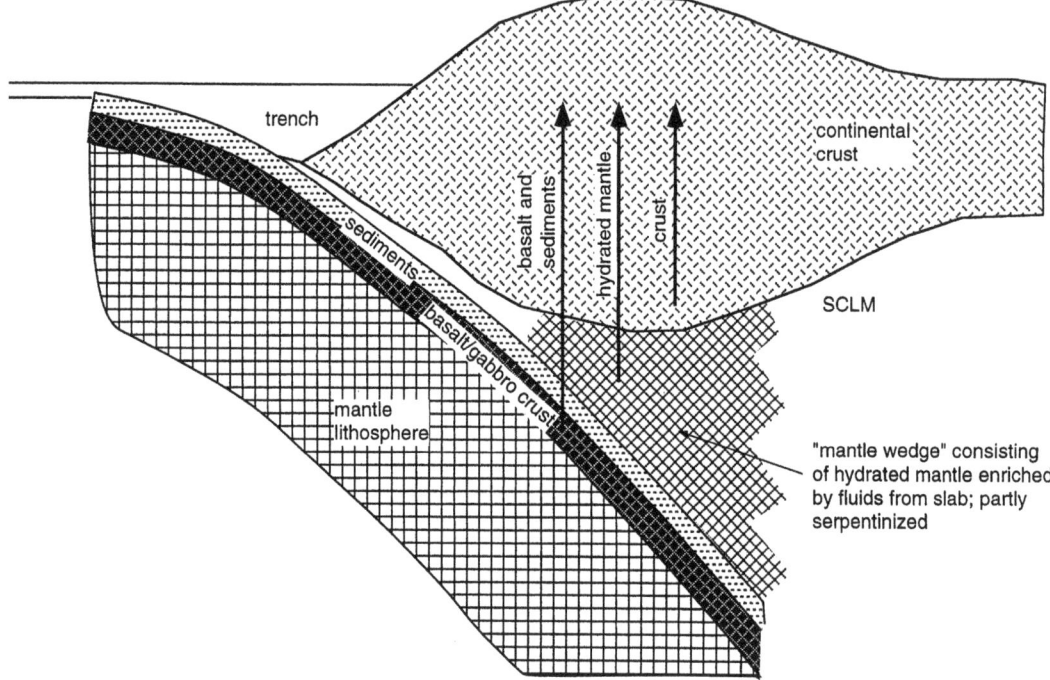

Figure 5.7. Sources of subduction-zone magmas.

magma that constitute the continental-margin magmas. This process is plausible, but verifying it for individual magmatic suites requires detailed balancing of the abundances of elements and isotopes in the suites and all possible source rocks. Thus far, all efforts have been highly controversial, and we cannot reach a conclusion here.

- A third, more recent, proposal is that the magmas are formed from a "mantle wedge" of old subcontinental lithospheric mantle (SCLM) that underlies the sialic crust on the margin of the continent. Water to cause this melting is provided by dehydration of the underlying slab, and it is likely that this water also carries LIL and other elements from basalts and sediments in the subducted lithosphere. Magmas formed by this melting apparently have the composition and isotopic properties needed to produce typical continental-margin igneous suites, and the rocks that formed from them would be new additions to the volume of continental crust. The problem with this proposal is that the volume of SCLM directly beneath continental margins is not sufficient to produce all of the magmas that form along the margins. As a solution to this problem, Glazner and Brandon (2002) proposed that continental-margin batholiths are underlain by thrusts that move the mountain chains toward the continental interior. This process provides a continual supply of new SCLM to be partially melted throughout the generation of the magmatic suite.

### Passive-Margin Sediments

Rifting creates passive continental margins composed of thin crust that subsides as the continents move away from new spreading ridges (chapter 2). Shallow-water clastic and carbonate sediments accumulate progressively on these margins and locally attain thicknesses of more than 10 km where subsidence is rapid. The marginal sediments grade into thinner sequences deposited in shallow seas on continental platforms and locally are thrust over the continental interior. As examples of the involvement of passive-margin sediments in orogeny, we discuss the Zagros Mountains here and the Himalayas in the next section.

The Zagros Mountains on the western border of Iran provide an excellent example of the incorporation of passive-margin sediments into orogenic belts (fig. 5.8; Alavi, 1994). Stabilization of the Arabian–

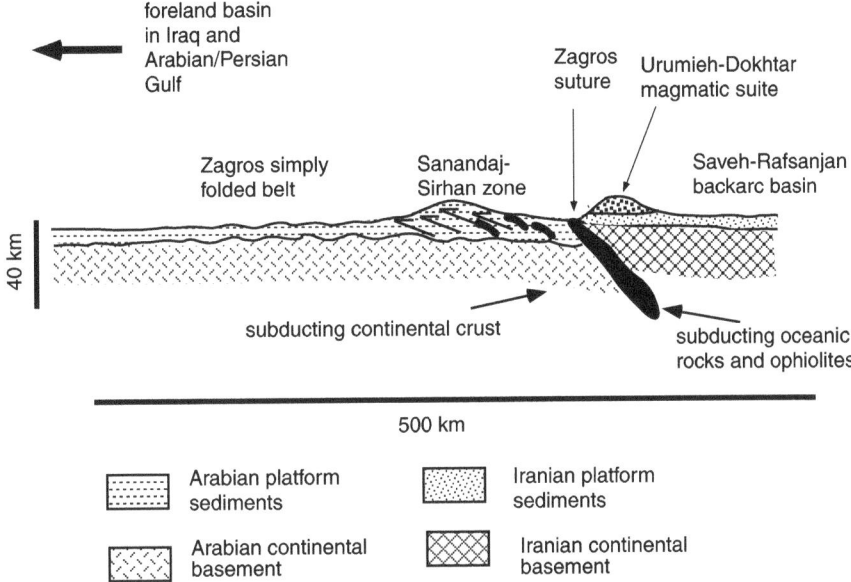

Figure 5.8. Diagrammatic cross-section of Zagros Mountains (terminology of the different zones is from Alavi, 1994).

Nubian craton near the end of the Proterozoic (chapter 4) developed a rapidly subsiding continental margin that accumulated as much as 10 km of shallow-water sediments throughout most of the Phanerozoic. In the Cretaceous the Arabian part of the craton began to subduct under the southwestern margin of Iran, which had accreted to Asia only a short time earlier.

This subduction created three major tectonic zones that trend NW–SE. Farthest to the northeast is a magmatic arc that has been active since the Cretaceous and is very similar to the Andes (see above). It contains a variety of intrusive rocks from gabbro to granite and extrusive rocks dominated by andesite and rhyolites, some of which were erupted as ignimbrites and other types of pyroclastics. Southwest of the magmatic arc is an intensely deformed zone that consists largely of slightly metamorphosed sediments of the Arabian margin. This zone also contains numerous ophiolites that indicate it was the site of an ocean that formerly existed between Arabia and Iran. Because of deformational complexity, it is commonly referred to as the "Zagros crush zone."

Southwest of the crush zone is a region designated as the "Zagros simply folded belt." It consists of unmetamorphosed sediments of the Arabian margin that have been slightly folded and thrust back onto the Arabian craton. This thrusting has at least doubled the thickness of the sedimentary sequence in local areas and greatly increased overall crustal thickness. The Zagros now acts as a crustal load that creates a foreland basin in Iraq, where recent sedimentation has filled the trough, and in the Arabian (Persian) Gulf, which is not yet filled to sealevel.

## Processes of Accretion

Terranes collide with a continental margin only after an intervening ocean basin has been destroyed by subduction. In ancient orogens, subduction of oceanic lithosphere beneath the continent is demonstrated by the presence of igneous rocks that penetrate the continent in a narrow zone along its margin (magmatic arc). Conversely, subduction beneath the approaching block leaves a zone of magmatism within the exotic terrane instead of the continent. In complex areas, such as the present southwestern Pacific, some small oceans are closing because of subduction on both sides, and some subduction zones have flipped from one side to another beneath small crustal blocks. This type of complexity may have occurred in ancient orogens, but it would be extremely difficult to decipher. Most collisions involve some component of strike-slip movement (they are "transpressional"), and a few ancient collisional belts seem to have been dominated by such movements.

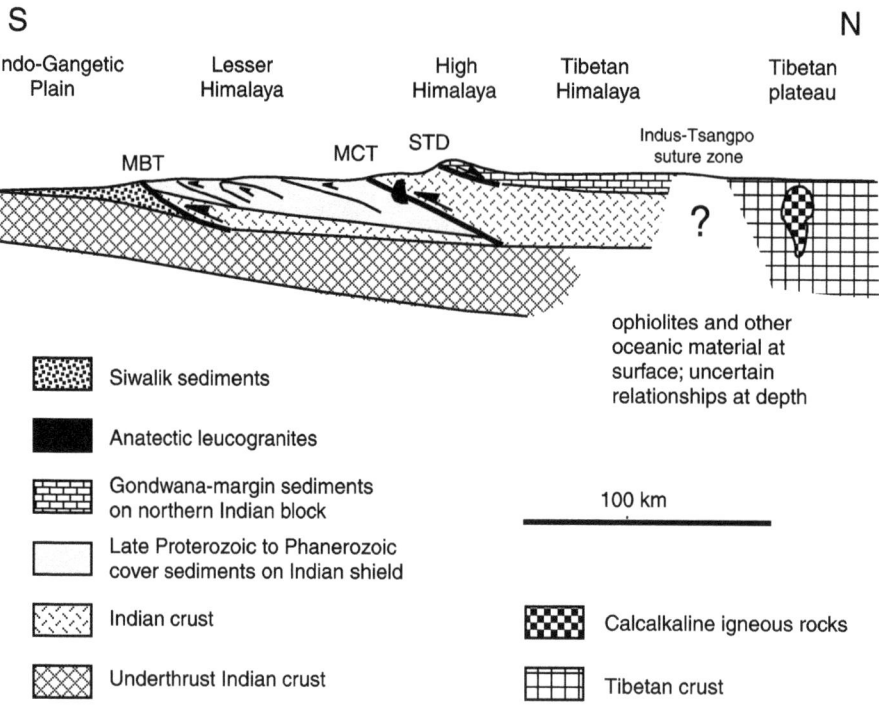

Figure 5.9. Generalized cross-section of Himalayas (based partly on Hauck et al., 1998). Relationships between mantle and crust below the mountains are unclear, and the diagram does not attempt to show them in any detail. MBT, Main Boundary Thrust; MCT, Main Central Thrust; STD, South Tibetan Detachment.

We describe examples of collision caused by subduction beneath a continental margin and also beneath approaching terranes.

## Subduction Beneath a Continental Margin—The Himalayas

Continental fragments rifted away from Gondwana and collided with Eurasia throughout much of the Phanerozoic (chapters 9 and 10). In the Mesozoic, these terranes moved northward across an ocean referred to as "Tethys," and the term "Tethyan" is used to refer to the collision along southern Eurasia, to ophiolites and other ocean material in the orogens, and to sediments that accumulated along the formerly passive margin of the Gondwana fragments.

The Mesozoic–Cenozoic collisions resulted in the Alpine–Himalayan mountain chain that stretches from the western edge of Europe into Southeast Asia, a distance of ~10,000 km. In the Alps, Europe was subducted under the advancing terrane that contains Italy and part of the Balkans (Apulian plate or peninsula; chapter 2). In southeastern Europe and southwestern Asia, subduction directions appear to have been variable from place to place and possibly from time to time. In the Himalayas, however, subduction has occurred under the Asian continental margin since about 50 Ma (review of Tibetan tectonics by Hodges, 2000).

Initially the subducting slab consisted of Tethyan oceanic lithosphere that caused magmatism in southern Tibet. At about 40 Ma, the major Kohistan island arc was sutured in the Pakistan part of the orogen, and shortly afterward closure of Tethys brought the entire Indian subcontinent into collision with Eurasia (fig. 5.9). The collision was so severe that the Indian subcontinent moved northward along major strike-slip zones that now occur in the Indo-Burman ranges of Southeast Asia and the Baluchistan region along the border between Pakistan and Afghanistan. Collision intensified at about 25 Ma, when major uplift of the Himalayas began, the Indian plate was warped downward, and rapid sedimentation was initiated in the Arabian and Bengal Seas.

Figure 5.9 shows a transect across the present Himalayas between India and Tibet. South of the Main Boundary Thrust are Miocene to Recent sediments (Siwaliks) eroded from the Himalayas. The MBT uplifts metasediments of the Indian plate to form the Lesser Himalayas. Some of them have been approximately dated, and they seem to correlate with Late Proterozoic sediments that overlie crystalline basement of the Indian shield. Metamorphic grade increases upward in the metasediments to the Main Central Thrust, which brings crystalline rocks of the Indian shield upward to form the Greater Himalayas. Some of the rocks at the highest elevations in the Greater Himalayas are Tethyan sediments deposited on the northern margin of the Indian plate, but most of the Tethyan deposits occur at lower elevations north of the South Tibetan Detachment, a series of down-to-the-north normal faults. The northern margin of this Tethyan belt is the Yarlung–Tsangpo (Indus–Tsangpo) suture zone, which is a tectonic melange of ophiolite, radiolarian chert, and other material from the Tethyan Ocean plus blocks of continental-margin sediments.

North of the suture zone is a series of plutonic rocks (Gangdese batholith) emplaced through Asian continental crust (Debon et al., 1986). Subduction of oceanic crust produced early plutonism in the Late Cretaceous, and the first stages of the collision of India and Asia caused additional magmatism in the Eocene. Initial $^{87}Sr/^{86}Sr$ ratios of 0.704–0.707 indicate that source regions of the magmas did not include any sediments or Indian continental crust. Magmatism appears to have ended at about 25 Ma when collision was complete and continental India began to subduct.

One of the major problems of Himalayan geology is the cause of crustal thickening both in the mountains and in the Tibetan plateau to the north (Harrison et al., 1992). The only magmatic rocks in the Himalayas are anatectic granites derived by melting of crustal rocks, so there has been no addition of juvenile (mantle-derived) material to any part of the area south of the Yarlung–Tsangpo suture. Thus, the 70-km crustal thickness of the Himalayas must have been caused by compression and thickening of the Indian continental crust and underlying mantle. Some of this thickening may have resulted from thrust stacking of both crust and mantle, but whether this mechanism is effective at depths of several tens of kilometers is presently controversial.

The 50-km depth of the crust beneath Tibet north of the Himalayas has been explained in several ways, including an early suggestion that the Indian plate was underthrust all the way to the northern edge of the plateau. The mechanism of this thrusting is unclear, however, and the elevation may be primarily the result of a thermal anomaly that reduces the density of the mantle and crust.

## Accretion by Subduction under Arriving Terranes—Paleozoic North America

Fragmentation of Rodinia between about 800 Ma and 600 Ma (chapter 7) left North America almost completely surrounded by rifted margins. These margins accumulated rift sediments and shelf deposits until the Middle Paleozoic. During this time, shallow ocean water encroached over most of the midcontinent, forming the basal deposits of sedimentary suites that accumulated until near the end of the Mesozoic.

Orogeny began on all margins of North America in the Middle Paleozoic by collision of exotic blocks (fig. 5.10). Along the east coast, the Avalonian terranes collided as oceanic lithosphere was subducted beneath them (see above). In the south, the Ouachita Mountains were formed in the Late Carboniferous (Pennsylvanian) by collision of a terrane that is now completely buried beneath coastal plain sediments (Viele, 1989). Subduction beneath this terrane instead of under North America is shown by a complete absence of magmatism in and north of the Ouachitas and by preservation of volcanic rocks in Paleozoic sediments beneath the coastal plain to the south.

Orogenies in western North America in the Devonian and Permian are known mainly because thrust sheets of those ages bring volcanosedimentary sequences of island arcs and backarc basins eastward over platform sediment (Burchfiel et al., 1992). The absence of magmatism in the North American platform indicates that orogeny was not caused by subduction of lithosphere beneath North America but by subduction of the intervening ocean basin beneath the approaching arcs. Similarly, the Innuitian area of northern Canada contains sediments in a fold and thrust belt generated by collision of the offshore terrane Pearya in the Middle Paleozoic (Trettin, 1991). The absence of magmatic rocks and metamorphism in the orogenic belt and the presence of both in Pearya demonstrate subduction of North America under the colliding terrane.

With the partial exception of Late Paleozoic subduction under the continental margin in New England and the maritime area of Canada, no subduction occurred beneath North America until the Late Triassic. At this time, opening of the Atlantic

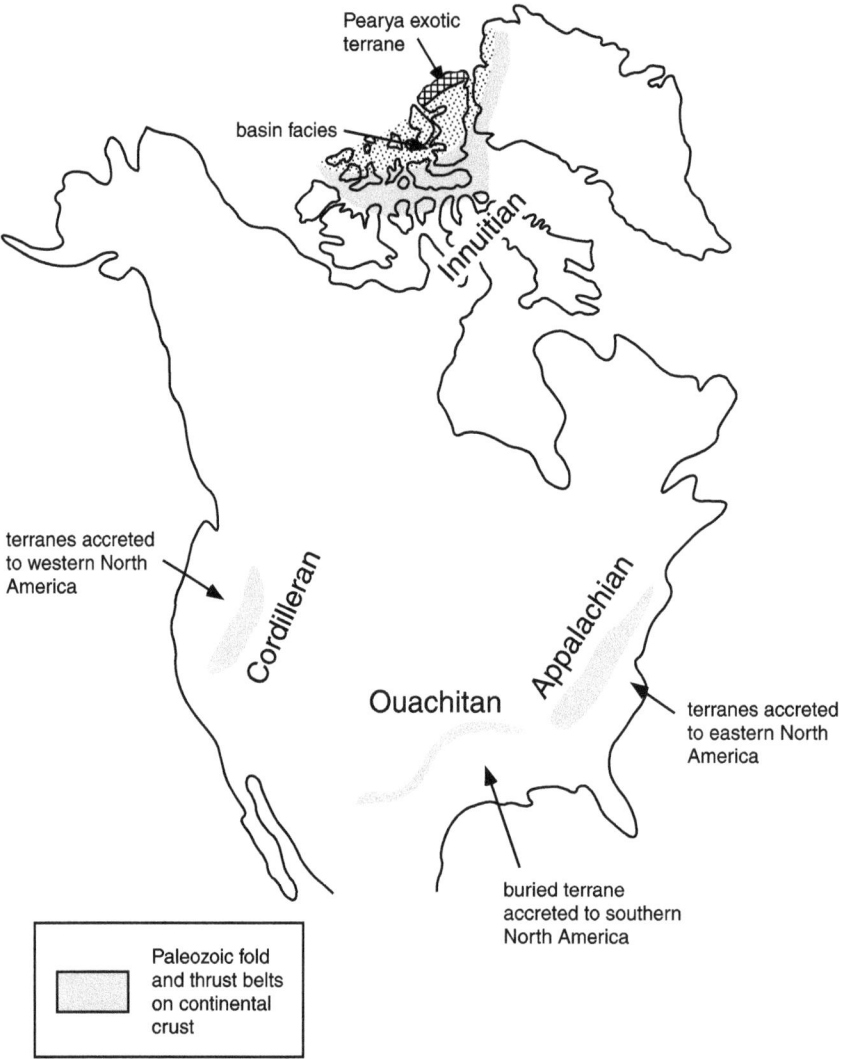

Figure 5.10. Paleozoic orogenies around North America. Accreted Avalonian terranes in eastern North America are shown in more detail in fig. 5.3.

Ocean and rearrangement of plates in the Pacific Ocean forced oceanic lithosphere under the western margin of North America from Canada to Mexico. Before this happened, however, North America was almost completely surrounded by passive margins for approximately 500 million years.

## Welding Continents in the Lower Crust and Upper Mantle

Discussions in chapters 3 and 4 demonstrate complex histories for both the lower continental crust and the mantle beneath it. Numerous investigations have shown that the mafic part of the lower crust and the SCLM have younger ages than the ages of the overlying upper crust. In chapter 3 we demonstrated that these lower ages were caused partly by metamorphism of older mafic rocks and partly by addition of new material to the base of the crust ("mafic underplating"). In chapter 4 we found that general correlations between the ages of cratons and ages and compositions of their underlying SCLM indicate that cratons have been linked to their SCLM since they formed. Ages of mantle xenoliths and information about continental heat flow, however, also show metasomatism and other

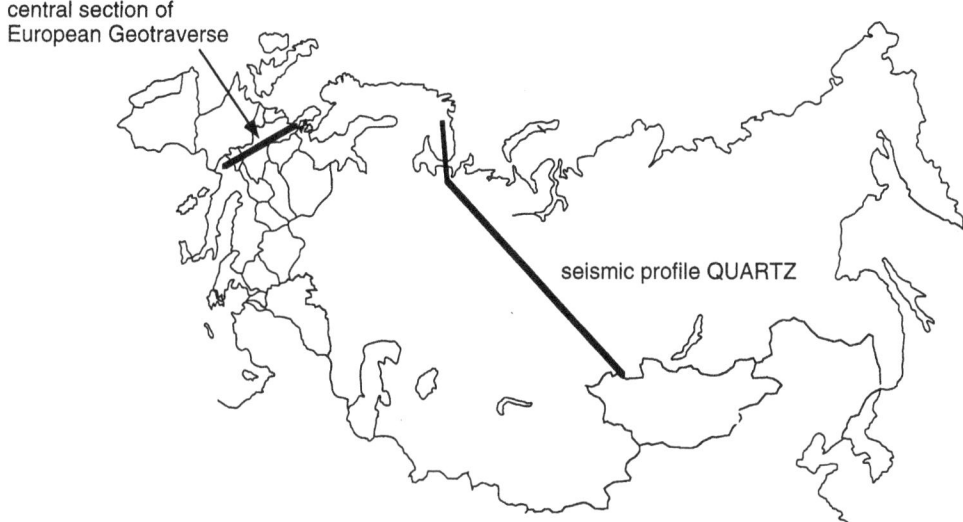

Figure 5.11. Map of Eurasia showing locations of central section of European Geotraverse and Quartz seismic profile across Russia and Siberia.

reworking of the SCLM long after stabilization of overlying cratons.

The reprocessing and addition of new material to the lower crust and SCLM are probably necessary for the development of continents, because assembly of terranes merely develops a group of separate blocks. These blocks must be "fused" or "welded" together before they become a coherent continent, and much of this welding clearly occurs in the lower crust and SCLM following assembly. This reprocessing and welding apparently develop continent-wide domains by processes that "blur" (homogenize) the original differences within the lower crust and also within the SCLM. Some major structures recognized on the surface have been followed seismically into the lower crust and SCLM, but the variation from place to place is far smaller than in the separate terranes that can be recognized in the upper crust.

We begin this section with an investigation of the properties of the lower crust and then describe the SCLM, mostly using the results of two long seismic surveys (fig. 5.11). Then we finish with a discussion of changes in the depth of the Moho through time, including the distinction between seismic Moho and petrologic Moho (chapters 2 and 3).

*Lower Crust*

The process of forming a homogeneous lower crust is well shown in fig. 5.12 (Morozova et al., 1999).

The transect crosses the exposed Baltic shield, sedimentary basins on subsided shield, the Ural Mountains collisional belt, the West Siberian Basin, and the Altay Mountain Range. The Urals contain a major suture between colliding terranes and several smaller sutures between island arcs and microcontinents within the suture zone (see above). The West Siberian Basin contains a thick sequence of sediments deposited on basalt that was erupted into a series of rifts that may locally have separated to form a short-lived ocean basin. The Altay Range is bounded on the north by one or more major sutures that separate it from the Siberian Precambrian crust.

The various crustal blocks clearly had separate geologic histories before assembly into the Asian continent. Despite these original differences, however, the present lower crust shows only small variations from place to place and is now a coherent block that extends through the entire transect. Some of this welding and the blurring of terranes may have occurred during assembly, but the lateral continuity of different crustal regions suggests that complete establishment of the crust did not occur until after all of the crustal blocks had accreted. Part of the blending must have involved changing the lower crust of island arcs and other oceanic terranes into continental crust. The method is unclear, although it probably involved intracrustal melting and other methods for separating LIL elements upward and leaving a more mafic and depleted crust at depth (chapter 2).

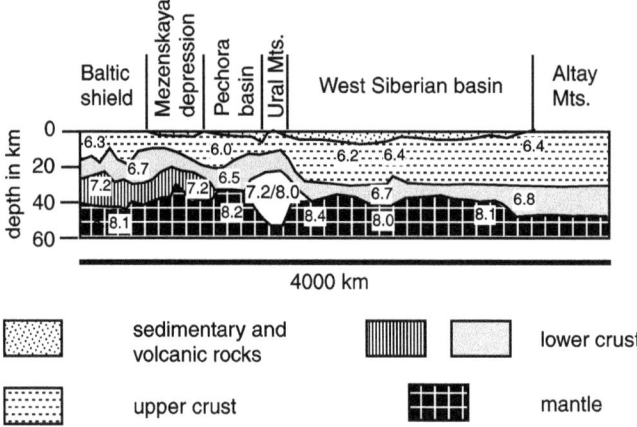

Figure 5.12. Lithologic interpretation of seismic properties of crust in Quartz geotraverse located in fig. 5.11. The mountain root under the Urals is shown with P-wave velocities of 7.2–8.0 km/sec, with velocities generally higher under the eastern part of the range. The two types of lower crust shown have distinctly diffferent seismic characteristics. Discussion of the history of the Ural Mountains earlier in this chapter (figs. 5.1 and 5.2). Morozova, E.A., Morozov, I.B., Smithson, S.B., and Solodilov, L.N. (1999). Heterogeneity of the uppermost mantle beneath Russian Eurasia from the ultra-long-range profile QUARTZ. J. Geophys. Res., v. 104, 20,329–30,348. Copyright 1999 American Geophysical Union. Modified by permission of American Geophysical Union.

Depths to the Moho shown in fig. 5.12 average about 40 km in most regions and reach 50 km beneath the Ural and Altay ranges. These differences, however, do not show a relationship to variations in P-wave velocities in crust just above the Moho. Velocities of 7.2 km/sec (almost certainly mafic rock) are present beneath the Baltic shield, the Urals, and basins between them, but velocities of only 6.7 to 6.8 km/sec occur below both the West Siberian Basin and the Altay Range.

Similar conclusions can be drawn from seismic refraction studies in the central-European section of the European Geotraverse, from an investigation of the European crust along a transect from the northern tip of Norway across the Mediterranean to Tunisia (fig. 5.13; Prodehl and Aichroth, 1992). The section crosses an area that contains numerous Paleozoic fold belts and accreted terranes, with seismically recognizable sutures throughout the upper crust and locally extending into the lower crust. There is some indication of a Conrad discontinuity in several places at depths of 20–25 km, and part of the lower crust appears to be mafic. The Moho is nearly flat at depths of ~30 km, and typical upper-mantle velocities are present below it. Although there is some seismic variability in the lower crust, it is much smaller than in near-surface terranes, indicating that blurring and fusing of original terranes has occurred.

## Subcontinental Lithospheric Mantle and the Low-velocity Zone

Seismic data amplify the information about the SCLM derived from studies of xenoliths and heat flow (see above; chapters 3 and 4). As a generalization, velocities of seismic waves in the upper mantle beneath continents are a few percent higher than at comparable depths below oceans. Because rock density and seismic velocities are higher where temperatures are lower, this difference is consistent with the likelihood that thermal gradients in the mantle below continents are lower than below ocean basins. Within the SCLM, small variations in wave velocities provide further detail about its development (Mechie et al., 1997).

Figure 5.14 shows seismic velocities in the SCLM along the same transect that we used to discuss crustal structure from the Baltic shield to the Altay Mountains (Morozova et al., 1999). Velocities tend to increase downward from the Moho until low-velocity zones (LVZs) are reached. The highest LVZ has velocities only slightly lower than in the mantle above it, and the LVZ at depths centered around 200 km is probably the low-velocity lid at the top of the asthenosphere (lithosphere–sthenosphere boundary). Morozova et al. recognize two smaller LVZs within the asthenosphere, and their origin and significance are less clear.

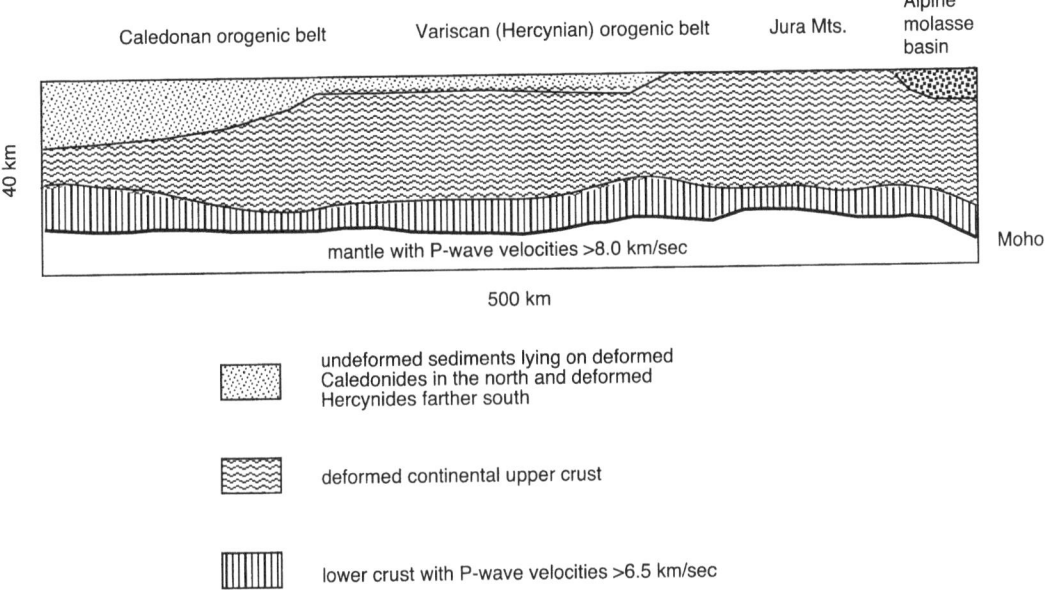

Figure 5.13. Lithologic interpretation of seismic features of European geotraverse located in fig. 5.11 (interpreted from information in Prodehl and Aichroth, 1992).

Similar differences in seismic-wave velocities occur under other continental regions. As an example, lateral velocity differences under Australia are mostly restricted to mantle above a depth of ~200 km, perhaps locally deeper under the Archean cratons of the west. The zone of variations is generally 150 km deep along the eastern margin of the continent, but is locally as shallow as 80 km. These depths appear to represent the base of the lithosphere, with more lateral uniformity in the underlying asthenosphere. Investigations of the uppermost SCLM here by Simons et al. (1999) show velocities that are a few percent higher under the Precambrian regions of the western part of the continent, decrease toward the east, and are particularly low under the areas of Phanerozoic accretion and recent volcanism along the eastern margin of the continent.

Determination of the thickness of the SCLM is important because the LVZ presumably is the base of the rigid lithosphere that moves with its overlying crust on top of a more mobile asthenosphere. It is also the base of the "tectosphere," defined by Jordan (1975) as the part of the earth rigid enough

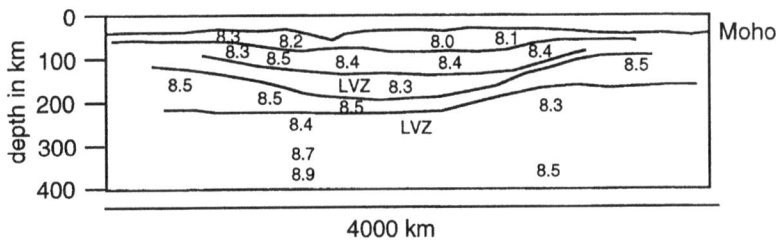

Figure 5.14. Seismic properties of subcrustal lithospheric mantle (SCLM). The lowermost low-velocity zone (LVZ) shown in this diagram is probably the top of the asthenosphere. Morozova et al. (1999) propose two LVZs in the lithosphere, and we show only the one that is most apparent. Modified from Morozova, E.A., Morozov, I.B., Smithson, S.B., and Solodilov, L.N. (1999). Heterogeneity of the uppermost mantle beneath Russian Eurasia from the ultra-long-range profile QUARTZ. J. Geophys. Res., v. 104, 20,329–30,348. Copyright 1999 American Geophysical Union. Modified by permission of American Geophysical Union.

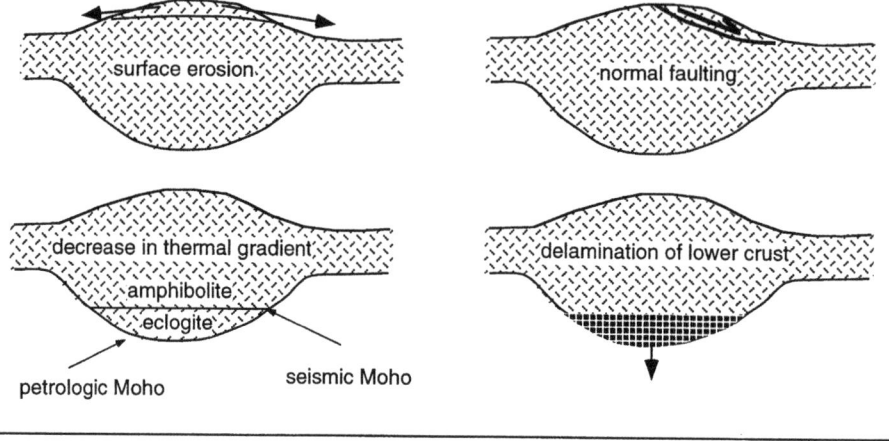

processes that increase depth to Moho

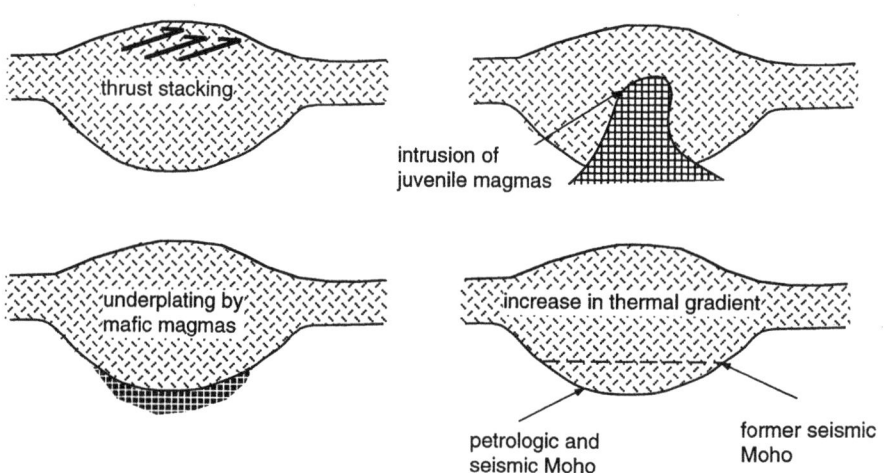

Figure 5.15. Changes in depth to Moho.

to preserve deformational features for long periods of time (perhaps indefinitely). Various theoretical arguments have been used to suggest that this rigid zone might be more than 500 km deep in some areas, but current seismic information suggests that it varies between about 100 km and 200 km in stable continental areas.

*Depths to the Moho and their Changes Through Time*

Because figs. 5.12 and 5.13 are based on seismic data, the Moho shown on both of them is the seismic Moho, underlain by material with a P-wave velocity of 8 km/sec or higher. This discontinuity, however, may not be the petrologic Moho, defined as the top of the peridotitic mantle. Where the lower crust is granulite or eclogite, it can have P-wave velocities of 8 km/sec or higher, thus causing the seismic Moho to be higher (shallower) than the petrologic Moho. Several processes can cause vertical movements and complex relationships between these two types of Moho through time, and we use fig. 5.15 to summarize them.

Depths to the Moho can be decreased (the Moho made shallower) by the following processes.

- Erosion. High elevation of mountain ranges causes them to erode rapidly. If the seismic and/or petrologic Moho beneath a mountain range was 70 km deep when it was formed,

20 km of erosion would reduce the depth to 50 km, which is normal for many ranges developed in the Paleozoic. The problem with this explanation is that 20 km of erosion would bring the level of exposure into granulite facies. Because granulites are rarely exposed in the higher elevations of mountain ranges, however, surface erosion can be responsible for only part of the reduction in Moho depth.

- Stretching and normal faulting. Normal faulting in very high mountain ranges (see Himalayas above) reduces the depth to the Moho by decreasing the load imposed by the range. Similarly, lateral stretching thins the crust, commonly by developing one or more detachment faults (discussion of rifting in chapter 2). Both processes reduce the thickness of the crust and thereby cause the Moho to rise.
- Conversion of gabbro to eclogite. In some places the Moho measured seismically may partly represent a zone of transition between gabbro, with a density of ~3.0 g/cc, and mafic rocks in eclogite facies, which have densities ranging from ~3.3 g/cc for garnet granulites to as much as ~3.6 g/cc for eclogites consisting solely of garnet and omphacite. Mafic lower crust converts to eclogite facies at depths that depend on the thermal gradient, deeper where the gradient is high and shallower where it is low. As an average, eclogite-facies rocks are stable below depths of about 40 km (fig. 2.2), and conversion between gabbro and eclogite moves the Moho up or down as the thermal gradient changes. Thus, a deep mountain root should become shallower as heat flow decreases following the peak of orogeny. This explanation for changes in Moho depth requires large parts of the region below the seismic Moho to consist of rocks in eclogite facies, causing the top of the petrologic Moho (mantle peridotite) to be deeper.
- "Delamination." Depth to the Moho may be decreased by a process of "delamination," which is separation ("peeling off") of SCLM and lower crust downward into the mantle. This occurs because the density of lithospheric mantle is slightly higher than the density of the uppermost asthenosphere, and lithosphere is buoyant only because it is attached to crust of lower density. Under conditions that are not fully understood, this mantle lithosphere can be detached from lighter crust above it and drop into the mantle. Where the lower crust is dense rocks in eclogite or granulite facies, it can also be detached from lighter overlying crust, causing the Moho to become much shallower as new mantle flows into the space left by the delaminated material.

Several processes can increase the depth to the Moho.

- Thrust stacking. Where several thrust sequences overlap each other in a mountain belt, the crustal thickening causes a deepening of the Moho. This process can construct a thick orogenic belt by compression without addition of new crust from the mantle.
- Addition of juvenile magmas. If the magmatic rocks of mountain belts represent juvenile additions from a mantle source, they can cause crustal thickening without lateral shortening by compression. Whether this process is a significant part of orogeny is unclear because of uncertainty about the source of magmatic suites in mountain belts (discussion of Andes above).
- Mafic underplating. Intrusion of mafic magmas at or near the base of the crust has commonly been regarded as a source of heat for such processes as metamorphism and intracrustal melting (chapter 3). Unless these new mafic rocks are converted to granulite or eclogite facies, they can cause thickening of crust above the seismic Moho, which is underlain by rocks with a P-wave velocity of 8 km/sec.
- Increase in thermal gradient. This process causes the opposite effects of a decrease in thermal gradients discussed above. Higher temperatures, perhaps accompanied by infiltration of water, reduce the density of granulites and thus move the zone of 8-km/sec P-wave velocities to greater depth.

## Summary

Continents form by the assembly of allochthonous terranes and become larger by growth on continental margins. Far-traveled terranes include large and small continental blocks, island arcs, and small amounts of oceanic crust. Marginal growth takes place largely where subduction of oceanic lithosphere forms mountain belts characterized by deformation and magmatism, but sediments of passive margins are incorporated in continental interiors when they collide with another continent.

Assembly requires closure of oceans by subduction either under the margin of the accreting continent or under the colliding block. All of the collisions are commonly accompanied by strike-slip movements approximately parallel to the margin of the growing continent.

Welding of terranes into coherent continents probably occurs both during and after collision. This develops mafic lower crusts and underlying mantles that are somewhat more homogeneous than the diverse felsic terranes that overlie them. Some of this continued activity locally causes the Moho to move up or down because of intrusion of juvenile magmas, structural thickening and thinning of the crust, delamination of dense crust, and the effects of changing temperatures on the density of the lower crust.

# 6

# Assembly and Dispersal of Supercontinents

Supercontinents are assemblies that contain all, or nearly all, of the earth's continental blocks. The concept arose with the recognition of Gondwana in the late 1800s (chapters 1 and 8), and it has been greatly expanded since then. In this chapter we build on the ideas developed in chapters 2 through 5 to discuss the origin and dispersal of supercontinents. The first section considers various mechanisms for the accretion of supercontinents, and the second section considers the reasons for their assembly. The last two sections consider evidence that former supercontinents have broken up and the reasons for their dispersal.

We emphasize that the processes of accretion and dispersal overlap, with rifting of some parts of a supercontinent occurring at the same times as suturing in other areas. This overlapping produces a time when the supercontinent has its largest coherent area, which we refer to as the time of "maximum packing."

## Models for the Assembly of Supercontinents

All supercontinents contain the same types of terranes that occur in individual continents (chapter 4). All models of assembly recognize that some terranes accreted as small individual blocks and some as continental-sized masses that contained several individual blocks that had been previously sutured together. Differences between models involve the area of the supercontinent that consisted of previously sutured large blocks and the area formed by accretion of small individual blocks. Resolution of this problem requires an understanding of the nature of orogenic belts developed during assembly, and we discuss this issue first.

## Types of Orogenic Belts

All orogenic belts have many similarities. They all underwent lateral compression that led to rock deformation and crustal thickening. Thickening pushed some rocks down into realms of higher temperature and pressure, causing metamorphism, and magmatic intrusion locally raised temperatures even higher in some areas. Almost all orogens contain magmatic rocks from various sources, including rocks partially melted within the orogen and magmas from subducted lithosphere and mantle below the deformed belt.

Despite these similarities, different orogens contain features that enable us to distinguish different environments of formation. We discuss them here in three categories: (1) "intercratonic" orogens formed by closure of ocean basins; (2) "intracratonic" orogens developed within continental areas where there was no pre-existing ocean basin; and (3) "confined" orogens formed by closure of small ocean basins that existed as indentations into continental blocks that had not been completely split apart. All of the different rock suites mentioned here have been discussed in previous chapters.

- The classical view of an orogenic belt is an intercratonic one formed by collision of two or more continental blocks previously separated by an ocean basin, and here we use fig. 6.1 to summarize the characteristic features of this type of orogen. The presence of ophiolites and other oceanic material is a clear indication of ocean closure and formation of a suture zone. Many suture zones also contain volcanosedimentary suites of island arcs, and some include microcontinents. Syn- to late-tectonic magmatism commonly produces calcalkaline batholithic assemblages of gabbro–diorite–granodiorite–granite. They commonly have $T_{DM}$ ages only slightly greater than the age of emplacement, and initial $^{87}Sr/^{86}Sr$ ratios are between ratios for average continental crust and average oceanic crust (~0.705–0.708 in recent batholithic suites). Shortly after the close of defor-

86  Continents and Supercontinents

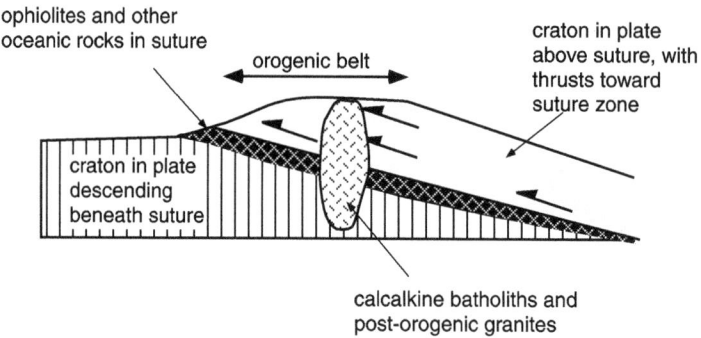

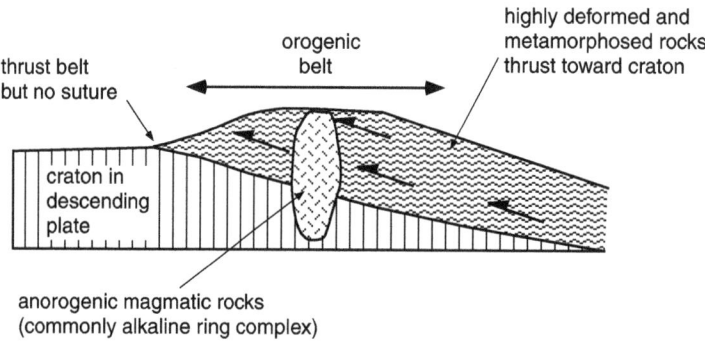

Figure 6.1. Characteristics of intercratonic and intracratonic orogens.

mation, the orogenic belt is penetrated by numerous bodies of post-orogenic granite.

- Intracratonic orogenic belts are also illustrated in fig. 6.1. Eroded belts consist largely of granulites thrust over one continental block and overlain by other continental rocks that range from gneisses to unmetamorphosed sediments. Ophiolites or other oceanic remnants are not present in the orogenic zone, but because these materials are squeezed upward and easily eroded in intercratonic belts, their absence is not by itself sufficient proof of intracratonic activity. Intracratonic orogeny is also shown by an absence of batholithic suites and postorogenic granites, with magmatism commonly restricted to local anatectic granites and post-tectonic alkaline suites. The anatectic granites show their derivation from within the orogen by having high initial $^{87}Sr/^{86}Sr$ ratios typical of old continental crust and $T_{DM}$ values that indicate much greater source ages than the ages of emplacement. Figure 5.2 shows an oceanic suture outside of the orogenic belt because the lateral stresses needed to cause an orogeny presumably developed by subduction of oceanic lithosphere on one or both sides of the orogen.

A typical intracratonic belt is the Rocky Mountains of western North America. The Laramide orogeny that was responsible for development of much of the Rocky Mountains was clearly caused by subduction of Pacific oceanic lithosphere beneath the western margin of North America in the Late Cretaceous/Early Tertiary. This compression was transmitted to the continental interior, where it caused east-vergent thrusting of supracrustal rocks in the northern Rockies, and upward movement of Proterozoic rocks along high-angle faults in the southern Rockies.

The Rocky Mountains are regarded as intracratonic because no ocean closed beneath the mountain range, and the suture at the source of compression was on the continental margin hundreds of kilometers to the west. The difference from intercratonic orogenic belts is well shown by Laramide and younger magmatism. Subduction beneath the continental margin produced batholiths along the entire western edge of North America, but none within the Rockies. Tertiary granite intrusions into the Proterozoic uplifts have $T_{DM}$ ages centered around 1000 Ma and initial $^{87}Sr/^{86}Sr$ ratios generally greater than 0.707 (Stein and Crock, 1990). Some of the intrusions were accompanied by volatiles that produced topaz, as in other magmas commonly regarded as anorogenic (chapter 4).

Intracratonic belts occur in other areas of the earth. One is the Tien Shan Mountains of central Asia (Bullen et al., 2001), where Late Cenozoic and current uplift are caused by collision of India with an Asiatic margin well to the south of the Tien Shan (see Himalayas, chapter 4). An older belt is the Eastern Ghats of India (chapter 7; appendix H), which underwent repeated orogeny at ~1.6 Ga, 1 Ga, and 0.5 Ga.

- The concept of confined orogens has been developed very recently, with the type example being the Aracuai belt of eastern Brazil (fig. 6.2; Pedrosa-Soares et al., 2001). The Aracuai belt indents the southern margin of a cratonic block stabilized at ~2 Ga and now separated by the opening of the Atlantic Ocean to form the Sao Francisco craton of eastern South America and the Congo craton of western Africa. The former union of these cratons is shown by correlation of geologic suites between the two blocks, and we discuss below the possibility that the combined Sao Francisco–Congo craton was part of a much broader continental assembly (see Atlantica below) that developed at ~2 Ga.

The Aracuai belt began when rifting developed a small triple junction within the Sao Francisco–Congo craton at approximately 1.0 to 0.9 Ga. The three arms of this junction include an ocean basin extending into the craton from its southern margin (present orientation) and two aulacogens, one in South America and one in Africa. Ocean opening is shown by the preservation of an ophiolite and a calcalkaline magmatic arc in the present Aracuai belt. Closure of the basin was centered around 600 Ma, when subduction closed the small basin, produced syn- and late-tectonic granitic rocks, and thrust metamorphosed supracrustal and basement suites onto continental blocks on both sides of the belt. As in other collisional orogens, compressional orogeny was followed by intrusion of post-orogenic granites about 50 million years after the end of deformation.

During rifting, the Aracuai basin was very similar to the present Red Sea (chapter 10). The Red Sea extends northward from a very large triple junction at its southern end, with creation of ocean crust in the southern part and extreme stretching that may develop into an ocean basin in the north. Two rifts (Suez and Aqaba) extend from the northern end of the Red Sea just as the Paramirim and Sangha rifts did from the northern end of the Aracuai ocean basin. If the Red Sea ever closes back on itself in the future, it seems likely that it will develop an orogenic belt similar to Aracuai.

The concept of confined orogen may explain the evolution of other orogens than Aracuai. One example may be the Ordovician–Silurian Caledonide belt of northern Europe and Greenland (Hurich, 1996; Kalsbeek et al., 2000). Correlation of orogenic and magmatic events shows that North America, Greenland, and Baltica became a coherent block during the Middle Proterozoic (we discuss the supercontinent Nena below), much earlier than the Caledonide orogeny. The presence of ophiolites in the upthrust terranes in Europe and subduction-generated magmatism in Greenland, however, clearly indicates closure of ocean basin. Therefore the Caledonides cannot have formed in an intracratonic belt that did not contain an ocean basin, and they show all of the characteristics of a confined orogen.

## Accretion of Small or Large Continental Blocks to Form Supercontinents

All known supercontinents contain numerous orogenic belts developed during the time range in which the supercontinents accreted. If all of these

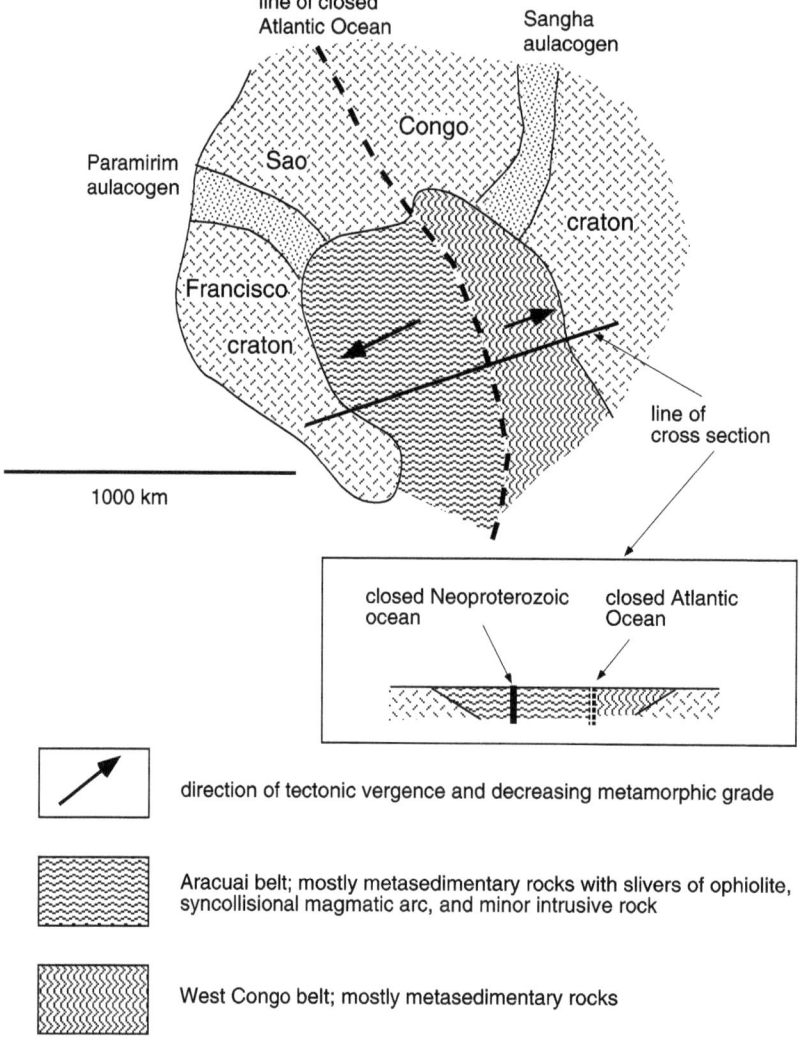

Figure 6.2. Map of Aracuai belt, a confined orogen (modified from Pedrosa-Soares et al., 2001).

belts resulted from the closing of ocean basins, then the accreting blocks were only as large as the areas between the orogenic belts (Unrug, 1992). If, however, some of the orogenic belts were intracratonic or confined, then the accreting blocks may have contained the belts and been much larger.

Three large continental blocks with ages of 2.0 Ga and older may have been involved in the accretion of all younger supercontinents (Rogers, 1996). They are named Atlantica, Arctica, and Ur and are recognized primarily by the pattern of cratonic stabilization ages within Pangea (fig. 6.3; discussion of stabilization in Chapter 3). The existence of these large blocks requires younger orogenic belts within them to be intracontinental or confined, which is highly controversial, with the degree of controversy increasing from youngest to oldest.

- Atlantica (2.0 Ga; fig. 6.4). The numerous ~2-Ga cratons of South America and western Africa seem to have formed a coherent continent that did not fracture until the breakup of Pangea. Creation of the continent occurred from 2.1 Ga to 2.0 Ga, an orogenic episode referred to as the Transamazonian in South America and Eburnian in Africa (Boher et al., 1992, in Africa; Hartmann, 2002, in South America). The continent is named Atlantica because the Atlantic Ocean opened through it.

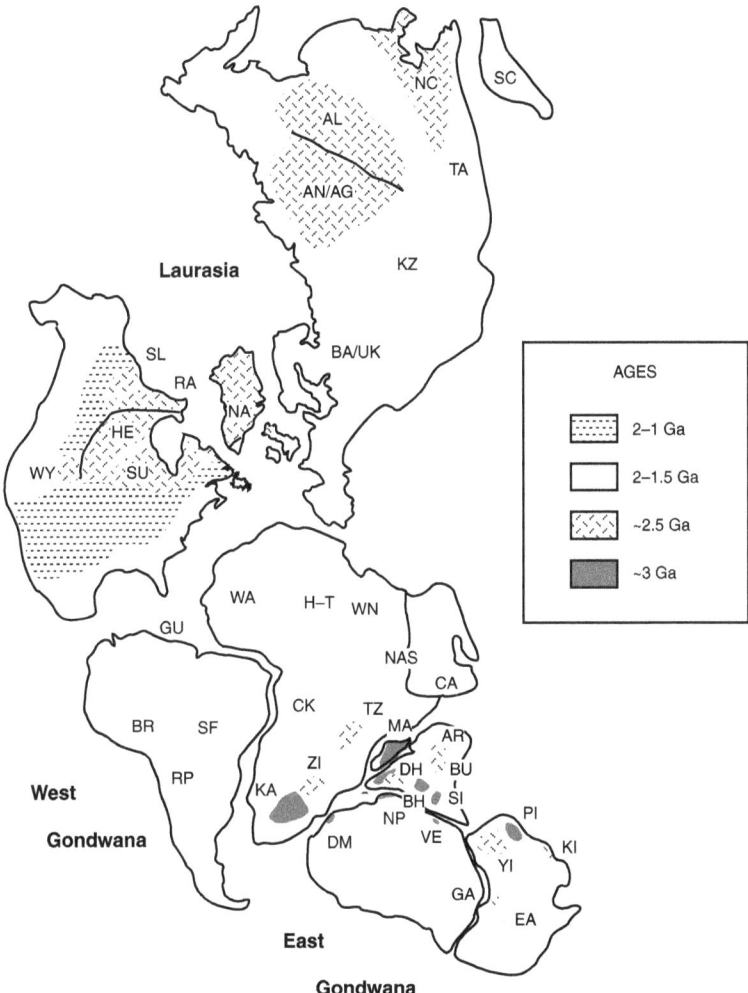

Figure 6.3. Ages of cratons in Pangea (from Rogers, 1996). Symbols designate the following cratons: AL, Aldan; AR, Aravalli; AN/AG, Anabar/Angara; BA/UK, Baltic/Ukraine; BH, Bhandara (Bastar); BR, Brazil (Guapore); BU, Bundelkhand; CA, Central Arabia; CK, Congo/Kasai; DH, Dharwar (west and east); DM, western Dronning Maud Land; EA, an assemblage of Early to Middle Proterozoic areas in eastern Australia; GA, Gawler; GU, Guiana; HE, Hearne; H–T, terranes, including Archean, assembled in the Hoggar and Tibesti areas of North Africa in the Pan-African; KA, Kaapvaal; KI, Kimberley; KZ, Kazakhstan; MA, Madagascar; NA, North Atlantic, including Nain, Greenland, and Lewisian; NAS, mostly juvenile Pan-African crust in the Nubian–Arabian shield; NC, North China (Sino-Korean); NP, Napier; PI, Pilbara; RA, Rae; RP, Rio de la Plata; SC, South China (Yangtze); SF, Sao Francisco, including Salvador; SI, Singhbhum; SL, Slave; SU, Superior; TA, Tarim; TZ, Tanzania; VE, Vestfold; WA, West Africa, including the small Sao Luis craton now on the coast of South America; WN, West Nile (Nile–Uweinat); WY, Wyoming craton; YI, Yilgarn; ZI, Zimbabwe. The 2–1-Ga crust in North America is mostly juvenile growth on the southern and western sides of the Archean shield.

The most important evidence for the existence of a coherent Atlantica is correlation of ~2-Ga fluvio-deltaic sediments among five of the cratons (Ledru et al., 1994). This correlation seems impossible if the cratons were together 2 billion years ago and rifted apart later. If they became widely separated, the probability seems vanishingly small that these five cratons would have come back together during the growth of Gondwana

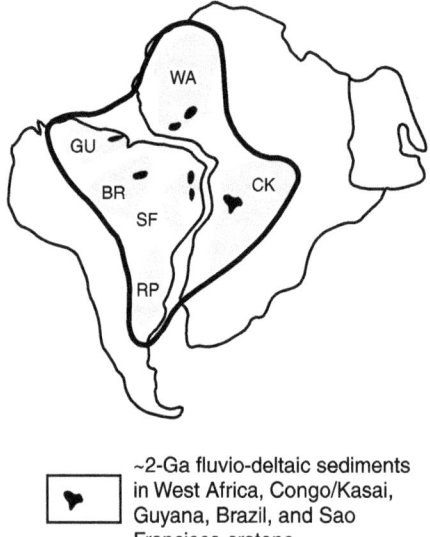

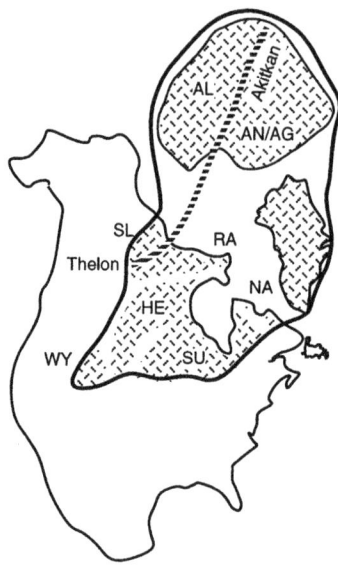

Figure 6.4. Map of Atlantica (from Rogers, 1996). Fluvio-deltaic sequences correlated by Ledru et al. (1994) include: Tarkwaian in West Africa; Francevillian in the Congo craton; Corrego and Rio do Ouro in the Sao Francisco craton; and variously named sequences in the Brazil and Guiana cratons. Symbols for cratons are as shown in fig. 6.3.

Figure 6.5. Map of Arctica (from Rogers, 1996). Arctica includes Kenorland (fig. 6.6) plus Siberia, Baltica, and the Wyoming craton. Symbols for cratons are as shown in fig. 6.3.

and Pangea in the same relative positions that they had occupied 1.5 billion years earlier, and it is far more likely that they would have been dispersed throughout younger supercontinents.

Accretion of Atlantica to Gondwana as a coherent block requires that the numerous 600–500-Ma orogenic belts within it were either intracratonic or confined orogens. The evidence that the Aracuai belt in the Sao Francisco–Congo craton was confined is very clear (see above), but it is less certain for other belts. In chapter 3 we used the relationships between zircon U/Pb ages and $T_{DM}$ ages in these belts as an example of the development of new crust from pre-existing crust instead of production of juvenile crust from the mantle. This lack of juvenile crust supports the concept that these orogenic belts did not form by the closure of large ocean basins but by intracontinental orogeny, and we discuss the issue in more detail in chapter 8.

- Arctica (2.5 Ga; fig. 6.5). Cratons of ~2.5-Ga stabilization age occur primarily in North America and Siberia, and the continent containing them was named Arctica because the Arctic Ocean opened through them. The existence of Arctica at 2.5 Ga faces two challenges. One is the position of Siberia, which Condie and Rosen (1994) and Pelechaty (1996) place against the northern margin of North America, but which Sears and Price (2000, 2002) suggest was positioned against the western edge of North America through much of the Proterozoic. The second problem is the history of the cratons in Canada. H. Williams et al. (1991) proposed the name "Kenorland" for a 2.5-Ga continent that included the Canadian cratons of Arctica (fig. 6.6). Numerous ages ranging from 2.4 Ga to 2.0 Ga occur in the region between the Superior and Slave provinces. These ages suggest that Kenorland underwent widespread rifting during this time and was then reassembled at 1.9–1.8 Ga by collision along the Trans-Hudson orogen and Taltson magmatic belt.

Reassembly of rifted fragments from Kenorland raises the same question that we discussed for Atlantica—the problem of bringing widely separated blocks back to the same positions they occupied before dispersal. For this reason, Aspler and Chiarenzelli

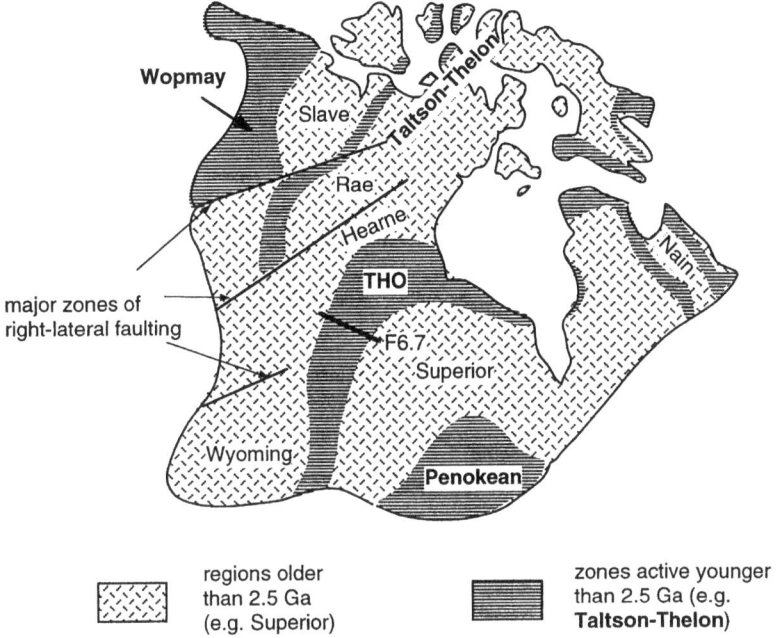

Figure 6.6 Map of Kenorland (based on Williams et al., 1991, and Aspler and Chiarenzelli, 1998). Major post-Archean orogenic belts are named in bold type. The line marked F6.7 shows the approximate location of the cross section of the Trans-Hudson orogen (THO) shown in fig. 6.7.

(1998) proposed that the Trans-Hudson and Taltson orogens are "internal" orogens that underwent only limited rifting and then closed back upon themselves (the same concept discussed above for confined orogens). To clarify the issue, we discuss both the Trans-Hudson and Taltson orogens in more detail.

Figure 6.7 shows a generalized section across the Trans-Hudson orogen from the Hearne craton to the margin of the Superior craton (Lucas et al., 1993; Lewry et al., 1994). The Hearne province consists of Archean rocks that underwent high-grade metamorphism and local anatexis in the Early Proterozoic (Bickford et al., 1994; Orrell et al., 1999). The Wathaman (Wathaman–Chipewyan) batholith is a typical calcalkaline suite formed on the margin of the Hearne province from 1.87 Ga to 1.85 Ga (Meyer et al., 1992). The Reindeer zone is a general term for an assemblage of oceanic and continental rocks thrust southeastward (present orientation) over Archean crust at the same time as production of the Wathaman magmas (Lewry et al., 1994; Theriault et al., 2001). The Reindeer zone is separated from the Archean rocks of the Superior province by a thrust that Lewry et al. (1994) suggest is intracratonic. Because Archean crust underlies the Reindeer zone, the only part of the Trans-Hudson orogen that clearly resulted from oceanic closure is the Wathaman batholith. Chiarenzelli et al. (1998) propose that it resulted from subduction on the margin of a small ocean that opened and then closed around the same pole of rotation.

The Taltson magmatic zone has also been proposed as an intracratonic orogen. Chacko et al. (2000) and De et al. (2000) investigated the geochemical and isotopic properties of the 2.0–1.9-Ga intrusive rocks that dominate the zone and found diagnostic features that include: almost all rocks with $SiO_2$ contents greater than 64%, crustal $\varepsilon_{Nd}$ values of $-3.8$ to $-9.8$, average $T_{DM}$ of 2.8 Ga, and Pb and O isotopes that were probably derived from metasedimentary wall rocks of the intrusions. These compositional properties are incompatible with subduction of oceanic lithosphere, showing that the Taltson zone did not contain an ocean basin and was not a plate boundary at 2.0–1.9 Ga.

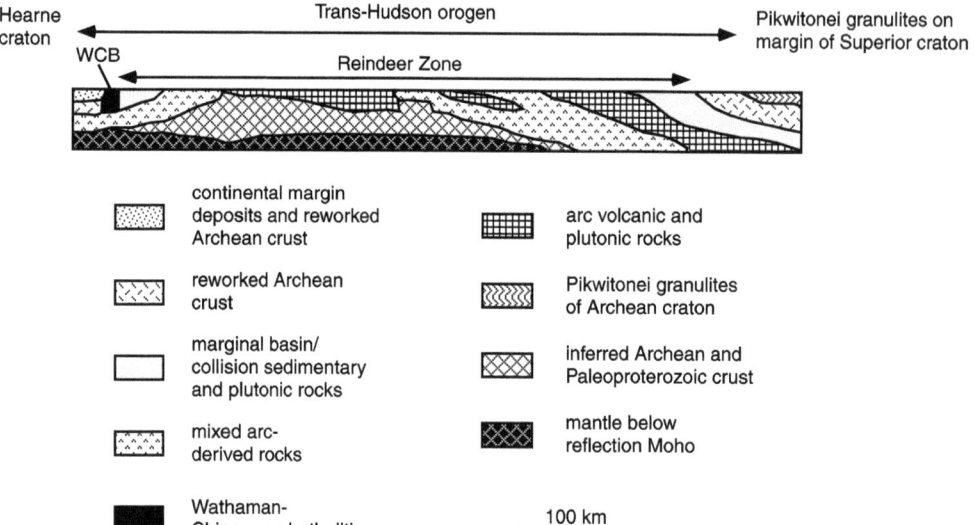

Figure 6.7. Generalized lithologic cross-section of Trans-Hudson orogen (based on Lucas et al., 1993; Lewry et al., 1994; and Chiarenzelli et al., 1998). The section unites three separate seismic profiles, and this figure simplifies relationships in complex terranes. Discussion in text includes names of tectonic regions.

Whether the Kenorland part of Arctica has been coherent since 2.5 Ga or was assembled at 1.9–1.8 Ga, it is clear that it became significantly larger by the Middle Proterozoic. Part of this growth was accretion on the western margin (Wopmay orogen, Bowring and Grotzinger, 1992), and part consisted of continued accretion on the southern and eastern margins from ~1.8 Ga to the formation of Rodinia at ~1 Ga (chapter 7). Accretionary belts in eastern North America correlate so well with similar belts on the southern margin of the Baltic shield, that Gower et al. (1990) named the combined North America, Greenland, and Baltica as the supercontinent Nena (an acronym for Northern Europe North America; fig. 6.8). Nena apparently remained a coherent block from ~1.8 Ga to the breakup of Pangea.

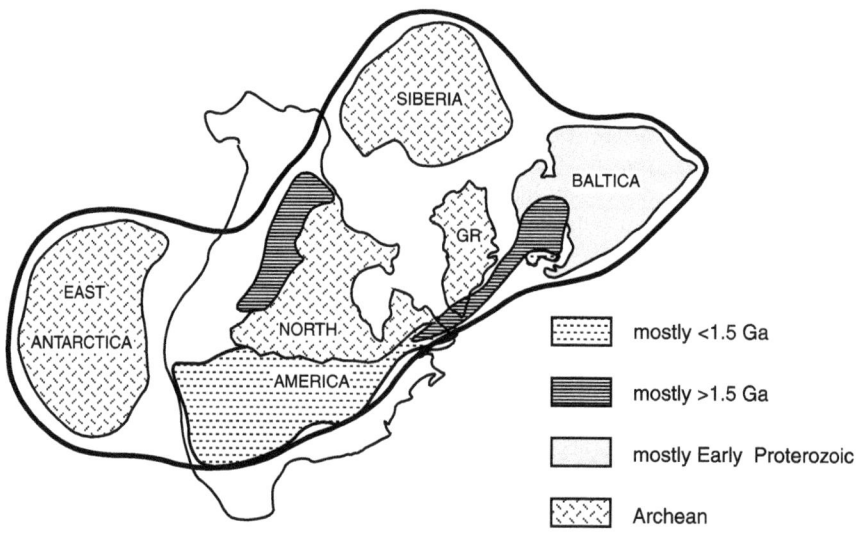

Figure 6.8 Map of Nena. GR is Greenland (from Rogers, 1996; based on work by Gower et al., 1990).

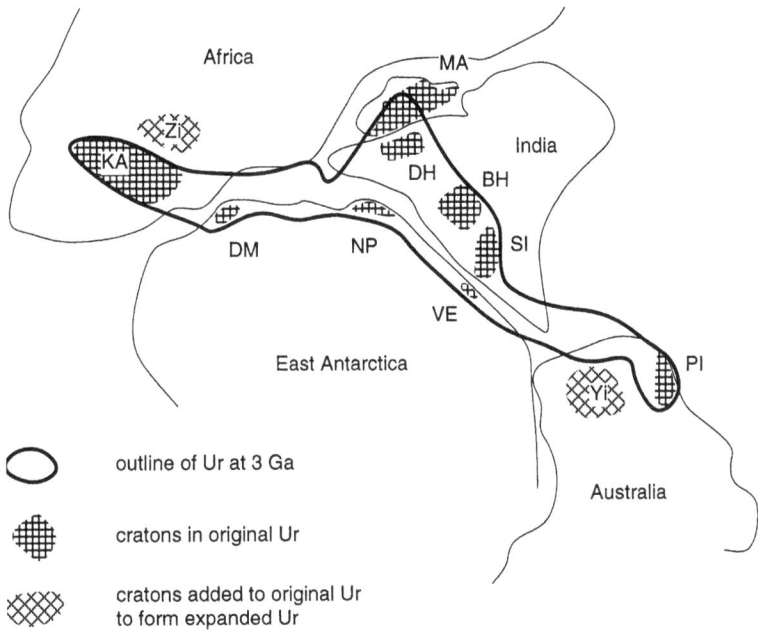

Figure 6.9 Maps of Ur and expanded Ur (from Rogers, 1996). Symbols for cratons are as shown in fig. 6.3.

- Ur (3.0 Ga; fig. 6.9). The only large cratonic areas stabilized as old as 3.0 Ga are all in one part of Pangea (fig. 6.3). They include the Kaapvaal craton (appendix E), the Western Dharwar craton of India, the Singhbhum craton of India (recently dated as 3.2 Ga by Mukhopadhyay, 2001), and the Pilbara craton of Australia (Barley, 1993; Kloppenberg et al., 2001; Blewett, 2002). Small exposures of 3.0-Ga cratons also occur on the margin of East Antarctica, where they are known to have been joined to India and Australia no later than Middle Proterozoic and probably earlier. As with Atlantica and Arctica, the probability of all of these old cratons assembling into one area of Pangea is extremely small if they had been separated at any time between 3 Ga and the end of the Paleozoic. Because they are the oldest cratons known, the supercontinent created by them was named Ur, from a German word meaning "original."

Tests of the existence of Ur are very difficult because of its age. Specific correlations of rocks are impossible, although D. Nelson et al. (1999) suggested that both the Kaapvaal and Pilbara cratons have roughly similar histories. The existence of Ur as old as 3 Ga also requires that numerous younger orogenic belts be intracratonic or confined orogens. One uncertain area is the Eastern Dharwar craton (Jayananda et al., 2000; Mazumder et al., 2000).

Regardless of whether Ur became a coherent continent as old as 3 Ga, it may have formed a core for a continental block that became coherent no later than 1.8 Ga. At this time, northern India was sutured to the older cratons of southern and eastern India, and Australia expanded by attachment of the Yilgarn craton and several terranes in central and eastern Australia (fig. 6.9). We refer to this larger block as expanded Ur. Expanded Ur may have included the "Mawson continent" proposed by Peucat et al. (1999), but relationships are unclear.

Our discussions of supercontinent assembly in chapters 7 and 8 use both of the models of accretion that we describe above—one in which all orogenic belts were active during accretion of small individual blocks and one in which much of the supercontinent was formed by accretion of one or more of these three old blocks.

## Causes of Supercontinent Assembly

Processes in the mantle exercise fundamental controls over the accretion and movement of conti-

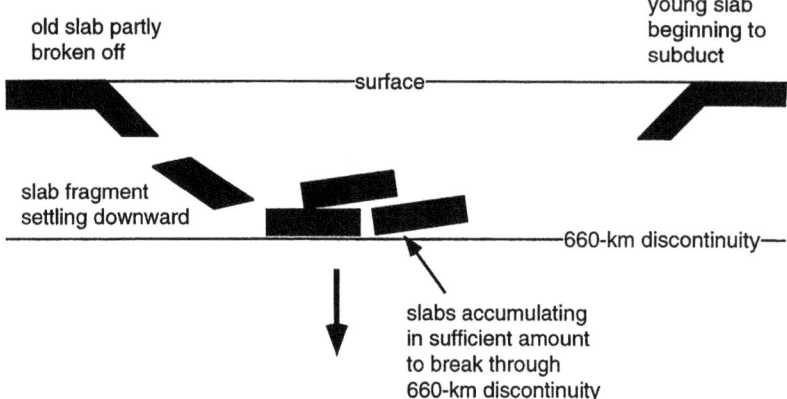

Figure 6.10. Diagrammatic representation of slab avalanches.

nents. Information about the present mantle can be obtained from a variety of geophysical methods, but ancient mantle processes must be inferred from the records that they left in rocks that we are able to study. In this section we discuss relationships among plumes, superplumes, and slab avalanches in the mantle and crustal processes such as the rate of production of continental crust and the times of supercontinent assembly.

A fundamental seismic discontinuity in the mantle at a depth of 660 km is caused by downward phase change to denser material (chapter 2). Mantle processes associated with plate tectonic movements normally occur above this discontinuity, which forms the base for convection cells that rise at spreading centers and also the lowest depth reached by subducting slabs (hence the deepest earthquakes). Mantle below the discontinuity affects surface processes largely through plumes that may be generated as deep as the base of the mantle, which is seismically designated the $d''$ layer.

At various occasions in earth history, the 660-km discontinuity seems to have been disrupted (fig. 6.10). Subducted slabs of oceanic lithosphere tend to accumulate at the discontinuity because they are denser than the mantle above and around them. Their low temperature and lack of LIL elements, removed during subduction-zone magmatism, also make them slightly denser than the mantle beneath the discontinuity. Thus, places where numerous slabs have accumulated can become unstable, with the slabs dropping into the lower mantle as a "slab avalanche" (Condie, 2001). This avalanche displaces material in the lower mantle, causing it to rise elsewhere on the earth as a plume or, if large enough, as a "superplume."

Material rising from the lower mantle has important effects on the composition of the upper mantle. Because continental crust is derived solely by extraction of material from the upper mantle, that part of the mantle becomes rapidly depleted in LIL elements, including those important in analysis of isotopic systems (Rb, Nd, Re; appendix D). The upper mantle may not contain enough of these LIL elements to generate all of the continental crust in the earth and to account for isotopic changes. Consequently, many earth scientists propose that compositional evolution of the upper parts of the earth can be explained only if plumes, superplumes, or other infusions from the lower mantle periodically replenish the upper mantle in LIL elements.

In addition to compositional effects, plumes cause plate movements on the surface (Gurnis, 1988). Both the geoidal low created by a slab avalanche and the high created by rising large plumes/superplumes should move surface plates toward the area of the avalanche. Plates that are purely oceanic lithosphere would presumably disappear into the mantle, but plates that contain continents would be too light to subduct. Thus, where this process continues for a few hundred million years, it is likely that virtually all continental blocks would accumulate in one area to form a supercontinent.

Because the breakup of one supercontinent overlaps the assembly of the next one, it is possible that mantle movements that cause breakup automatically begin the accretion process elsewhere. Figure 6.11 shows how rising mantle beneath a supercontinent sends continental blocks toward areas of accretion above a geoid low. Presumably this was a continuing process during most of earth history.

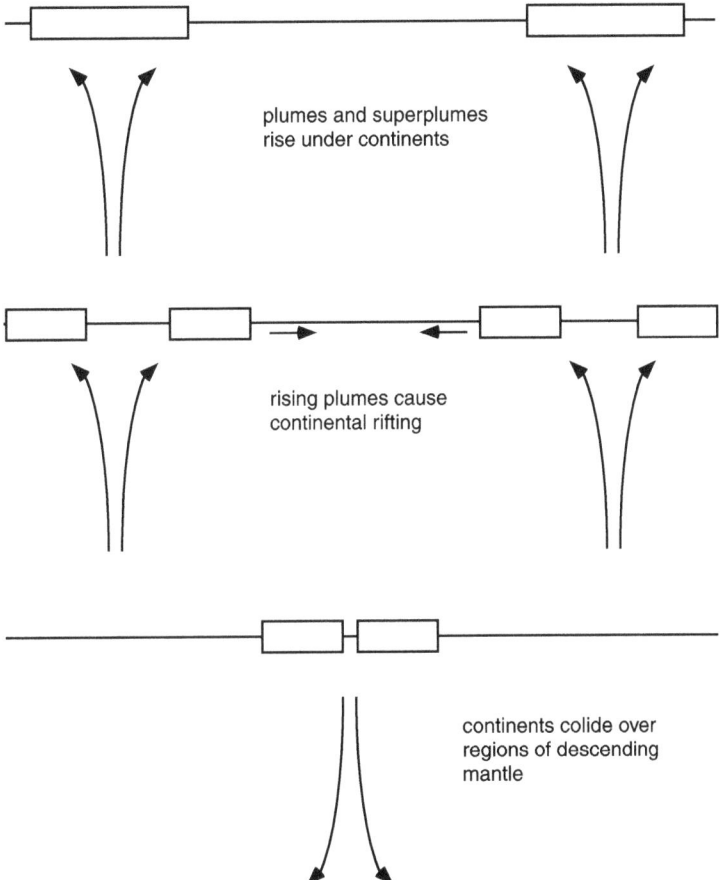

Figure 6.11. Effect of mantle plumes on breakup and assembly of supercontinents.

## Evidence for Dispersal of Former Supercontinents

Evidence for the dispersal of Pangea is clear. Breakup of Pangea created the present pattern of continents and ocean basins, and the histories of ocean opening can be determined by the patterns of magnetic stripes that they exhibit. We discuss this process in chapter 9, and here we merely summarize the conclusions that can be drawn about the mechanism of fragmentation.

### Breakup of Pangea

Fragmentation of Pangea provides two important insights on the dispersal of supercontinents. One is that dispersal of Pangea began with rifting of fragments from Gondwana while the supercontinent was still being assembled and continues to the present. Presumably older supercontinents underwent progressive dispersal over a time period at least as long. The second observation is the number and sizes of blocks formed during dispersal. In addition to present continents, blocks rifted from Pangea include Cimmeria (as one or a few blocks), Greenland, India, Arabia, Madagascar, and the Seychelles–Mascarene ridge. Most of these blocks are large, and if this pattern was followed during the dispersal of an older supercontinent, then most of the blocks available for accretion to the next younger supercontinent should also have been large. This supports the view that supercontinents grew mostly by assembly of large blocks, plus a few smaller blocks that accreted around their margins, instead of by accretion of numerous small blocks.

### Evidence of Breakup of Supercontinents Older than Pangea

Because all oceanic lithosphere older than ~200 Ma has disappeared (chapters 1 and 2), the breakup of

96  Continents and Supercontinents

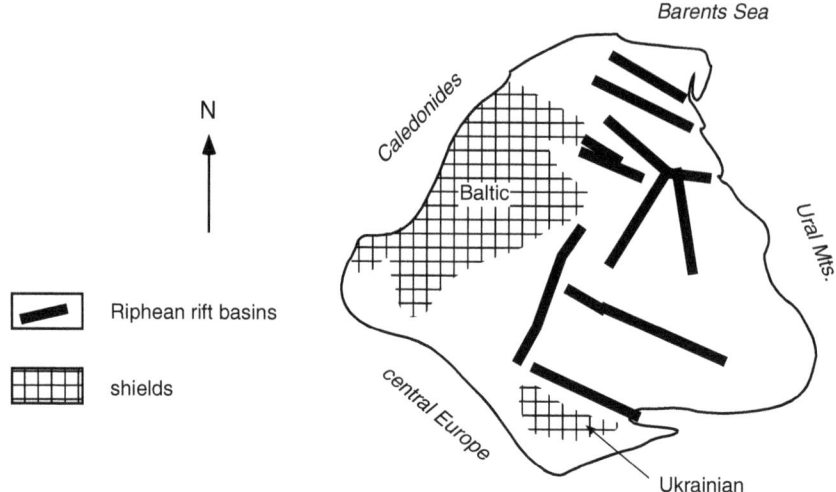

Figure 6.12. Riphean rifts in East European craton (interpreted from map of the depth of platform cover on basement; Aleinikoff et al., 1980). Further information in Nikishin et al., 1996.

continents older than Pangea cannot be investigated by studying modern oceans. Details of the fragmentation can be inferred for individual supercontinents by correlating geologic features between continents that are now separated. We leave those for chapters 7 and 8, however, and here we concentrate only on the general evidence that shows that some supercontinent was dispersed during some period of time. This evidence consists of geologic features that indicate extension over a broad area by relaxation of stress throughout a supercontinent. We classify them into four categories: systems of rift valleys; granite–rhyolite terranes; anorthosite (AMCG) complexes; and mafic dike swarms.

Systems of Rift Valleys  Linear belts of rift valleys commonly develop into spreading oceans, and evidence of their existence is preserved only along the margins of the new continents and in some aulacogens that extend into the continental interiors. Many continents, however, preserve broad areas of rifts with numerous different orientations. They indicate a general relaxation of lateral compression, and some apparently were developed during the breakup of supercontinents.

An example of widespread rifting is the Riphean basins of the East European craton (fig. 6.12; Nikishin et al., 1996). As we discussed in chapter 5, the East European craton formed by the fusion of its Baltic and Ukrainian areas at about 2 Ga. Rifting of the stabilized area began about 1650 Ma and continued to approximately 650 Ma as sedimentation continued in old rifts and new ones developed. Throughout this time, the rifts were filled almost entirely by shallow-water sediments consisting mostly of sandstones, siltstone, and shales, with minor conglomerate and carbonate rocks. The only igneous rocks are very minor bimodal assemblages of basalt and rhyolitic tuff. The Riphean rifts became active at the time that the proposed supercontinent Columbia was reaching maximum packing (chapter 7), and it is possible that the rifting was an early result of extension of part of the supercontinent.

Rifts in India initially formed at about the same time as the Riphean rifts, but their later history is very different (fig. 6.13; Naqvi et al., 1974; Chaudhuri et al., 2002). They began to develop shortly after suturing of different parts of India converted it into a single continental block at ~1.8 Ga. The Indian rifts are unique among all of the earth's known rift systems because they remained active as basins of subsidence continually from the Middle Proterozoic to the present, a time range of at least 1.5 billion years. This longevity clearly indicates some thermal or compositional anomaly associated with the Indian lithosphere, but no satisfactory explanation is now available.

Other rift systems were associated with breakup of other supercontinents. One of the oldest is the Statherian rifts of Amazonia, which began at ~1.8 Ga, shortly after stabilization of the supercontinent Atlantica at ~2 Ga (Brito Neves, 2002). These rifts remained episodically active throughout the proposed assembly of the supercontinent Columbia

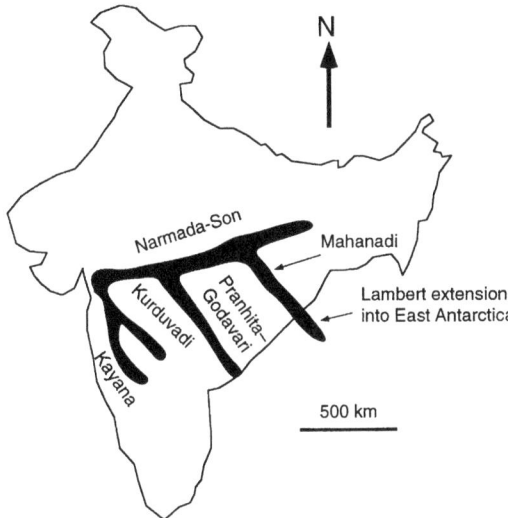

Figure 6.13. Rift valleys in India (from Rogers and Santosh, 2002).

(chapter 7), possibly demonstrating synchronous extension and compression in different parts of the assembling supercontinent. Other major examples include aulacogens and related rifts developed in North America shortly after the breakup of Rodinia at ~800–700 Ma (chapter 7), and early stages of rifting in Africa followed the breakup of Pangea (Lambiase, 1989).

Granite–Rhyolite Terranes  Granite–rhyolite terranes are large areas of rhyolitic flows and pyroclastics locally penetrated by granitic plutons emplaced during the same period of time as eruption of the volcanic rocks. Many of the rocks are peralkaline, with alkali concentrations high enough to generate riebeckite and aegerine. Intrusive rocks occur in ring complexes and as individual plutons. The types of rocks and their compositions are characteristic of areas of continental rifting.

An example of granite–rhyolite terranes is the Malani suite of northwestern India (Bhushan, 2000; Pandit et al., 2001; Roy, 2001). It covers ~20,000 sq km with extrusive suites up to 300 m thick formed from ~900 Ma to ~700 Ma. The oldest volcanic rocks are minor tholeiitic and alkali basalts, but almost all of the rest of the suite consists of peraluminous or peralkaline rhyolite plus some trachyte. Granites that intrude the suite locally have $SiO_2$ concentrations greater than 70%, total alkali between 5% and 10%, and extremely low MgO and FeO. Most of the granites contain hydrothermal veins and other evidence of fluid activity, including local development of topaz and tin minerals. The age range of the Malani suite is very similar to the range proposed for rifting of Rodinia (chapter 7), and it may have been erupted because of widespread extension of the supercontinent at that time.

Older granite–rhyolite terranes with large areal extent have been recognized in three places. The largest underlies much of the Phanerozoic platform of North America, where it is known from drill cores throughout the area and outcrops in the St. Francois Mountains (Bickford and Anderson, 1993). It formed in an age range centered around ~1.3 Ga and probably overlies terranes accreted to the southern margin of the Canadian shield. A terrane of similar lithology and age has been proposed in eastern India (Chaudhuri et al., 2002) based on rhyolite outcrops preserved in rift basins. The suites in both North America and India have been regarded as the result of widespread extension during the breakup of Columbia (chapter 7). Similarly the Uatuma suite in South America followed consolidation of northern South America (Brito Neves, 2002) and may be related to extension of the supercontinent Atlantica (see above).

AMCG Suites  Massif anorthosites mostly occur in rock associations known as AMCG complexes (Emslie and Hunt, 1990). The A signifies anorthosite, composed almost entirely of plagioclase (labradorite to andesine) plus minor oxides and a few mafic minerals. The rocks are granulated because of intense shearing during and after emplacement. The M refers to mangerite, and the C stands for charnockite formed by igneous processes, in contrast to the many charnockites developed by anhydrous metamorphism of gneisses (chapter 4). The G refers to granite, rich in quartz and K feldspar and commonly exhibiting rapakivi texture.

The AMCG complexes probably developed initially from basaltic magma ponded near the base of the crust or in the upper mantle. Starting from this basalt, separation involved several processes, generally involving both crystal accumulation and magma mixing (Wiebe, 1992). Production of the initial basalt and subsequent processing almost certainly occurred during crustal extension, which also enabled the resulting magmas to move upward in the crust.

Mukherjee and Das (2002) summarized the features of 72 massif anorthosites worldwide. A few are Phanerozoic and are commonly associated with ring complexes and emplaced into near-surface rocks. More than 90% of the dated complexes, however, were emplaced into granulite-facies wall rocks between 2000 Ma and 1000 Ma, and half of them are restricted to an age range of 1500 Ma to 1200 Ma.

All of the AMCG complexes and other massif anorthosites indicate crustal extension during the time when they were formed. For some suites, this extension may have been related to the end of a localized orogeny, but the concentration and widespread distribution of complexes with ages from 2000 Ma to 1000 Ma indicate one or more periods of widespread extension during this time. We discuss the possibility that the 1500–1200 Ma suites are related to dispersal of the supercontinent Columbia in chapter 7, but the cause of emplacement at other times is unclear.

Mafic Dyke Swarms  Mafic dike swarms develop when basaltic magma penetrates (and partly creates) cracks near the earth's surface and crystallizes to gabbros, diabases, dolerites, and other rocks with a variety of textures. Dike swarms of any orientation are clear indicators of crustal extension and occur in numerous parts of the earth (review by R. Ernst and Buchan, 1997), but the most useful are swarms that radiate from a small area. The center of radiation is generally inferred to be a plume that was responsible for at least some of the horizontal stress relaxation (see discussion of "large igneous provinces" in Condie, 2001).

The best studied swarm is the Mackenzie dikes emplaced at about 1.4 Ga in northern Canada (LeCheminant and Heaman, 1989). They radiate from the Coppermine River Basalts, formed near the end of orogeny on the continental margin of North America. The dikes were intruded during the same time range as AMCG complexes and the granite–rhyolite terrane were developed farther south in North America, and all of these suites may be related to dispersal of the supercontinent Columbia (chapter 7).

No correlatives of the Mackenzie dikes have been found in other continents, but several dike swarms have been proposed to occur in two continents that were rifted apart after the dikes were formed. For example, Correa-Gomes and Oliveira (2000) proposed that 1.0-Ga dikes in eastern Brazil and western Africa were emplaced as the supercontinent Rodinia was breaking up (chapter 7). Similarly, Park et al. (1995) suggested that a 780-Ma radiating dike swarm in Rodinia was separated by rifting between Australia and North America (fig. 6.14).

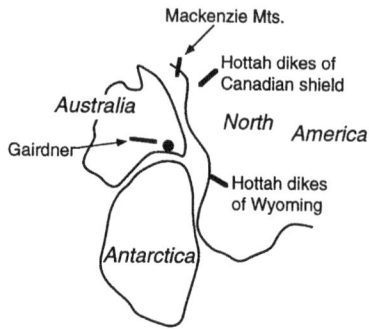

Figure 6.14. Orientations of ~800-Ma mafic dikes in Australia and North America (modified from Park et al., 1995). The black circle shows the location of a possible plume in southeastern Australia.

## Causes of Dispersal of Supercontinents

Supercontinents break up because they are subjected to lateral extension throughout most of their area. This requires uplift that can be caused only by an accumulation of heat under the supercontinent. Part of this accumulation results from the comparatively low conductivity of continental crust, which prevents escape of heat across the earth's surface as rapidly as it does across oceanic lithosphere. Some of the increase in temperature may also result from the inability of subducting slabs to reach the interior of the supercontinent and cool the upper mantle beneath it.

It is unclear whether uplift of a supercontinent is related to the development of plumes and superplumes (Condie, 2001). The problem is complicated by the uncertain relationship between plumes/superplumes and basalt plateaus. Many geologists assume that large areas of basaltic eruption are the "heads" of plume (hotspot) tracks, although this assumption has been challenged by examination of sequences of ages along the tracks (chapter 2).

Some basalt plateaus are clearly related to the breakup of Pangea. Hawkesworth et al. (1999) and Storey et al. (2001) discussed the possibility that a broad region of Gondwana was underlain by hot mantle (superplume?) that generated basalt provinces during different ages of fragmentation in at least three locations (further discussion in chapter 9). Courtillot et al. (1999) also suggested that much of the breakup of Pangea was caused by plumes that developed at different times. Broad uplift of Pangea before rupturing is also shown now by a geoidal high in most of the Atlantic Ocean, which is almost certainly the remains of an axis of uplift beneath Pangea.

The relationship of plumes (and basalt plateaus) to fragmentation of supercontinents, however, is not always clear (Storey and Kyle, 1997; Condie, 2001). Some parts of Pangea seem to have rifted apart without any associated magmatism, either of basalts or other magmatic rocks. Furthermore, some basalt plateaus clearly have no relationship to supercontinents. Examples include the ~15-Ma Columbia River basalts in western North America and, less clearly, a possible superplume in the Pacific Ocean from about 120 Ma to 80 Ma (Larson, 1991). This superplume apparently caused so much reorganization of the earth's interior that no magnetic reversals occurred for 40 million years, referred to as the Cretaceous "magnetic quiet zone." The plume also formed numerous oceanic plateaus that are still preserved in the ocean and may have initiated some of the Pacific hotspot tracks.

## Summary

Models for the formation of supercontinents depend highly on interpretation of the history of orogenic belts within the continents. Intercontinental belts formed by closure of oceans within the belts demonstrate assembly of terranes on either side. Intracontinental belts that formed by reworking of older crust within the belt, however, probably developed within a large continental block that accreted intact to the supercontinent. Some confined orogens opened briefly and closed without dispersing the continental blocks that contain them, thus permitting those blocks to accrete intact to a supercontinent.

Supercontinents probably accrete where mantle processes bring most of the earth's terranes into one area, probably above the downgoing limbs of very large convection cells. Different interpretations of the nature of orogenic belts lead to different models for the assembly of supercontinents. If most belts are intercontinental, then supercontinents have been formed by accretion of numerous small terranes. If many belts are intracontinental or confined, however, then supercontinents formed by accretion of a few large blocks and accumulation of smaller terranes around their margins.

Dispersal of supercontinents is probably caused by accumulation of heat under the supercontinent, possibly above rising convection cells. The dispersal of Pangea can be traced by patterns of ocean opening, but the breakup of older supercontinents must be inferred from evidence of widespread continental extension. Indicators of extension include rift systems, granite–rhyolite terranes, anorthosite (AMCG) complexes, and dike swarms.

# 7

# Supercontinents Older than Gondwana

The configurations of Gondwana and Pangea are well known because the histories of oceans that opened to disperse Pangea can be reconstructed from their patterns of magnetic stripes (chapters 1 and 9). The configurations of older supercontinents cannot be easily determined because the oceanic lithosphere formed when they dispersed is so old that it has been completely subducted and destroyed. Thus the histories, and even existence, of these older continents must be inferred from indirect evidence.

The four most widely used techniques for reconstructing old supercontinents are: paleomagnetic data; correlation of orogenic belts that developed during accretion of the supercontinent: correlation of extensional features that developed when the supercontinent fragmented; and recognition that sediment in one present continent was derived from a source now in another continent.

Paleomagnetic information can be used in two ways. One is to compare APW curves for different continental blocks to determine whether there were periods of time when two or more blocks seem to have been joined (appendix C). If similar movements are found for several continental blocks that are now separated, then we can infer that they formed a single block, perhaps a supercontinent, during the period when they had identical APW paths. Another method of using paleomagnetic data is simply to compare the apparent latitudes of numerous continental blocks. Even though longitudes cannot be specified, latitudes can be used to infer proximity of different blocks, thus supporting other information that suggests the configuration of a supercontinent.

Correlation of orogenic belts starts with identification of belts of different ages in present continents. Belts of the same age are now scattered all over the earth's land surface because of fragmentation of supercontinents and movement of modern continents to their present positions. The configurations of older supercontinents can be inferred by placing modern continents into positions in which these orogenic belts line up to form a pattern that would be expected to develop during accretion of a supercontinent. We demonstrate this technique below in our discussion of the configurations of Rodinia and Columbia.

Correlation of extensional features is possible because fragmentation of a supercontinent is generally accompanied by widespread extension throughout much of the former landmass (chapter 6). This extension results in some features that can be correlated among the newly formed fragments, including orientations of mafic dike swarms and positions of aulacogens and other rifts. Features such as granite–rhyolite terranes and AMCG suites, however, indicate broad extension but are not readily correlatable from place to place.

Determination of the sources of sediments is most commonly done by measuring the ages of zircons in the clastic suites. In many sediments, the ages correspond to the ages of nearby crystalline rocks, but some sediments contain zircons that could not have derived from adjacent rocks, suggesting that their source has now rifted away during fragmentation of a supercontinent.

Other, less direct, methods have been employed in efforts to infer configurations of old supercontinents. In chapter 6 we described how the grouping of cratons with similar stabilization ages in different parts of Pangea suggests the existence of three old large continents. We discuss below how the distribution of different types of sediments has been used to infer the existence of two supercontinents in the Late Archean.

All of these methods become less certain as the age of the proposed supercontinent increases. For this reason, we discuss supercontinents older than Gondwana in order of their increasing age. The youngest of these is Rodinia, which apparently reached its maximum packing (chapter 6) at about 1 Ga. The configuration of Rodinia is highly controversial, partly because the entire concept of this supercontinent was initially suggested as recently as 1990. Columbia, with maximum packing at about 1.6 Ga, was originally proposed by Piper (2003) and is even more controversial. The oldest is an unnamed supercontinent that was proposed in 2003 to have existed in the Late Archean and Early Proterozoic.

Because of the newness of all of these proposals and their highly controversial nature, we cannot present a timely discussion of them in this book, and we use a website at www.gondwanaresearch.com/csbook to maintain up-to-date information on new data and ideas. This information supplements the discussion of the configurations of supercontinents in this chapter and of related orogenic belts in appendices G (belts of Grenville age) and H (belts of 2.1–1.3 Ga age).

## Rodinia

The transition from Precambrian to Cambrian is marked by the sudden appearance of skeletal organisms (chapter 12). In stratigraphic sections that cross this boundary without interruption in sedimentation, lower sequences with enigmatic soft-bodied organisms gradually change upward to sediments that contain organized burrows, small shells of controversial origin, and then shelled invertebrates that represent most of the phyla that exist today. The development of skeletal structures apparently occurred within a few million years between 540–530 Ma, and McMenamin and McMenamin (1990) proposed that it followed the evolution of various groups of soft-bodied metazoans that occurred in large areas of continental shelves in the latest Precambrian. This suggested a supercontinent of some uncertain configuration, and McMenamin and McMenamin named it Rodinia, derived from a Russian word for "beget."

The search for a configuration of Rodinia was assisted by the observation that a time of 1 Ga and slightly older was a period of major orogeny in many parts of the world. Named for the type Grenville area of eastern Canada, correlative belts with various names occur in linear orogens in all continents (fig. 7.1). Events of this time, however, are absent from many broad regions, suggesting that the Grenville-age belts were linked along worldwide zones of orogeny that may have created the supercontinent proposed by McMenamin and McMenamin.

Grenville-age orogens are shown in fig. 7.1 and are summarized in appendix G. Figure 7.1 distinguishes orogens regarded as "interior magmatic belts" from those classified as "exterior thrust zones." The criteria for these two types of orogen are discussed in our description of the type Grenville belt in appendix G.

Fitting Grenville-age belts together has led to numerous proposed configurations for ~1-Ga supercontinents, most of which are referred to as Rodinia but also include an early proposal for Palaeopangaea.

### Proposed Configurations of Rodinia

The first configurations of a 1-Ga supercontinent to use the name Rodinia were published in the early 1990s. Following them, numerous modifications have been suggested throughout the past decade, but here we discuss only four of them and leave many proposals to our website.

All proposals begin with the concept that much of North America ("cratonic North America") had achieved nearly its present shape by ~1.3–1.2 Ga. Whether it had reached this configuration earlier, perhaps 1.8 Ga or 2.5 Ga, is unclear and is discussed in chapter 6.

The configuration of Rodinia most commonly referred to was published by Hoffman (1991). Figure 7.2a shows blocks surrounding Laurentia and attached East Antarctica along a series of Grenville-age belts. Baltica is attached to southern Greenland and Amazonia to eastern North America. Next to Amazonia, Grenville-age belts continue around blocks east of the Congo craton and attach them to Congo. Grenville-age belts almost completely surround the Kalahari craton and follow around the coast of East Antarctica, and all of East Antarctica is assumed to have been a coherent block at ~1 Ga. India and Australia are shown sutured to East Antarctica in the same relative positions they occupied in East Gondwana (chapter 8). The positions of numerous blocks now in Asia are suggested but not shown on the map. This configuration is similar to one proposed by Meert (2001) based on paleomagnetic evidence (fig. 7.2b).

The configuration of Rodinia proposed by Hoffman (1991) assumes that all Grenville belts are zones of oceanic closure between continental blocks. It also accepts a juxtaposition of western North America and East Antarctica (Dalziel, 1991), referred to as the SWEAT (Southwest North America East Antarctica) proposed by Moores (1991) rather than a position of Australia against North America (AUSWUS connection; see below). Siberia is located adjacent to northern Canada, as proposed by Condie and Rosen (1994). West Africa is shown related to both Congo and Amazonia but without Grenville-age attachment.

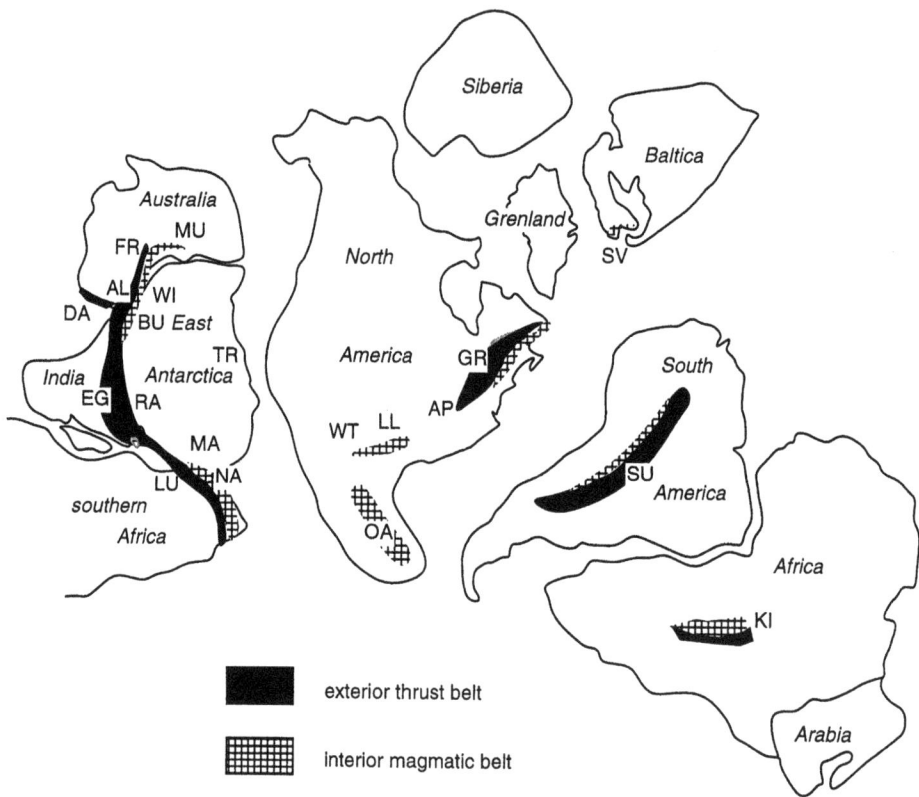

Figure 7.1. Locations of Grenville belts. For clarity, southern Africa is shown at two locations. DA, Darling; MU, Musgrave; FR, Fraser; AL, Albany; WI, Windmill Islands; BU, Bunger Hills; EG, Eastern Ghats; RA, Rayner; TR, Transantarctic; MA, Maudheim; NA, Natal; LU, Lurio; GR, Grenville; AP; scattered outcrops in the Appalachians; LL, Llano; WT, West Texas; OA, Oaxaquia; SV, Sveconorwegian; SU, Sunsas; KI, Kibaran. Further discussion in text and appendix H.

On the basis of an early suggestion by Brookfield (1993), Karlstrom et al. (1999) and Burrett and Berry (2000, 2002) proposed a configuration for part of Rodinia that is different from the ones discussed above (fig. 7.2c). The key relationship is an "AUSWUS" connection, named for a position in which eastern Australia adjoins the western United States. This configuration correlates the Grenville/Sveconorwegian belts with the Musgrave and Albany/Fraser belts of Australia and further into adjoined belts in East Antarctica. It also requires Oaxaquia (appendix G) to be a link between North America and Australia. By placing Australia south (present orientation) of its position in other models of Rodinia, the AUSWUS configuration leaves room for Siberia to be joined to western Canada, as proposed by Sears and Price (2000, 2002).

A configuration of a ~1-Ga supercontinent quite different from Rodinia was suggested by Piper (1982, 2001; fig. 7.2d). He referred to it as "Palaeopangaea" because it had a slightly curved shape similar to that of Pangea. The principal evidence for Palaeopangaea is paleomagnetic data supplemented by correlation of Grenville belts. Piper compared APW curves (chapter 1) for different blocks in both Palaeopangaea and a Rodinia configuration, and the better correlation of curves within Palaeopangaea has been used as evidence to support that configuration. The configuration proposed by Piper (2001) accepts the coherence of the three old continental assemblies proposed by Rogers (1996; chapter 6).

Many alternative configurations have been proposed for Rodinia. We discuss them in the website and mention only a few examples here. Li et al. (1995) placed Baltica off the coast of Greenland close to the position it occupied in Pangea, South China between Australia and North America, and North China adjacent to Siberia along a continuation of the join between Baltica and Greenland.

Dalziel et al. (2000b) placed the Kalahari block against southern North America. Torsvik et al. (2001) proposed that India and East Antarctica were separated in Rodinia, and Hartz and Torsvik (2002) place the eastern side of Baltica (present orientation) instead of the western side against Greenland. Wingate et al. (2002) proposed an AUSMEX connection, in which Australia was adjacent to Mexico instead of to the western United States.

## Breakup of Rodinia

Separation of blocks in Rodinia probably occurred mostly between 800–600 Ma. In this section we discuss the evidence for dispersal and briefly describe movements of continental blocks from Rodinia to Gondwana. We leave a more complete discussion of the movements of blocks between 750 Ma and 250 Ma to chapter 8.

Rifting   Because it was at or near the center of Rodinia, the consequences of rifting are well displayed in North America (fig. 7.3). The earliest development is the midcontinent rift system, which began during the last stages of accretion in other parts of Rodinia. The only outcrop is the lopolithic Duluth gabbro with an age of ~1.1 Ga, and the rest of both the western and eastern branches are buried by flat-lying Paleozoic sediments. This rift system remained active throughout the Paleozoic, forming depocenters for sedimentary accumulations thicker than on the midcontinent platform.

Rifting of the western margin of Precambrian North America may have begun at ~750 Ma, when APW curves between Laurentia and Australia began to diverge (Powell et al., 1993). Subsidence analysis of latest-Proterozoic and Paleozoic sequences in western North America shows that accumulation of sediments in this area began at ~625 Ma and continued without interruption until docking of arc terranes in the Devonian (Bond et al., 1984; chapter 5). The similarity of this series of sediments in North America with sequences preserved in the Transantarctic Mountains led Moores (1991) to propose the SWEAT connection (see above).

Eastern and southern North America developed a complex margin as the continent rifted away from some part of South America and attached blocks. The margin is highly irregular because of offset along numerous transform faults (fig. 7.3; Thomas, 1977). The largest fault causes the present separation of the Appalachian and Ouachita orogens, and smaller offsets along the Appalachians controlled the development of salients where the Appalachian fold and thrust belt was deformed farther west during the Alleghenian orogeny (chapter 10). Several rifts extend into the continent from the margin, and some remained active from the latest Proterozoic through most of the Paleozoic.

Complex patterns of rifting associated with the breakup of Rodinia are known in continental areas outside of North America. In two of them, the Riphean rifts of Russia (chapter 6) and the rift valleys of India, post-Rodinia sedimentation simply continued in rifts that had already been established in the Middle Proterozoic. In South America and West Africa, however, new patterns of rifts developed shallow ocean basins floored by continental crust throughout a broad continental area that had been stable since ~2 Ga (Brito Neves, 2002; fig. 7.4).

Movement after Breakup   Rodinia broke up into continental blocks whose histories are very controversial (fig. 7.5). Because Laurentia was at the center, it was almost completely surrounded by rifted margins that remained passive until the Paleozoic (chapter 5). South America moved away from North America by a counterclockwise rotation as it traveled to the position south of North America that it would occupy in Pangea (chapter 8). As it moved, fragments of western South America rifted away and traveled to the east coast of North America at the same time as rifted terranes from eastern North America were attached to western South America (Thomas and Astini, 2002; see discussion of Andes and Avalonian terranes in chapter 5). Brief collisions between North and South America may also have occurred during this period of migration, possibly contributing to compressional stresses generated during Paleozoic orogenies (Dalziel, 1997).

The western margin of Africa in Rodinia had apparently been passive during most of the accretion of the supercontinent. This condition may have remained unchanged if South America and west-central Africa moved largely as a coherent block, but if all of Africa consisted mostly of separate blocks, they would have followed different paths until their accretion to Gondwana. Much of eastern and southern Africa became involved in compressive orogeny by latest Proterozoic time as West and East Gondwana were fused together (chapter 8).

Numerous blocks separated from western North America when Rodinia broke apart. North and South China probably moved separately from controversial positions in Rodinia until their accretion to Pangea. Australia, India, Madagascar, at least

104    Continents and Supercontinents

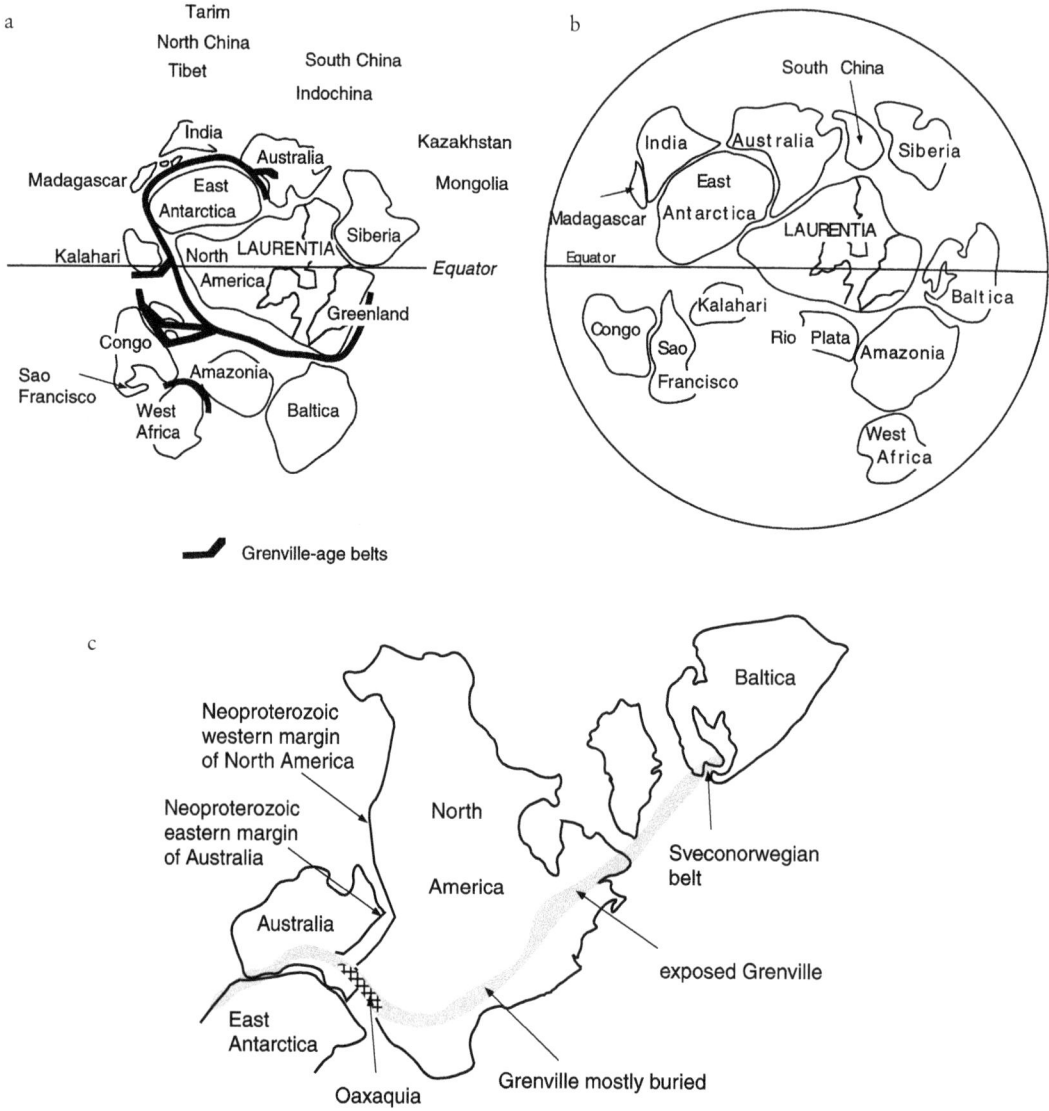

Figure 7.2. a. Configuration of Rodinia proposed by Hoffman (1991). Terranes not specifically located are shown only by their names. b. Configuration of Rodinia proposed by Karlstrom et al. (1999). South America and Africa are not shown in their diagram. c. Configuration of Rodinia proposed by Meert (2001); also see Weil et al. (1998).

parts of coastal East Antarctica, and possibly the Kalahari–Maudheim block became the nucleus of East Gondwana, either moving separately or as a single block (chapter 8). In order to join with West Gondwana they had to move across much of the earth. Depending on the movement of South America–Africa, this might have required more than 10,000 km of movement in some 200 million years. This rate is high in comparison with current rates of plate movements and much more rapid than would have been required if the ~1-Ga supercontinent had a configuration of Palaeopangaea rather than any of the proposed Rodinia configurations (Piper, 2001).

## Columbia

Very little orogenic activity is known anywhere in the world between the Late Archean and about 2.1

Figure 7.2. d. Interpretation of configuration of Palaeopangaea proposed by Piper (2001), including the three old continental assemblages discussed in chapter 6.

Ga. This period of approximately 500 million years seems to have been a time when the earth was tectonically quiet, with the only major activity consisting of initial phases of growth of cratons in eastern South America and western Africa (see Atlantica in chapter 6). Numerous orogenic belts, however, became active during the period of ~2.1–1.8 Ga, suggesting that the earth had entered a new phase in its history (fig. 7.6; Condie, 2002; Zhao et al., 2002b) The geology of most of these orogenic belts is summarized in appendix H, which discusses tectonic activity from 2.1–1.3 Ga.

The possibility that this new phase of earth history involved assembly of a supercontinent was suggested by the abundance of orogenic activity between about 2.1 Ga and 1.6 Ga and the evidence of widespread extension beginning at approximately 1.5 Ga. Figure 7.7a shows the configuration of a supercontinent named Columbia that began to accrete at ~1.9–1.8 Ga (Rogers and Santosh, 2002). This configuration is very similar to the shape of Pangea, with one margin undergoing active orogeny and the other margin divided into two parts, one that underwent rifting and one where the rifted blocks accreted (chapter 8). Figure 7.7b shows an alternative configuration proposed slightly later by Zhao et al. (2002b).

The principal evidence supporting the configuration of Columbia shown in fig. 7.7a consists of the pattern of Middle Proterozoic rifting in eastern India and western North America, the history of orogenic zones along the proposed suture, and the nature of continental outgrowth along one margin of the supercontinent.

Mesoproterozoic rifting in India developed several rifts that have been episodically active from the Mesoproterozoic to the present (fig. 7.8; Naqvi et al., 1974; Chaudhuri et al., 2002). This rifting

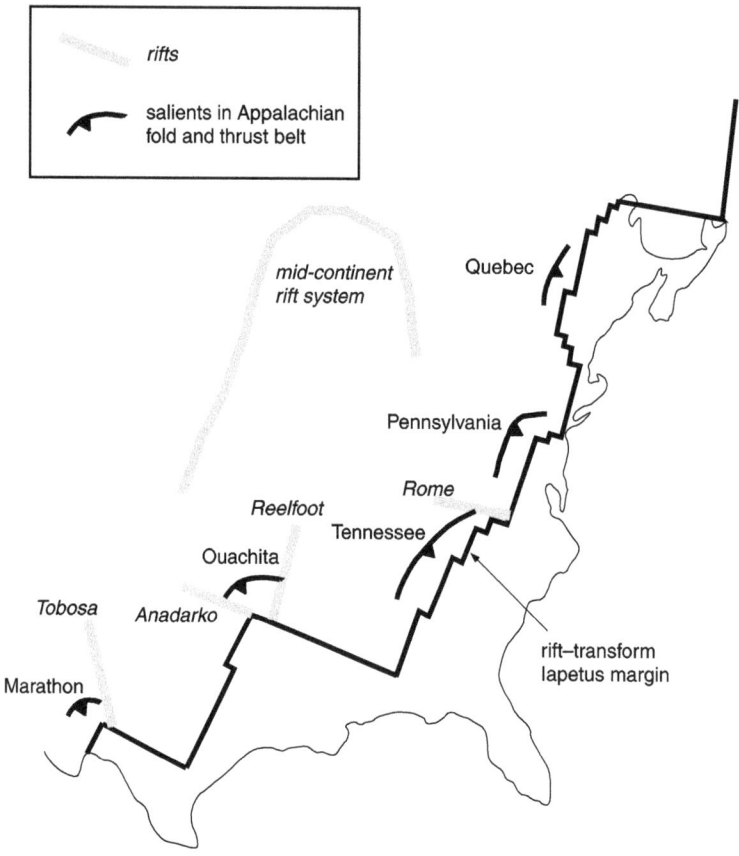

Figure 7.3. Neoproterozoic rifting along the eastern margin of North America. The complex pattern of transforms that intersect the margin is from Thomas (1977).

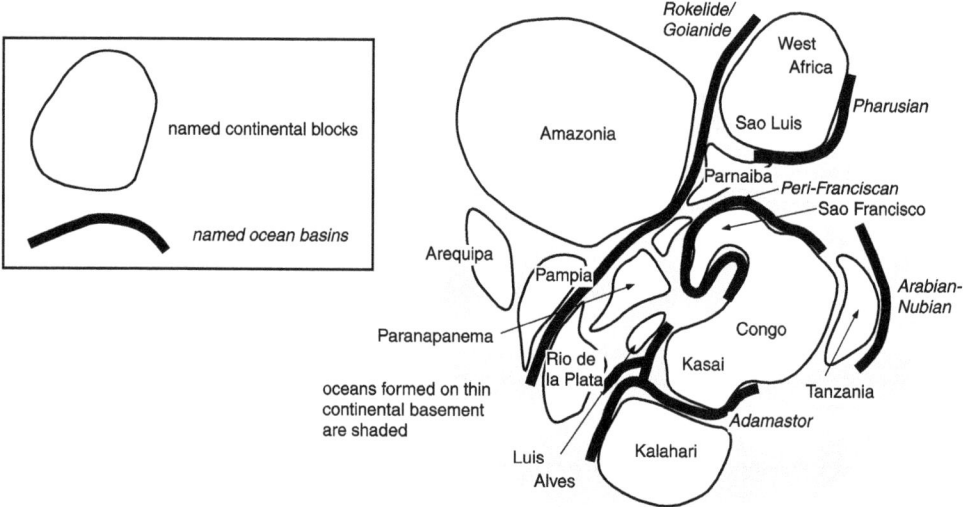

Figure 7.4. Rift patterns in Amazonia and western Africa following the breakup of Rodinia (modified from Brito Neves, 2002).

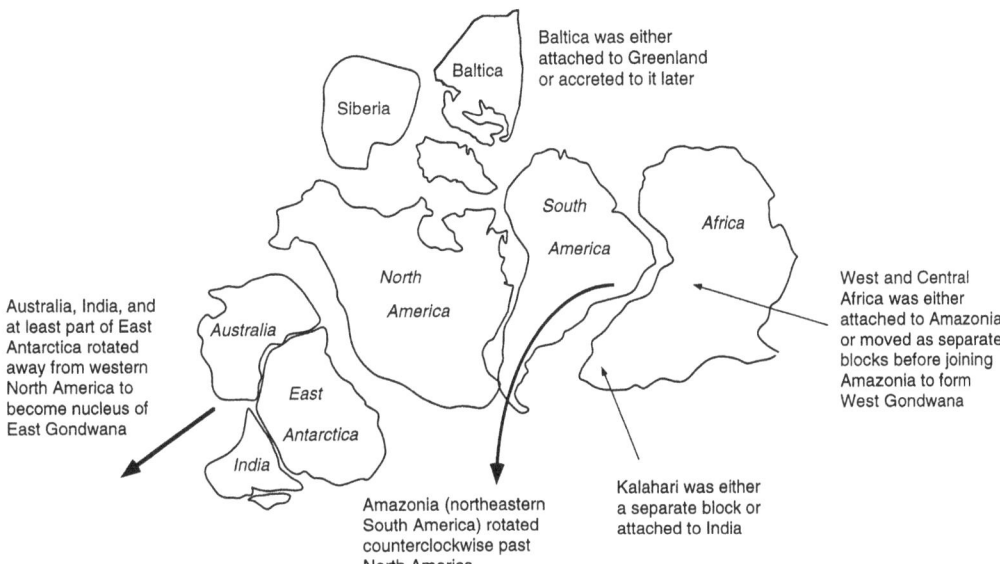

Figure 7.5. Breakup of Rodinia.

Figure 7.6. Locations of orogenic belts with ages from 2.1–1.3 Ga. For clarity, southern Africa is shown at two locations. Tn, Trans-North China; Ak, Akitkan; Ag, Angara; Kk, Kola–Karelia; Vo, Volhyn; Pa, Pachemel; Sv, Svecofennian; Kg, Konigsbergian–Gothian; Fo, Foxe; Wo, Wopmay; Tt, Taltson–Thelon; Ri, Rinkian; Ng, Nagssugtoqidian; To, Torngat; Un, Ungava; Nq, New Quebec; Ke, Ketilidian; Mk, Makkovikian; La, Labradorian; Th, Trans-Hudson; Gf, Great Falls; Pe, Penokean; Yv, Yavapai; Ma, Mazatzal; Rj, Rio Negro–Juruena; Ro, Rondonian; Te, Transamazonian–Eburnian; Ca, Capricorn; Af, Albany–Fraser; Wb, Windmill Islands–Bunger Hills; Ra, Rayner; Ci, Central Indian; Eg, Eastern Ghats; Ad, Aravalli–Delhi; Ln, Lurio–Namama; Li, Limpopo. Further discussion in text and Appendix H.

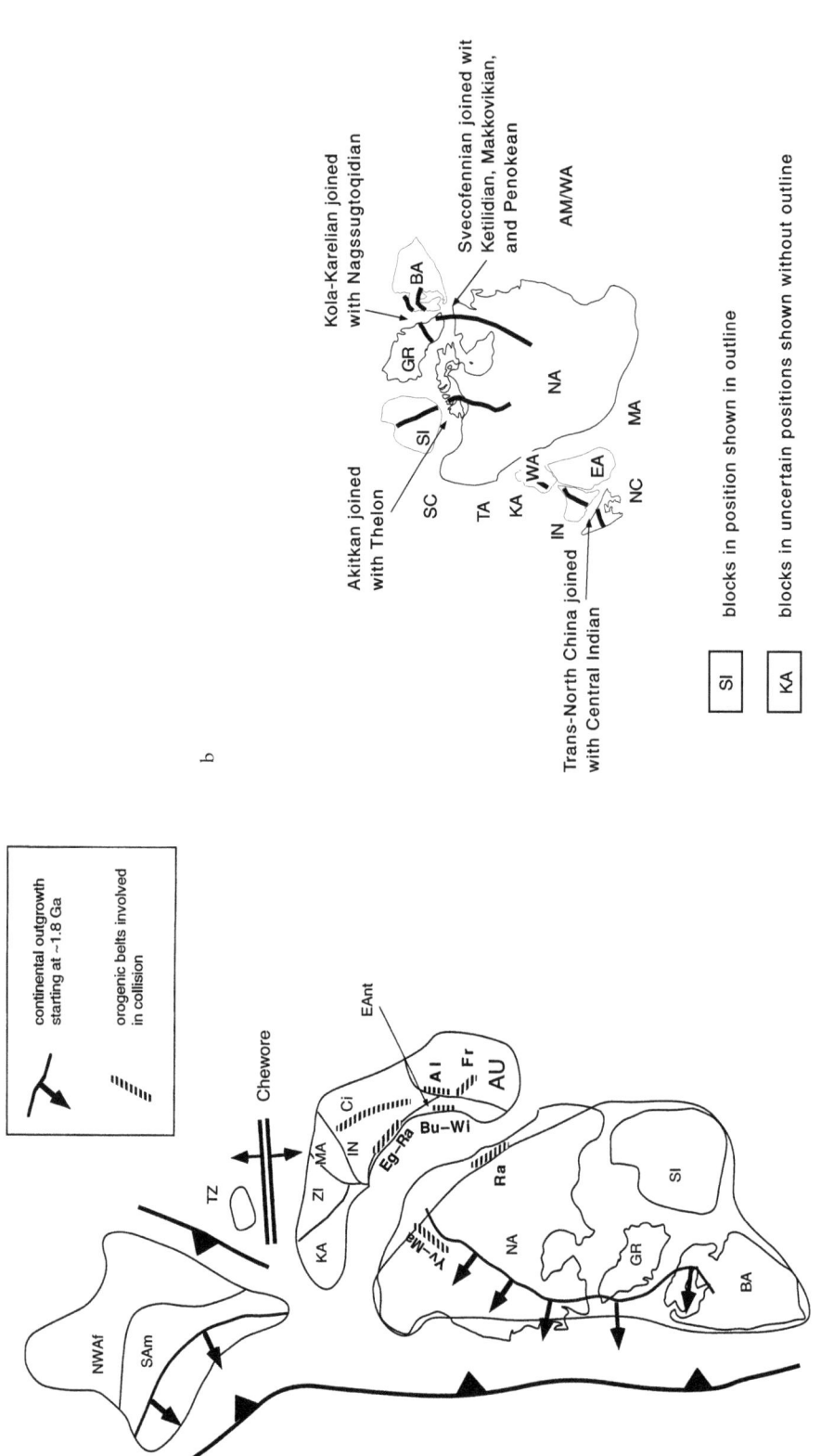

Figure 7.7. a. Configuration of Columbia at ~1.5 Ga proposed by Rogers and Santosh (2002) (references in text; conventional symbols for subduction zones and rifts). SI, Siberia; BA, Baltica; GR, Greenland; EAnt, coastal zone of East Antarctica correlative with marginal orogenic belts in Australia, India, and South Africa but separated during post-Pangea rifting; AU, Australia; NA, North America; IN, India; MA, Madagascar; ZI, Zimbabwe; KA, Kalahari; SAm, South America; NWAf, northwestern and central Africa; TZ, Tanzania. Chewore is the Chewore Ocean discussed in the text. b. Configuration of Columbia proposed by Zhao et al. (2002a). This diagram is drawn from information on orogenic belts of 2.1–1.8 Ga age discussed by Zhao et al. (2002a) and further described in Appendix H. Arrows point to principal joins that support this configuration. Blocks in uncertain positions are shown without outlines. Abbreviations are SI, Siberia; GR, Greenland; BA, Baltica; NA, North America; WA, Western Australia; IN, India; NC, North China; SC, South China; EA, East Antarctica; TA, Tarim; KA, Kalahari; MA, Madagascar; AM/WA, joined Amazonia/West Africa.

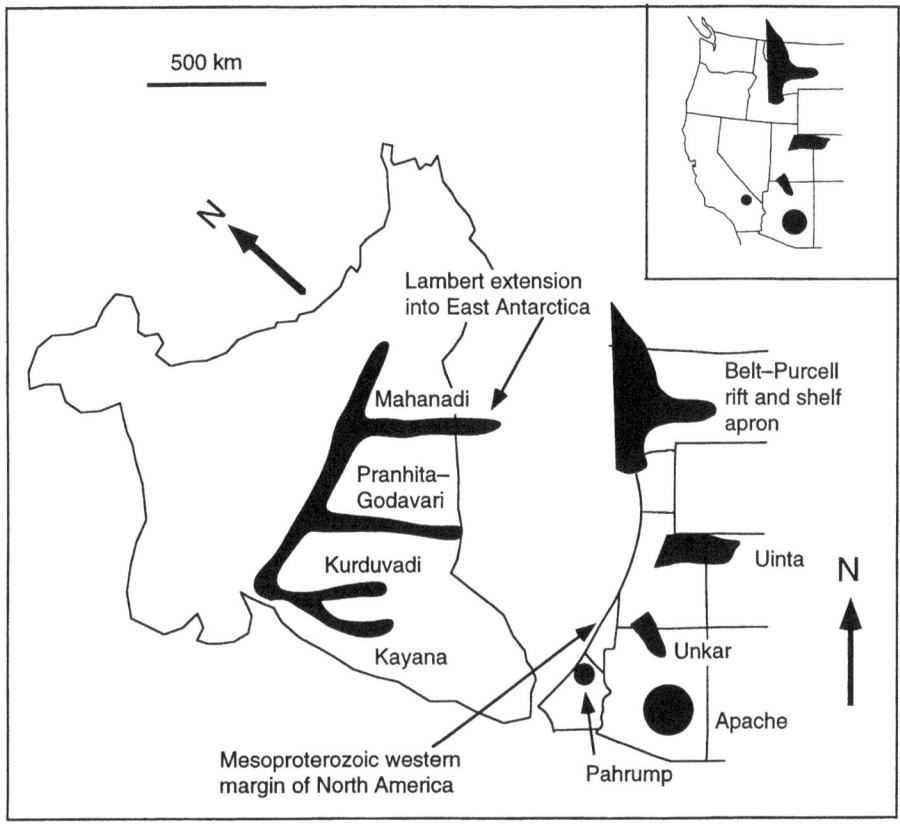

Figure 7.8. Rifts in eastern India (left) and western North America (right). Arrows show present north for each area.

also created broad basins containing sediments that thicken from the Indian interior toward the east, apparently indicating an eastern continental margin formed by rifting at the same time as the basins were initiated. Similarly, at least three rifts intersect the Precambrian western margin of North America (fig. 7.7a; summarized from Link et al., 1993). The Belt–Purcell trough started to rift before 1.47 Ga, when sediments near the base were intruded by mafic sills (Ross et al., 1992; Sears et al., 1998; Luepke and Lyons, 2001). The Uinta trough is at least as old as Neoproterozoic, and its unexposed base could be Mesoproterozoic. The Unkar Group is a rift sequence intruded by 1.1-Ga mafic sills, and the age of initial sedimentation could be as old as 1.35 Ga (Timmons et al., 2001). Rogers and Santosh (2002) correlated the Mahanadi–Lambert and Godavari rifts of India with the Belt and Uinta rifts of North America, respectively, thus placing eastern India against western North America in Columbia.

- Several orogenic belts are aligned along the suture zone shown in fig. 7.8 (details and references in appendix H). In western North America they include the 1.95–1.84 Wopmay orogen and Great Bear intrusive suites of the Racklan orogen, with ages apparently ranging from >1.7 Ga to <1.3 Ga, and possibly part of the Great Falls tectonic zone. Mesoproterozoic orogenic activity in Australia, India, and coastal East Antarctica includes the older stages (~1.7–1.3 Ga) of the Albany and Fraser orogens of Australia, which extend into the Bunger Hills and Windmill Islands of coastal East Antarctica, and the Eastern Ghats granulite terrane of eastern India and the correlative Rayner terrane of East Antarctica, which underwent metamorphism at ~1.7–1.6 Ga. Several of the orogenic belts show evidence of right-lateral movement, suggesting that the suture zone was dominated by dextral transpression, with subduction beneath North America.

The subduction that began at 2.1–1.8 Ga along the southern margin of Baltica, eastern margin of North America, and western margin of Amazonia continued as a broad zone of continental outbuilding that continued into Grenville time (appendix H). In Baltica these belts include the Svecofennian orogen, Western Gneiss region, the Southwest Scandinavian gneiss complex, and the Konigsbergian–Gothian magmatic suite. The Ketilidian orogen of Greenland and Makkovik province of Labrador developed between approximately 1.8–1.7 Ga, and further outbuilding continued in North America through the early phases of the Grenville orogeny. This outgrowth is well shown by the Yavapai and Mazatzal provinces of the southwestern United States, which developed from ~1.8–1.5 Ga. These orogens apparently continue northward to the Canadian shield beneath Phanerozoic platform sediments. The western part of the Amazon craton also underwent marginal growth from ~1.8–1.3 Ga, when subduction beneath Amazonia progressively formed the Rio Negro–Juruena and Rondonian suites (appendix H).

The configuration of Columbia proposed by Zhao et al. (2002b; fig. 7.7b) compares with the one proposed by Rogers and Santosh in the following ways.

- The major part of the supercontinent in both proposals is Nena, consisting of North America, Greenland, Siberia, and Baltica, with Siberia in the same position in both configurations.
- South America and northwestern Africa are joined in both configurations, referred to as the continental assemblage Atlantica in chapter 6, but Zhao et al. do not place this region in contact with other continental areas in the supercontinent.
- South China is shown as a separate block by Zhao et al. and omitted by Rogers and Santosh.
- Zhao et al. show North China attached to India, with the Trans-North China Orogen connected to the Central Indian Tectonic Zone, but North China/India are shown only questionably attached to the rest of the supercontinent. Rogers and Santosh omit North China from Columbia because of uncertainty about its location.

- A major difference between the configurations proposed by Zhao et al. (2002b) and by Rogers and Santosh (2002) centers around East Antarctica. Zhao et al. group all of East Antarctica with Australia, South Africa, and the Tarim craton of central Asia. This assembly is sutured against the western margin of North America, with East Antarctica opposite the United States and Australia opposite Canada. In contrast, Rogers and Santosh propose an expanded Ur (chapter 6) that contains India, South Africa, Australia, and coastal Antarctica, with a southern margin (present orientation) at an unknown location under the Antarctic ice-cap. The configuration proposed by Rogers and Santosh omits the rest of East Antarctica (including the "Mawson continent" proposed by Peucat et al., 1999) because Fitzsimons (2000) indicates that East Antarctica may not have been a coherent block before Pan-African time.

Paleomagnetic confirmation of the configurations of Columbia is difficult to find because of the age of the proposed supercontinent. Meert (2002) found that the configuration suggested by Rogers and Santosh (2002) was consistent with paleomagnetic data but could not be supported by them. Similarly, Zhao et al. (2002b) found paleomagnetic support for parts of their proposed configuration but no complete confirmation.

## Evidence for Breakup of Columbia

In addition to the rifting between India and North America discussed above, abundant evidence shows widespread fragmentation and extension of continental blocks beginning at ~1.5 Ga. It applies equally well to either configuration of Columbia proposed above, but for simplicity we show it in fig. 7.9 on the map of Columbia proposed by Rogers and Santosh (2002; fig. 7.7a). The general characteristics of extensional features described here are described in chapter 6.

Rifting occurred along all margins of Columbia from about 1.6–1.4 Ga. A rift along one margin is shown by the ~1.4-Ga Chewore ophiolite in the Zambezi orogenic belt between Zimbabwe and central Africa (fig. 7.7a; Oliver et al., 1998; S. Johnson and Oliver, 2000). It may indicate that the Tanzania craton rifted away from the Zimbabwe craton at this time and then moved across an intervening ocean basin to collide with the Congo craton

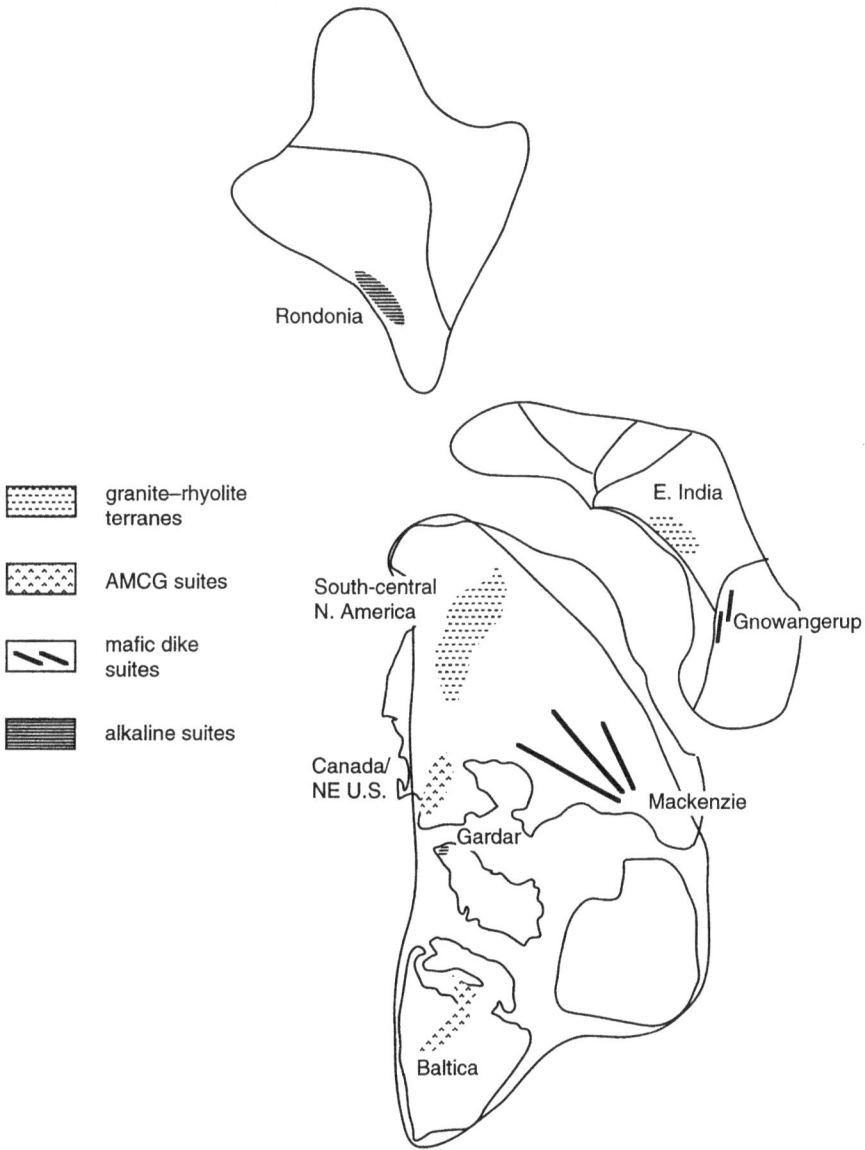

Figure 7.9. Large areas of Mesoproterozoic post-orogenic and anorogenic magmatism in Columbia. Granite–rhyolite terranes occur in south-central North America and eastern India; ~1.4-Ga post-orogenic granites occur in and around the North American terrane. Dike swarms include Mackenzie and Gnowangerup. The AMCG suites of North America and Baltica were connected before opening of the Atlantic Ocean. Alkali granites and associated anorogenic activity occur in South America and southern Greenland. References and further descriptions in text.

during the Kibaran orogeny at ~1.3–1.2 Ga (Meert et al., 1994; Fernandez-Alonso and Theunissen, 1998). Rifting at ~1.4 Ga was widespread in orogenic belts that formed the long margin of Columbia from Baltica to Amazonia (appendix H). In all of these areas, a period of stability and/ or backarc rifting interrupted the subduction and outgrowth that appear to have begun about 1.8 Ga and extended to Grenville time.

Figure 7.9 shows the locations of post-orogenic and anorogenic magmatism developed during ~1.6– 1.3 Ga. One area is the granite–rhyolite terrane, a widespread suite of ~1.4-Ga rhyolites and local granite that underlies much of midcontinent

United States (Bickford and Anderson, 1993; Van Schmus et al., 1996). Granites of ~1.4-Ga age occur within the granite–rhyolite terrane and also are spread throughout most of the pre-1.5-Ga orogenic belts of the central and southwestern United States (Anderson and Morrison, 1992; Anderson and Cullers, 1999). They are a typical post-orogenic suite, consisting almost exclusively of granite without any associated mafic or intermediate rocks and few alkaline rocks (chapter 4). A rhyolite province similar to the granite–rhyolite terrane is now preserved only as fragments in the rifts and marginal basins of eastern India that began to form at approximately 1.5 Ga. Rocks consist partly of flows and ash flows but are predominantly air-fall tuffs deposited in water and partly mixed with other sediments.

Several dozen plutonic assemblages in northern North America and southern Baltica contain a mixture of rock types broadly classified as anorthosite complexes (AMCG; Emslie and Hunt, 1990; Haapala and Ramo, 1990; Puura and Floden, 1999; Ahall et al., 2000; Lu et al., 2002). Almost all of the massifs have ages ranging from about 1.6–1.4 Ga, and all of them are clearly related to relaxation of horizontal stress shortly after orogenic crustal growth. Minor AMCG suites in North China have ages of ~1.6 Ga, similar to ages of older suites in Baltica, suggesting this age of separation of the two cratons.

Rapakivi granites and related rocks occur in numerous plutonic suites in the Rondonian orogenic region of the southern Amazonian craton (Bettencourt et al., 1999). The massifs consist primarily of rapakivi granite with minor associated mafic rocks, and some alkaline intrusives are so metalliferous that the area is commonly referred to as the Rondonian tin province. Ages range from ~1.6 Ga to slightly greater than 1.3 Ga and center between 1.5–1.4 Ga. They either follow, or are associated with, the last stages of the Rondonian orogeny, presumably as a result of post-orogenic crustal extension.

The Mackenzie dike swarm consists of mafic dikes that radiate from a small area of northwestern Canada and extend east and south across much of Canada. Widely spaced samples have ages closely centered at ~1.3 Ga (LeCheminant and Heaman, 1989), and the suite may include the Coppermine basalts of the Racklan orogen (appendix H). The continent-wide extent of these dikes indicates extension associated with asthenospheric uplift, possibly a broad mantle plume. Similarly, the Gnowangerup dikes are aligned along the southern margin of the Yilgarn block in the western Australian craton (Myers, 1993). Their intrusion at ~1.3 Ga is essentially synchronous with intrusion of the Mackenzie dike swarm of North America and presumably indicates extension throughout much of Columbia at this time.

Magmatic rocks in the Gardar province of southernmost Greenland are an approximately 1.3–1.2-Ga assemblage of mafic dikes and highly alkaline plutonic suites. Many of them have compositions similar to those of the ultrapotassic and other alkaline rocks of the western branch of the East African rift zone (Macdonald and Upton, 1993), and this similarity presumably shows that the Gardar suite occupies a Mesoproterozoic rift.

The approximate age equivalence of all of these suites with the Mesoproterozoic rift valleys of India and North America suggests widespread relaxation of horizontal stress starting at about 1.5 Ga throughout much of the earth's continental crust. Presumably this relaxation was associated with the processes that caused the breakup of Columbia.

## A Neoarchean Supercontinent

Piper (2003) used paleomagnetic data to study the positions of continental blocks during the period 2.9–2.2 Ga. He suggested the presence of a supercontinent that had a linear to slightly curved shape similar to that of both Palaeopangaea and Pangea (fig. 7.10). The two limbs of the supercontinent were also proposed to contain approximately the same continental blocks as Palaeopangaea and Pangea. One limb, located at high latitudes, contained older cratons now in India, Australia, and the Kaapvaal region of southern Africa (discussed as the continental assembly Ur in chapter 6). The second arm was in low latitudes and consisted mostly of North America and Baltica (see Arctica and Nena in chapter 6).

Piper (2003) proposed that the similarity of the shapes of this Neoarchean supercontinent, Palaeopangaea, and Pangea required only small amounts of movement to form one supercontinent from its predecessor. The slow rate during the Proterozoic particularly suggests that the Proterozoic was generally a time of slow mantle convection, which affected both tectonic and surface processes (chapter 11).

Support for the configuration of the supercontinent shown in fig. 7.10 comes from the distribution of Paleoproterozoic sediments. Aspler and Chiarenzelli (1998) suggested the presence of two Neoarchean supercontinents that were essentially

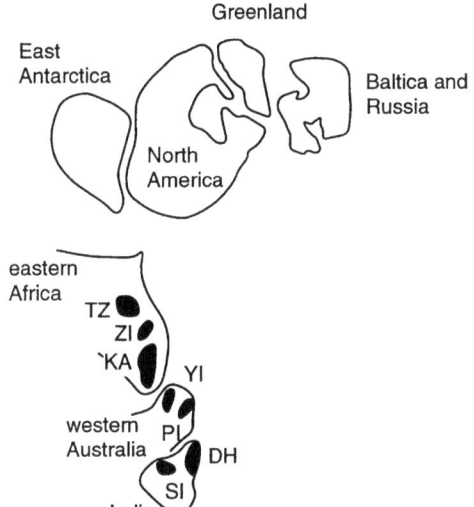

Figure 7.10. Configuration of Neoarchean supercontinent proposed by Piper (2003). Principal cratons in the southern arm of the supercontinent are: TZ, Tanzania; ZI, Zimbabwe; KA, Kaapvaal; YI, Yilgarn; PI, Pilbara; DH, Dharwar; SI, Singhbhum.

the two arms of the one supercontinent proposed by Piper (2003). One supercontinent (northern at present) consisted largely of the Canadian shield and possibly included Baltica. Rifting in this area started at ~2.5 Ga, and the oldest Paleoproterozoic deposits include quartz sandstone and glacigenic materials that developed almost completely above sealevel. Deposition of widespread iron formations did not begin in this supercontinent until the end of glaciation and the supercontinent subsided below sealevel. The other supercontinent (southern) contained India, parts of western Australia, and one or more blocks in southern Africa. Rifting in this area began at ~2.65 Ga, earlier than in the Canadian shield, and most of the region was submerged by the start of the Paleoproterozoic. This permitted subsidence below sealevel in the earliest Paleoproterozoic and deposition of large sequences of iron formation before glaciation.

## Summary

Three or more supercontinents may have existed before the assembly of Gondwana at ~500 Ma. A supercontinent named Rodinia or Palaeopangaea almost certainly accreted at ~1 Ga and broke up a few hundred million years later. An earlier supercontinent, named Columbia, may have accreted by ~1.6 Ga and broken up ~100 million years later. One or more supercontinents may also have existed in the latest Archean, although they have not yet been named.

This sequence of supercontinents, ending with Pangea, suggests a rough periodicity of approximately 750 million years between times of accretion to maximum packing. There is also the possibility that this cycle includes approximately 500 million years of assembly followed by 250 million years of dispersal.

# 8

# Gondwana and Pangea

Pangea, the most recent supercontinent, attained its condition of maximum packing at ~250 Ma (fig. 8.1). At this time, it consisted of a northern part, Laurasia, and a southern part, Gondwana. Gondwana contained the southern continents—South America, Africa, India, Madagascar, Australia, and Antarctica. It had become a coherent supercontinent at ~500 Ma and accreted to Pangea largely as a single block. Laurasia consisted of the northern continents—North America, Greenland, Europe, and northern Asia. It accreted during the Late Paleozoic and became a supercontinent when fusion of these continental blocks with Gondwana occurred near the end of the Paleozoic.

The configuration of Pangea, including Gondwana, can be determined accurately by tracing the patterns of magnetic stripes in the oceans that opened within it (chapters 1 and 9). The history of accretion of Laurasia is also well known, but the development of Gondwana is highly controversial. Gondwana was clearly a single supercontinent by ~500 Ma, but whether it formed by fusion of a few large blocks or the assembly of numerous small blocks is uncertain. Figure 8.1 shows Gondwana divided into East and West parts, but the boundary between them is highly controversial (see below).

We start this chapter by investigating the history of Gondwana, using appendix SI to describe detailed histories of orogenic belts of Pan-African age (600–500-Ma). Then we continue with the development of Pangea, including the Paleozoic orogenic belts that led to its development. The next section summarizes the paleomagnetically determined movement of blocks from the accretion of Gondwana until the assembly of Pangea, and the last section discusses the differences between Gondwana and Laurasia in Pangea. The patterns of dispersal and development of modern oceans are left to chapter 9, and the histories of continents following dispersal to chapter 10.

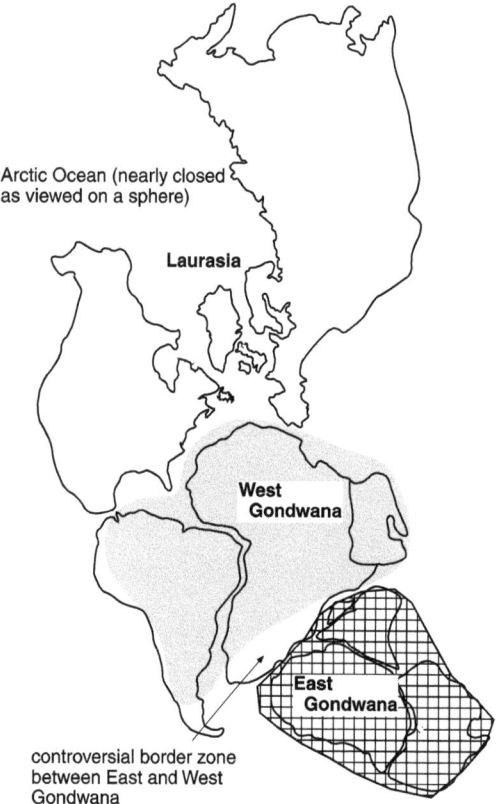

Figure 8.1. Pangea, showing division into Laurasia, West Gondwana, and East Gondwana. The line between East and West Gondwana is controversial (see discussion in text).

## Gondwana

By the later part of the 1800s, geologists working in the southern hemisphere realized that the Paleozoic fossils that occurred there were very different from those in the northern hemisphere. They found similar fossils in South America, Africa, Madagascar, India, and Australia, and in 1913 they added Antarctica when identical specimens were found by the Scott expedition. When geologists wanted a

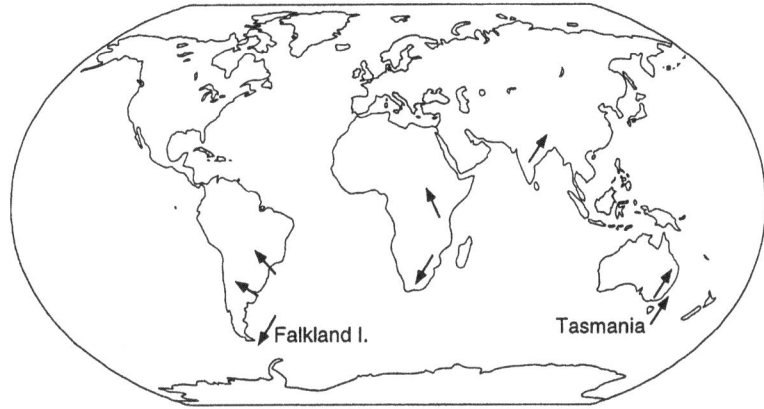

Figure 8.2. Areas that contain evidence of Late Paleozoic (Gondwana) glaciation are shown by arrows in South America, Falkland Islands, southern Africa, India, continental Australia, and Tasmania. Evidence consists of glacial deposits, glacial scours, and cold-climate organisms such as Glossopteris. Arrows show predominant direction of ice flow.

name to refer to all of these areas together, they chose to call it after the Gonds, a group of people who dominated parts of central India until about the 18th century. Thus Gondwana, the empire of the Gonds, became Gondwana Land, simplified to Gondwanaland in the classic book "Das Antlitz der Erde" (The Face of the Earth) by Eduard Suess in 1881 (see Suess, 1904–1909).

At first, geologists assumed that the parts of Gondwana in different continents had similar fossils because there were paths of migration between the fragments, perhaps "land bridges" that are now sunk beneath the oceans. After publication of the concept of continental drift, the idea that the Gondwana fragments were pieces of a former single continent—a "supercontinent"—gradually began to take hold. Putting all of the present Gondwana fragments into one supercontinent solved another problem raised by early studies. All of these fragments had been glaciated at approximately the same time in the Permian, and numerous types of evidence showed that the glaciers had flowed mostly from the general area of the Indian Ocean (fig. 8.2). Synchronous glaciation was supported by the finding that all of the Gondwana fragments contained a characteristic type of plant, Glossopteris, that apparently developed in periglacial environments.

The shape of Gondwana that resulted from placing all of the fragments into positions compatible with glaciation in one region led to the configuration of fig. 8.3, which shows Gondwana with the major fragments assembled into one supercontinent. The configuration can be inferred from patterns of opening of modern oceans, but it does not represent the full size of Gondwana. Many of the outer parts of Gondwana rifted away before the formation of the present Indian and Atlantic Oceans and accreted to North America and Eurasia in the Paleozoic and Early Mesozoic. Because these parts cannot be located precisely in Gondwana, we show only their approximate positions in fig. 8.3 and also designate those fragments whose attachment to Gondwana cannot be demonstrated on the basis of available evidence.

Just as the Grenville orogeny is an essential part of the history of Rodinia, the Pan-African orogeny is the defining event in a large part of Gondwana. We begin this chapter by discussing it, with details provided in appendix I. Then we continue with an investigation of the assembly of Gondwana, including apparent differences between "West Gondwana" and "East Gondwana," and the locations of sutures that brought the entire supercontinent together.

*Pan-African Orogeny*

In 1964 W.Q. Kennedy identified a major period of deformation and metamorphism in Africa that occurred at approximately 500 Ma. Orogeny at this time was soon recognized to be widespread throughout Africa and was referred to as "Pan-African." Geologists in South America also found numerous orogenic belts of the same age and named them "Brasiliano." Now it is understood that belts of this age occur throughout most of Gondwana and were highly involved in assembly of the super-

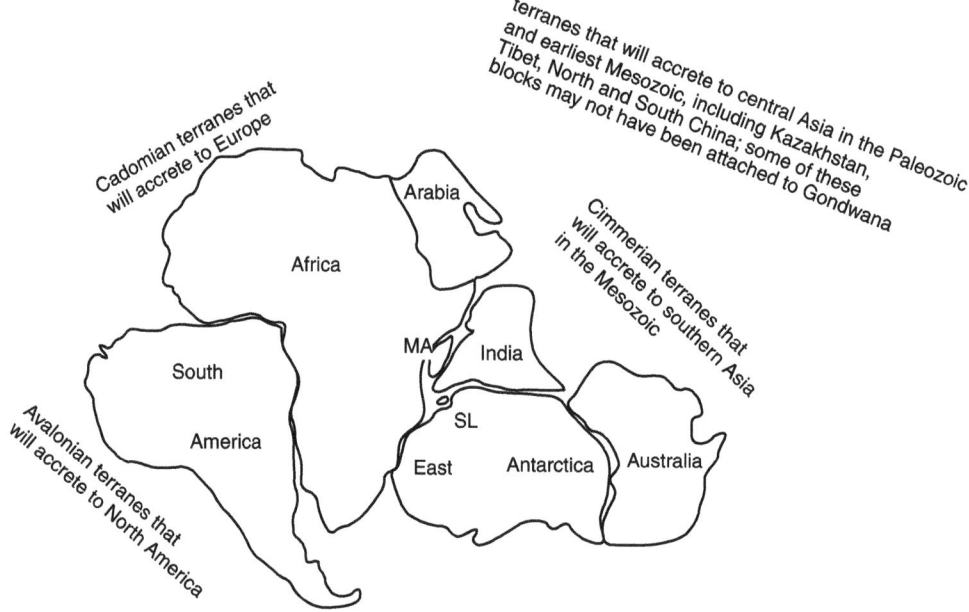

Figure 8.3. Gondwana at ~550 Ma. Blocks that rifted before the Middle Mesozoic are approximately located. MA is Madagascar; SL is Sri Lanka.

continent (fig. 8.4). We also know that the Pan-African–Brasiliano orogeny was restricted to Gondwana, because extensive search in Laurentia, northern Europe (Baltica), and Precambrian shields of Asia has demonstrated that the only parts of Laurasia affected by Pan-African orogeny are in exotic blocks from Gondwana.

Continued work during the past several decades has now shown that activity regarded as Pan-African or Brasiliano spans almost the entire Neoproterozoic and Early Cambrian (from ~1000 Ma to ~500 Ma). Meert (2003) proposed that the Pan-African could be subdivided into three different orogenic episodes in eastern Africa and farther eastward into East Gondwana (see below). The earliest activity, from ~900 Ma to 600 Ma, consisted largely of compression and consumption of oceanic lithosphere, primarily in the Nubian–Arabian shield and farther south along the East African Orogen (appendices E and I). This episode was followed by a period of extension and intrusion of postorogenic granites during a period of ~50 million years centered around ~570 Ma, designated as the Kuunga Orogeny (from a Swahili word meaning "coming together"). The final episode consisted largely of thermal resetting of many areas and lasted as young as 450 Ma.

Although the age of the Pan-African orogeny is clear, the significance of individual deformational belts is highly controversial. Traditionally, they have been regarded as zones of ocean closure, where individual cratons were sutured to form Gondwana (Unrug, 1992). An alternative viewpoint suggests that many areas of Pan-African activity are overprints on zones of older deformation, signifying that the belts may be intracratonic or confined (see terminology in chapter 6). If they are not zones of closure of Pan-African oceans, then Gondwana may have been formed by accretion of a few large blocks rather than of many small ones.

The controversy can be investigated in two principal ways. One is to determine whether the rock suites in the belts of Pan-African age were extracted from the mantle in Pan-African time, including development in ocean basins, or whether they were formed by reworking of older continental rocks. Juvenile rocks normally include ophiolites and continental-margin batholiths, whereas reworking of pre-existing orogenic belts produces granites formed by crustal anatexis and metamorphic rocks with structures and textures inherited from older rocks. Isotopic information is extremely useful in making this distinction, and we discuss it for many belts in appendix I.

A second method for distinguishing between zones of ocean closure and zones of intracontinental reworking is paleomagnetic data from stable areas on opposite sides of the orogens. If these

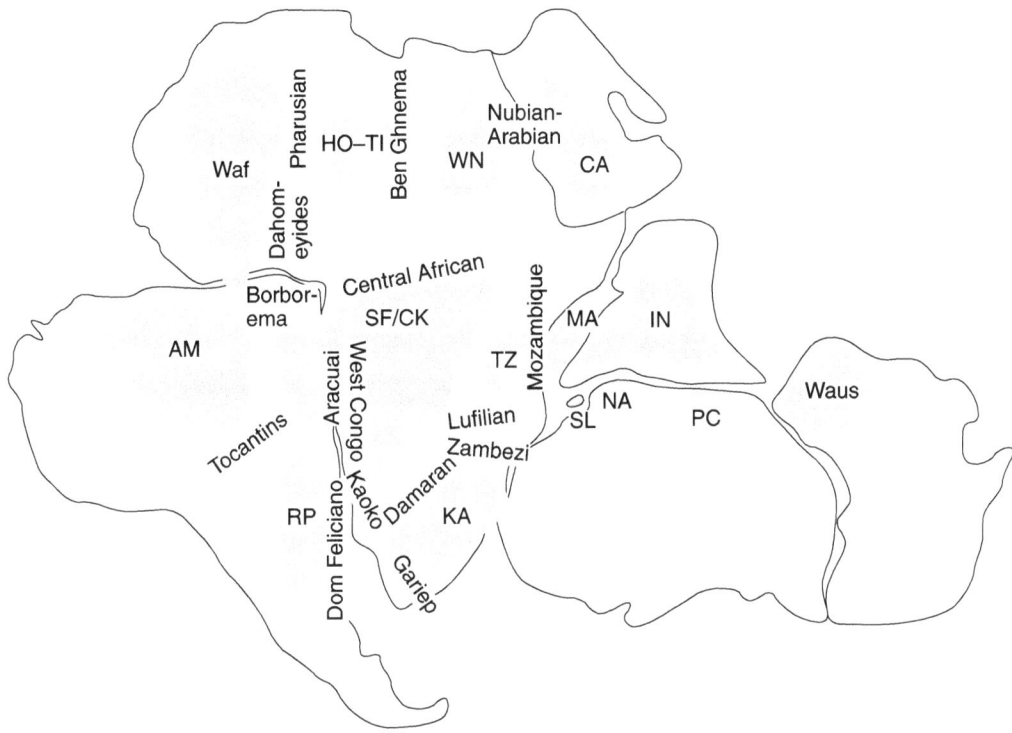

Figure 8.4. Sketch of Gondwana showing names of major belts and areas of Pan-African activity (further information in Fig. I.1 and appendix I). Cratonic blocks (shaded) are: Waf, West Africa; HO–TI, Hoggar–Tibesti; WN, West Nile; CA, Central Arabian; AM, Amazonia; SF/CK, Sao Francisco joined with Congo–Kasai; TZ, Tanzania; MA, Madagascar; IN, India; Waus, western Australia; RP, Rio de la Plata; KA, Kalahari, including Kaapvaal of southern Africa and Maudheim region of East Antarctica; SL, Sri Lanka; NA, Napier; PC, Prince Charles.

areas have nearly identical APW curves before and during Pan-African time, then they presumably formed a coherent block that accreted to Gondwana intact. Conversely, if they have different APW curves, then the Pan-African belt between them was a zone of ocean closure.

## Differences Between East and West Gondwana

East Gondwana traditionally consists of Australia, India, East Antarctica, Sri Lanka, and Madagascar. West Gondwana has generally been regarded as consisting of South America and all or most of Africa. Figure 8.5 displays four types of evidence that have been used to distinguish between East and West Gondwana:

- Ages of cratonization (from Rogers, 1996) show a clear difference between the eastern and western parts of Gondwana. Cratons of 3.0–2.5-Ga age are concentrated in Australia, India, coastal East Antarctica, and parts of southern Africa. They are completely absent, however, from central and northern Africa and all of South America.
- With the exception of the Sunsas belt in western South America, Grenville-age (~1-Ga) orogenic belts are restricted to the same part of Gondwana as older cratons (appendix H). They are common in western Australia, where there is very little evidence of younger activity, and in India, Sri Lanka, and adjacent East Antarctica. Grenville–Kibaran orogenic activity also appears to surround the Kalahari craton of southern Africa and its attached Maudheim region of Antarctica. The northernmost extent of Grenville-age orogeny in Africa appears to be the type Kibaran belt between the Tanzanian and Congo cratons.

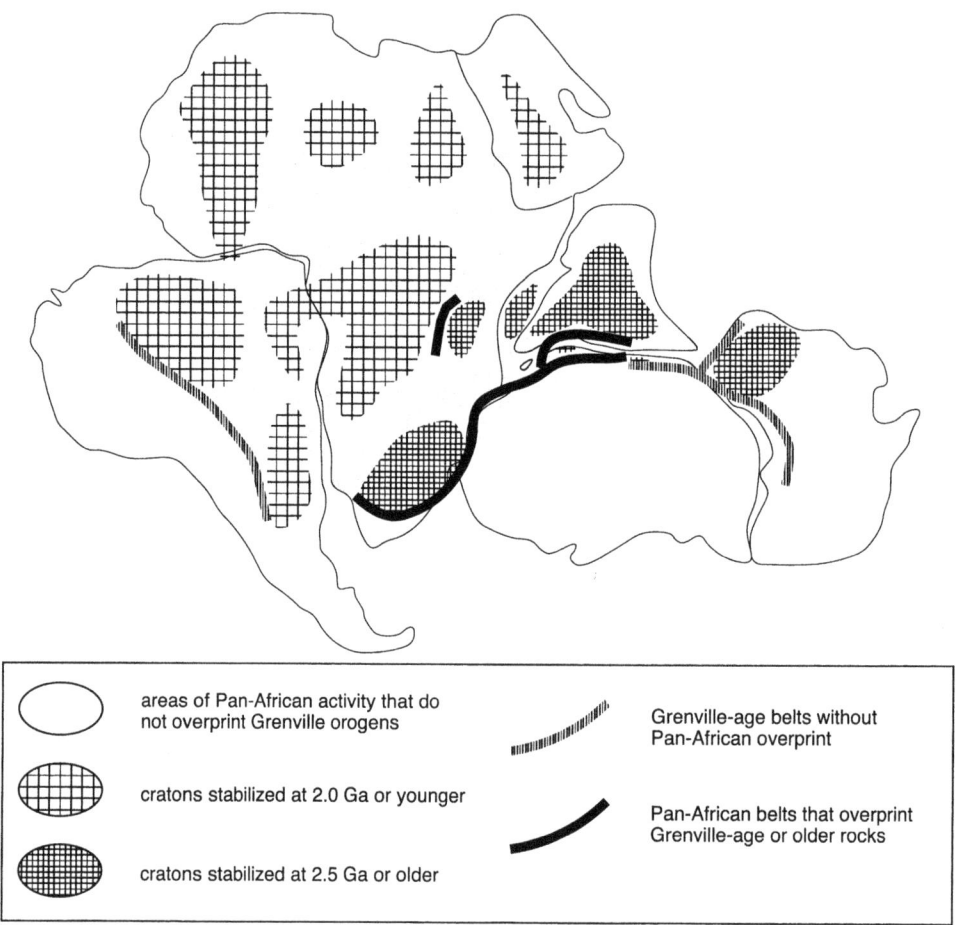

Figure 8.5. Relationships between ages of cratonization and Grenville and Pan-African activity in Gondwana.

- Pan-African activity occurs from India throughout all areas of Gondwana to the west and south, including Sri Lanka, Madagascar, southern and central Africa, and adjoining parts of Antarctica. Pan-African orogenies mostly overprint older areas of Grenville-age orogeny. Some of these belts may be reworked intracontinental terranes, and some may represent Pan-African oceans that opened and then closed within older orogenic belts. Our discussion of individual Pan-African orogens (appendix I) suggests that different tectonic models apply to different belts.
- Pan-African orogeny in northern and western Africa and in all of eastern South America does not overprint rocks of Grenville age except in the Sunsas belt along the western margin of Amazonia. Some areas, such as north–central and northeastern Africa, are clearly regions of both growth of juvenile crust and extensive reworking of older crust (appendix I). Igneous rocks in some Pan-African belts may have been derived from crust formed as early as the Transamazonian orogeny (~2 Ga). Some belts contain evidence of both the existence of small ocean basins and inheritance from rocks older than Pan-African, which probably indicates that they were confined orogens (appendix I; chapter 6).

On the basis of the information in fig. 8.5, many workers believe that East Gondwana was a coherent block by no later than Grenville time (Yoshida, 1995), although numerous doubts have been expressed recently. West Gondwana, consisting of South America and northern and western Africa, clearly became a coherent block during the Pan-African. Between regions that are clearly East or West Gondwana, however, is central and southern Africa, which was affected by both Grenville and Pan-African events. The presence of both orogenic events indicates that at least one suture between the two halves of Gondwana is somewhere within Africa or along its border, and we investigate this before considering the possibility that multiple Pan-African sutures developed throughout Gondwana.

## One Zone of Suturing in Gondwana During the Pan-African Orogeny

The northern end of any individual area of suturing (present orientation) between East and West Gondwana must begin somewhere in the broad region where Pan-African oceans closed between the Arabian peninsula and the West African craton (fig. 8.6; appendix I). Most investigators look for a suture that originates in the Nubian–Arabian shield (appendix E) and then follows at least part of the Mozambique belt (appendix I). The Mozambique belt is defined in Africa but is only the western part of a broad zone of Pan-African activity that includes all of Madagascar (Shackleton, 1996; A. Collins and Windley, 2002), plus southern India and parts of East Antarctica (Jacobs et al., 1998; Jacobs, 1999). Between ~650–500 Ma, this entire zone was subjected to thermal activity and complex shearing that resulted in a net left-lateral displacement between rocks to the west, in Africa, and rocks to the east in stable parts of India. Total displacement is the sum of displacements on numerous individual shears, although some of them may have undergone both sinistral and dextral movements at different times. Heating was sufficiently intense that it reset isotopic systems in rocks outside the area as well as within the deformational zone.

Shears within the Nubian–Arabian shield are clearly also suture zones, but shears farther south show either little or no evidence of ocean closure, and the absence of evidence of ocean closure is the basic problem in locating a suture between the two parts of Gondwana. Bodies of ultramafic rocks, possibly pieces of ophiolite, occur in a few locations, but they do not form a consistent belt of oceanic lithosphere that might indicate a suture zone. Similarly, some parts of the Mozambique belt and adjacent areas appear to be assemblages of terranes with different histories and lithologic/compositional characteristics (Appel et al., 1998; Muhongo et al., 2002a,b), but joins between them have not been shown to be former ocean basins.

Other sutures between East and West Gondwana are even more difficult to locate than one through the Mozambique belt because they cannot penetrate regions in central Africa that have been stable since Grenville time or earlier. The Central African Fold Belt and the northern margin of the Congo craton, have formed a stable region since ~2 Ga, and the southern (Kasai) part of the Congo craton was fused to the Tanzanian craton earlier than 1.2 Ga (appendix H). One possibility is that the Zambezi and Damaran belts form a zone of both ocean closure and left-lateral shearing extending westward from the Mozambique belt. This suture would connect on its western end with the known zone of ocean closure between the Gariep and Dom Feliciano belts (appendix I), but Hanson et al. (1994) proposed that cratons on opposite sides of the Zambezi belt had been fused much earlier than Pan-African time.

A second possibility for a suture through central Africa stems from increasing evidence of a Pan-African subduction zone that extended southwestward from the Nubian–Arabian shield into the poorly known area shown by question marks in fig. 8.6 (Schandelmeier et al., 1994; Kuester and Liegeois, 2001). If this trend continues farther southwest, then a possible suture may extend under the Congo basin, which has many of the stratigraphic characteristics found in other "successor" (post-collisional) basins. A connection with the Gariep–Dom Feliciano ocean would complete a zone of ocean closure along this trend.

## Multiple Zones of Suturing in Gondwana During the Pan-African Orogeny

Instead of one suture between a coherent East Gondwana and a coherent West Gondwana, numerous proposals have been made for a variety of sutures active in the age range of 600–500 Ma. All of them generally assume that India was a coherent block at that time, and most regard western and central Australia as a single block. Investigators looking for sutures in West Gondwana also accept paleomagnetic and tectonic evidence for the long-continued unity of three blocks: the Congo part of the Congo craton with the Sao Francisco craton; the Tanzanian craton with the Kasai part of the Congo craton (Borg and Shackleton, 1997); and northeastern South

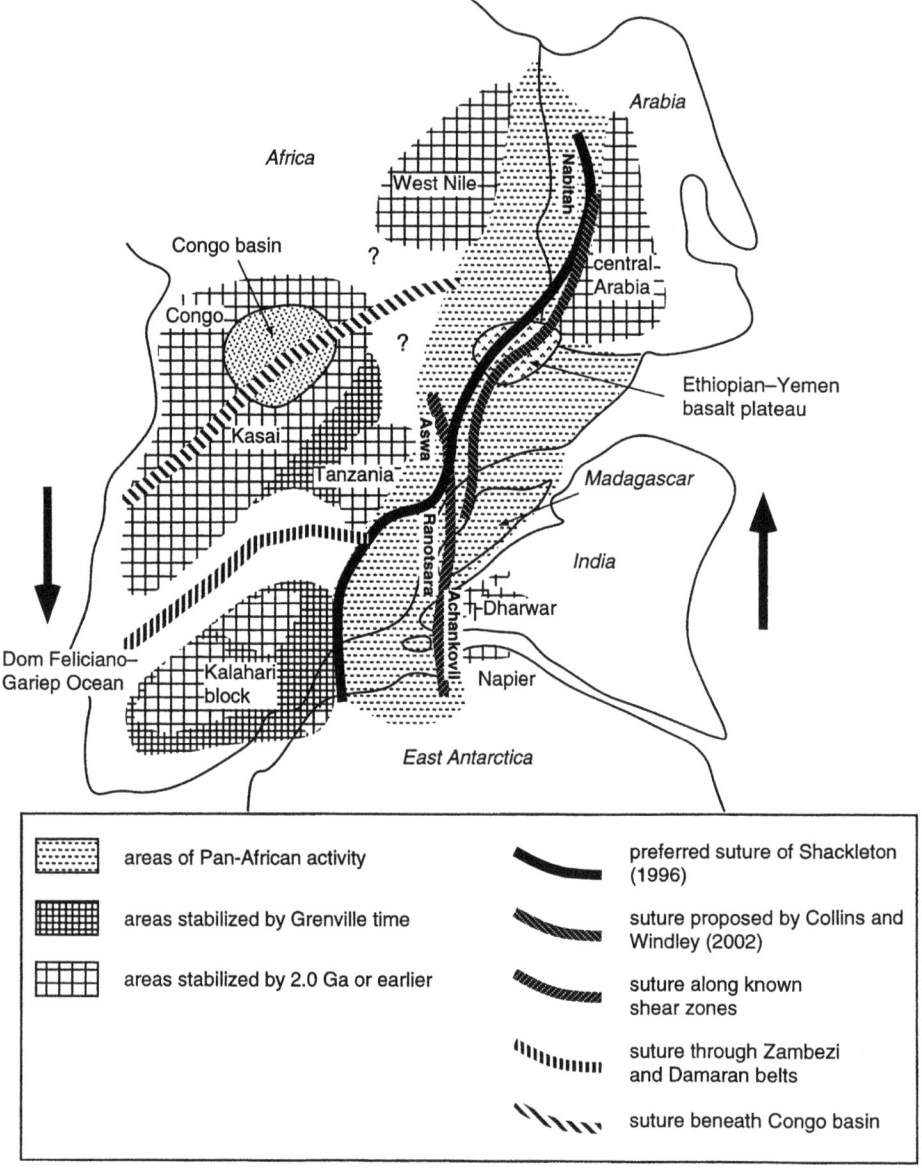

Figure 8.6. Region of possible suturing between East and West Gondwana. Question marks indicate area of poor exposure previously regarded as Mesoproterozoic or older but recently found to have Pan-African activity.

America (Amazonia) with West Africa. On the basis of the stability of these blocks, fig. 8.7 shows several possibilities for Pan-African sutures throughout Gondwana.

The two major sutures within East Gondwana both extend into East Antarctica. Fitzsimons (2000) proposed that East Antarctica contains three separate Grenville belts of slightly different ages and, consequently, was not assembled as a single continent until the accretion of Gondwana (further discussion in Boger et al., 2001). Torsvik et al. (2001) and Powell and Pisarevsky (2002) supported this proposal with paleomagnetic data that showed that India and Australia did not join until the accretion of Gondwana. These proposals suggest that Pan-African assembly of East Gondwana occurred partly along a zone between eastern India and western Australia (present orientations) and also along

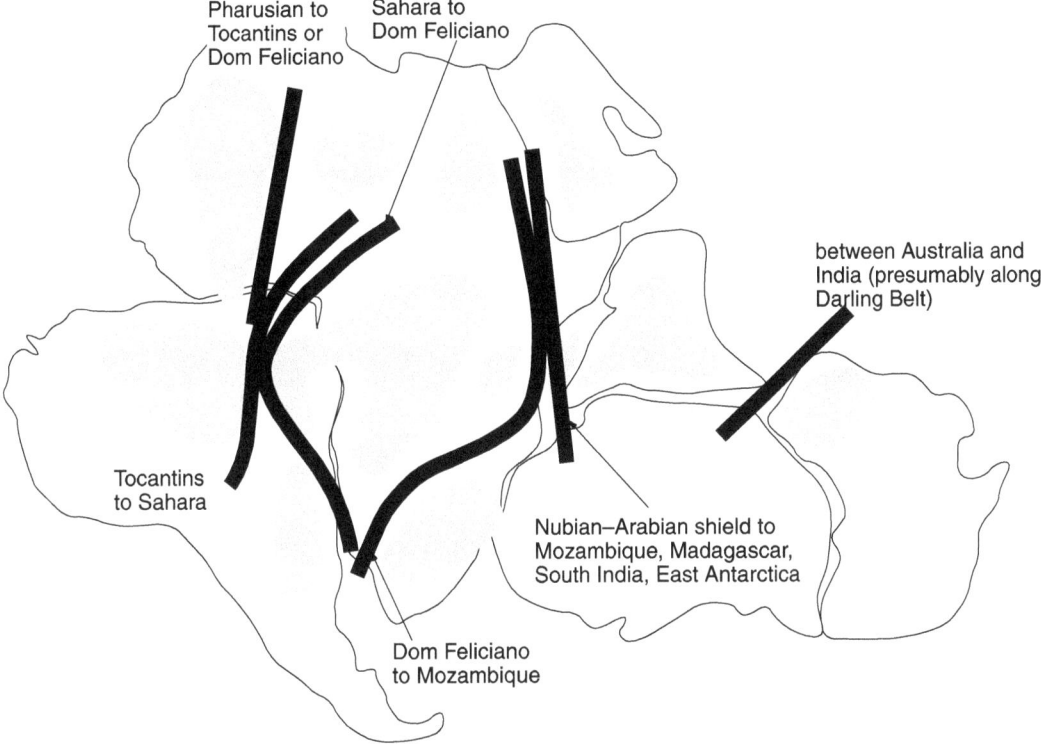

Figure 8.7. Possible zones of Pan-African suturing in Gondwana shown by dark lines.

a zone between western India and eastern Africa. The suture east of India is presumably along the Darling belt (Wilde, 1999; appendix H), and the suture west of India is probably within the broad zone of suturing and/or shearing along the east coast of Africa and through Madagascar (see above).

In West Gondwana, numerous Pan-African sutures exist in the Nubian–Arabian shield (appendix E) and on the western and southern edges of the largely unknown area that extends from the Nubian–Arabian shield to the eastern margin of the West African craton (appendix I). This broad area almost certainly contains additional Pan-African sutures, but the lack of exposure has prevented accurate location. As discussed above, any suture between East and West Gondwana would have to extend southward from this area of Pan-African ocean closure.

Sutures of Pan-African age may exist elsewhere within West Gondwana, but their locations are controversial. The controversy stems from the difficulty of distinguishing between Pan-African belts that formed by ocean closure and those that overprint older rocks (see above). Trompette (1997) and Alkmim et al. (2001) accepted the existence of both types of orogen but proposed somewhat different zones of suturing caused by the closure of oceans that existed in Pan-African–Brasiliano time. The conclusion that these zones are Pan-African sutures conflicts with the proposal of Brito Neves (2002) that Brasiliano oceans formed by rifting of Rodinia were shallow and floored by thin continental crust (chapter 7).

Possible sutures in West Gondwana all extend from the Sahara region of ocean closure to either of two oceanic regions on the opposite side of the supercontinent (fig. 8.7). One is the recognized zone of ocean closure between the Dom Feliciano belt of South America and the Gariep belt of southern Africa (appendix I). The other is the Tocantins region on the southern margin of the Amazonian block, where evidence of ocean closure is controversial (appendix I).

## Summary

The assembly of Gondwana was completed by 550–500 Ma, generally referred to as Pan-African time. The assembled supercontinent is commonly regarded as comprising East Gondwana, which consists mostly of Archean cratons and Grenville-age belts, and West Gondwana, which consists of Paleoproterozoic cratons and shows little evidence of Grenville orogeny. The two major tectonic problems involving the assembly of Gondwana are the nature of Pan-African belts and the location of sutures within East and West Gondwana and between them.

Possible sutures between East and West Gondwana must include at least one zone in Africa and adjacent regions. Most investigators consider one suture to be somewhere within the Mozambique belt, extending southward along a zone between southern India and the Kalahari block of southern Africa. Possible other sutures include one between Australia and India and at least two in eastern South America and western/central Africa.

## Pangea

Figure 8.8 shows the major regions that we discuss below in order to illustrate the assembly of Pangea. It does not, however, show all of the continental blocks that existed when Pangea reached its time of maximum packing. At all times since Gondwana was assembled, fragments of various sizes have rifted away from it and accreted to North America, northern Europe, and the Precambrian core of Asia. This rifting continued as Pangea was assembled and continues to the present. In addition, some blocks that may never have been attached to Gondwana also were not yet accreted to Pangea in the configuration shown in fig. 8.8.

The accretion of Pangea took place during the Paleozoic following the assembly of Gondwana. We begin this section with a discussion of the assembly of Pangea along Paleozoic orogenic belts and then continue with the movements of blocks that became parts of Pangea by the end of the Paleozoic. These movements can be traced paleomagnetically, and we show not only relative movements but also absolute movements across latitudes. The assembled Pangea consisted of Gondwana and the newly accreted Laurasia, and we describe the differences between these two regions.

### Paleozoic Orogeny that Led to the Assembly of Pangea

Compressive orogeny during the Paleozoic occurred on one side of Gondwana (western as shown in fig. 8.8), on all sides of Laurasia, and possibly within Laurasia. Orogeny caused by collision between North America and northwest Africa completed the assembly of Pangea into its configuration of maximum packing.

Outer (Western) Margin of Gondwana  Compressive orogeny along the western margin of Gondwana began as the supercontinent was assembling in the latest Proterozoic (fig. 8.9). The western margin of the Amazon shield (Brazil and Guiana cratons) was created by post-Rodinia rifting after the Grenville orogeny, but blocks with Grenville-age basements occupy most of the area between the margin of the shield and the present Pacific coast (chapter 6).

The origin of many of the exposures of rocks with Grenville basement is uncertain, but at least one of the blocks, the Precordillera of Argentina, can be shown to have formed by rifting from the Ouachita embayment of eastern North America (Astini and Thomas, 1999; chapter 7). The Precordillera contains a Grenville basement that was rifted in the latest Neoproterozoic or Early Cambrian, forming a thin crust throughout the block and syn-rift graben deposits locally. These rift deposits were covered by Late Cambrian and Early Ordovician limestones that contain fauna different from indigenous South American forms. Apparently the Precordillera drifted across the newly developed Iapetus Ocean and collided with the South American (Gondwana) margin in the Middle Ordovician, causing deformation, metamorphism, and magmatism.

Extensive cover by Mesozoic–Cenozoic rocks in the Andes makes interpretations of Paleozoic activity difficult. The deformed Paleozoic rocks are mostly quartzites, shales, and carbonates, implying deposition along a passive continental margin. In one area, deformation has been dated as Late Carboniferous to Permian (Jacobshagen et al., 2002), and there probably was almost continuous orogeny along the continental margin throughout the Paleozoic. Subduction directions during this time are unknown, with no evidence for subduction either beneath the continent or beneath colliding arcs or microcontinents.

The Ross orogen is displayed throughout the Transantarctic Mountains and smaller ranges on strike toward the east, giving it a total length of 3500 km (summaries by Laird, 1991; Stump, 1995; Encarnacion and Grunow, 1996; Storey et al., 1996; Goodge et al., 2001). The Ross orogeny began in the latest Neoproterozoic and continued in different areas until the Late Cambrian or Early Ordovician. It affected sediments that accumulated

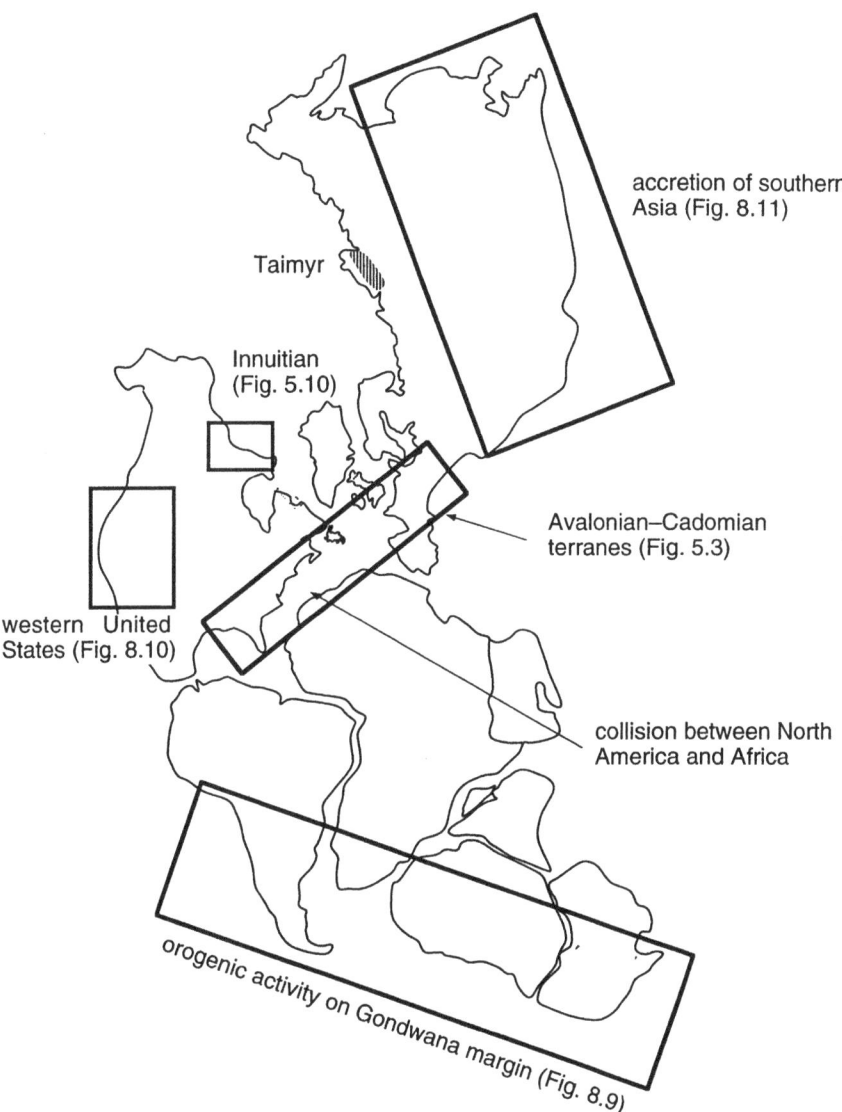

Figure 8.8. Major events in the accretion of Pangea.

on the rifted margin of East Antarctica after it separated from the North American part of Rodinia, produced highly deformed metasedimentary rocks, and generated batholithic magmas by subduction of oceanic lithosphere beneath the Antarctic continent. By the Middle Paleozoic most of the deformed rocks of the Ross orogen were covered by laterally extensive platform sediments that extend inland under the East Antarctic icecap (Collinson et al., 1994; ten Brink et al., 1997).

The Tasman orogen is a general term that includes the series of orogenic belts that formed along the eastern margin of Australia during the Paleozoic. Outgrowth of the continent began with the Cambrian Delamerian orogeny in the Adelaide trough, which is probably correlative with the Ross orogeny. Some of the compression was caused by eastward subduction under a colliding microplate. Outgrowth continued eastward in the Lachlan orogen (Vandenberg, 1999) and New England/Yarrol orogens (Leitch et al., 1994), all of which were accompanied by westward subduction of the Pacific plate beneath Australia (Muenker and Crawford, 2000; summary by Veevers, 2001).

The Lachlan orogen consists mostly of Ordovician–Silurian sediments derived from the

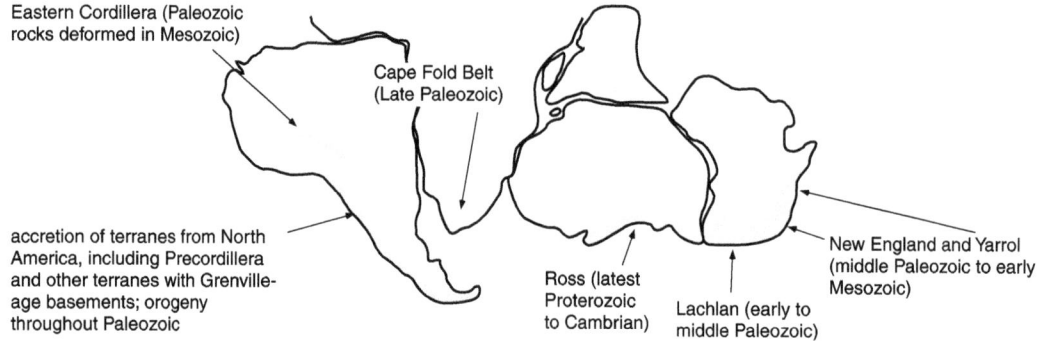

Figure 8.9. Orogeny along the margin of Gondwana during the Paleozoic.

west (Vandenberg, 1999; Foster and Gray, 2000). Final orogeny in the Early Carboniferous caused intense deformation, low-rank metamorphism, and injection of large volumes of granitic magma. The amount of granite increases toward the east, and Chappell et al. (1988) used rocks in the Lachlan orogen to establish their concept of I-type and S-type granites. The I-type granites contain compositional and isotopic properties that suggest derivation largely from oceanic lithosphere subducted under the continental margin, whereas S-type granites appear to have formed by remelting of pre-existing sedimentary rocks. The boundary between these two suites is known as the I–S line, and it may mark the boundary between older Australian crust and oceanic crust.

As subduction occurred along the Australian margin throughout the Paleozoic, intraplate deformation took place throughout much of the continental interior (Scrimgeour and Close, 1999; Braun and Shaw, 2001; Scrimgeour and Raith, 2001). This formed several belts of deformation, some of which now expose rocks of granulite facies. The weakness that permitted this type of deformation may have resulted from lithospheric thinning associated with development of Neoproterozoic sedimentary basins during and after separation of Australia and North America.

Outer (Western) Margin of Laurasia  The western edge of North America during the Paleozoic was the passive margin of a continent that was gradually being covered by a sequence of shelf sediments (fig. 8.10; Burchfiel et al., 1992). The inferred continental margin at the beginning of the Paleozoic illustrates the outgrowth that has occurred during the Phanerozoic by some combination of accretion of exotic terranes and incorporation of sediments eroded from North America into a continental wedge along the border of the continent. Although no subduction occurred beneath western North America during the Paleozoic (chapter 5), evidence of compressive orogeny is well displayed in the United States. Orogeny in the western United States was created by collision of island arcs that were generated by westward subduction beneath the arc. Ocean closure caused thrust sequences to overlap the continental margin by a few hundred kilometers during the Devonian (Antler orogeny) and Permian (Sonoma orogeny). The thrust sheets contain sediments derived from continental erosion, sediments washed from volcanic arcs into the ocean basins between the arcs and the continent, and volcanic and volcaniclastic rocks of the arc.

Within Laurasia (?)  Paleozoic orogeny occurred in the Innuitian belt on the northern margin of Canada (chapter 5), and Late Neoproterozoic to Early Paleozoic deformation affected the Taimyr belt on the northern margin of Siberia (Vernikovsky and Vernikovskaya, 2001). Because the Arctic Ocean opened between North America and Siberia during the Cenozoic, the two continents must have been joined at some earlier time. If this suturing occurred in the Proterozoic (chapter 6), then the Innuitian and Taimyr belts developed within an ocean that extended into Laurasia during the Late Neoproterozoic and Early Paleozoic.

Inside (Eastern) Margin of Laurasia  Orogeny occurred almost continuously throughout the Paleozoic along the continental margins of North America and Europe. Dalziel (1997) attributed some of the orogeny in North America to repeated collision with South America as the two continents rotated past each other from their positions

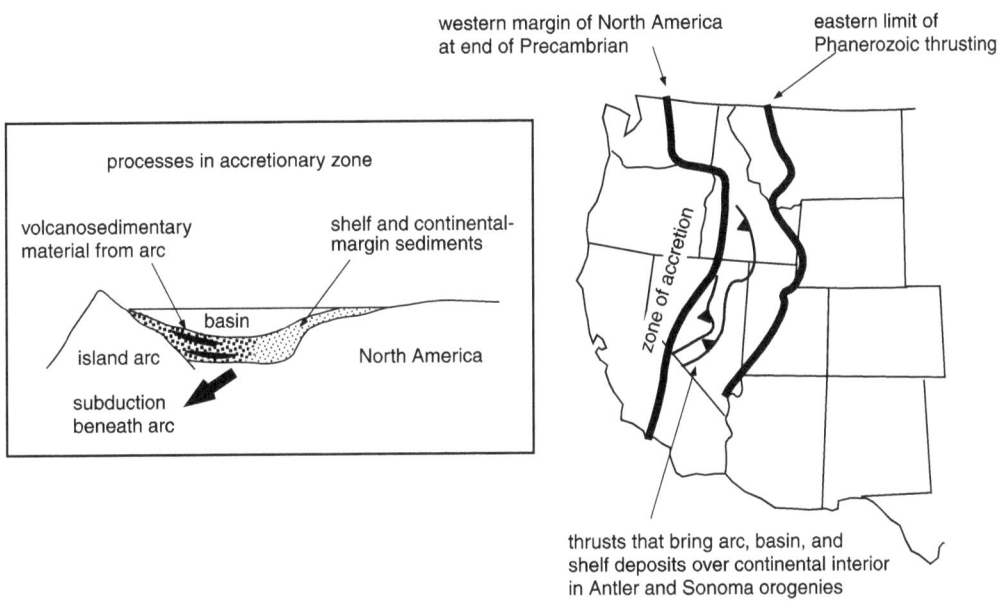

Figure 8.10. Paleozoic orogeny along the western margin of North America.

in Rodinia (chapter 7). An alternative interpretation is that the orogenic compression was caused mostly by accretion of exotic blocks and oceanic juvenile terranes to the eastern margin of North America and southern margins of Europe. This accretion formed a collage of Avalonian–Cadomian exotic blocks that we discussed in chapter 5.

Deformation in North America began in the Late Ordovician and Silurian, commonly referred to as the Taconic orogeny. The approximately synchronous Caledonian orogeny occurred largely between present Norway and Greenland, and whether it is collisional or intracratonic is discussed in chapter 6. Most of the accretion of Avalonian and Cadomian terranes occurred during the Devonian–Carboniferous orogenies referred to as Acadian in North America and Variscan (Hercynian) in Europe (Tait et al., 2000). In Europe these blocks are incorporated as a collage in a large region of oceanic materials, but in North America the amount of juvenile crust appears to be small.

The growth of Asia began in the Middle to Late Paleozoic (fig. 8.11; review by Sengor, 1985; Sengor et al., 1993; papers in Hendrix and Davis, 2001). The Ural Mountains formed when the Kazakh block, possibly already joined to Siberia, collided with the East European platform by subduction toward the east (chapter 5). Following this emplacement, the southern margin of Asia was nearly linear and continued to grow southward throughout the rest of the Phanerozoic, both before and after the collision of North America and West Africa that created the supercontinent Pangea (see below). Accretion of these blocks appears to have been accompanied by subduction in various directions, and we show tentative estimates of these directions in fig. 8.11. Much of the accreted material consists of juvenile crust, including island arcs and oceanic material. These rocks now occur as ophiolites, blueschists, eclogites, and a complete sequence of high $T/p$ suites ranging from greenschist to granulite facies. The largest accumulation of juvenile material is in the southwestern Chinese flysch basin, which separates old continental blocks in China from those in Tibet and other regions to the west.

Although the ages of accretion clearly become younger farther south in Asia, the exact times are uncertain for most terranes. The line that shows the outer margin of Pangea is arbitrarily drawn south of the North China craton and north of the South China craton, although the two may have fused before the end of the Paleozoic (Li, 1998). This problem is complicated because the accretion of North China to Siberia probably occurred over a period of time, with the western end of North China attached to Siberia in the Permian, followed by rotational closure of the two blocks that may not have been complete until the Triassic. The margin is drawn north of terranes regarded as Cimmerian (chapters 9 and 10), which may have rifted from

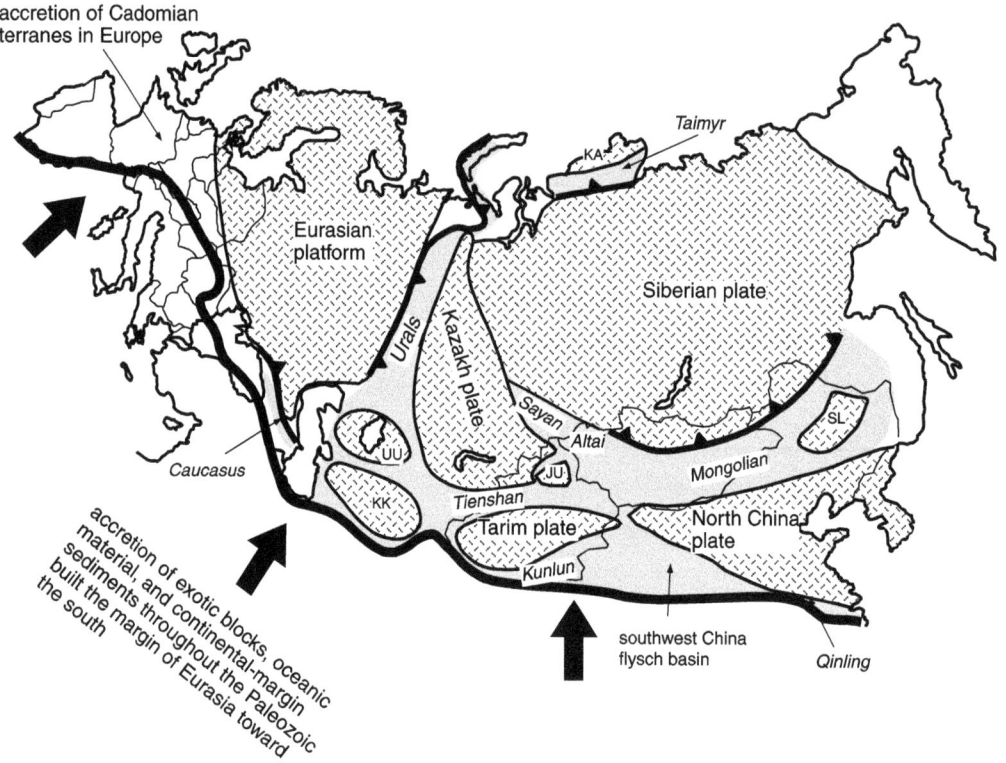

Figure 8.11. Assembly of Asia during the Paleozoic. Discussion in text. Major orogenic belts are shown in italics. Abbreviations of continental blocks are: UU, Ust-Urt; KA, Kalahari; KK, Karakum; JU, Junggar; SL, Songliao.

Gondwana before the end of the Paleozoic but had not yet accreted to Asia.

The exotic blocks shown in fig. 8.11 are probably all either fragments of Gondwana or evolved near the Gondwana margin. This origin is demonstrated both by their Gondwanan fossils (Metcalfe, 1996) and by paleomagnetic data (Li, 1998). The exact locations of the blocks within Gondwana, however, is highly uncertain. None of them were affected by the Pan-African orogeny that is characteristic of West Gondwana, which means that they presumably came from the northern margin of East Gondwana. The South China craton, however, is the only block that contains evidence of the Grenville-age orogenies that are typical of East Gondwana, and it is the only block whose position in Rodinia and Gondwana can be inferred (Yang and Besse, 2001; chapter 7). All of the other blocks must have originated in some part of Gondwana that escaped both the Grenville and Pan-African orogenies, perhaps along the northernmost margins of India and Australia or as isolated terranes near the northern margin of East Gondwana.

Collision of North America and northwest Africa

Collision of North America and northwest Africa in the Permian brought continental fragments together in the maximum packing identified as the supercontinent Pangea. At this time, the Alleghanian orogeny in North America and the Mauritanide and Rokelide orogenies in Africa caused final closure of the Iapetus Ocean that had been formed by rifting of Rodinia (chapter 7). Subduction beneath Africa created Mauritanide fold and thrust belts that were intruded by Late Paleozoic granites (Dallmeyer, 1990), and a complex suite of rocks was thrust over eastern North America. In the Blue Ridge of the eastern Appalachians, basal thrusts carry older thrust complexes that contain continental-margin sediments and some oceanic materials, all of which were metamorphosed to high-amphibolite or granulite grade (Hatcher, 2002). The western Appalachians is primarily a zone of folding and thrusting of unmetamorphosed shelf sediments, some of which were derived by erosion of older Taconic and Acadian orogens farther east (Castle, 2001; Tull, 2002).

## Movements of Continental Blocks from Rodinia to Pangea

Different configurations and breakup patterns of Rodinia require different paths of movement of blocks from Rodinia to Gondwana. Because all of them are highly controversial, we begin to describe some absolute movements starting with positions of continental blocks at 580 Ma, prior to the assembly of Gondwana (fig. 8.12a). The diagram shows eastern Laurentia just beginning to separate from the western margin of the Precambrian region of eastern South America along a rift that crossed the South Pole. Rifting between these two blocks formed the Iapetus Ocean, which closed at the end of the Paleozoic during the suturing of Pangea (see above). The diagram also shows a block containing the Sao Francisco, Congo, and Kalahari cratons separated by the Brasiliano Ocean from eastern South America and by the Mozambique Ocean from a block containing India, Australia, and East Antarctica. The position of Siberia relative to North America is uncertain, and we show its general latitude in the diagram. Complete opening of the Iapetus Ocean may not have occurred until after the assembly of Gondwana at ~550 Ma (fig. 8.12b).

Figure 8.12b shows the positions of major continental blocks at 550 Ma, the time at which the assembly of Gondwana was nearly complete. The South American part of Gondwana had drifted only a short distance from the South Pole since 580 Ma, and the assembled continent extended into the northern hemisphere. Laurentia and South America had separated since 580 Ma to form the Iapetus Ocean, and Baltica–Russia is an independent block at approximately the same latitude as 30 million years earlier. It is shown in an orientation that requires rotation through 180° before its collision with Greenland approximately 150 million years later (Caledonian orogeny). Siberia is not shown in the diagram and may have been in the northern hemisphere, although some evidence suggests that Siberia and North America were joined until the Mesozoic (chapter 6).

The positions of major blocks at the end of the Ordovician are significantly different from their locations at the beginning of the Cambrian, approximately 100 million years earlier (fig. 8.12c). By this time, Gondwana had begun to move across the South Pole, creating glaciations that we discuss in chapter 11. This movement continued through the Paleozoic, at which time the South Pole was located in East Antarctica at approximately the same position that it now occupies. Avalonian and Cadomian terranes had completed their separation from Gondwana and were about to collide with North America (which is partly in the northern hemisphere) and northern Europe, causing a series of Middle Paleozoic orogenies. Baltica had completed its rotation from its orientation at 550 Ma, and its western margin would soon collide with eastern Greenland. Siberia and Tarim were in the northern hemisphere, and North China, South China, and the southern part of the future Indochina were all straddling the equator. Various Cimmerian terranes (chapter 9) were apparently still attached to Gondwana.

Figure 8.12d places Pangea at ~250 Ma in a paleomagnetic framework (simplified from fig. 8.1; Scotese and McKerrow, 1990). North America and the Precambrian regions of Eurasia had moved completely into the northern hemisphere, and Avalonian and Cadomian blocks from Gondwana had also moved northward to collide with eastern North America and southern Europe. The collision of West Africa with eastern North America closed the Iapetus Ocean and completed the Paleozoic assembly of Pangea. At this time, Cimmerian blocks may still have been attached to Gondwana, and the remaining part of Gondwana had moved almost completely across the South Pole. North and South China are shown as still separated from Asia, moving into position to collide with the southern margin of the continent in the Early Mesozoic, although the times of these collisions are very controversial (see above).

## Differences Between Gondwana and Laurasia in Pangea

The fully assembled Pangea was divided approximately equally between Laurasia in the north, and Gondwana in the south. These two regions had very different histories before the formation of Pangea and were significantly different within Pangea. Our summary of these differences is based partly on information discussed above and partly on Veevers (1995) and Veevers and Tewari (1995).

Because Gondwana was completely assembled by ~500 Ma, it was more than 200 million years old when it was incorporated into Pangea. Conversely, Laurasia consisted partly of the very old regions of North America, Greenland, and Baltica (referred to as Nena in chapter 6), but most of Asia formed during the accretion of

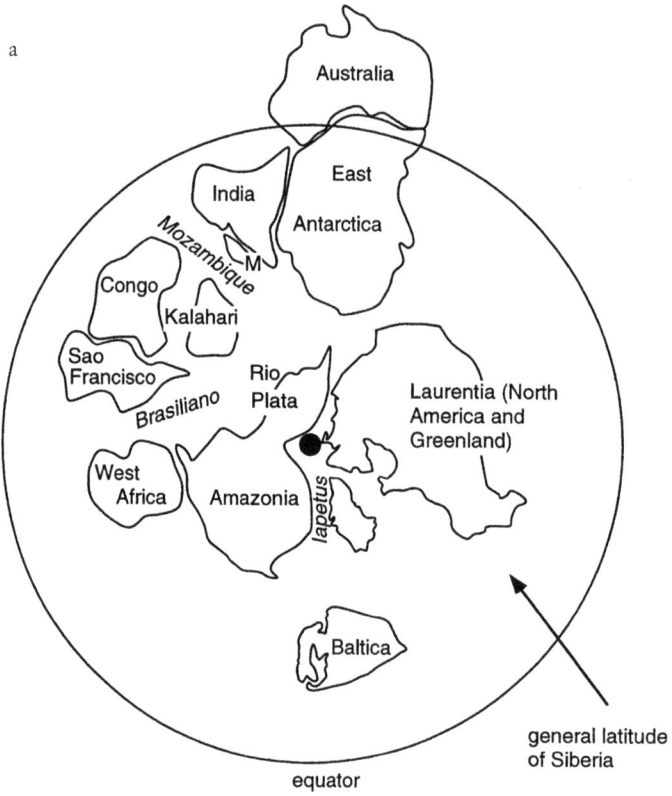

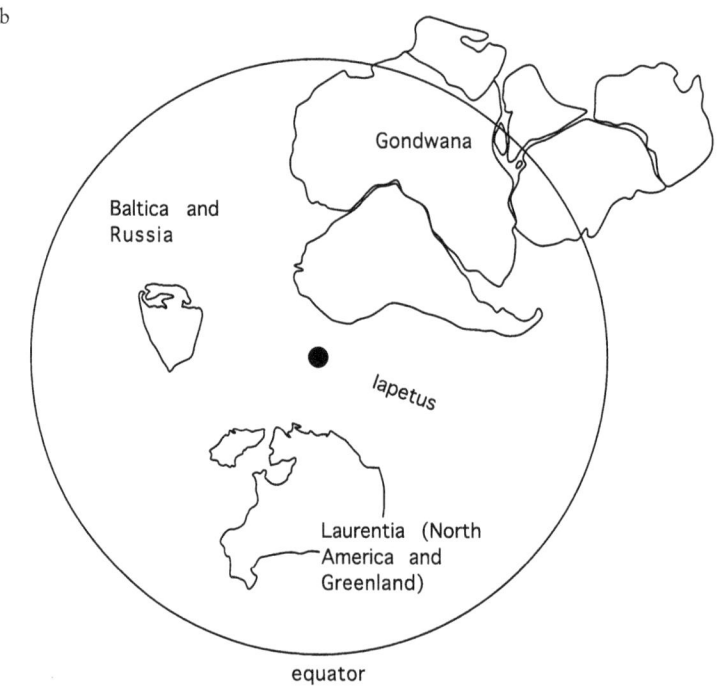

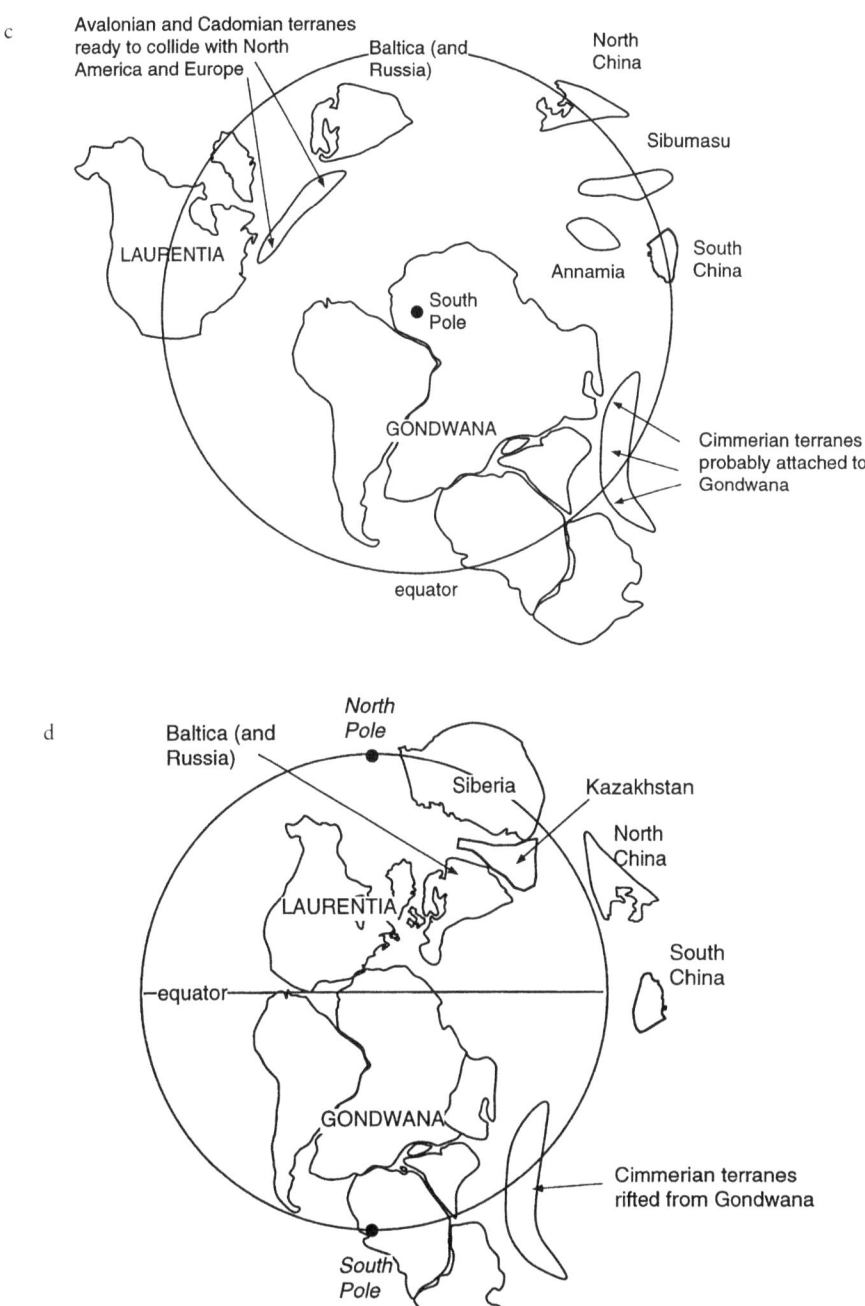

Figure 8.12. Paleomagnetically determined locations of major continental blocks. Diagrams a, b, and c are views of the southern hemisphere; diagram d is a view of the hemisphere (pole to pole) that contained Pangea. In all diagrams, continental blocks shown outside the circle of the hemisphere were on the opposite hemisphere. a. 580 Ma (modified from Meert, 2001). Major oceans are shown in italics, with the Iapetus just beginning to rift. M is Madagascar. b. 550 Ma (modified from Meert, 2001). The Iapetus had become wider since 580 Ma. c. 450 Ma (based on positions of blocks proposed by Cocks and Torsvik, 2002). Annamia is a block currently between South China and the major part of Indochina; Sibumasu is an acronym for Siam, Burma, Malaysia, and Sumatra. d. 250 Ma. Information supporting this configuration is in Scotese and McKerrow (1990).

Pangea. This difference in average age between Gondwana and Laurasia largely controlled broad patterns of basin formation and also affected the transport of sediment into those basins.

Pangea began to break up in the Late Triassic and Early Jurassic from ~225–175 Ma, a time that followed the Permo-Triassic extinction event and also the maximum glacial episode (chapters 11 and 12). During the period before rifting, marine water encroached only along the margins of the supercontinent, leaving the interior almost completely as exposed land. The highest area at this time seems to have been in East Antarctica, which provided sedimentary debris that washed outward over much of East Gondwana. Sequences of terrestrial sediment from this source are particularly well exposed in the "Gondwana" rifts of India, that had been active since the Middle Proterozoic but were now filled by clastic debris and developed abundant coal (Veevers and Tewari, 1995). Sediment derived from this Antarctic upland did not reach areas of Laurasia, where Early Mesozoic terrestrial sequences were thin and derived from local sources.

Erosion of sediment from the interior toward its northern margin apparently began in the Cambro-Ordovician, shortly after the fusion of Gondwana. North Africa and Arabia are covered by fluvio-deltaic sediments of this age, both in the Nubian–Arabian shield (chapter 4) and in the area of North Africa converted into continental crust by accretion of microcontinents and subduction of oceanic lithosphere (see above).

Differences in average elevation of continental surfaces apparently both preceded and followed the assembly of Pangea. During the Paleozoic, shallow marine water covered ~40–30% of the cratons and other stabilized continental areas that would later accrete to Laurasia. At this time, however, only about 20–10% of Gondwana was submerged below shallow epeiric seas (Veevers, 1995). In the Mesozoic and Cenozoic, after dispersal of Pangea, Laurasian fragments continued to show much higher percentages covered by shallow seas than fragments of Gondwana (Algeo and Wilkinson, 1991).

Differences between continental blocks in Laurasia and those in Gondwana throughout the Phanerozoic imply that some process that operated during the formation of Gondwana was fundamentally different from processes that caused assembly of Laurasia. Veevers (1995) proposed that this process developed a lower crust with P-wave velocities of ~7.5 km/sec throughout most of Gondwana but only in limited parts of Laurasia, such as the Colorado plateau. The exact nature of this process is unclear, but recent finding of similar P-wave velocities beneath much of the Rocky Mountains of North America (Dueker et al., 2001; CD-ROM Working Group, 2002) suggests that it is a characteristic of intracontinental activity.

## Summary

Pangea grew by continental-margin orogeny along one entire side (western in present orientation), by accretion of exotic and juvenile material along the eastern margin of North America and Eurasia, and ultimately by collision of eastern North America with western Africa. At its time of maximum packing (~250 Ma), Pangea had western and northeastern margins undergoing subduction and a southeastern (Gondwana) margin undergoing rifting. The rifted fragments from Gondwana accumulated on the northeastern margin of Pangea, a process that continues to the present. Even at its time of maximum packing, Pangea did not contain all of the world's continental blocks. At the end of the Paleozoic separate blocks probably included various Cimmerian terranes that had already rifted from Gondwana and may have included both North and South China.

# 9

# Rifting of Pangea and Formation of Present Ocean Basins

At the end of the Paleozoic the supercontinent Pangea was surrounded by the "superocean" Panthalassa (all ocean). We have no way of knowing what islands, island arcs, spreading ridges, and other features most of the ocean contained, because all of it has now been subducted. We can, however, be somewhat more specific about continental fragments and spreading ridges in the small region of Panthalassa directly adjacent to the eastern margin of Pangea. This part of the ocean, known as "Tethys," left a record of its history as continental fragments continued to rift from the Gondwana (southern) part of Pangea and move across Tethys to collide with the Laurasian (northeastern) margin of Pangea (chapter 8).

During the Mesozoic and Cenozoic the positions and configurations of continents and ocean basins gradually attained their present form. Major continental reorganization resulted from movements of fragments across Tethys and the opening of the Atlantic, Indian, Arctic, and Antarctic Oceans and associated smaller seas. The size of Panthalassa, now known as the Pacific Ocean, gradually decreased as other oceans opened and small seas formed by a variety of processes in the western Pacific. Separation and collision of continental plates in what had been the center of Pangea formed the Gulf of Mexico–Caribbean and the Mediterranean.

By creating new spreading centers, the breakup of Pangea generated a larger volume of young ocean lithosphere both in the new ocean basins and in the Pacific than the volume occupied by spreading centers in Panthalassa. By filling more of the ocean basins, these ridges forced seawater to rise eustatically onto continental platforms, creating shallow seas and filling cratonic basins where the crust was tectonically depressed.

We begin this chapter by discussing the successive changes in Tethys and then the origin of the world's major ocean basins. This is followed by an investigation of the smaller seas of the western Pacific region and the specific histories of the Gulf of Mexico–Caribbean and the Mediterranean. We continue with a discussion of the causes and locations of rifts that break up supercontinents and finish with a description of eustatic sealevel changes.

## Tethys

The process of fragments rifting from Gondwana and moving across Tethys to accrete with Laurasia continued without interruption across the time boundary between the Paleozoic and Mesozoic. Some of these fragments had almost certainly rifted away from Gondwana but not yet reached Eurasia at the beginning of the Mesozoic, thus yielding a map similar to the one shown in fig. 9.1.

The shape of Tethys began to change when the Atlantic and Indian Oceans started to open at ~180 Ma (see below). Patterns of drift established by spreading centers in the new oceans caused Eurasia to rotate toward an east–west orientation from its position in Pangea, where Europe was near the equator and Siberia near the north pole. At the same time, Africa, India, and Australia began to move northward because of spreading in the newly forming Indian and Antarctic Oceans. This caused a Tethyan Sea that had been elongated north–south along the eastern margin of Pangea to become elongated east–west as Eurasia and parts of the former Gondwana rotated toward each other like the blades of a pair of scissors.

These movements shortened the distance between the southern and northern margins of Tethys and decreased the time that it took Gondwana fragments to cross to the Eurasian margin. We show this process diagrammatically in fig. 9.2, which uses only one of the several terminologies that have been applied to the parts of Tethys

132   Continents and Supercontinents

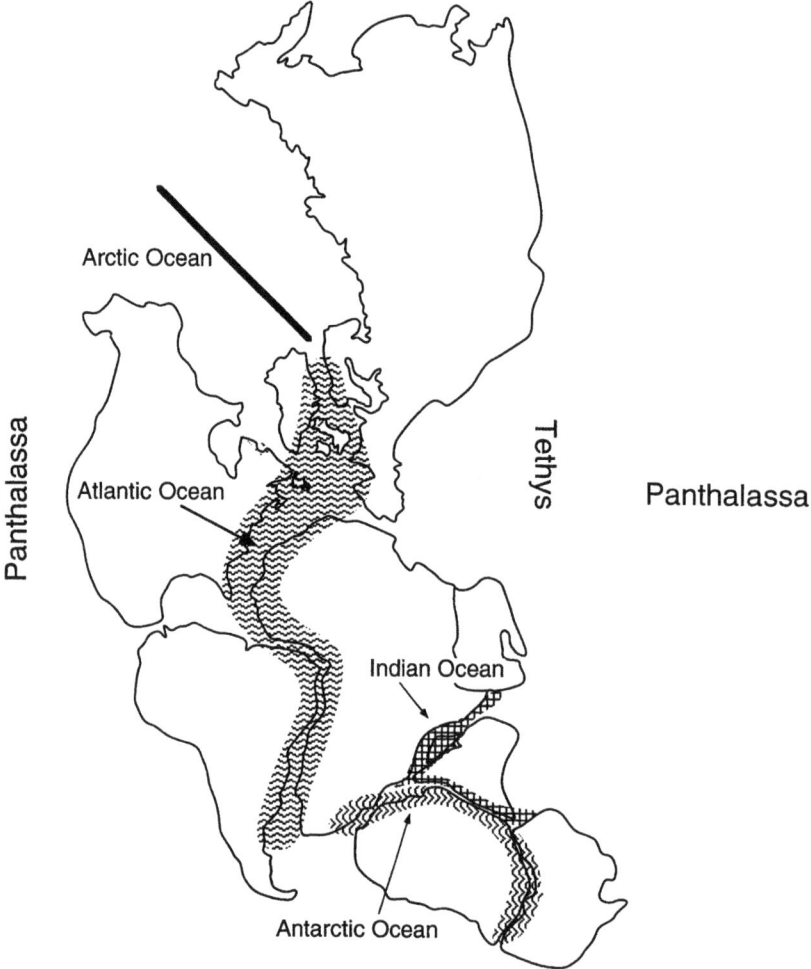

Figure 9.1. Pangea at ~250 Ma, showing oceans around it and oceans that caused fragmentation by opening within it. The border between the Indian and Antarctic Oceans is arbitrary.

that closed northward of the moving fragments and opened behind them. The closure of Paleotethys occurred in the Mesozoic as Cimmerian blocks accreted to Eurasia, with similar closure of Neotethys and opening of the Indian Ocean in the Cenozoic. Both of these accretions generated widespread orogenic events, all of which are described more completely in chapter 10.

## Formation of Modern Oceans

Almost all of the breakup of Pangea and movement of continents to their present positions can be described in detail because of the preservation of magnetic stripes in modern ocean basins. This wealth of information is well beyond the scope of this book, and we present only the outline of these movements and refer readers who want more detail to the website operated by C.R. Scotese at www.scotese.com. This site shows configurations of continents and ocean basins at numerous times throughout the Phanerozoic, including animations of many of the movements.

### Indian Ocean

The Indian Ocean opened behind (south of) the last continental blocks to rift from Gondwana and travel northward across Tethys to Eurasia. In contrast to the linear opening in the Atlantic (see below), the Indian Ocean exhibits an extremely complicated

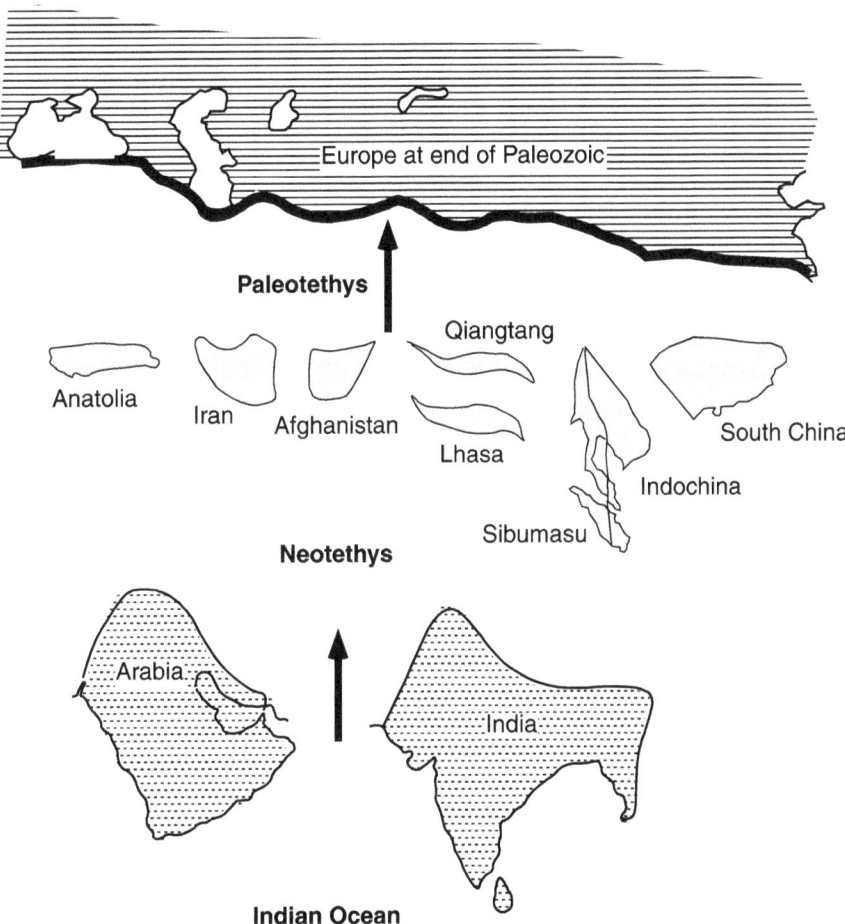

Figure 9.2. Diagrammatic representation of movement of blocks across Tethys. Lhasa and Qiangtang are blocks in Tibet. Sibumasu is an acronym for Siam–Burma–Malaysia–Sumatra. A separate part of Southeast Asia is labeled Indochina.

pattern (fig. 9.3a; Norton and Sclater, 1979; Besse and Courtillot, 1988). The reason for this difference is not well explained, but it is possible that it results from the difficulty of fracturing the old crust of East Gondwana (Ur) to form the Indian Ocean versus the greater ease of fracturing the younger crust of other parts of Pangea to create the Atlantic Ocean (chapter 8).

The first rifting in the Indian Ocean occurred between Madagascar and the Somali coast of Africa at ~175 Ma. Fragmentation at ~100–90 Ma broke up most of the rest of East Gondwana, forming spreading ridges that are nearly perpendicular to each other. This rifting separated East Antarctica from Africa, India, and Australia, but separation of the Antarctic and South America did not fully develop until about 25 Ma. The final major rifting in the Indian Ocean separated the Seychelles–Mascarene ridge from India at ~65 Ma, which some investigators have associated with the early stages of eruption of the massive Deccan basalts in western India.

Major reorganization of movements in the Indian Ocean began at approximately 30 Ma (fig. 9.3b). At that time, Neotethys was completely destroyed as the last part of its oceanic lithosphere was subducted beneath Asia and the Indian subcontinent collided with the mainland. India began to move separately from Australia–New Guinea, which started to rotate counterclockwise and compress areas in the southwestern Pacific (see below). The Red Sea also began to open, sending the Arabian peninsula northward to create greater compressive stresses in southwest Asia.

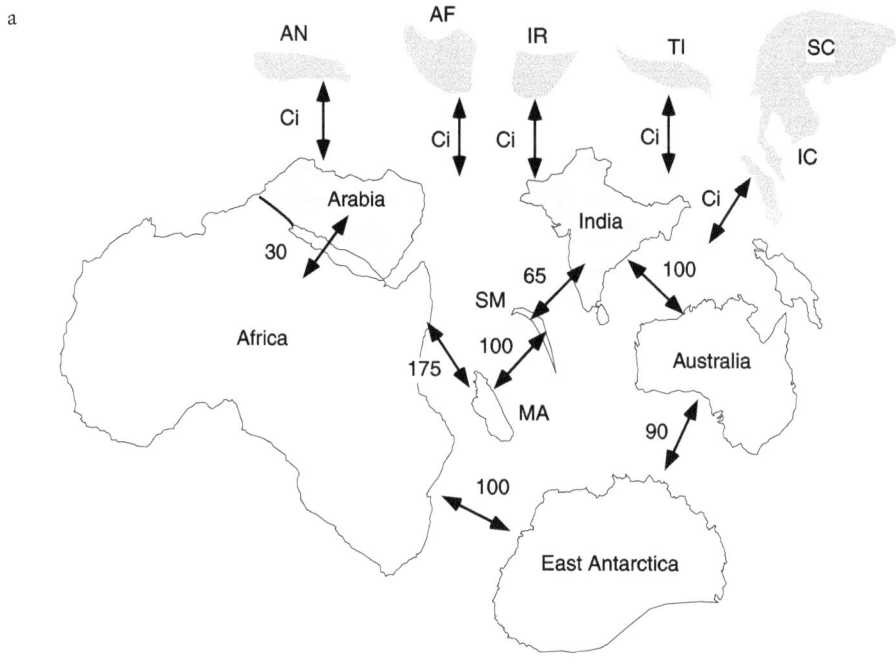

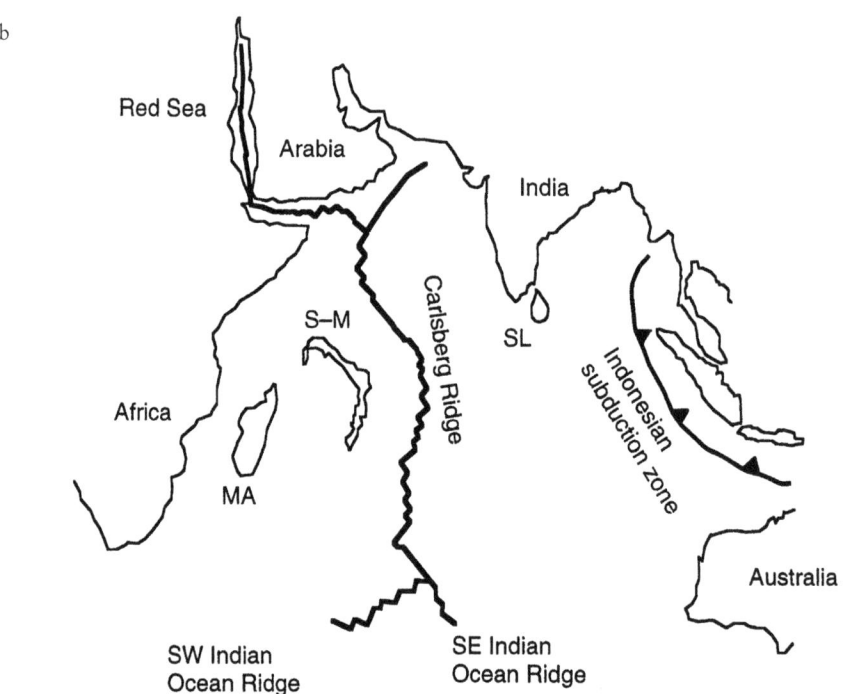

Figure 9.3. a. Opening of Indian Ocean. Ages of initiation of spreading are shown in Ma. Ci indicates Cimmerian rifting of uncertain age. Abbreviations for Cimmerian terranes are: AN, Anatolia; AF, Afghanistan; IR, Iran; TI, Tibet (two separate terranes); SC, South China; IC, terranes in Indochina. Abbreviations for terranes in Indian Ocean are: SM, Seychelles–Mascarene; MA, Madagascar. b. Present Indian Ocean. Abbreviations are: S–M, Seychelles–Mascarene; MA, Madagascar; SL, Sri Lanka.

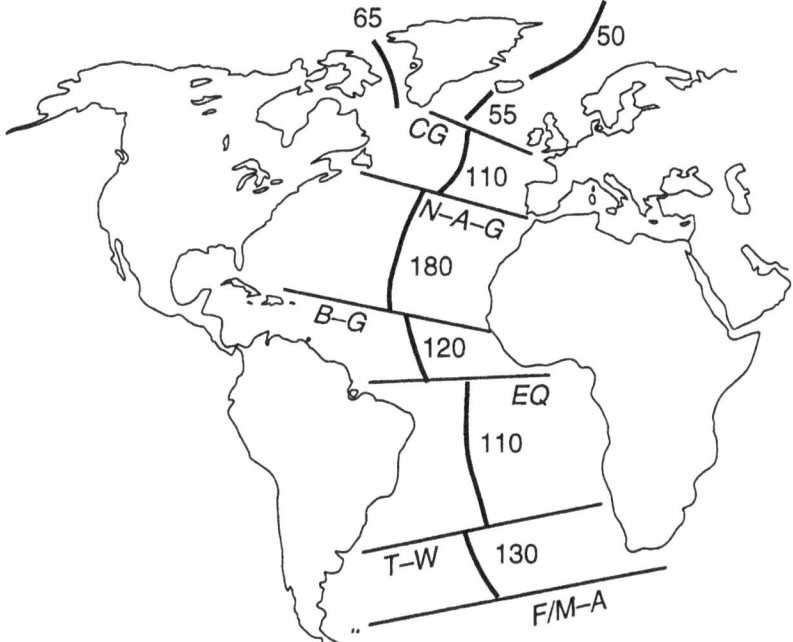

Figure 9.4. Opening of Atlantic Ocean. Ages of initiation of spreading are shown in Ma. Abbreviations of fracture zones are: CG, Charlie Gibbs; N–A–G, Newfoundland–Azores–Gibraltar; B–G, Bahamas–Guinea; EQ, Equatoria (Romanche); T–W, Torres and Walvis; F/M–A, Falkland (Malvinas)–Agulhas.

## Atlantic Ocean

The Atlantic Ocean opened along a series of spreading centers separated into segments of different ages by very large transform faults (fig. 9.4; Emery and Uchupi, 1984). The oldest segment began in the North Atlantic at approximately 180 Ma, bordered by the Bahama–Guinea fracture zone to the south and the Newfoundland–Azores–Gibraltar zone to the north. The spreading center formed toward the eastern side of this suture, leaving a series of aborted rift basins and a broad continental shelf along eastern North America but only a very thin shelf in northwestern Africa.

The Atlantic Ocean south of the Bahama–Guinea fracture zone began to open some 60 to 70 million years later than the North Atlantic. It was broken into three segments of slightly different ages by: the very long offset along the Equatoria (Romanche) zone; the combined Torres–Walvis Ridge zone; and the Falkland/Malvinas–Agulhas zone, which forms the northern border of the Antarctic Ocean in this area. The Equatoria zone was originally a strike-slip fault that developed as northeastern South America moved westward past the southern bulge of West Africa (chapter 1).

Spreading in the northernmost part of the Atlantic Ocean and the Arctic Ocean generally began at a younger time than farther south. A spreading segment started at about 110 Ma between the Newfoundland–Azores–Gibraltar fracture zone on the south and the Charlie Gibbs zone on the north. The final separation in the Atlantic Ocean occurred between Greenland and Norway and was almost certainly related to the development of the broad North Atlantic igneous province that is now a plume located beneath Iceland (Johnston and Thorkelson, 2000). An offshoot of this spreading opened the Labrador Sea between Greenland and Labrador at some time in the early Cenozoic.

## Arctic Ocean

Opening of the Arctic Ocean separated North America from Siberia, but the process is very uncertain (fig. 9.5; Kristofferson, 1990). Magnetic anomalies are absent in many parts of the ocean, and their interpretation is clear only in the spreading center that extends north from Iceland into the Eurasian basin and separates the Lomonosov Ridge from thinned crust of the Siberian mainland. The Canada basin contains a few magnetic anomalies

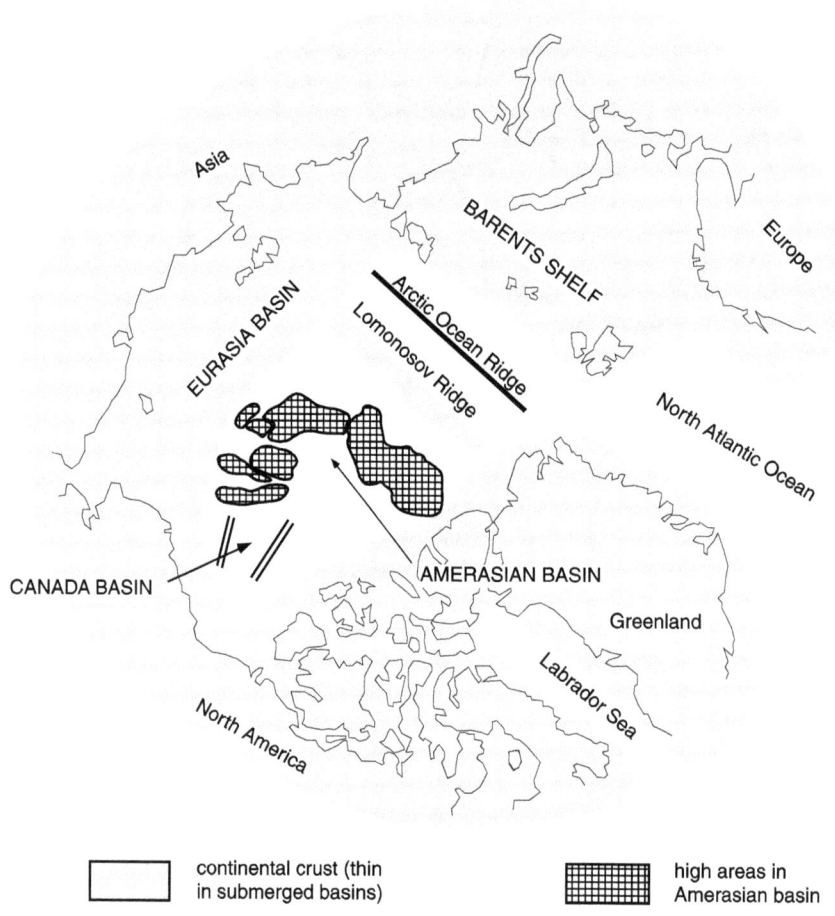

Figure 9.5. Opening of Arctic Ocean.

that show how it swung open between Canada and the Lomonosov Ridge. Formation of the Amerasian basin has been tentatively attributed to rotation of Alaska away from Canada, but evidence is not conclusive (Grantz et al., 1998).

## Antarctic (Southern) Ocean

Opening of the Antarctic (Southern) Ocean left the Antarctic continent at the South Pole and isolated it from all other continents as they moved northward throughout the Mesozoic and Cenozoic (fig. 9.6; J. Anderson, 1999). The present Antarctic plate is bounded by spreading ridges and fracture zones that developed in the past 100 million years. The Falklands/Malvinas–Agulhas fracture zone terminates the southern Atlantic Ocean and connects eastward with a spreading ridge between Antarctica and Australia. This ridge connects farther eastward with a series of ridges and short fracture zones that extend to the East Pacific ridge. A similar series of fracture zones and short spreading ridges at the southern margin of the Nazca plate completes the isolation of the Antarctic continent.

The most recent separation was between South America and the Antarctic Peninsula at ~25 Ma, at the time the Scotia arc was initiated in the new oceanic lithosphere. This final separation established a world-encircling ocean around Antarctica, leading to glaciation of the continent and other climatic effects that we discuss more completely in chapter 11.

Only the East Antarctic part of the Antarctic continent rifted from Pangea, with the area known as West Antarctica formed by post-rift accretion (chapter 10). The rifted edges are exposed where East Antarctica adjoins the Indian Ocean,

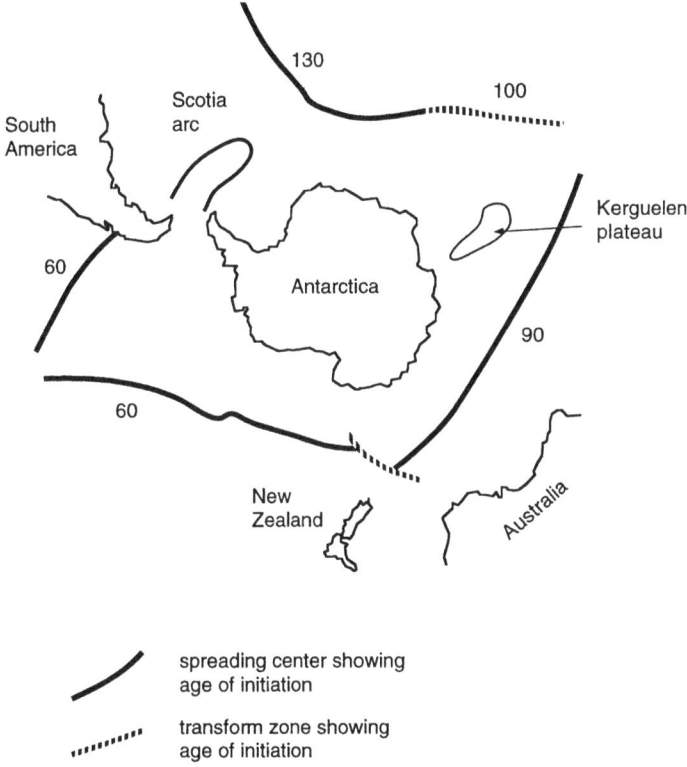

Figure 9.6. Opening of Antarctic (Southern) Ocean. Ages of initiation of spreading are shown along ridge segments in Ma.

but rifts along the Transantarctic Mountains are now obscured because this area is also the join between East and West Antarctica. In Pangea, the Transantarctic Mountains and eastern Australia were probably inland from areas of continental crust that now include New Zealand and submerged plateaus around it (Muir et al., 1996). This region around New Zealand probably rifted away from the Transantarctic Mountains at about the time that rifting began in the Tasman Sea in the Late Cretaceous.

## Pacific Ocean

At the end of the Paleozoic, Panthalassa covered about three fourths of the earth's surface. The earth's ocean basins together still cover the same total area, but the Pacific now occupies half of that area, with the remainder divided mostly among the Atlantic, Indian, Arctic, and Antarctic Oceans plus small seas on the western margin of the Pacific.

Figure 9.7 shows a capsule history of the Pacific beginning in the Cretaceous. It is pieced together from magnetic stripes on both sides of the East Pacific Rise, which is still active, and from stripes preserved on one side of ridges that have now been subducted. Subduction of ridges also destroyed plates on the subducted side of the ridges and led to significant reorganizations of plates through time in the Pacific basin.

Several unknown plates in the Pacific may have been subducted and destroyed in the Early Mesozoic, but the oldest one for which there is significant evidence is the Izanagi plate. It was produced by spreading on the northern side of the Izanagi–Pacific spreading ridge, which has been subducted and is now represented only by magnetic stripes that become younger toward the northern margin of the Pacific (the opposite of the pattern for spreading oceans; see chapter 1). After the Izanagi plate and its spreading center were subducted, the Kula plate was generated on the northern side of the Kula–Farallon spreading center and was separated from the Pacific plate by a long fracture zone. Subduction of the Kula ridge and plate and destruction of most of the Farallon plate left magnetic stripes that become younger toward the margin of Alaska and also toward the Canadian margin.

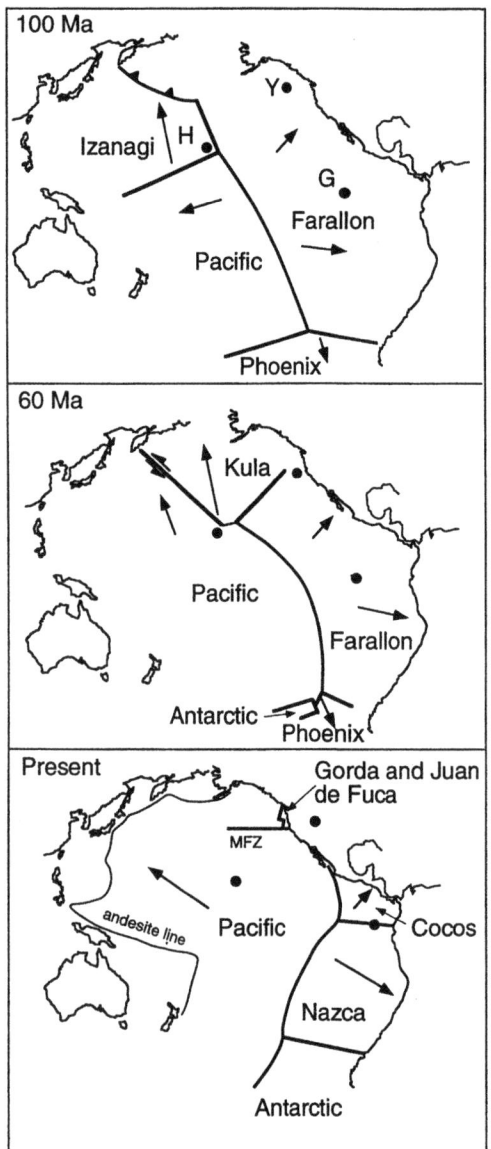

Figure 9.7. Summary of changes in Pacific Ocean since the Cretaceous. Arrows show direction of movement of plates. MFZ is the Mendocino Fracture Zone. Movements are shown relative to three hotspots in their present positions: Y, Yellowstone; H, Hawaii; G, Galapagos. Andesite line is explained in text.

Progressive encroachment of North and South America into the Pacific Ocean and the recent development of the Galapagos spreading ridge destroyed much of the Farallon plate by subduction and separated the remnants into three parts. A very small section of the Farallon plate remains off the coast of the northwestern United States, where it is referred to as the Gorda and Juan de Fuca plates. They are separated by the Mendocino fracture zone and the San Andreas fault from the Sea of Cortez, which contains the northern tip of the East Pacific Rise. The Cocos and Nazca plates are separated by a transform north of the Galapagos spreading center and subduct beneath their adjacent continents in slightly different directions. The Phoenix plate has now completely disappeared beneath Antarctica, leaving the Antarctic plate separated from the Nazca plate by a series of small spreading ridges and transforms.

The direction of movement of the Pacific plate may have changed abruptly at ~40 Ma. The best evidence for the shift in direction of the Pacific plate is the path of the Hawaii–Emperor seamount chain, which changes from a mostly N–S orientation of seamounts older than Midway Island to a more westerly direction of the younger part of the chain, including active volcanic centers in Hawaii (chapter 2). Northward movement of the Pacific plate before 40 Ma caused subduction along the northern margin of the Pacific basin, thus generating the Aleutian Island arc and contributing to southward growth of the Alaskan margin. Movement after 40 Ma, however, was more toward the west, converting the western part of the Aleutian subduction zone into a strike-slip fault and terminating volcanism.

The western Pacific is a complex region of small subduction zones, backarc basins, and fracture zones (fig. 9.8; Ballance, 1999). It contrasts strongly with the eastern margin of the Pacific, where subducting oceanic lithosphere descends beneath continental margins without any development of offshore island arcs. The difference between the eastern and western Pacific may result from the much older lithosphere in the west than the east, but the question is not resolved.

The small basins of the western Pacific are west of a feature that geologists have known about for more than 100 years. This "andesite line" was originally recognized by differences between the volcanic rocks on islands west of the line and those within the main Pacific Ocean basin. Islands inside the Pacific basin consist almost entirely of basalts, whereas islands west of the line contain a complete "calcalkaline" suite dominated by andesite but ranging from basalt to rhyolite. The reasons for this difference were unclear when the andesite line was first discovered, but the distinctions remain valid after a century of research has now provided explanations. We now know that the andesite suite is generated in island arcs above subducting slabs of

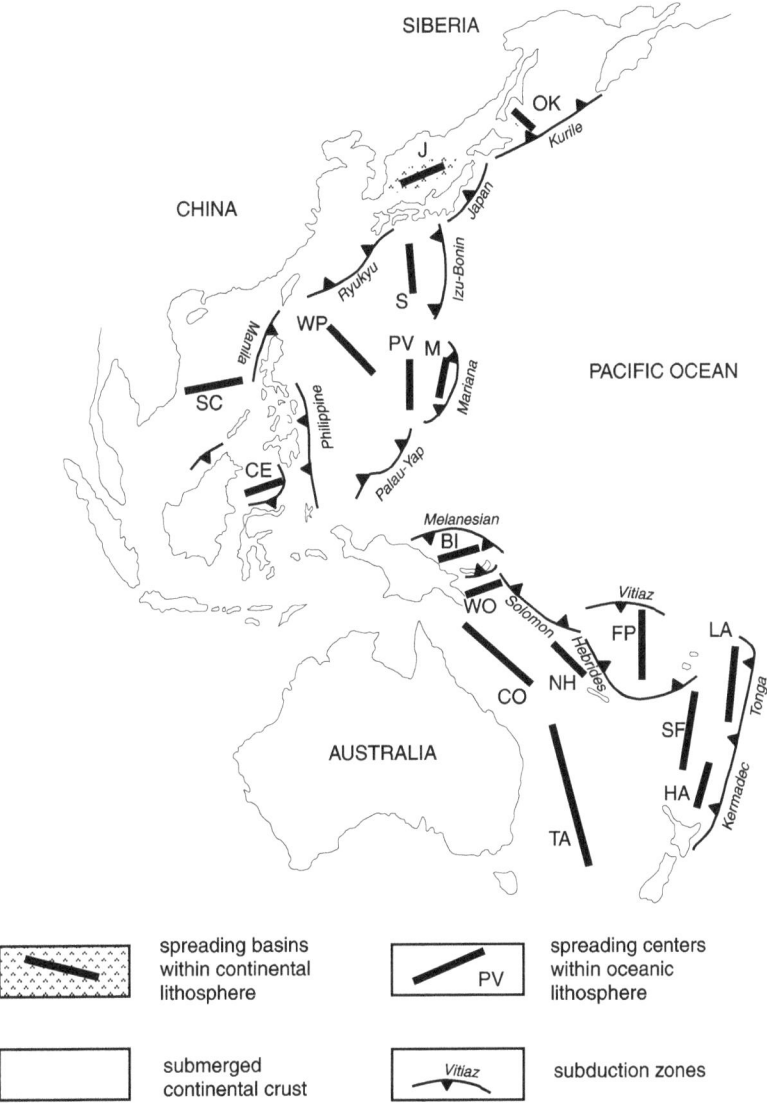

Figure 9.8. Arcs and backarc basins in the western Pacific Ocean. Arcs are shown by conventional symbols for subduction zones. Spreading centers are shown by thick black lines and named on the diagram. Abbreviations of spreading centers are: OK, Okhotsk; J, Japan; S, Shikoku; WP, West Philippine; PV, Parece Vela; M, Mariana; SC, South China; CE, Celebes; BI, Bismarck; WO, Woodlark; CO, Coral; NH, New Hebrides; FP, Fiji plateau; LA, Lau; SF, South Fiji; TA, Tasman; and HA, Havre.

oceanic lithosphere, that the central-Pacific islands are hotspots above mantle plumes and contain a variety of rocks in addition to basalt, and that basalt formed by seafloor spreading occurs both in the Pacific basin and in backarc basins behind spreading centers (see chapters 1 and 2 for more information).

The Tasman spreading center was active almost entirely in the Late Cretaceous, but all of the other backarc basins shown in fig. 9.8 were formed in the Cenozoic. Ages of initiation are generally younger in the north than the south, and Miyashiro (1986) suggested that the rifting was caused by northward movement of a region of hot mantle. Most of the rifting is within oceanic lithosphere, but a few of the marginal basins were initiated by rifting within the Asiatic continent (see chapter 10 for further discussion of Eurasian extension). The principal example

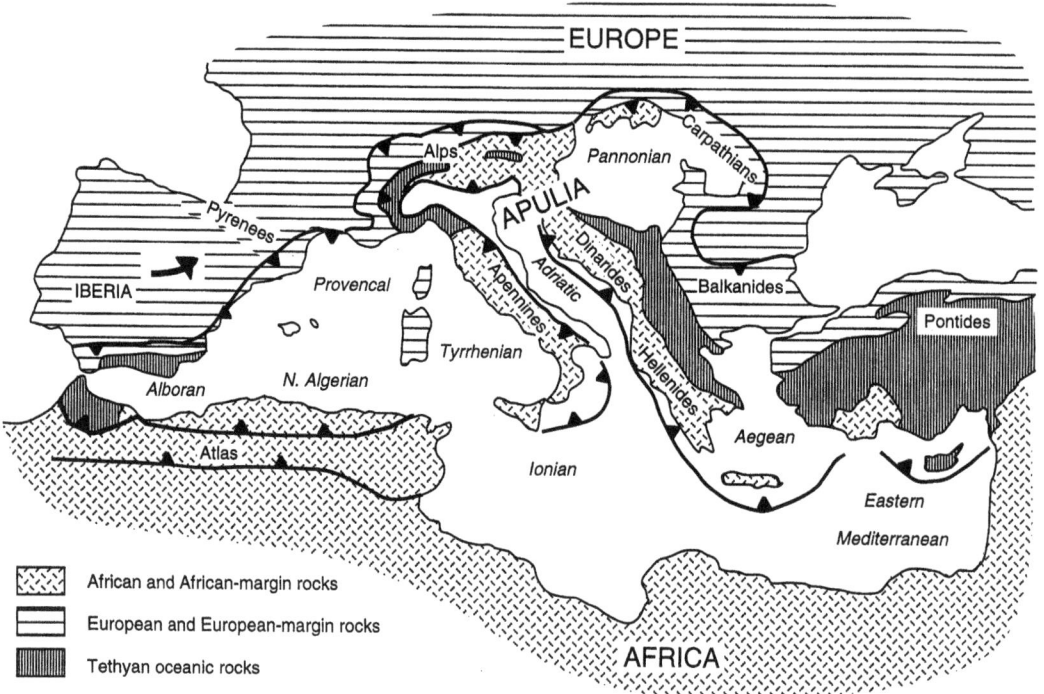

Figure 9.9. Evolution of Mediterranean Sea. Subduction zones are shown by conventional symbols. Basins are in italics and mountain ranges in normal font. Iberia is shown rotating counterclockwise because of opening of the Bay of Biscay to the north. The Alps are further discussed in chapter 2.

of this continental fragmentation is the Sea of Japan, which began to separate Japan from the Asiatic mainland in the Cenozoic.

Subduction zones associated with these Pacific basins have a complex pattern that changed throughout their history (Ballance, 1999). The easternmost ones (along the andesite line) all lie above descending slabs of Pacific Ocean lithosphere, but various directions of subduction occur farther west. Directions vary not only from place to place, as at present, but from time to time at individual places. For example, subduction under the northern Philippines (Manila trench) is now down to the east, but it has "flip-flopped" between the east and west sides of the island at least twice during the Cenozoic (Yumul et al., 1998). Similarly, the Hebrides Islands now lie above a westward-descending slab east of the Hebrides arc, which has recently replaced subduction beneath their western side.

The complex developments in the western Pacific were caused not only by spreading within the Pacific Ocean but also by stresses from spreading in the Indian and Antarctic Oceans. The Indonesian island arc lies on thin continental crust with ages as old as the Proterozoic, and subduction of Indian Ocean lithosphere that began in the Mesozoic caused construction of the emergent part of the arc (Hamilton, 1979; Katili and Reinemund, 1982). Further complication was caused when Australia–New Zealand separated from the Indian plate at ~25 Ma (chapter 10) and began to rotate counterclockwise. This rotation compressed the entire region of Indonesia and the Philippines and established several long systems of strike-slip faults.

## Mediterranean Sea

The Mediterranean Sea occupies the zone where the collision of Africa, Europe, and possibly the Apulian plate closed the part of Tethys that formerly separated the two continents (fig. 9.9). With the exception of the Atlas Mountains along the northern margin of West Africa, almost all of the orogeny produced by this collision occurred in southern Europe (chapter 10). Both in Europe and North Africa, the Mediterranean plate was thrust up and over the surrounding continents, and many of the features of the Mediterranean result from extensional stresses above the descending slabs.

The initial crust of the Mediterranean area was probably thin continental crust of the North African plate. Extension in the western Mediterranean stretched this crust so much that oceanic crust broke through in several areas (Gueguen et al., 1998; Robertson and Comas, 1998). Extension has been less severe in the eastern Mediterranean, possibly because compressive orogeny occurred only on the European margin (Garfunkel, 1998). For this reason, the eastern Mediterranean is floored almost completely by thin continental crust, with only a few areas where oceanic lithosphere is exposed.

Three seas on the northern margin of the Mediterranean have somewhat different histories. The Aegean is being stretched north–south because of east–west compression between Greece and Turkey and also because of extension above the slab that is subducting beneath Crete and generating magmatic activity throughout the southern Aegean (Dinter, 1998). The Tyrrhenian Sea is extending north–south because of a slab that is subducting beneath the Apennines, and arc magmatism on its southern margin results from descent of a slab beneath the Calabrian arc. The Adriatic, conversely has no associated magmatism and is simply a foreland basin to mountain belts on its northeastern shore.

Closure of both the eastern and western ends of the Mediterranean resulted from rotation of continental blocks that began at ~30 Ma. The opening of the Red Sea (see above) and later translation along the Aqaba–Dead Sea transform isolated the eastern end of the Mediterranean from the Indian Ocean (chapter 10; Guiraud and Bosworth, 1999). Rotation of the Iberian peninsula caused by spreading in the Bay of Biscay nearly closed the western end of the Mediterranean, leaving only the Straits of Gibraltar as a connection to the world's major oceans.

## Gulf of Mexico and Caribbean

The Gulf of Mexico opened when South and North America began to drift apart in the Triassic (fig. 9.10). The initial rifting developed a shallow and rapidly subsiding basin that accumulated a thick sequence of Middle Mesozoic evaporites before the crust stretched far enough for oceanic lithosphere to break through (Pilger, 1978). This stretching took place at the same time that blocks formerly on the western side of North America began to move counterclockwise to fill the void left by South America.

The Caribbean Sea developed as South America and North America moved farther apart in the Late Mesozoic and Early Cenozoic (fig. 9.10). The nature and origin of the Caribbean plate have never been fully resolved, but at least part of it is probably a remnant from the Pacific region that became isolated as both North and South America moved westward when the Atlantic opened (Donnelly, 1989). In the Mesozoic, lithosphere of the Atlantic Ocean underthrust the northern Caribbean, but this compression was changed to strike-slip faulting when subduction began under the Lesser Antilles in the Cenozoic. Similarly, a formerly compressive regime along the northern coast of South America was changed to strike-slip movement that connects in an uncertain pattern with the subduction zone along western South America. Subduction beneath Central America constructed a land corridor between North and South America that closed circulation between the Caribbean and Pacific at about 5 Ma. This closure started the cooling of the northern hemisphere that ultimately resulted in episodic Pleistocene glaciations, which we discuss more fully in chapter 11.

## Origins and Locations of Rifts that Cause Breakup

Heat rising under ocean basins is convected to the surface and easily dispersed (chapter 1), but the only method of heat loss through rigid continental lithosphere is the relatively slow process of conduction. Consequently, it seems likely that supercontinents break apart because heat from the earth's interior accumulates at the base of their lithospheres. This accumulation of heat may cause generalized fracturing of the lithosphere, but plumes and superplumes may also create local extensional stresses.

As soon as continental fragmentation and drift were recognized as a major part of earth history, geologists began to wonder what determined the location of the new ocean basins. When the concept of plumes was developed in the 1970s, triple junctions radiating from them were regarded as potential sites of new spreading centers (chapter 1). This led to efforts to determine whether plumes developed under supercontinents in definable patterns or whether they were apparently random. In addition to looking for patterns of plume development, geologists also attempted to correlate new rifts with

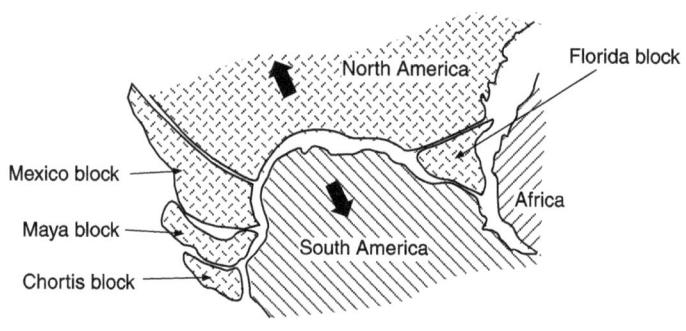

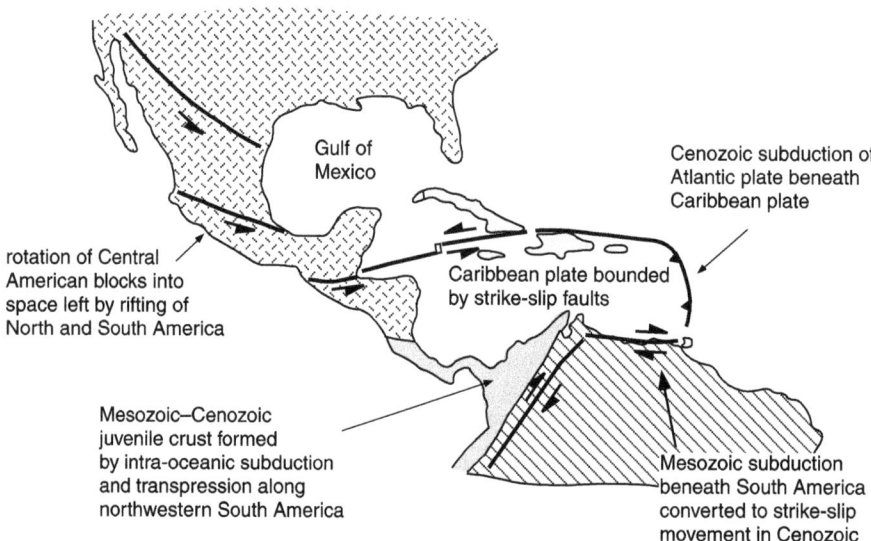

Figure 9.10. Evolution of Gulf of Mexico and Caribbean Sea.

former suture zones. Figure 9.11 shows the nature of the crust where rifting caused fragmentation of Pangea and also the locations of plumes or possible plumes that were active in Pangea fragments during rifting.

An enormous variability in precursor history is shown by the numerous rifts that created the Indian Ocean. If India and Australia accreted separately to Gondwana (chapter 8), then the rift between them probably followed the suture zone along the Darling belt. Also, if a suture between East and West Gondwana followed the east coast of Africa and included Madagascar and southern India, then the rift between Madagascar and Africa may have been along one part of that suture (chapter 8). The rift between Madagascar and India, however, split Precambrian terranes that had developed together at least since the Middle Proterozoic. Similarly, none of the rifting along the coast of East Antarctica followed earlier sutures (see below).

Rifts in the Atlantic Ocean also follow numerous types of precursor history. The earliest rifting in the North Atlantic follows the Paleozoic suture between the Appalachians of North America and Mauritanides of Africa (chapter 8). The spreading

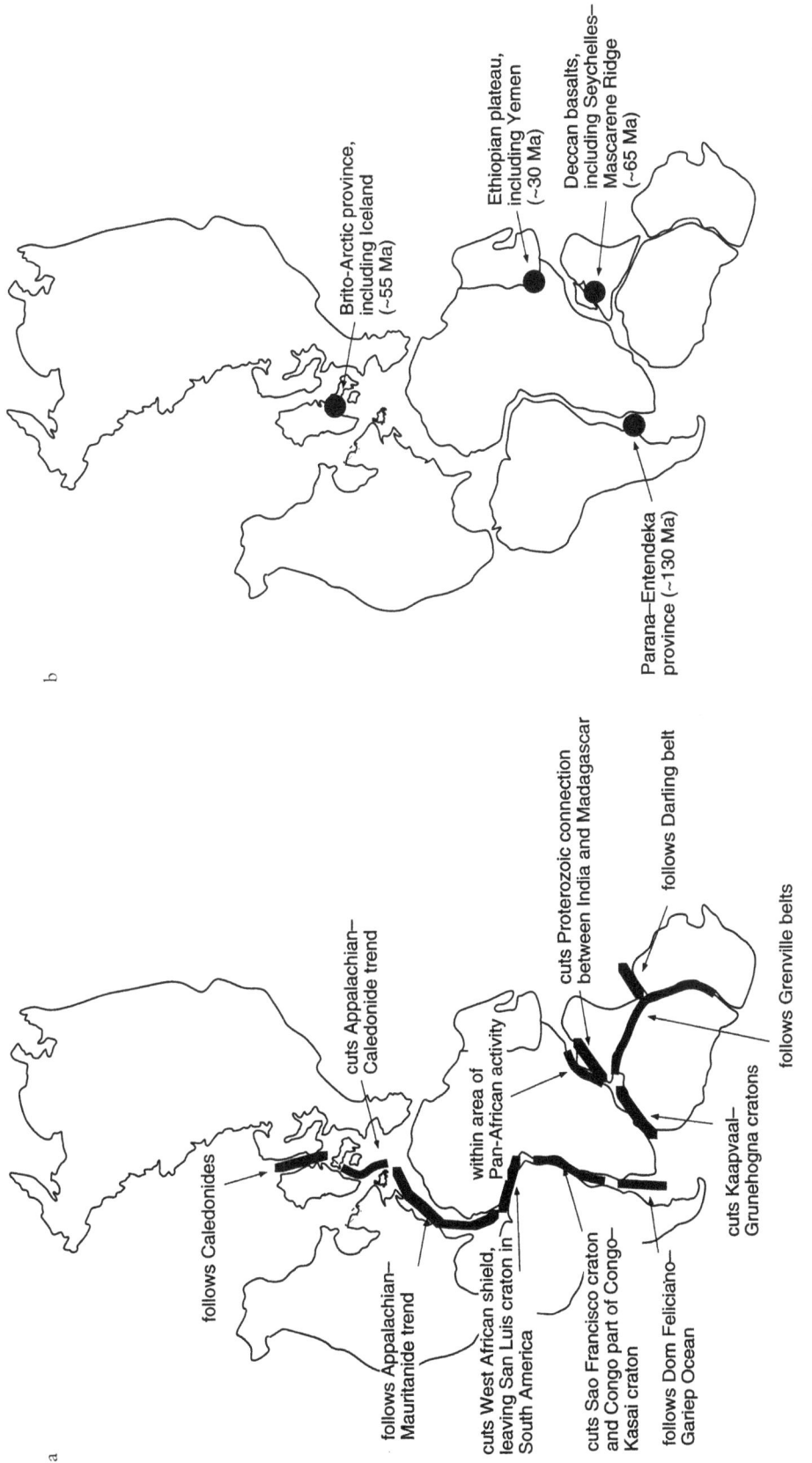

Figure 9.11. Locations of rifts and plumes during the breakup of Pangea. a. Rifts. The diagram shows the type of continental crust that was fractured by each rift. Each of the features is discussed in the following chapters and appendices: Caledonides, chapter 6; Appalachian–Mauritanide trend, chapter 8; West African shield and strike-slip fault between West Africa and eastern South America, chapter 2, appendix I; Sao Francisco and Congo–Kasai craton, chapter 6; Dom Feliciano–Gariep Ocean, chapter 8, appendix I; Kaapvaal–Grunehogna cratons, chapter 7, appendix G; Grenville belts in East Antarctica, chapter 7, appendix G; Darling belt, chapter 8, appendix G; Madagascar and India, chapter 4; Madagascar and Africa, chapter 8. b. Plumes.

center between Greenland and Norway follows the Caledonide orogen, but whether this was an intercontinental suture or an intracontinental orogen is uncertain (chapter 5). Rifting between the Newfoundland–Azores–Gibraltar and Greenland–Scotland zones cuts across Middle Paleozoic orogenic belts that formerly extended from North America into Europe. Rifting south of the Bahama–Guinea fracture zone also appears to show no relationship to older structures. The Equatoria fracture zone cut across the southern tip of the West African craton and left a small part as the San Luis craton of South America. Spreading centers in the South Atlantic cut the formerly united Congo–Sao Francisco craton (chapter 6) until they reach the former Dom Feliciano–Gariep Ocean (chapter 8) at the southern end of the Atlantic.

Whether rifts in the Arctic Ocean follow former zones of tectonic weakness is unclear. The Arctic Ocean ridge apparently cut across ancient Siberian continental crust, but it is possible that this submerged area might have contained an older suture zone. The Canada and Amerasian basins are so filled with fragments of continental crust, and the opening pattern is so unclear, that the location of initial rifting cannot be determined from present information.

The rifts that formed around East Antarctica do not appear to follow any former suture. Between Africa and East Antarctica the rift cuts through the Archean Kaapvaal–Grunehogna craton (chapter 7), which had been coherent since the Paleoproterozoic or earlier. Farther east, separation between East Antarctica and India/Australia is within a set of Meso-/Neoproterozoic orogenic belts in India and Australia that correlate so well with belts in East Antarctica that the sutures associated with these orogens are presumably farther south under the Antarctic icecap (appendix G). Rifting along the Transantarctic Mountains followed an Early-Paleozoic orogen, but it is not clear whether it developed in a suture zone.

Most, but not all, geologists believe that large basalt plateaus were developed above plumes (chapter 1). Figure 9.11b shows the locations of four basalt plateaus, presumably caused by plumes, that were clearly involved in the breakup of different parts of Pangea. Iceland, which is still astride the mid-Atlantic ridge, is part of a much broader Brito-Arctic basalt province that is now preserved in both eastern Greenland and the Hebridean Islands of northern Scotland (Kent and Fitton, 2000). Because this province began to form at ~55 Ma, the same age as the oldest magnetic stripes between Greenland and Europe, it is clear that the Iceland plume either initiated the spreading or was able to rise because the area was undergoing extension.

Similar relationships between age of basalt plateaus and initiation of spreading are present in three other locations. The Parana basalts and associated volcanic rocks of South America and the similar Entendeka suite of southern Africa developed as a single volcanic province at ~130 Ma, when this part of the Atlantic Ocean began to open (Kirstein et al., 2001; Marsh et al., 2001; Ernesto et al., 2002). The island of Tristan da Cunha, on the mid-Atlantic ridge, is still volcanically active and may be the present site of the plume that formed the Parana–Entendeka suite. Tristan da Cunha is connected to the Entendeka province by the Walvis ridge and to the Parana province by the Torres ridge, but whether these ridges are both hotspot tracks is unclear (fig. 9.4).

Spreading in the northern Indian Ocean at ~65 Ma split the 65-Ma Deccan basalts in eastern India from their apparent continuation in the Seychelles–Mascarene Ridge. The Deccan basalts were apparently the head of a plume that left a trail across part of the Indian Ocean and is now located at Reunion Island (Lenat et al., 2001). The most recent example of spreading associated with a basalt plateau is at the southern end of the Red Sea, where the Ethiopian–Yemen basalt plateau developed at the time all three arms of the triple junction began to spread (Ebinger and Casey, 2001; Menzies et al., 2001; chapter 2).

In addition to Tristan da Cunha, two other volcanic islands are located along or near the mid-Atlantic ridge in the southern part of the Atlantic Ocean. Ascension and St. Helena have both been proposed to be the tails of plume tracks that were developed during the opening of the ocean. Efforts to relate them to volcanic activity in Africa, however, have been mostly unsuccessful, and we discuss the issue more completely in chapter 10.

## Eustatic Changes in Sealevel

Since the beginning of the Phanerozoic, global sealevel probably has not varied by more than a few hundred meters. The two major processes that control the changes that have occurred operate over very different periods of time. Glacial–interglacial variations typically have periodicities of less than a few hundred thousand years, and we discuss them in chapter 11. Variations with periodicities of tens of millions of years are determined by the volume of ocean basins that is filled by spreading ridges, and we discuss them here because the breakup of Pangea triggered a major rise in global sealevel.

Variations in global sealevel during the Phanerozoic are controversial in detail but fairly well known over intervals of tens of millions of years. A generalized illustration of variations in sealevel shows highstands in the Ordovician and Cretaceous with lowstands near the end of the Paleozoic and at present (fig. 9.12). The cause of the Ordovician highstand is unclear, but the one in the Middle Cretaceous is almost certainly the result of the fragmentation of Pangea and development of numerous spreading centers in the Atlantic and Indian Oceans at about 100 Ma (see above). Sealevel rise was enhanced in the Cretaceous by development of a superplume in the Pacific Ocean (Larson, 1991).

The extent of continental submersion during the Cretaceous was controlled not only by global sealevel but also by local tectonic activity. Thus the amount of continental submergence or emergence varied between continents and within continents. Figure 9.13 shows the principal areas of Cretaceous flooding, distinguishing between areas submerged because of tectonic subsidence and those flooded only by sealevel rise and subsidence of rifted margins.

Rifting causes continental margins to become thinner and susceptible to encroachment by marine water, with the extent of submersion dependent on global sealevel, post-rift tectonic activity, and the rate of sedimentary outbuilding of the continental margin (chapter 2). The Atlantic Ocean covered the continental shelf of North America and penetrated far into the interior but encroached only slightly on Africa because the initial rifting was on the eastern side of the extended region (see above). Encroachment similarly covered shelves in South America but did not extend into the interior except in areas of tectonic subsidence. Large regions of the Arctic Ocean are floored by thin continental crust (see above), and they have been submerged continually since rifting.

Several continental margins that were not formed by rifting of Pangea are now submerged and/or show the effects of Cretaceous sealevel rise. One of the thickest sedimentary suites accumulated on the northeastern margin of the Arabian peninsula before it collided with Asia, and remnants of sediments from the northern margin of India are locally preserved in the Himalayas. The margins of eastern Asia, the Sunda shelf, and northern Australia/New Guinea are all thin and covered by seawater, either because of stretching of old con-

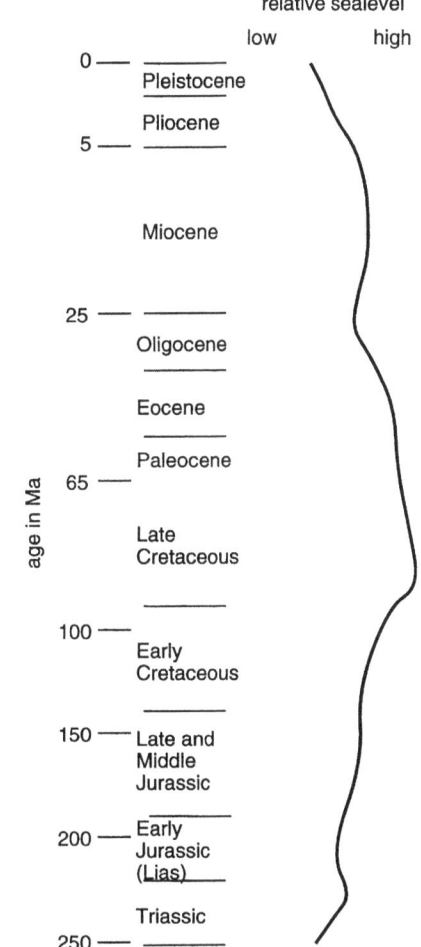

Figure 9.12. Eustatic sealevels during the Phanerozoic. Adapted from Haq et al. (1987).

tinental crust or because the crust is now forming and has not yet developed normal thickness.

The general emergence of Gondwana and submergence of Laurasia within Pangea continued into the Early Mesozoic (chapter 8). The principal area affected was northern Europe, the Russian platform, and areas toward the southeast. This region accumulated widespread suites of Mesozoic rocks before becoming emergent in the Cenozoic.

Figure 9.13 shows some of the large areas where Cretaceous flooding was caused partly by tectonic downwarp. The Western Interior Seaway of North America was a foreland basin depressed by loading of the Sevier orogenic belt to the west. The Mississippi, Amazon, and Parana basins all formed along major river systems, possibly indicating that the rivers followed long-term zones of weakness. A

146   Continents and Supercontinents

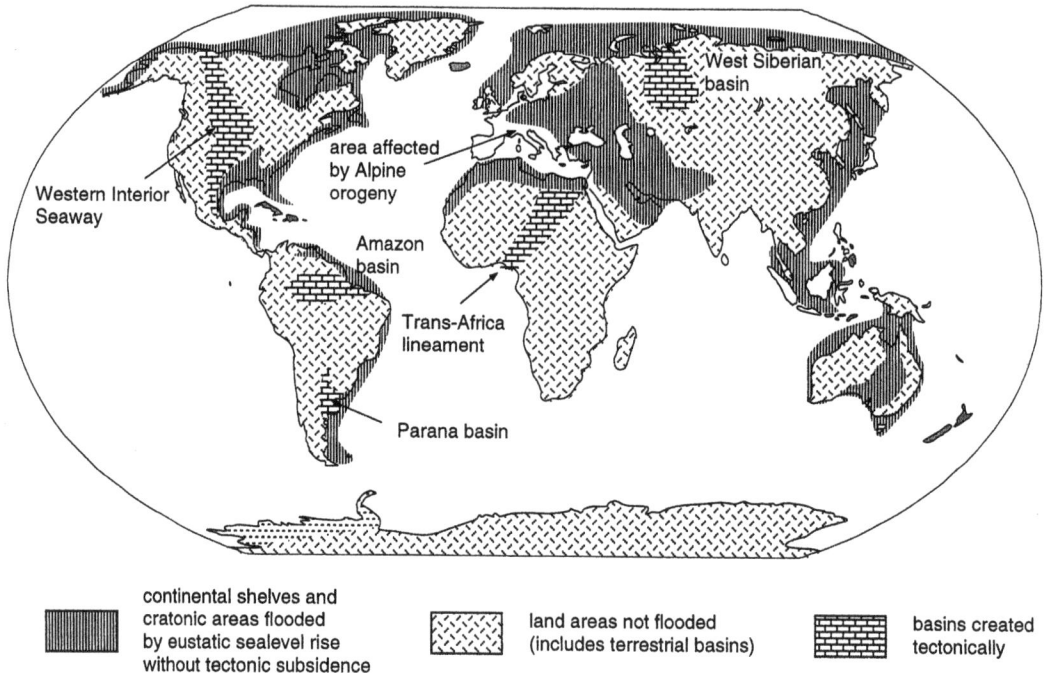

Figure 9.13. Areas submerged during Mesozoic and Cenozoic.

similar zone of weakness follows the Transafrican lineament (Nagy et al., 1976), and the West Siberian basin represents a major area of rifting that formed as collision occurred along the southern margin of Asia.

The major Precambrian cratons of Africa are largely covered by marine sediments that formed before and during the Cretaceous sealevel highstand, with Cenozoic sedimentation being primarily intracontinental (Kogbe and Burollet, 1990). All or some of these basins may lie above rifts, and the Congo basin may be the successor to Pan-African suturing of its two parts (chapter 8).

## Summary

Pangea began to break up at ~175 Ma, with the largest pulse of seafloor spreading starting at ~120 Ma. Formation of the Atlantic, Indian, Arctic, and Antarctic Oceans caused continents to encroach on the Pacific Ocean, which is a remnant of the Panthalassa that formerly surrounded Pangea. Fragmentation was presumably caused by buildup of heat beneath Pangea and may have been augmented by stresses produced by a few superplumes. Extension created rifts throughout the supercontinent that show little or no relationship to the precursor geology of the zones that were fractured. Development of new spreading centers in the Cretaceous filled so much of the ocean basins that seawater was forced out of areas underlain by oceanic lithosphere and onto continents.

# 10

# History of Continents after Rifting from Pangea

As continents moved from Pangea to their present positions, they experienced more than 100 million years of geologic history. Compressive and extensional stresses generated by collision with continental and oceanic plates formed mountain belts, zones of rifting and strike-slip faulting, and magmatism in all of these environments. In this chapter we can only provide capsule summaries of this history for each of the various continents, but many of their salient features have been discussed as examples of tectonic processes in earlier chapters. The final section analyzes the breakup of Pangea as part of the latest cycle of accretion and dispersal of supercontinents. Because it involves continuation of this cycle into the future, it is necessarily very speculative.

## Histories of Individual Continents

Figure 10.1 shows approximate patterns of movement of each continent from its position in Pangea to the present. The dominant feature of this pattern is northward movement of all continents except Antarctica, which has remained over the South Pole for more than 250 million years. Shortly after geologists recognized the concept of continental drift, this movement was referred to by the German word "Polflucht" (flight from the pole) because all of the continents were seen to be fleeing from the South Pole. The only continent that did not simply move northward was Eurasia, which essentially rotated clockwise and changed its orientation from north–south to east–west. Comparison of fig. 10.1 with fig. 8.12a (locations of continents shortly before the assembly of Gondwana) shows that the net effect of the last 580 million years of earth history has been a transfer of most continental crust from the southern hemisphere to the northern hemisphere.

### Eurasia

Accretion and compression against the southern margin of Eurasia constructed a series of mountain belts from the Pyrenees in the west to the numerous ranges of Southeast Asia in the east. This collision generated extensional and transtensional forces that opened rifts and pull-apart basins. Tectonic loading created foreland basins with sediment thicknesses of several kilometers. Opposite the area where the collision of India caused the most intense compression, the extensional basins are interspersed with mountain ranges that were lifted up intracontinentally. We divide the discussion of Eurasia into a section where compression dominates to the south (present orientation) of the former margin of Pangea and a section that describes processes within the landmass to the north.

*The Collision Zone* Compression in southern Europe produced the Alps and related mountain belts farther east (chapter 2; fig. 10.2). The orogenic belts contain passive-margin sediments deposited on both the African and European margins plus a broad range of supracrustal rocks formed in the Tethyan Ocean that separated the two continental margins. These Tethyan rocks include ophiolites, various rocks of island arcs, and thick sequences of turbidites and other flysch. They were all thrust in a general northerly direction as the European continental lithosphere descended beneath Tethyan lithosphere and, ultimately, beneath Africa or continental fragments north of Africa (Frey et al., 1999; Schmid and Kissling, 2000).

Most of the compression began in the Cretaceous and continues to the present. The complex pattern of tectonism was largely caused by a region known as Apulia (Adria) that was either a promontory of North Africa or a separate microcontinent (fig. 10.2). The Apennines, Alps, and Dinarides formed

148  Continents and Supercontinents

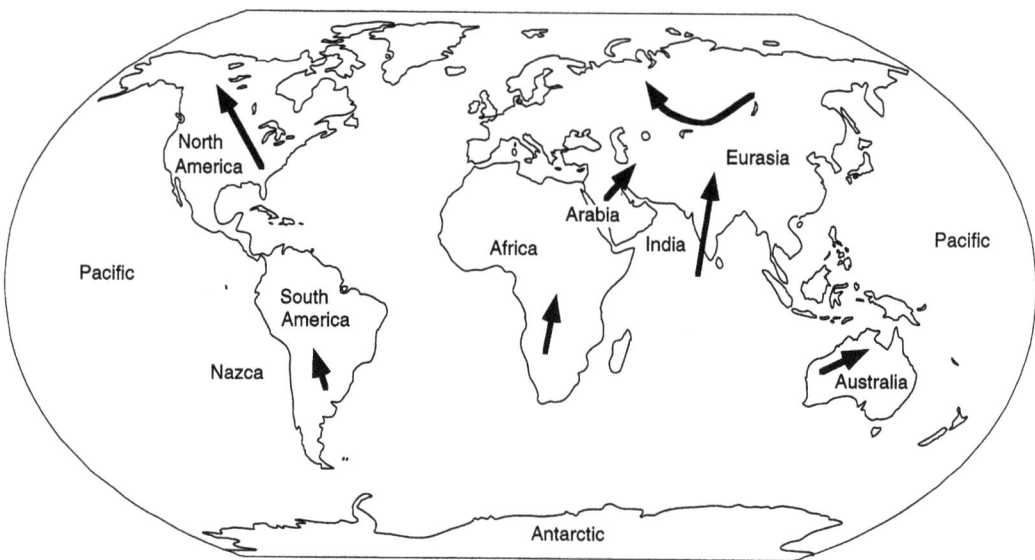

Figure 10.1. Movement of continents from Pangea to the present. Arrows show general directions relative to other fragments of Pangea, and lengths of arrows indicate the relative amounts of movement.

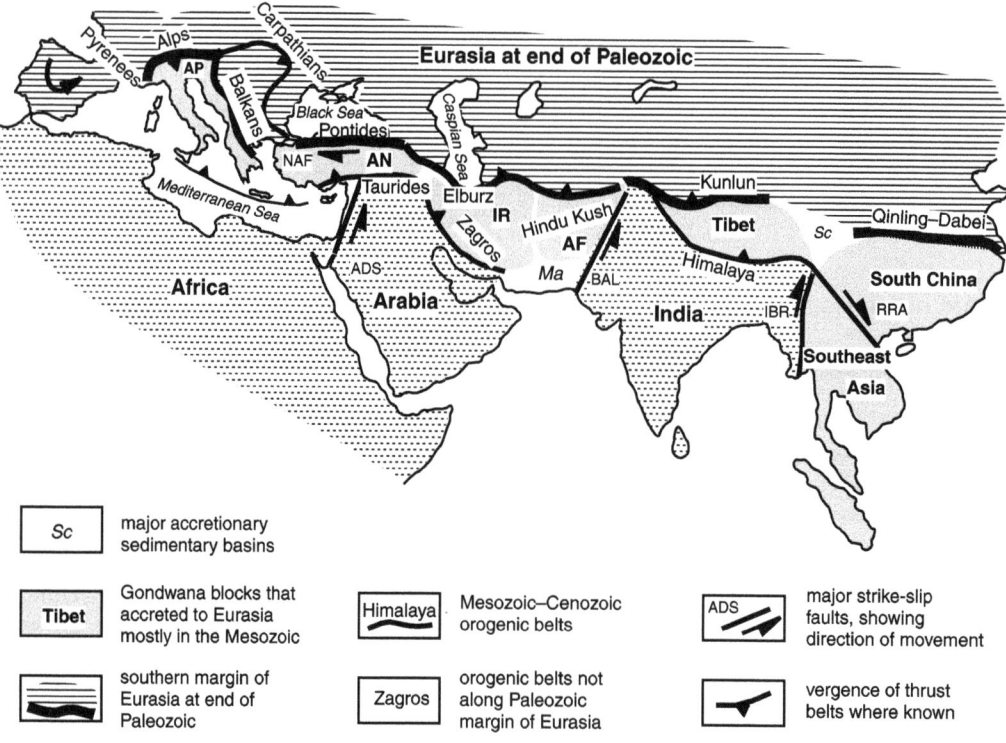

Figure 10.2. Compression along the southern margin of Eurasia. The Iberian peninsula is shown rotating counterclockwise. Abbreviations are: AP, Apulia plate; NAF, North Anatolian fault zone; AN, Anatolia; IR, Iran; AF, Afghanistan; Sc, Southwest China flysch basin; ADS, Aqaba–Dead Sea fault zone; Ma, Makran accretionary prism; BAL, Baluchistan fault zone; IBR, Indo-Burman ranges; RRA, Red River–Ailongshan fault zone. Three mountain ranges are discussed in more detail elsewhere: Alps, chapter 2; Zagros, chapter 5; Himalaya, chapter 5.

around the margins of Apulia, and the Carpathians are separated from that margin by an unknown amount of extension in the Pannonian basin (fig. 9.1; Peresson and Decker, 1997).

Orogeny along the southern margin of Europe seems to have been nearly continuous since the Late Mesozoic and is commonly regarded as a single "Alpine" event. Farther east in Asia, however, geologists commonly separate the waves of collision into two separate events. The older one resulted from closure of Paleotethys (chapter 9) in the Middle to Late Mesozoic and is referred to as "Cimmerian" (Sengor, 1985, 1987; Sengor et al., 1986, 1988, 1993). Orogenic belts caused by the closure of Neotethys developed mostly in the Cenozoic and are referred to by a variety of local names.

Cimmerian blocks include: Anatolia, bordered on the north by the Pontides (Adiyaman et al., 2001); Iran, bordered by the Elburz Mountains (Besse et al., 1998; Stampfli, 2000); Afghanistan, dominated by the Hindu Kush; two terranes that now form Tibet; and South China, separated from North China by the Qinling–Dabei orogen (Meng and Zhang, 2000; Yang and Besse, 2001; Sun et al., 2000). North China is not shown on fig. 10.2 as a Cimmerian terrane, and the margin of Pangea is drawn along its southern border even though the accretion of North China to Siberia along the Mongol–Okhotsk orogenic belt was only partly completed by the end of the Permian (chapter 8). Based on their content of Gondwana fossils and paleomagnetic evidence of their positions before collision, all of the Cimmerian blocks are clearly from Gondwana (Metcalfe, 1996).

The process of Cimmerian accretion is controversial. One possibility is that all of the Cimmerian terranes now in Asia were once part of a microcontinent that formed the northern border of Gondwana and moved across Paleotethys to collide with Asia as a single block. If this happened, the present isolation of these terranes resulted from disruption by strike-slip faulting and other processes after collision. An alternative explanation is that the terranes collided as separate blocks and later underwent further disruption. This problem is virtually identical to the question about collision of Avalonian–Cadomian terranes with North America and Europe in the Paleozoic, which we discuss in chapter 5.

The blocks that form the basement of southeastern Asia are also clearly from Gondwana (Schwartz et al., 1995; Li et al., 1996; Metcalfe, 1996; Besse et al., 1998; Lan et al., 2001; Morley, 2002). Part of their accretion to Asia occurred in the Mesozoic, but at least two post-Mesozoic processes have made it extremely difficult to decipher their early history. One was the collision of India with Asia, which pushed terranes already attached to Asia away from the collision zone along a series of strike-slip faults with displacement of hundreds of kilometers. In southeastern Asia, movements along these and similar faults disrupted original terranes and juxtaposed formerly separate terranes in an enormously complicated pattern (X. Wang et al., 2000). The geology of Southeast Asia has been further complicated by subduction of Indian Ocean lithosphere that began in the Mesozoic and still continues (Hamilton, 1979; Katili and Reinemund, 1982). This caused further deformation and movement of older blocks and created new crust by injection of juvenile magmas throughout the area.

Collision of Arabia and India completed the assembly of Asia. Arabia began to move northward when the Red Sea and Gulf of Aden opened at about 30 Ma (see above), and the pressure increased when the Aqaba–Dead Sea transform was initiated at ~18 Ma. This collision formed the Tauride Range along the southern margin of the Cimmerian Anatolian block as it closed the Mediterranean from the Indian Ocean and thus completed the destruction of Tethys (see above). Similarly, rotation of Arabia into Iran created the Zagros Mountains, which we discuss more completely in chapter 5. Collision of Arabia with southern Asia was partly responsible for development of the North Anatolian fault zone, one of the world's largest dextral strike-slip faults.

Collision of continental India with Asia occurred after island arcs and Tethyan oceanic lithosphere had been descending northward during most of the Cenozoic. The Himalayas started to rise at ~25 Ma when Indian continental lithosphere began to descend (chapter 5), and stresses were distributed throughout all of southern Asia. Much of this stress resulted because India penetrated deeply into the Asiatic mainland along zones of strike-slip faults both east and west of India. To the west, movement of northwestern India in the Baluchistan region created the enormously complicated "Pamir syntaxis," from which several mountain ranges radiate. Similar northward movement along the Indo-Burman ranges formed the "Assam syntaxis," generating both mountain ranges and strike-slip faults that disrupted much of eastern Asia (see above). Spreading in the Indian Ocean continues to push India northward, resulting in uplift of southern India to expose the extensive granulite terrane throughout the area (chapter 3).

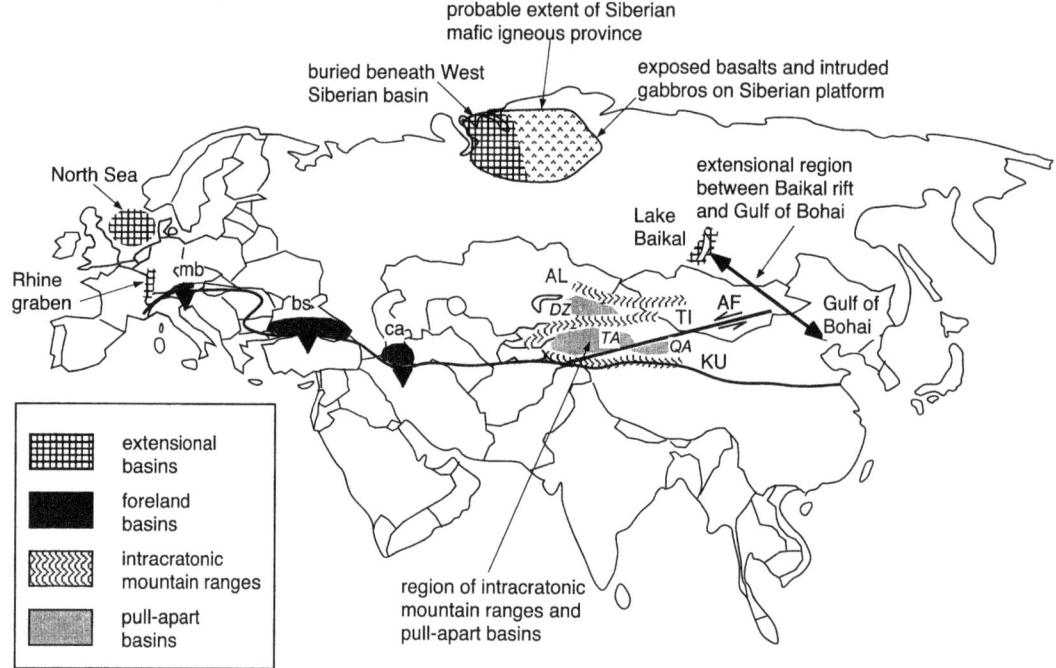

Figure 10.3. Extensional features in Eurasia. Abbreviations are: mb, Molasse Basin; bs, Black Sea; ca, Caspian Sea; AL, Altai Range; DZ, Dzunggar Basin; TA, Tarim; TI, Tien Shan Range; KU, Kunlun Range; QA, Qaidam Basin; AF, Altyn Tagh fault.

Eurasia North of the Collision Zone  Collision of India and other blocks on the southern margin of Eurasia distributed stresses throughout the entire landmass, forming foreland basins, extensional basins, and a region of intracratonic mountain ranges and pull-apart basins. They are shown in fig. 10.3, and we discuss each type of deformation briefly in this section.

Foreland basins develop where the weight of a thickened orogen causes subsidence of an adjacent block (chapters 1 and 2). The concept was originally developed from studies in the Alps, which yielded the terms "flysch" and "molasse" for sediments deposited in different tectonic environments. Flysch included sediments deposited within an orogen undergoing active deformation, and molasse referred to sediments that were deposited either after active deformation had stopped or in areas outside the zone of deformation. The Alpine origin of the term is shown by designation of the foreland basin north of the Alps as the "Molasse Basin," which accumulated as much as 10 km of sediment in the Late Mesozoic and Early Cenozoic (chapter 2; Homewood et al., 1986).

Figure 10.3 shows two other foreland basins. The Pontide orogen on the northern edge of the Anatolian plate caused subsidence of the European foreland to create the Black Sea. This depression caused the floor of the Black Sea to tilt southward, forming a shallow shelf on the north and a deep-water basin on the south (Meredith and Egan, 2002). Similar depression of the southern Caspian Sea by the Elburz Range caused accumulation of a thick sequence of sediment in the southern part of the sea, and the depression may be so severe that Caspian lithosphere is subducting beneath the Iranian region (Axen et al., 2001).

East–west extension in many areas north of the collision zone was caused by north–south compression along the southern margin of Eurasia In Europe, the Rhine graben extends northward almost perpendicular to the Alps (Laubscher, 2001; Schumacher, 2002). A much broader region of extension began to form in the North Sea and thinned the crust sufficiently that the area is now submerged. Extensional development of the Pannonian region is within the collision zone and is mentioned above.

Extension of the very large West Siberian basin began shortly after development of the Siberian basaltic province at 250 Ma (Reichow et al., 2002). Sediments accumulated most rapidly during active extension in the Mesozoic and then more slowly during the entire Cenozoic (Aleinikov et al., 1980; Rudkevich and Maksimov, 1988). The duration of activity in the basin suggests that it was affected not only by collision along the southern margin of Eurasia but also by subduction of Tethyan lithosphere at times when continental collision was not occurring.

Broad upwarp and depression occurred at several places in Asia in the Cenozoic. Upwarp formed a dome with ~300 m of uplift in Mongolia at the same time as the development of Lake Baikal (Cunningham, 2001). The Aral Sea and Lake Balkash are very shallow bodies of water that apparently formed in slightly downwarped basins.

Lake Baikal is the world's deepest lake and contains the largest volume of freshwater. It lies in a volcanically active rift zone with a NE–SW orientation that was presumably generated by northward-directed compression from the colliding Indian plate during the Cenozoic (Delvaux et al., 1997; Ionov, 2002). Lake Baikal is on the western edge of an extensional region that has a width of more than 1000 km from central Siberia to northeastern China (Liu et al., 2001; Ren et al., 2002). This area makes it even larger than the comparable Basin and Range province of western North America (see below).

When India collided with central Asia it encountered a region that had been assembled largely during the Late Paleozoic (chapter 8; Heubeck, 2001). This collage of microcontinents, arcs, and oceanic suites was intensely reactivated by renewed compression in the Cenozoic. Areas that were uplifted ("rejuvenated") during this compression became mountain ranges, locally raising Paleozoic rocks that had been eroded almost to sealevel to form ranges such as Altay and Tien Shan, which did not begin to rise until the Late Cenozoic. The southernmost of the rejuvenated ranges is Kunlun which formed the southern margin of Eurasia before accretion of the Cimmerian blocks that constitute Tibet (Jolivet et al., 2001).

The Dzunggar, Tarim, Qaidam, and smaller basins developed among the rejuvenated ranges (Vincent and Allen, 2001), with their depression caused partly by loading of adjacent mountain ranges but probably more significantly by pull-apart movements along fault zones. Figure 10.3 shows only the largest of these fault zones as an extension of the Altyn Tagh fault, which forms the northern margin of Tibet and distributes left-lateral movement through much of eastern Asia (Yin et al., 2002). The pull-apart motions created basins that are locally very deep, with the Turfan depression being more than 100 km below sealevel.

## North America

Rifting of Pangea left North America with passive margins on three sides of the continent and an active margin only on the west. The rate of convergence between North America and the Pacific increased when the North Atlantic opened and westward subduction beneath colliding island arcs (chapter 8) changed to eastward subduction of Pacific lithosphere beneath continental North America. The interaction of sequences of plates in the Pacific, plus the steady northward movement of the Pacific relative to North America, created a complex history of the western part of the continent that we can summarize only briefly here (fig. 10.4a and b). Ward (1995) provides a general summary of the history of western North America since the beginning of subduction.

Infilling of the broad embayment in the Precambrian margin of North America began in the Paleozoic (chapter 8) and was probably completed by the middle of the Mesozoic. The largely unknown mix of exotic terranes and continental-margin suites that accreted in this region, plus rotation of some terranes southeastward toward the Gulf of Mexico, produced a western continental margin that was an approximate straight line by some time in the Jurassic. At this time, the eastern Pacific consisted solely of the Farallon plate (fig. 9.7), which subducted beneath North America at a rate equal to the combined spreading velocities of the East Pacific Rise and mid-Atlantic ridge. This subduction generated batholithic suites along almost the entire western margin of North America from the Jurassic until the Early Cenozoic.

While batholiths were intruded through continental crust, North America built outward by accumulation of a collage of terranes west of the zone of intrusion. This collage contains an enormous variety of suites, many of them with controversial origins. Some may be exotic fragments from unknown continents. Others appear to be island arcs that formed offshore of North America before collision. A few have been proposed to be oceanic plateaus that had densities too low for them to be subducted under the continental margin. All of these terranes are interspersed through a mixture of continental-margin sediments, ophiolitic melanges, and marine sediments such as chert and deposits of marine fans.

Intrusion and outbuilding were greatly affected by changes in configuration of plates in the Pacific Ocean (chapter 9; Atwater, 1989; Atwater and Stock, 1998). During the Late Cretaceous the northern margin of the Farallon plate was a spreading center on the southern edge of the Kula plate,

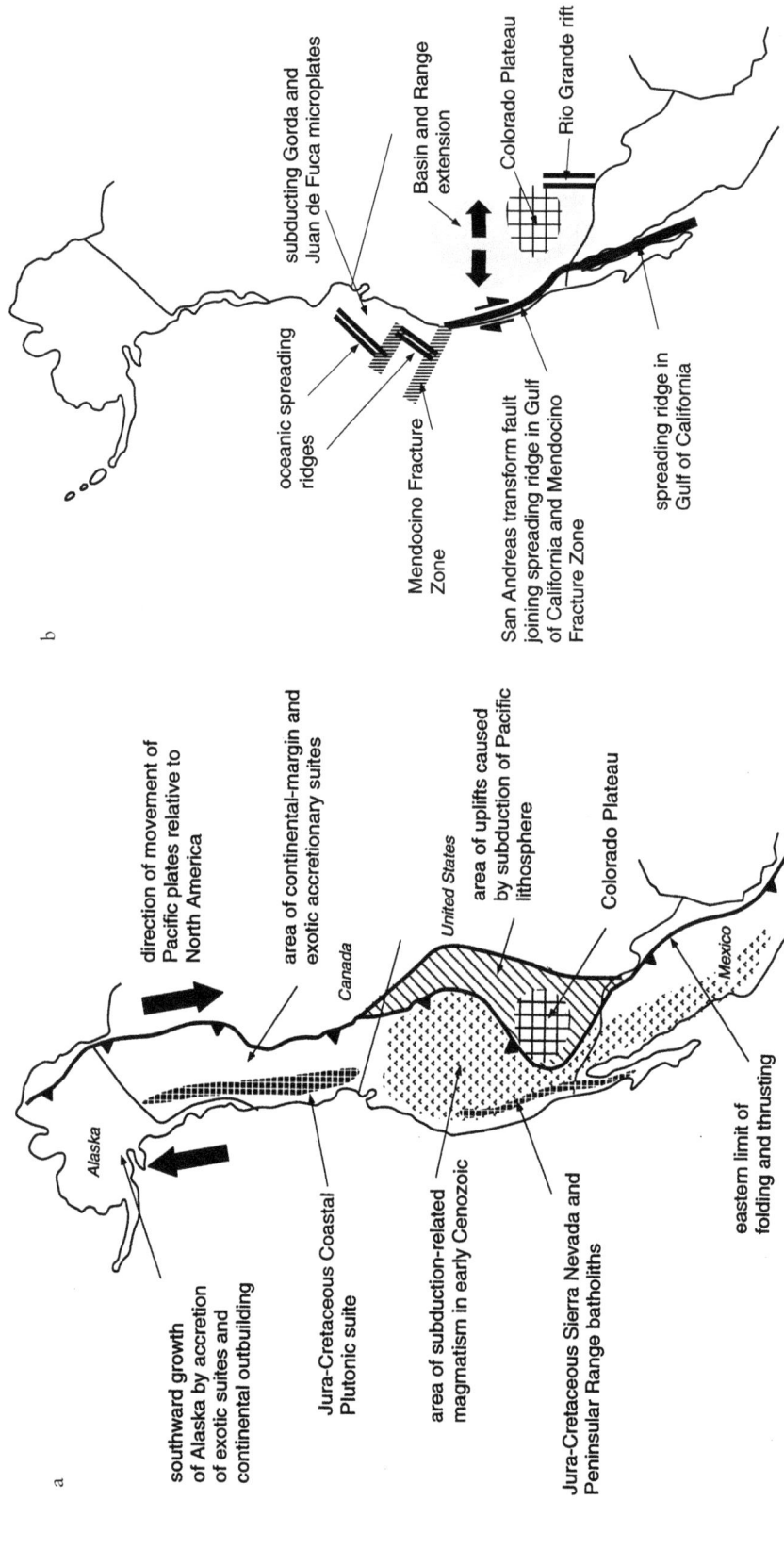

Figure 10.4. Mesozoic–Cenozoic tectonics of western North America. a. Compressional features in western North America before collision with the East Pacific Rise. b. Extensional features in western North America after collision with the East Pacific rise.

which directed the Kula plate almost directly northward under the margin of Alaska and the Aleutians. This movement caused southward growth of Alaska as oceanic materials subducted under the advancing continental margin, and subduction beneath areas of thin continental crust or oceanic crust developed the Aleutian island arc. This growth continued until ~40 Ma, when the Kula plate was completely destroyed and the Pacific plate changed its direction of movement (chapter 9).

The Farallon plate underwent further fragmentation as North America moved farther westward, and parts of the Farallon–Pacific ridge intersected the margin of North America. This process has now left only a remnant of the Farallon plate (Gorda and Juan de Fuca plates) subducting to form the Cascade Range in Oregon and Washington and another remnant referred to as the Cocos plate to cause continuing volcanism in Mexico and Central America. The Gorda and Juan de Fuca plates are terminated southward by the Mendocino fracture zone, which extends to the northern end of the San Andreas fault. The San Andreas fault lies within the broad zone of right-lateral displacement between the Pacific and North American plates and shows more than 100 km of dextral movement in the Late Cenozoic. The southern end of the San Andreas connects with the part of the East Pacific Rise that extends into the Sea of Cortez and forms the eastern margin of the Cocos plate.

Mesozoic and Cenozoic activity affected North America up to 1000 km east of its Pacific margin. During the Cretaceous, construction of the Sevier thrust belt produced a load that created the Western Interior Seaway (chapter 9). In the Late Cretaceous and Early Cenozoic (Laramide time), areas to the east of the Sevier orogen were uplifted along faults of uncertain origin to form a series of blocks that now constitute much of the southern Rocky Mountains. Orogeny during Laramide time north of these uplifts developed more conventional orogenic features, moving thrust sheets hundreds of kilometers eastward over "miogeoclinal" rocks to form the Rocky Mountains in the northern United States and Canada.

Compression changed largely to transtension along the central (U.S.) part of the continental margin as the continent began to override the East Pacific Rise. The disappearance of the rise and Farallon plate in this area occurred at about the same time as the Basin and Range extensional province began to open (chapter 2). This series of N–S-trending fault basins and intervening ranges has stretched to a width nearly three times as much as the original area, partly resulting in the westward bulge of North America along the U.S. coastline. The Rio Grande rift began to form at ~30 Ma as a result of dextral rotation between the Colorado plateau and continental crust to the east (Wawrzyniec et al., 2002).

Both compressive and extensional activity in western North America largely bypassed the Colorado plateau (Selverstone et al., 1999; D. Smith, 2000). The plateau consists of Precambrian basement overlain by flat-lying sedimentary rocks that range from Late Proterozoic to the present. Older suites are mostly marine and younger ones almost completely terrestrial, showing a long-term emergence of the plateau. The almost complete absence of rifting or compression throughout the interior of the plateau even while it was surrounded by intense deformation make this area one of the world's most enigmatic features, with no present explanation for its ability to remain a bulwark against surrounding activity.

## South America

The Precambrian crust of South America extends to the western coastline of most of the continent (chapter 8; fig. 10.5). The South Atlantic Ocean began to open at ~110 Ma (see above), and the increased rate of subduction of Pacific oceanic lithosphere beneath the continental margin began to construct the modern Andes. Construction resulting from subduction of the Farallon plate continued without interruption when the Farallon plate broke up, and now the Andes lie wholly above the Nazca plate. We provide further information about the Andes in chapter 5, where they are used as the type example of orogenic activity on continental margins not accompanied by collision of exotic terranes.

The only places along the western margin of South America not underlain by Precambrian crust are the northwestern corner and the southern tip of the continent. The northwest has a basement of mostly Cenozoic arc and continental-margin suites and is dominated by the northeasterly subduction of the Cocos plate and by dextral strike-slip faults that form the margin between the South American and Caribbean plates. Modern volcanism is intense in the southern part of this area but nonexistent to the north, where the orogenic trends swing toward an E–W orientation along the southern margin of the Caribbean.

The southern tip of South America is underlain by a suite of Paleozoic and Early Mesozoic arcs and accretionary wedges. This terrane was apparently

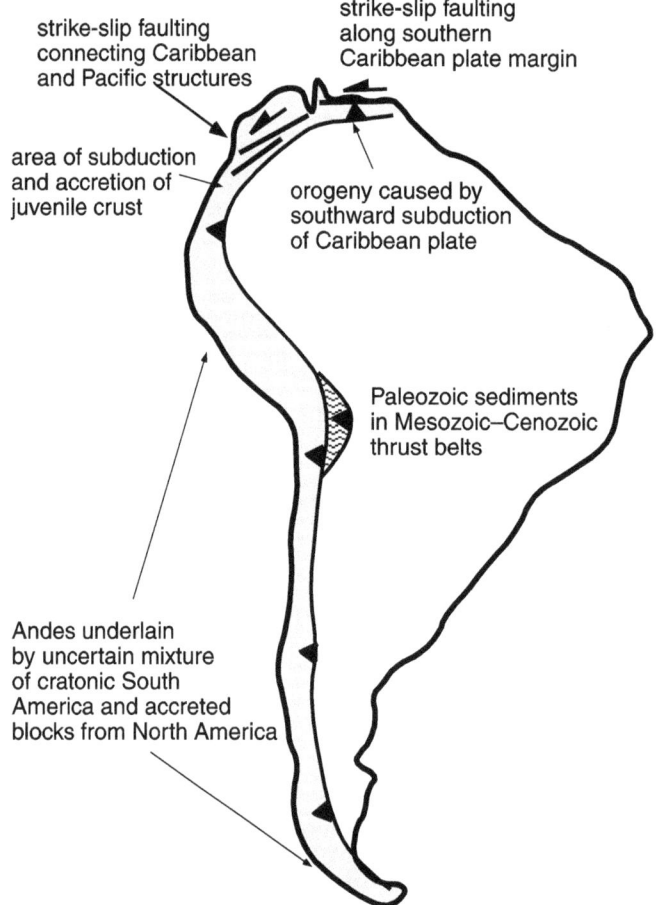

Figure 10.5. Major Mesozoic–Cenozoic tectonic features of South America. The border between South America and the Caribbean is also discussed in chapter 9. The Andes are discussed in chapter 5.

continuous into similar suites in the Antarctic Peninsula before rifting opened the last land barrier to creation of the Antarctic Ocean (chapter 9). This rifting terminated subduction in southernmost South America and led to development of the recent subduction zone beneath the Scotia arc.

## Africa

Rifting of Pangea left Africa with passive margins on all sides except the Mediterranean (chapter 9). Relaxation of horizontal stress started in Africa as early as the Late Paleozoic, when Karoo rift basins began to accumulate sediments commonly referred to as "Gondwana" sequences (chapter 8; fig. 10.6). Progressive detachment from the rest of Pangea then produced a series of rifts of different ages and a variety of orientations (Lambiase, 1989).

Most of these rifts show very little relationship to older structures, reinforcing observations (see below) that rifting of supercontinents is not controlled by pre-existing sutures or other structures.

The most recent rifts include the "Great Rift Valley," of East Africa, where investigators such as Gregory (1921) established many of the concepts of continental rifting. These youngest rifts are the failed arm (chapter 2) of the triple junction where the Red Sea and Gulf of Aden began to open at ~30 Ma (Ebinger and Casey, 2001; Menzies et al., 2001). The East African rift valleys separate into two branches around the Lake Victoria stable region, with the eastern branch regarded as "wet" because much of it is filled with volcanic rocks, whereas the western branch is "dry" because the only volcanism associated with it produces a small amount of highly alkaline lava (Ebinger and Sleep, 1998; Macdonald et al., 2001).

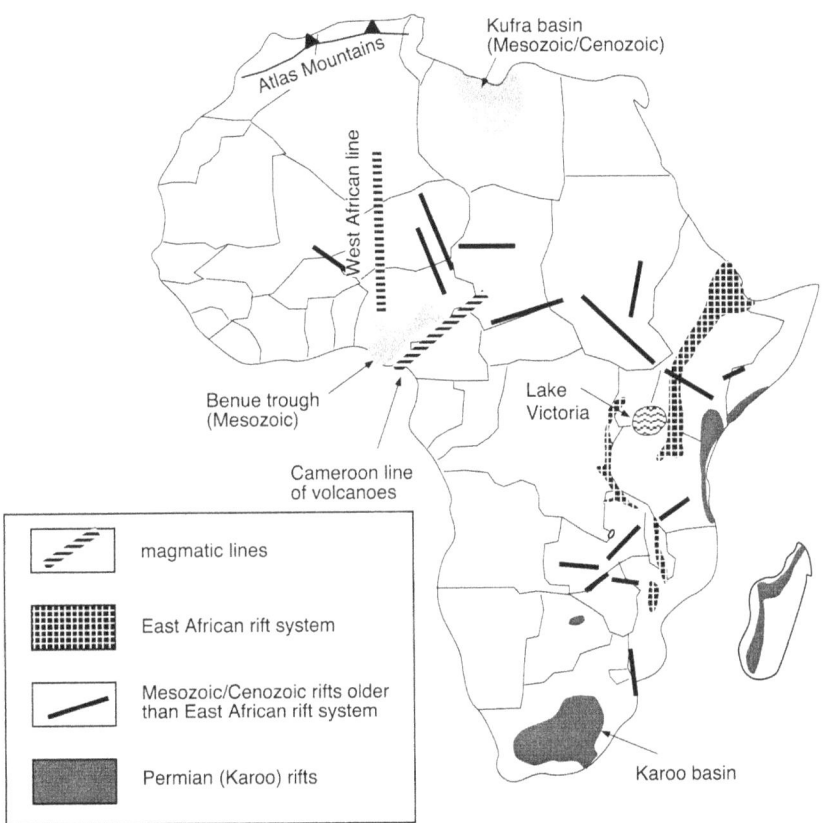

Figure 10.6. Major features of Africa in the Permian, Mesozoic, and Cenozoic. Major basins of subsidence are named, and small linear rifts older than the present East African rift system are shown as lines. Lambiase (1990) discusses the sequence of rifting in more detail. Hotspot tracks are shown through West Africa and along the Cameroon line. The Atlas Mountains formed by compression on the southern margin of the Mediterranean.

Another major feature of Africa in the Mesozoic and Cenozoic is widespread distribution of anorogenic magmatism that appears not to be associated with rift valleys. At least two zones may be hotspot traces associated with opening of the Atlantic Ocean, but many of the more recent areas have been nearly stationary through most of the Cenozoic. One proposed hotspot track forms a line of alkaline magmatic rocks that extends approximately N–S across West Africa before drifting moved Africa away from the plume that may have produced it (Meyers et al., 1998). The track is directed approximately toward the island of St. Helena, near the mid-Atlantic ridge, but the island is not now active, and no clear hotspot trace extends from it across the Atlantic Ocean. A second track may be the Cameroon line, which is a set of alkaline volcanic suites that extends partly offshore in the direction of Ascension Island. The only current volcanic activity along this zone is within Africa, particularly including the Nyos volcano.

A hotspot track has also been proposed to extend from the Tristan da Cunha island group toward the African mainland along the Walvis ridge, but there is no indication of an onshore extension. Tristan da Cunha is regarded as the present site of a plume that developed during the opening of the South Atlantic and created the Parana igneous province of South America and Entendeka province of Africa, which we discuss more in chapter 9.

Recent areas of concentrated volcanism that show very little movement are numerous throughout most of Africa except those areas that were fully

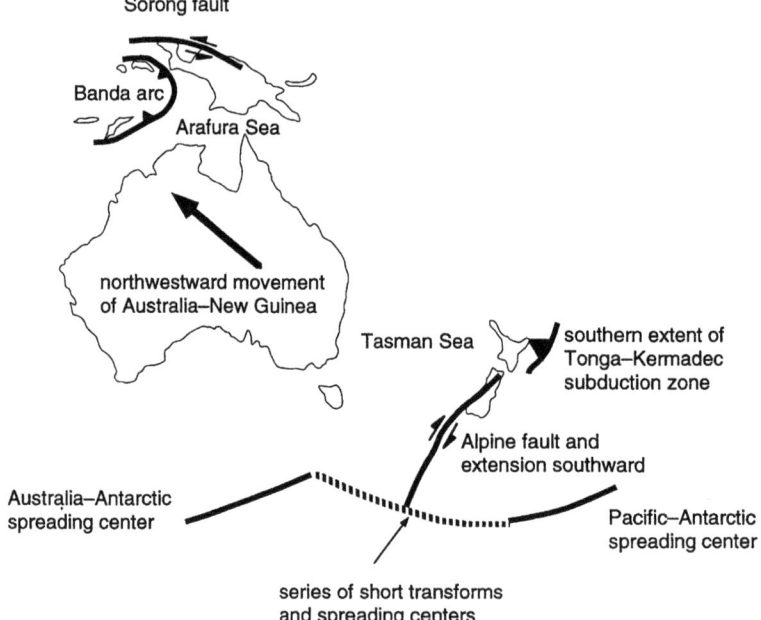

Figure 10.7. Major Mesozoic and Cenozoic tectonic features of Australia, New Guinea, New Zealand, and adjacent seas. Australia and New Guinea were a single joined block during this time. Subduction zones are shown by conventional symbols.

stabilized prior to Pan-African time (chapter 8; Ritsema and van Heijst, 2000). They erupted a variety of rocks that include large volcanic plateaus, local centers of alkalic suites, and mixtures of both types of rocks. This typical anorogenic magmatism probably results from some combination of the lack of compressive stress and a thick lithosphere that appears to characterize much of the continent.

### Australia–New Guinea–New Zealand

Australia and New Guinea have been part of a single plate floored by continental crust throughout the Phanerozoic and possibly earlier (fig. 10.7). The rifting that disrupted Pangea left Australia with passive margins on the west and south, and only New Guinea has undergone compressive deformation since that time. Subduction beneath the New Guinea margin began in the Mesozoic and constructed the island as a melange of volcanic and intrusive suites, continental-margin sediments, obducted ophiolites, and oceanic plateaus (Hill and Raza, 1999).

This construction was terminated in the Middle Cenozoic when the Bismarck and other basins opened and new subduction zones were created (Hall and Spakman, 2002). Counterclockwise rotation of Australia–New Guinea began at ~25 Ma (chapter 9), and much of the deformation in New Guinea became transpressional. This movement formed the left-lateral Sorong fault, which extends from New Guinea to structures around the Philippines and is one of the world's longest fault zones.

New Zealand and at least some of the submerged continental plateaus around it apparently rifted away from East Antarctica in the Early Mesozoic. This crust then merged with an unknown amount of submerged crust east of the present margin of Australia to form a broad zone of thin continental crust. Opening of the Tasman Sea in the Late Cretaceous (chapter 9) separated the New Zealand area from Australia and isolated eastern Australia from all geologic activity except intraplate magmatism, which produced the voluminous Newer Basalts.

Both before and after opening of the Tasman Sea, lithosphere of the Pacific plate subducted under the North Island of New Zealand and the Tonga and Kermadec arcs to the north, creating a zone of magmatism that is still active. The subduction zone terminates near the south end of the North Island, and is replaced southward by a very long zone of

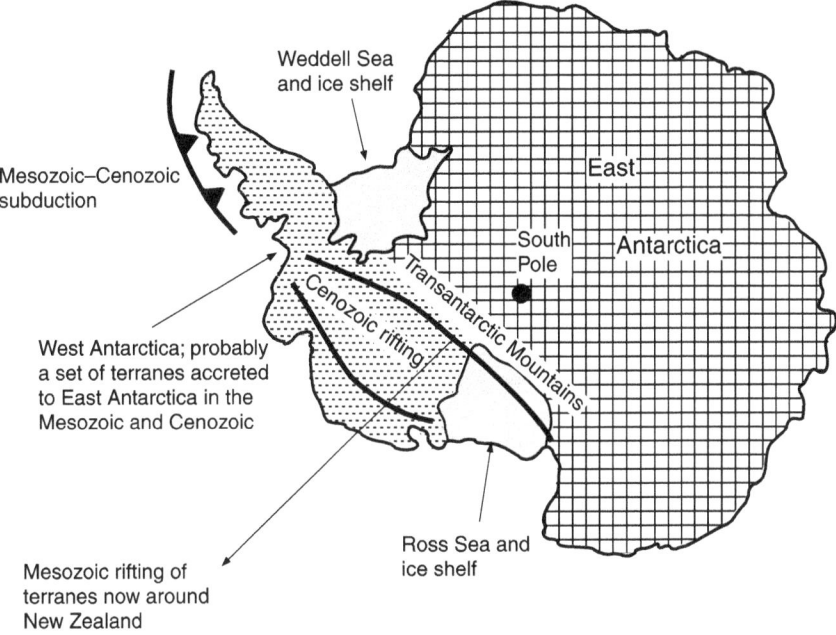

Figure 10.8. Major Mesozoic and Cenozoic tectonic features of Antarctica

dextral strike-slip faulting (Walcott, 1998). The Alpine fault of the South Island extends through the southern part of the Pacific Ocean until it meets the series of ridges and transforms that constitutes the Pacific–Antarctic plate boundary.

## Antarctica

Rifting of continental crust from the continental margin along the Transantarctic Mountains left a long zone of contact between oceanic lithosphere and East Antarctica (see above). Part of this margin was soon occupied by accreting terranes as West Antarctica began to assemble at some time in the Mesozoic (fig. 10.8; Dalziel et al., 2000a). The history of this area is poorly known because most of West Antarctica is buried by ice, but the exposed part in the Antarctic Peninsula is a magmatic arc that has been active since the Mesozoic as a series of plates in the Pacific Ocean subducted beneath it both before and after separation from South America (Millar et al., 2001).

The assembled terranes of West Antarctica were subjected to extensive rifting beginning in the Middle Cenozoic (J. Anderson, 1999; Behrendt, 1999; Woerner, 1999). The rift zone appears to be parallel to the Transantarctic Mountains, with crustal thinning sufficient to depress the surface of the rift below sealevel. This has caused open water to occupy the present Ross Sea, and at times when the West Antarctic ice sheet has disappeared, oceanic sediments as young as mid-Pleistocene were deposited throughout the rift.

## Cycle of Assembly and Dispersal of Supercontinents

Pangea began to disperse from its maximum packing about 200 million years ago. This fragmentation continues as the oceans that caused it still expand, with the six major continents produced by Pangea moving farther apart even as rifting of small fragments and accretion occur in local areas. This pattern may be similar to ones followed by older supercontinents, and we compare the assembly and breakup of supercontinents in this section. This section is highly conjectural, and we keep it very brief.

On the basis of the ages of maximum packing of Columbia, Rodinia (Palaeopangaea), and Pangea, the cycle of accretion and dispersal of supercontinents appears to be about 750 million years long. This period is divided into about 500 million years of accretion to maximum packing followed by about 250 million years of dispersal before continental blocks begin reassembling. If this periodicity is being followed by fragments from Pangea, then they are likely to spend the next 50–100 million years drifting farther apart until the succeeding supercontinent begins to form.

The name "Amasia" has been proposed for the next supercontinent (Hoffman, 1992). The North American and Asiatic plates are already fused along a zone of uncertain nature through eastern Asia (chapter 2), and it is possible that the next supercontinent will develop as more of the present continents collide around this Asia–North America nucleus. Part of this fusion might result from further accretion of Gondwana blocks to southern Asia, perhaps by closure of the Indian Ocean. If South America and Antarctica moved in some pattern that caused them also to accrete to Asia in about 500 million years, then the next supercontinent would reach its maximum packing at the same time in the cycle as previous supercontinents.

In chapter 6 we discuss two models for the formation of supercontinents. One model proposes that supercontinents develop through accretion of numerous small blocks along intercontinental orogenic belts formed by ocean closure. The second model suggests reorganization of large old continental blocks that had remained stable since they were originally formed at ~3 Ga (Ur), ~2.5 Ga (Arctica), and ~2 Ga (Atlantica). The breakup of Pangea mostly into large continental blocks and a few small ones seems to support the second model, which is consistent with the suggestion that the earth's lithosphere has always consisted of a relatively small number of plates (D. Anderson, 2002).

Despite the similarity that all supercontinents fragmented into large blocks, the breakup of Pangea may differ in one respect from that of the older supercontinents Columbia and Rodinia. According to the second model presented above, both Columbia and Rodinia fragmented around the three older continental blocks, thus preserving them for incorporation into Pangea, but only one of the three remained unfractured when Pangea broke up. The core of Arctica (Kenorland) remained intact, but Atlantica was cut into two halves (Africa and South America), and Ur was shattered into more than three fragments.

We close this chapter with two observations. One is that processes now occurring on the earth are fully consistent with operation of a cycle of supercontinent assembly and dispersal throughout earth history. The dispersal of Pangea appears to follow the same time schedule and to form approximately the same numbers of fragments as its predecessors. The second observation is that fragments formed from Pangea generally do not contain the same assemblies of cratons as those formed from Rodinia and Columbia. The reasons for both the similarity and difference are unknown and should constitute a basis for future research.

# 11

# Effects of Continents and Supercontinents on Climate

Continents affect the earth's climate because they modify global wind patterns, control the paths of ocean currents, and absorb less heat than seawater. Throughout earth history the constant movement of continents and the episodic assembly of supercontinents has influenced both global climate and the climates of individual continents. In this chapter we discuss both present climate and the history of climate as far back in the geologic record as we can draw inferences. We concentrate on long-term changes that are affected by continental movements and omit discussion of processes with periodicities less than about 20,000 years. We refer readers to Clark et al. (1999) and Cronin (1999) if they are interested in such short-term processes as El Nino, periodic variations in solar irradiance, and Heinrich events.

The chapter is divided into three sections. The first section describes the processes that control climate on the earth and includes a discussion of possible causes of glaciation that occurred over much of the earth at more than one time in the past. The second section investigates the types of evidence that geologists use to infer past climates. They include specific rock types that can form only under restricted climatic conditions, varieties of individual fossils, diversity of fossil populations, and information that the $^{18}O/^{16}O$ isotopic system can provide about temperatures of formation of ancient sediments.

The third section recounts the history of the earth's climate and relates changes to the growth and movement of continents. This history takes us from the Archean, when climates are virtually unknown, through various stages in the evolution of organic life, and ultimately to the causes of the present glaciation in both the north and the south polar regions.

## Control of Climate

The earth's climate is controlled both by processes that would operate even if continents did not exist and also by the positions and topographies of continents. We begin with the general controls, then discuss the specific effects of continents, and close with a brief discussion of processes that cause glaciation.

### Climate on an Earth Without Continents or Other Topographic Variation

The general climate of the earth is determined by the variation in the amount of sunshine received at different latitudes, by the earth's rotation, and by the amount of arriving solar energy that is retained in the atmosphere.

On an annual basis, the amount of solar radiation reaching the earth's surface decreases steadily from the equator toward both the North and South Poles. This variation causes air at the equator to be warmer, and consequently less dense, than colder air at the poles. The differences in density establish a series of "convection cells," bounded by rising and descending columns of air, that control much of the distribution of moisture in the atmosphere. Figure 11.1 shows the present distribution of convection cells in the earth's atmosphere, although it is not clear that this same pattern of three cells in each hemisphere has been in effect at all times in earth history. Warm air rising at the equator moves north and south until it has lost most of its moisture and cooled sufficiently in the upper atmosphere that it is dense enough to descend. This dry air reaches the earth's surface in the region of 20–30° N and S latitudes, causing most of the world's deserts to lie along these latitudes. After reaching the ground, the air is cool and dense enough to move toward higher latitudes until it encounters colder air moving

160  Continents and Supercontinents

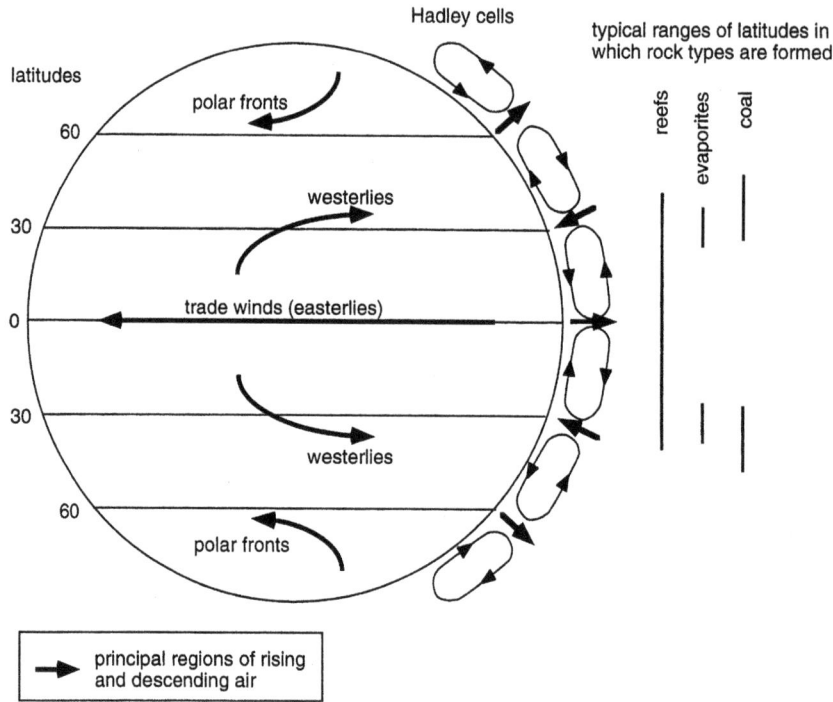

Figure 11.1. Convection cells, wind directions at different latitudes, and the typical latitudes in which different climatically sensitive rock types are deposited.

away from the North and South Poles. This meeting creates another region of rising air, which moves toward the poles until it descends as the coldest air on the earth.

Convection cells would exist even if the earth were not rotating, but the west-to-east rotation complicates the pattern of air movements by a process commonly referred to as the "Coriolis effect." The earth's radius of rotation decreases from the equator to the poles, causing a proportional decrease in the absolute rate of movement from highest at the equator (1511 km/hr) to the poles (where the absolute movement is zero). As the earth rotates, it drags the atmosphere with it at rates only slightly less than the ground underneath it. In equatorial regions, this slippage causes winds to blow from east to west, forming easterlies or "trade winds." As air rises at the equator and moves north or south in convection cells, however, it has a higher velocity than the ground beneath it when it drops to the surface at ~30° latitude. This higher velocity causes the air to move from west to east, forming "westerlies" (fig. 11.1). For similar reasons, cold air moving away from the poles is moving more slowly than the underlying ground and tends to move from east to west.

The heat that powers all of these air movements is available only because solar radiation is trapped in the atmosphere (fig. 11.2). This occurs because all bodies, such as the sun and earth, radiate energy away from themselves at frequencies that are proportional to the temperatures of their surfaces (known as the temperature of "black body" radiation). The black body temperature of the sun is ~6000C, and it produces energy with wavelengths mostly in the range of $10^{-6}$–$10^{-7}$ m, which is in the range that people and other animals can see. The earth has an effective temperature of ~25°C and radiates energy back into space at wavelengths centered around $10^{-5}$ m, known as "thermal infrared." This infrared radiation is much better absorbed by some of the gases in the atmosphere than the incoming radiation, thus retaining heat and keeping the earth's surface warmer than it would be without these gases.

The gases that retain heat are generally known as "greenhouse gases" because they operate in the same way as the glass of a greenhouse, letting solar radiation pass through and then holding some of the outgoing radiation. Nitrogen and oxygen, the two gases that constitute more than 99% of the present atmosphere, do not retain outgoing radia-

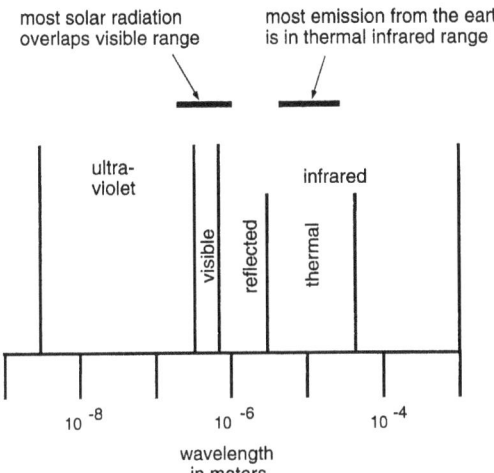

Figure 11.2. Greenhouse effect resulting from different wavelengths of incoming and outgoing radiation on the earth's surface.

tion, and the principal greenhouse gases consist of water, $CO_2$, and $CH_4$. The combination of solar radiation and atmospheric absorption keep the surface temperature of the present earth at ~25C, but it may have been very different in the past. One possible variable is the radiation output of the sun, with many solar astronomers proposing that this solar "irradiance" ("luminosity") has steadily increased from ~75% of its present value at the time the earth coalesced at about 4.5 Ga (Kuhn et al., 1989).

The principal reason for long-term variation in surface temperature probably is changes in the concentrations of greenhouse gases in the atmosphere. In the present atmosphere, $O_2$ reacts with $CH_4$ to form $H_2O$ and $CO_2$ at such a fast rate that the average methane molecule released into the atmosphere remains for fewer than 10 years. Before free oxygen began to accumulate in the atmosphere at ~2 Ga, $CH_4$ may have been a major component of the atmosphere, thus keeping the earth warmer than it is now. Another greenhouse gas that may have been much more abundant in the early earth than it is now is $CO_2$, which is known to be a major component of the atmosphere of Venus. Since 2 Ga, the principal variation in the concentrations of greenhouse gases has presumably been changes in the abundance of $CO_2$.

In addition to the long-term changes in the earth's temperature, short-term cyclical variations result from irregularities in the earth's orbit around the sun (fig. 11.3). These orbital effects are referred to as "Croll–Milankovitch cycles" and have three components. Slight changes in the shape of the earth's orbit, referred to as its "ellipticity," have a major periodicity of ~100,000 years. The angle that the earth's axis makes with its orbital plane (its "obliquity") varies through an angle of ~3° with a periodicity of ~41,000 years, and the rotational axis also moves with an approximately circular motion, forming cycles of "precession" with periodicities of ~23,000 years and ~19,000 years. Further information is provided by Clark et al. (1999) and Cronin (1999).

Orbital variations cause cyclical changes in the amount of solar radiation that reaches the earth. One principal effect is cyclical advance and retreat of glaciers during long periods of glaciation, caus-

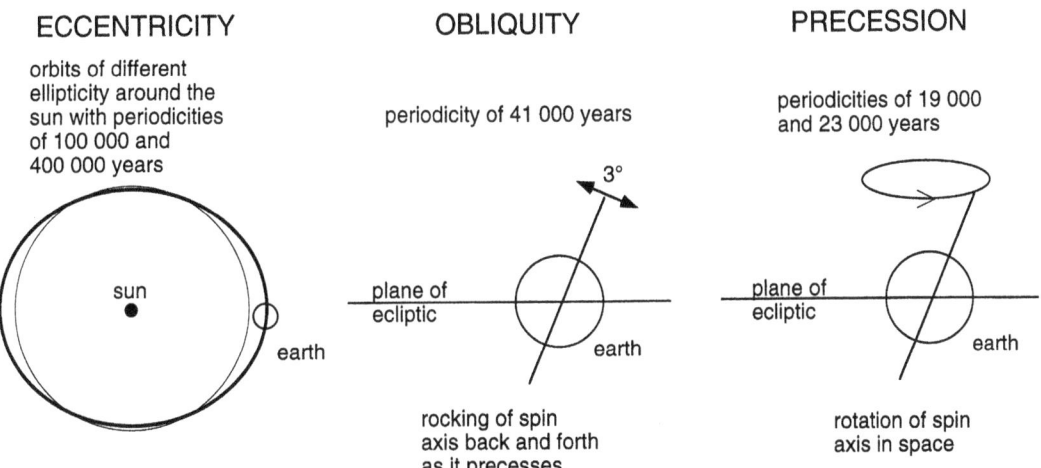

Figure 11.3. Orbital variations.

ing rise and fall of sealevel by as much as a few hundred meters. Even during times not regarded as glacial, cyclical variations in sealevel are generally presumed to have resulted from advance and retreat of small glaciers in elevated regions and small continental platforms where evidence of glaciation has not been found.

## Effects of Continental Location and Topography on Climate Within a Continent

The general orientation of the earth's axis of rotation probably has changed very little, if at all, during the entire history of the earth. Consequently, continents near poles of rotation have always received less solar radiation than continents at lower latitudes, and the position of a continent on the earth presumably has affected the continent's climate throughout geologic history. (We discuss below recent suggestions that some glacial periods occurred during times when the axis of rotation was tilted to be much closer to the plane of the earth's orbit than it is now.)

The presence of continents and their topography affect climates in four principal ways. One is that temperatures at high elevations are lower than at low elevations because the temperature of the atmosphere decreases with increasing altitude. This decrease results from the upward decrease in pressure at each altitude because the pressure is caused by the mass of air above that altitude. This decrease in pressure causes temperature to decline along an adiabatic path referred to as a "lapse rate" (see appendix B for a discussion of similar adiabatic processes in the earth's mantle). Because of the decrease in temperature upward, continents that are lifted up by some tectonic process have lower average temperatures than continents mostly at sealevel.

A second effect of continents is caused by mountain ranges, which deflect winds by forcing air to rise over the mountains into cool regions of the atmosphere. This causes rain to fall on the windward side of the mountains and leaves arid "rain shadows" on the leeward (back) side of the mountains. The Himalayas are so high that monsoonal winds flowing north–south cannot cross them, forming rain forests on the southern side of the mountains and leaving arid areas to the north in central Asia. A second example is the Cascade Mountains of the northwestern United States, which intercept the prevailing westerly winds and produce high rainfalls on their western slopes and semiarid conditions in the plains of Oregon and Washington to the east.

Another way in which continents affect climate is the onshore winds that they generate. The temperatures of ocean surfaces remain fairly constant, varying from near 0°C in polar regions to approximately 25°C near the equator. Continents, however, have surface temperatures that vary seasonally but are generally warmer than ocean surfaces. This difference causes air to rise over continents, and winds replace it by flowing "onshore" from the oceans. Thus, most coastal regions have at least moderate rainfall, whereas the interiors of large continents, such as present Asia, commonly receive only winds that are dry because they left their moisture along the coastline.

The dryness and coolness of large continents causes a condition known as "continentality." The centers of large continental masses commonly have relatively high elevations and are consequently dry and cool. This type of climate is currently found in Eurasia and, as we discuss below, also occurred in the middle of Pangea.

A fourth example of the effect of continents on climate is the production of tropical rain forests where continents are located in the zone of easterly trade winds. These winds form the Amazon rain forest by bringing moisture-laden air from the Atlantic Ocean all the way across equatorial South America until they reach the Andes. Similarly, the Congo rain forest forms where air from the Indian Ocean retains moisture as it flows across elevated regions of East Africa and forms rain mostly where it intersects moist air flowing onshore from the Atlantic Ocean. The rain forests of Southeast Asia develop because moisture-laden winds from the Pacific Ocean produce heavy rainfall where they intersect the numerous, but relatively low, mountains of the peninsula.

## Effects of Continental Location on Global Climate

The distribution of continents on the present earth creates separate sets of currents ("gyres") of marine water in each of the world's major oceans: Atlantic, Pacific, and Indian (fig. 11.4). The initial power for these currents comes from equatorial seawater that is moving westward because of slippage between the solid earth and the water above it, generating currents along the equator that flow continually from east to west (this slippage is the same reason for the more rapidly flowing trade winds described above). This equatorial water is deflected by the

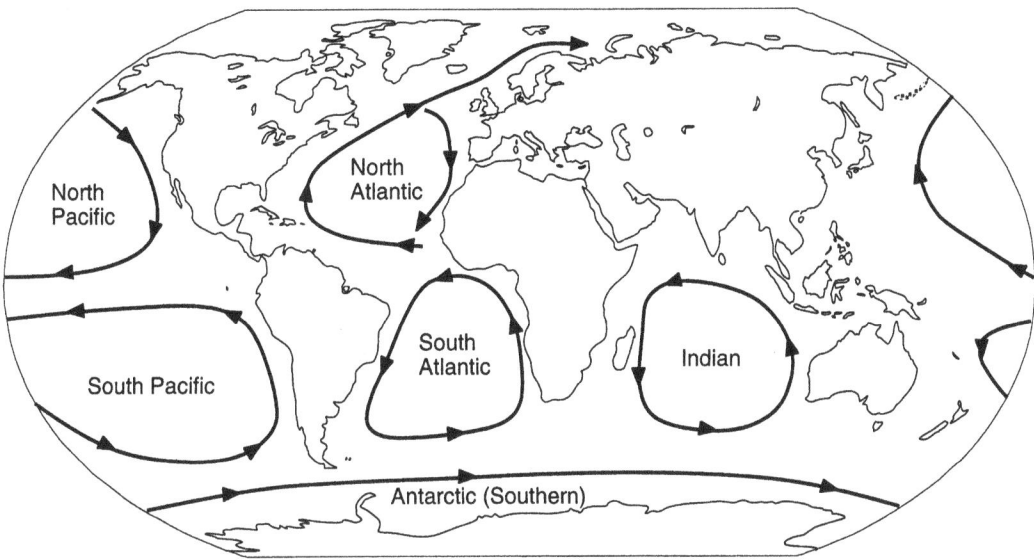

Figure 11.4. Major oceanic circulation on the present earth.

same Coriolis effect that controls wind movements, and in the North Atlantic this process creates a northeasterly flowing current that brings warm water around the north coast of Europe and a southerly return flow along the west coast. A similar current in the South Atlantic brings warm water southward along the east coast of South America, and the return flow along the coast of southwestern Africa carries water so cold that onshore winds produce so little rain that the area is a desert.

Currents similar to those in the Atlantic Ocean also form in both hemispheres of the Pacific Ocean. Both of them bring cold water along the west coasts of North and South America, with the current flowing northward along South America mixing with enough Antarctic water that it is so cold that it forms the same type of deserts as in southwestern Africa. The Indian Ocean is almost wholly in the southern hemisphere and has only the one major convection cell shown in fig. 11.4.

Water in the three major oceans of the present earth now meets only in the Antarctic (Southern) Ocean. This water continually circulates from west to east around the Antarctic continent, thus isolating it from warm water generated in the other oceans. We discuss below that Antarctica became glaciated only when the separation of South America and the Antarctic Peninsula allowed this isolation to occur at 30–25 Ma.

The positions of continents also affect global climate because of their ability to reflect solar radiation back into the atmosphere more effectively than ocean water, which tends to absorb radiation (i.e., land has a higher "albedo" than water). Because of this difference, global temperatures were probably lower in the past when continents were clustered near the equator than when they were broadly distributed or mostly in polar positions. This general cooling may have slightly reduced the high temperatures that land surfaces would have had because of their equatorial position.

## Possible Origins of Glacial Periods

During the past 2 million years, glaciation has clearly resulted from the position of the Antarctic over the South Pole, the proximity of North America and Europe to the North Pole, and the presence of high mountains that permit glaciation even in low latitudes. Ancient periods of glaciation may also have resulted from the presence of high mountains and the polar position of continents, but alternative explanations have recently been proposed. The three most widely known are a high obliquity of the earth in its orbit, a complete covering of the earth by ice when atmospheric $CO_2$ concentrations were very low, and formation of glaciers in orogenic belts and along uplifted rift margins developed during the breakup of supercontinents.

At present the earth's obliquity (see above) is 23.5°, causing the Tropics of Cancer and Capricorn to be located at 23.5° north and south, respectively, and the Arctic and Antarctic Circles at 66.5° north and south. Some investigators have proposed that the obliquity may have been much higher than 23.5° in the past (Williams, 1993). This higher value would have significantly changed the amount of solar radiation that the earth received at different latitudes. At the present low obliquity, the amount of radiation received declines continually from the equator to the poles. At high obliquity, however, the annual average radiation would have been almost constant at all latitudes. If the obliquity had been as high as 90° (axis of rotation in the plane of the orbit), polar regions would each have received six months of intense sunlight, with equatorial regions receiving only weak radiation throughout the year. In this situation, glaciation would have been more likely along the equator than near the poles.

The idea of a "snowball earth" (complete ice cover) was developed by several investigators, originally to explain Neoproterozoic glaciation that preceded the Cambrian development of skeletal organisms (chapter 12; Kirschvink, 1992; Hoffman et al., 1998; Hoffman and Schrag, 2002). The concept arose from presumed observations that continental (low-altitude) glaciation occurred simultaneously in several regions that paleomagnetic data showed were in low latitudes. The process envisions initial development of partial ice cover, which reflects solar radiation back into space (increases the earth's albedo). As less radiation reaches the earth, the ice cover expands eventually to cover the entire earth, both on continents and on ocean surface. The end of total ice cover comes only when volcanic outgassing releases enough $CO_2$ into the atmosphere to warm the earth back to normal temperatures. The cycle of onset of a snowball earth and return to normal climate may last about 15 million years, indicating that periods of glaciation as long as 200 million years in the Paleoproterozoic and Neoproterozoic (see below) required repeated cycles of snowball earth development and disappearance.

During periods of ice cover, normal life on earth should slow down because of the decrease in photosynthesis, and organic carbon should be buried in comparatively "dead" oceans. This possibility is supported by the observation that $\delta^{13}C$ values are high in oceanic sediments deposited at this time because of the removal of organic carbon rich in $^{12}C$ (further discussion in chapter 12). Further support comes from the observation that some glacial sediments are covered by "cap carbonates" that formed when warmer temperatures released an abundance of $CO_2$ into the atmosphere and oceans.

The proposal of a snowball earth has significant support, but many people oppose it. We note here that the proposal is valid only if: 1) paleomagnetic data indicating glaciation at low latitude are valid; 2) the glaciations can be shown to have occurred at low elevations; and 3) the glaciations can be shown to have been nearly synchronous around the world. Thus far, all of these requirements have been strongly challenged, and we discuss them briefly in our section on Neoproterozoic glaciation below.

Control of glaciation largely by tectonics was proposed by Eyles (1993). The concept was based partly on the distribution of 800–600-Ma glacial episodes along the margins of Rodinia (chapter 7). Because many rifted continental margins are uplifted (chapter 2), high elevations along these margins could have created glacial conditions regardless of the latitude where they occurred. Other glaciers may have formed where compressional orogeny created high mountains. We consider these concepts further in our discussion of the history of climate below.

## Methods for Inferring Past Climates

Inferring past climates becomes more difficult as the age being investigated increases (more complete discussion in Cronin, 1999). Some varieties of rocks can be deposited only under specific climatic conditions. Fossils that are similar to modern ones can be assumed to have lived in similar habitats, but ancient fossils that have no living relatives are less easily placed. Because of the scarcity or absence of fossils, Precambrian climates must be inferred from other types of evidence. Concentrations of stable isotopes, particularly $^{18}O$, are partly controlled by temperature, providing an opportunity to infer conditions of formation of ancient rocks.

### Special Types of Rocks

Numerous rock types provide information about the climates under which they were deposited, and here we discuss the four that are most widely used.

Reefs develop where calcareous animals are cemented together by calcareous ("coralline") algae. They form a solid mass that grows upward as either the sealevel rises or the reef subsides. Because all of the organisms live only in seawater and the algae need light for photosynthesis, reefs grow at sealevel and at latitudes that are low enough to provide adequate sunlight (fig. 11.1). The highest-latitude reefs at present are in

Bermuda, at 33°N, but it is possible that reef growth occurred at higher latitudes during times when the whole earth was warmer than it is now.

Reefs similar to present ones did not exist until the Phanerozoic because of the absence of skeletal organisms. Similar types of structures, known as "stromatolites," have been created by cyanobacteria since the Early Archean, but because of the ability of primitive bacteria to live in a wide range of temperatures they cannot be used as climate indicators (chapter 12).

Coal forms wherever trees, leaves, and other vegetative organic matter accumulate in thick masses that are later buried, compressed, and heated (L. Thomas, 2002). The compression and heating drive volatiles away from the organic matter and leave a carbon-rich residue that constitutes one of the several varieties of coal. Although this accumulation can occur at many geographic locations and over a broad range of latitudes (fig. 11.1), most of the world's coal resources were formed in coastal swamps during repeated incursions and withdrawals of seawater. Many of these deposits contain the remains of broad-leafed plants that apparently flourished in temperate latitudes, but coal developed over a broader range of latitudes in the Late Paleozoic (see below). Coal did not form until the development of land plants in the Silurian, with the oldest major coal deposits developing in the Late Carboniferous (Pennsylvanian).

Evaporites develop where large quantities of seawater can flow into a basin and continually evaporate (Warren, 1999). This replenishment is required in order to form thick sequences because of the low percentage of dissolved salts in seawater, and it can occur only under special tectonic and climatic conditions. Small deposits in environments such as playa lakes can develop at any latitude, but basinwide deposits generally form only in arid regions near 20–30° latitude (fig. 11.1). These large evaporite suites precipitate both in continually subsiding basins on continental platforms, such as the Permian deposits of west Texas, and in isolated basins formed along orogenic belts, such as the Miocene evaporites in the Mediterranean.

Evaporites similar to modern ones have formed only since the Neoproterozoic (A. Stewart, 1979). The oldest evaporites are exclusively sulfate (anhydrite and gypsum), and no salt deposits older than Cambrian have ever been found (Knauth, 1998). The absence of salt in Precambrian rocks may be the result of the ease of solution and removal of salt from original deposits, but it is also possible that Precambrian oceans were dominated by $Na_2CO_3$ instead of NaCl and did not deposit salt.

Both continental ice sheets and valley glaciers leave numerous types of evidence of their former existence. Much of the present topography of northern North America and Europe was produced by glaciers, including geomorphic features such as moraines and U-shaped valleys and minor effects such as sub-glacial scour marks (striations). Many rocks deposited by glaciers are "tillites" that consist of angular (unrounded) fragments ranging from large boulders to fine clay (fig. 11.5). Tillites occur in moraines and in deposits that are more widely distributed. In addition, glacial deposits include outwash plains, with angular gravel to silt, and varved sediment deposited in periglacial lakes. One rock type that is regarded as particularly associated with glaciers is "dropstone" deposits (fig. 11.6). Dropstones form where ice containing large pieces of glacial debris floats over deep water and drops its load when the ice melts.

Most of the rocks deposited by glaciers are difficult to distinguish from those formed in other environments (Eyles, 1993). Chaotic deposits of moraines are similar to landslide debris or subaqueous debris flows, distal deposits of streams originating in high mountains may look like the gravels of outwash plains, and varves can develop in nonglacial lakes. These similarities make it difficult to recognize ancient periods of glaciation unless the deposits are associated with dropstones or contain boulders and cobbles that show glacial striations or glacial faceting.

Rocks deposited by valley and continental glaciers are broadly similar, but the source of Pleistocene deposits can generally be distinguished by their relationships to modern topography. Ancient periods of glaciation, however, can be recognized only by glacial deposits, and the distinction between valley glaciation and continental glaciation is commonly difficult to make in outcrops of limited extent. This difficulty causes problems in distinguishing whether an ancient glacial episode occurred at low elevations over a broad continental area or only at high elevations in mountains, leading to arguments about the latitude at which glaciation occurred (see below).

## Fossils

The climatic ranges of modern animals and plants are sufficiently well known that they can be used to infer climates in which sediments were deposited up to 100 million years ago (possibly older). Modern organisms now living on some part of a continent can be compared with fossils in underlying sedi-

## D: Glacial deposits

Figure 11.5. Typical glacial tillite (photograph by Nick Eyles). Reproduced from Eyles, N., and Januszczak, N., Interpreting the Neoproterozoic glacial record; the importance of tectonics, in press. American Geophysical Union Monograph. Copyright 2004 American Geophysical Union. Reproduced by permission of American Geophysical Union.

ments, thus yielding a record of climatic changes that may have been caused by movement of the continent across latitudes. A prime example is the use of fossils to demonstrate the general northward movement of North America and Europe during the Cenozoic.

Determination of paleoclimates is more difficult for old fossils that do not have close living relatives. For animals, one general principle is that species diversity of a fossil assemblage is higher in low latitudes than in high latitudes. By contrast, the number of individuals of a single species preserved in a fossil assemblage is commonly higher in high latitudes. Another distinction can commonly be made for shelled organisms, such as bivalves and gastropods, because shells in low latitudes generally have more ornamentation than the simple shells of organisms that live at high latitudes.

## C: Glacially-influenced deposits

Figure 11.6. Typical dropstone (photograph by Nick Eyles). Reproduced from Eyles, N., and Januszczak, N., Interpreting the Neoproterozoic glacial record; the importance of tectonics, in press. American Geophysical Union Monograph. Copyright 2004 American Geophysical Union. Reproduced by permission of American Geophysical Union.

Plants also exhibit differences between different climatic regions. In tropical moist areas, plants show a high species diversity and have broad leaves and large flowers. Arid climates are characterized by plants with narrow leaves, spines, or other features that preserve water within the plant. Plants of high latitudes commonly have small flowers and show low species diversity.

### Stable Isotopes

Several light elements in sediments show variations in isotopic ratios that are partly related to temperature of deposition. The ones most widely used are the $^{18}O/^{16}O$ and $^{13}C/^{12}C$ ratios. Because the $^{13}C/^{12}C$ ratio is highly controlled by the varieties of global organic activity, we discuss it primarily in chapter 12, and here we consider only $^{18}O/^{16}O$ ratios.

Isotopes of all elements fractionate during geologic processes. For heavy isotopes the difference in masses between two isotopes (e.g., $^{235}U$ and $^{238}U$) is so small that the fractionation is not noticeable. For light elements, however, the mass difference is large enough to cause significant differences between different phases. Thus, the difference in masses of $^{18}O$ and $^{16}O$ is more than 10% of the mass of each isotope, leading to major fractionation during chemical reactions.

Determination of paleoclimates generally depends on the relationships between water and sediments deposited from it. These relationships are commonly measured in terms of the differences between $^{18}O/^{16}O$ ratios in sediments and the ratio in modern seawater (referred to as SMOW, or standard mean ocean water). These differences are recorded as $\delta^{18}O$, calculated as

$$\delta^{18}O = 1000 \times (^{18}O/^{16}O_{rock} - ^{18}O/^{16}O_{SMOW})/^{18}O/^{16}O_{SMOW}$$

Fractionation of $^{18}O$ and $^{16}O$ occurs during several processes. Evaporation removes more of the light isotope from seawater, leading to lower $^{18}O/^{16}O$ ratios in atmospheric moisture than in seawater. This further leads to low ratios in rain, snow, ice, and all freshwater, and the preferential retention of $^{16}O$ in glacial ice permits $\delta^{18}O$ to increase in seawater and fossils equilibrated with it during periods of glaciation and to decrease when glacial ice melts and flows back into the ocean.

Oxygen isotopes also fractionate when solids precipitate from water, either directly or by biological construction of the skeletal parts of organisms. In these reactions, $^{18}O$ is preferentially incorporated into the solids, with different degrees of fractionation for different minerals, such as calcite, aragonite, silica, etc. This fractionation makes $\delta^{18}O$ of all solids positive compared to $\delta^{18}O$ defined as zero for seawater. The degree of fractionation between all solids and seawater depends on temperature, with higher temperatures causing less fractionation. We use this relationship below to infer temperatures of ancient oceans.

## History of the Earth's Climate and the Effects of Continents and Supercontinents

As with most of the topics considered in this book, our information about the earth's climate is less clear as the age that we discuss increases. In this section we start with the very limited information available for the Archean and proceed to a summary of the more detailed knowledge available for the Cenozoic.

### Archean

The only organisms alive in the Archean were bacteria (chapter 12), and they provide virtually no information about climate. Modern bacteria, particularly the primitive Archaea, can live in cold Antarctic lakes and water boiling in geysers (chapter 12), and their Archean counterparts presumably could have existed under similar temperature ranges. Consequently, the traces of filamentous algae in chert and the stromatolites built by cyanobacteria (blue-green algae) merely demonstrate that liquid water existed on the earth's surface in the Archean because the surface temperature was between the freezing and boiling points.

The possibility that Archean temperatures were high because of high concentrations of $CO_2$ in the atmosphere is suggested by the scarcity of organisms that could precipitate calcite and the scarcity of limestones in Archean sediments (see discussion of greenstone belts in chapter 4). A partial confirmation of this high temperature comes from studies of $^{18}O/^{16}O$ ratios in 3.5-Ga cherts in the Kaapvaal craton of southern Africa (Knauth and Lowe, 2003; chapter 4). They found $\delta^{18}O$ of these cherts ranging from +11 to +20, compared to ranges of Phanerozoic cherts from +20 to +35. Because $\delta^{18}O$ decreases with increasing temperature (see above), these low values suggest very high temperatures of the Archean ocean, probably in the range of 55–85°C.

If temperatures were extremely high at 3.5 Ga, they may have decreased by 2.8 Ga. Two investigations suggest that glaciation occurred in the Kaapvaal craton at this time (Young et al., 1998; Modie, 2002). The evidence is controversial, but if glaciation did occur, it seems impossible for temperatures of surface water anywhere on the earth to have been much higher than they are at present. The cause of glaciation is also unclear because the latitude of the Kaapvaal craton at 2.8 Ga is completely unknown.

### Paleoproterozoic

Paleoproterozoic life provides no more information about the earth's climate than Archean life. An important indication that Paleoproterozoic surface temperatures were similar to the present is evidence of glaciation in the Canadian shield, Baltica, the Kaapvaal craton of southern Africa, and western Australia during a 200-million-year period from approximately 2.4–2.2 Ga (Eyles, 1993; Martin, 1999; Bekker et al., 2001; Young, 2002).

The causes of Paleoproterozoic glaciation are controversial. Evans et al. (1997) used paleomagnetic information to propose that the glaciations occurred at low latitudes and were caused by development of a snowball earth. The supercontinent that Piper (2003) suggested to have existed in the Paleoproterozoic also placed North America in low latitudes but indicated a high-latitude location of Australia and southern Africa (chapter 7). Eyles (1993) attributed glaciation in North America to formation of glaciers at high elevations on uplifted rift flanks formed during incipient breakup of Kenorland (chapter 6), which could have occurred at any latitude.

## Mesoproterozoic

The Mesoproterozoic atmosphere was clearly different from the atmosphere that had existed before ~2 Ga. Deposition of red beds instead of banded iron formations demonstrate the presence of free oxygen, which would have kept the $CH_4$ content extremely low. Thus, the only greenhouse gas whose concentration could vary significantly was $CO_2$. Except for a presumably very low concentration of $O_2$, the Mesoproterozoic atmosphere probably was similar to that of the modern earth.

Unfortunately, virtually no information is available about the climate of the Mesoproterozoic. Lithologic indicators such as coal and reefs did not exist at that time, evaporites are rare and not similar to Phanerozoic varieties (see above), and no glacial deposits have been found. The absence of glaciation can be explained in at least three ways.

One explanation is simply that glaciation existed but the evidence has been destroyed by later erosion. A second possibility is that the global climate was too warm because solar irradiance had increased since the Paleoproterozoic and atmospheric $CO_2$ had not decreased enough to permit glaciation. The third possibility involves the size and orientation of the supercontinent Columbia (chapter 7).

Although one proposed configuration of Columbia is similar to that of Pangea, Columbia may not have extended into latitudes high enough to cause glaciation. Using very limited paleomagnetic information, Meert (2002) proposed that Columbia at 1.5 Ga was primarily located in equatorial and temperate latitudes, although a preferred location of North America at 1.77 Ga was over the South Pole. Even if Columbia had a north–south orientation, it was smaller than Pangea and may not have extended into either the north or south polar regions.

## Neoproterozoic

Neoproterozoic climates are dominated by a series of glaciations that occurred between ~800 Ma and ~550 Ma. Although glaciation probably occurred almost continuously at different places throughout this time interval, conventional classifications divide the glacial periods into Sturtian from 750–700 Ma, Marinoan–Vendian from 625–580 Ma, and Sinian from 600–550 Ma. All of these glaciations occurred during the breakup of Rodinia, and younger ones were associated with the assembly of Gondwana. In addition to the lithologic evidence of glaciation at various locations, the effects of glaciation are found in nonglacial sediments deposited during variations in sealevel caused by Croll–Milankovitch cyclical advances and retreats of continental glaciers (Dehler et al., 2001).

The concepts of glaciation because of a snowball earth or high obliquity of the earth's axis originally developed from studies of Neoproterozoic glaciation. These explanations are opposed to the conventional explanations that glaciation occurs at high latitudes or at any latitude where elevations are sufficiently high. Distinction between these proposals must be based on the locations of the glaciations shown in fig. 11.7, which shows the locations and approximate ages of the major glaciations.

Eyles (1993) proposed that the major Neoproterozoic glaciations occurred where tectonic processes created high elevations regardless of latitude. This explanation signifies that the glacial deposits were formed predominantly by valley glaciers instead of broad continental ice sheets. Some early glaciation in South America and part of West Africa was attributed to formation of mountains during compressive orogeny between Amazonia and the combined Congo–Sao Francisco cratons. Collisions may also have been responsible for glaciation through much of West Africa in the latest Neoproterozoic. By contrast, the Sturtian glaciation beginning at ~750 Ma on the western margin of North America (present orientation) and younger glaciations on the eastern margin and in Baltica may have developed on uplifted rift flanks formed during the dispersal of Rodinia.

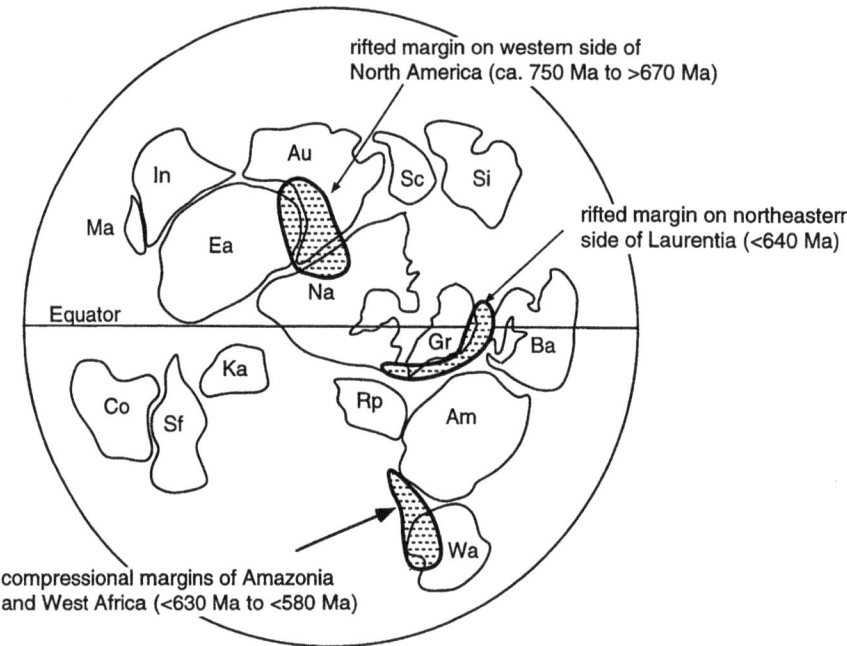

Figure 11.7. General locations of Neoproterozoic glaciations (based on figure 9.2 of Eyles, 1993). Au, Australia; In, India; Sc, South China; Si, Siberia; Ea, East Antarctica; Na, North America; Gr, Greenland; Ba, Baltica; Ka, Kalahari; Co, Congo; Sf, San Francisco; Rp, Rio de la Plata; Am, Amazonia; Wa, West Africa.

The possibility that the Neoproterozoic glaciations were formed at both high elevation and high latitudes has been suggested by numerous workers. Meert and van der Voo (1994) used paleomagnetic data to show that all glacial deposits had formed at latitudes greater than 25°, which meant that all of them could have developed at high altitudes, and some may have been continental ice sheets formed at near-polar positions. The concentration of most continents near the South Pole at the end of the Neoproterozoic (chapter 8, fig. 8.12b) strongly indicates that the latest Neoproterozoic glaciations formed in high latitudes. Evans (2000), however, disputed the paleomagnetic interpretations and suggested that the only viable data indicated glaciation in low latitudes.

The preceding discussion shows that the causes of Neoproterozoic glaciation are highly controversial. Further information and research are desirable not only to discover the causes of glaciation but also to understand the evolution of skeletal organisms in the Cambrian. In chapter 12 we show that the progressive development of soft-bodied organisms that preceded skeletal life took place at the same time as the Neoproterozoic glaciations, and many workers suggest that the two processes were closely related.

## Paleozoic

As we discuss in chapter 8, the general movement of continents and continental fragments during the Paleozoic was from southern latitudes toward the north. This movement included: passage of most of Gondwana across the South Pole; transfer of North America from near the South Pole to the northern hemisphere; movement of northern Europe and Siberia to northern latitudes; and northward transfer of blocks rifted from Gondwana to their collision with North America and Eurasia.

During the final breakup of Rodinia, the South Pole was located a short distance from the west coast of South America (chapter 8). The path of Gondwana across the South Pole is well defined for the Early Paleozoic but less certain for the later part of the era (fig. 11.8). A major ice sheet formed in the Ordovician when northwestern Africa was centered over the pole, but by Silurian time the pole was probably close to the southern parts of South America and Africa (Cocks, 2001). Further movement of Gondwana placed the South Pole near Antarctica in the latest Carboniferous and Permian (commonly referred to as Permo-Carboniferous), which coincided with the broadest development of

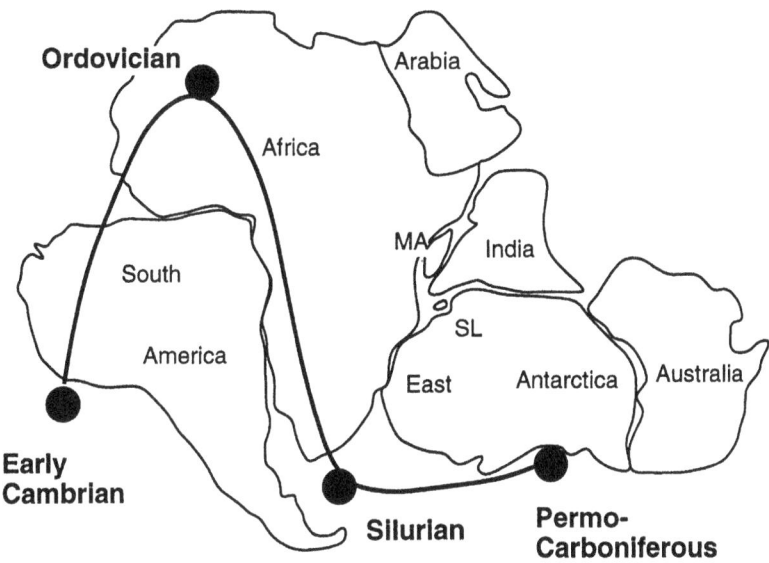

Figure 11.8. Path of Gondwana across the South Pole during the Paleozoic. Cambrian pole from Meert (2001); Ordovician and Silurian poles from Cocks (2001); uncertain path during the Late Paleozoic based on numerous proposals. Abbreviations are: MA, Madagascar; SL, Sri Lanka.

continental ice sheets in the history of Gondwana. This is widely referred to as the "Gondwana glaciation," and the evidence for its occurrence was important in developing the concept of the Gondwana supercontinent (chapter 8).

Permo-Carboniferous glaciation occurred as Pangea accreted to its condition of maximum packing. By interrupting the flow of equatorial currents that distribute warm water, the supercontinent contributed to and may have been essential for the development of this glaciation. Coal-producing vegetation flourished in much of Gondwana north of the glaciated region at latitudes of approximately 50–70° S (fig. 11.9a). These latitudes were higher than those normally occupied by coal-producing plants (fig. 11.1), and they demonstrate that the Gondwana glaciers were surrounded by regions much warmer than comparable latitudes in the present earth (Rayner, 1995). The reason for this warm area near the Gondwana ice sheets is unclear, and it may have resulted from some type of climate control exerted by the Pangean supercontinent (Crowley, 1994).

Terranes that rifted from Gondwana and moved to collision with North America and Eurasia separated throughout the Paleozoic. Avalonian and Cadomian terranes rifted before the development of land plants and provide little climatic information (chapter 5), but terranes that accreted to Asia generally rifted from Gondwana in the Middle to Late Paleozoic. One well-studied example is the North China craton, which carried a temperate to tropical flora northward to collide with Siberia in the Late Paleozoic/Early Mesozoic at a time when Siberia was close to the North Pole (Metcalfe, 1996; chapter 8).

Movement of continental blocks and global changes in climate affected the development of coal, reefs, and evaporites. Warm global climates were well established by the Early Carboniferous (Mississippian), and reefs grew in most epicontinental seas, possibly at latitudes much higher than they can in the present earth. Generally warm climates of the Late Carboniferous and low latitudes of many continental blocks caused a wide development of vegetation that was later converted to coal (fig. 11.9a), but by the Permian many of the continental blocks had moved out of the regions in which coal-producing vegetation could flourish. Large evaporite basins (fig. 11.9b) formed in continental blocks as they moved through arid latitudes, and the wide development of evaporites in North America was promoted when orogenic belts on the western margin of the continent created rain shadows in the interior.

At its time of maximum packing near the end of the Permian, Pangea extended from glaciated regions around the South Pole to cold, but apparently unglaciated, areas in Siberia at high northern latitudes. Consequently, Pangea spanned all lati-

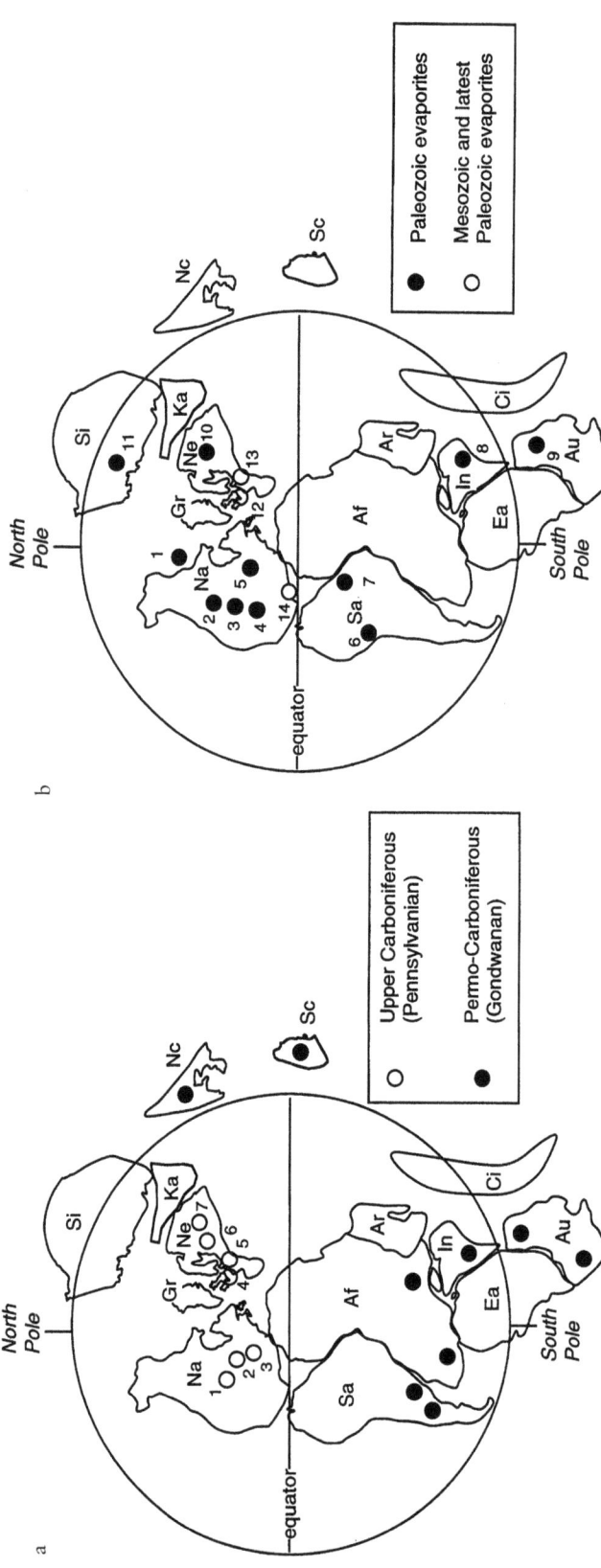

Figure 11.9. Coal and large evaporite basins plotted on a map of Pangea and blocks that were unattached during the Permian (from fig. 8.12d). The circle includes the hemisphere that contained most of Pangea, with continental blocks in the opposite hemisphere shown outside the circle. Abbreviations of continents and continental blocks are: Si, Siberia; Na, North America; Gr, Greenland; Ka, Kazakhstan; Ne, northern Europe; Sa, South America; Af, Africa; Nc, North China; Sc, South China; Ar, Arabia; In, India; Ea, East Antarctica; Ci, Cimmerian blocks; Au, Australia.

a. Locations of coal deposits (based on L. Thomas, 2002): 1, western interior basin; 2, Illinois basin; 3, Appalachians; 4 and 5, numerous deposits throughout British Isles and continental Europe; 6 and 7, numerous deposits in Russia and Ukraine. Permo-Carboniferous basins are not named because each location includes a large number of different deposits.

b. Locations of large evaporite basins of Paleozoic and Mesozoic age (based on Warren, 1999). Paleozoic basins that developed mostly during the accretion of Pangea are numbered: 1, Sverdrup Basin (Carboniferous); 2, Elk Point Basin (Devonian); 3, Paradox Basin (Pennsylvanian); 4, Delaware Basin (Permian); 5, Salina (Michigan) Basin (Silurian–Devonian); 6, Permian salt in Andes; 7, Parnaiba Basin (Lower Carboniferous); 8, Cambrian salt on northern margin of India; 9, Canning Basin (Ordovician–Silurian); 10, Moscow Basin (Devonian); 11, Eastern Siberian Basin (Cambrian). Mesozoic and latest Paleozoic evaporites include: the Triassic Cheshire Basin (12) and Permian Zechstein Basin (13) that formed in arid environments as Pangea was nearly completely assembled; and the Jurassic Gulf Coast Basin (14) that developed as South America and North America rifted apart.

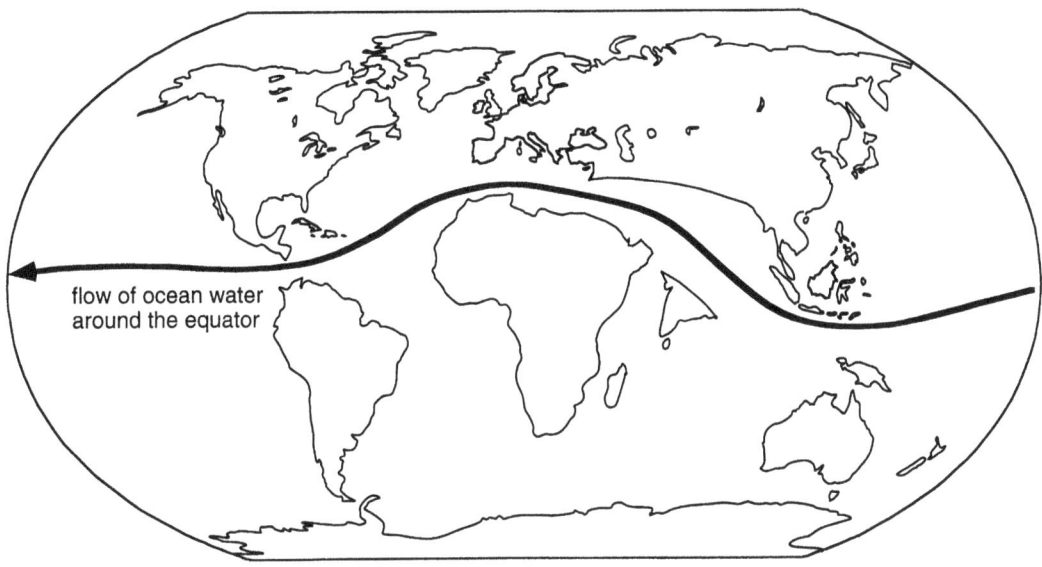

Figure 11.10. Worldwide equatorial circulation during the Cretaceous.

tudes, but it did not have the range of climates found on the present earth. Blockage of the flow of equatorial ocean currents and the presence of orogenic belts along almost the entire western side of Pangea created widespread arid conditions (Parrish, 1993). Permian coal is absent from almost all of Pangea and is known only in areas around the Gondwana ice sheet and in Chinese continental blocks that had not yet docked with Pangea. Permian evaporites are widespread, however, including large salt deposits in the western United States and in Europe.

## Mesozoic and Cenozoic

Rifting of Pangea affected global climates and caused climate change within individual continental blocks as they moved to new locations during the Mesozoic and Cenozoic. We discuss both of these processes in this section.

The breakup of Pangea caused global warming by permitting ocean currents to flow through the fragmented supercontinent and distribute heat more effectively around the earth. Warmer climates were established by the Triassic and almost immediately replaced glacial climates in Gondwana. Higher temperatures of the Triassic occurred even in Antarctica, which remained at high latitudes and was probably located over the South Pole (Retallack and Alonso-Zarza, 1998).

Slightly later in the Mesozoic, ocean currents flowed unimpeded along the entire equatorial region of the earth (fig. 11.10). These currents distributed heat so effectively that the Permo-Carboniferous ice sheet in southern Gondwana either disappeared or was greatly reduced in size. There is no direct evidence for its continued existence, but cycles of sealevel fluctuation in some Mesozoic sediments suggest control of seawater volume by advance and retreat of small ice sheets during Croll–Milankovitch cycles (Gale et al., 2002). A "greenhouse climate" was well established in the Cretaceous, when high sealevels were caused mostly by filling of ocean basins by spreading centers and partly by the lack of glaciation.

Much of the early fragmentation of Pangea occurred in low latitudes (chapter 9). During initial rifting of the Atlantic Ocean, reefs grew for many millions of years along the subsiding margin of North America until plate motions brought it too far north to support reef growth. Rifting in these hot regions formed shallow marine basins that caused precipitation of evaporites in basins now preserved on continental margins. An enormous thickness of salt was deposited in the area that ultimately became the Gulf of Mexico (fig. 11.9b; chapter 10).

Some plate motions brought continental blocks into temperate latitudes that promoted accumulation of coal-producing vegetation in coastal swamps. No coal was formed in the Triassic because the end-Paleozoic extinction destroyed vegetation capable of ultimately producing coal, but new varieties of coal-producing vegetation had developed by the Jurassic (chapter 12). They

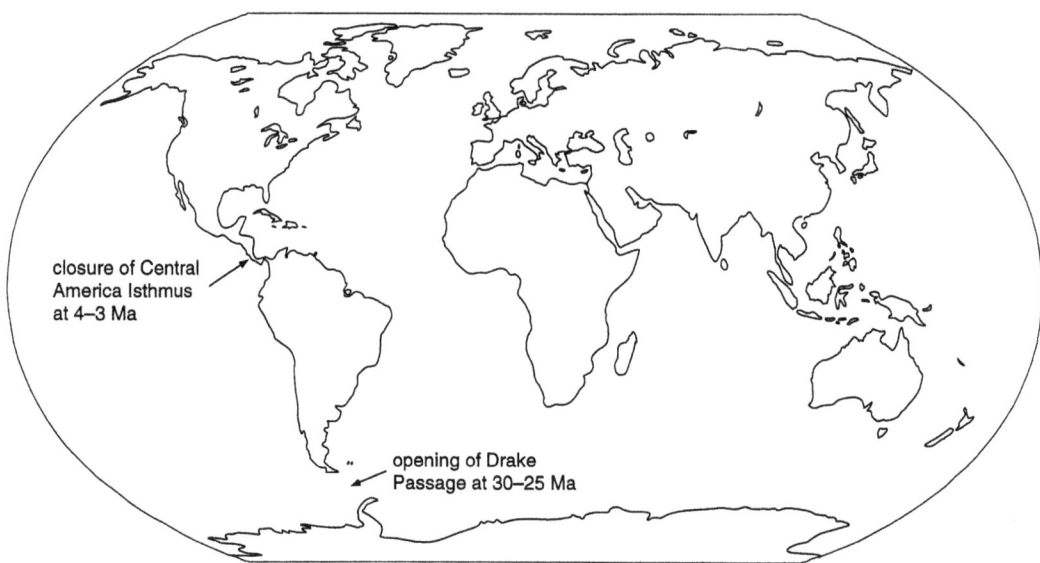

Figure 11.11. Changes in oceanic circulation that led to glaciation in Antarctica and the northern hemisphere.

formed major deposits along the Western Interior Seaway of North America (chapter 9), in South America, and in blocks that accreted to form eastern Asia. Because most of these coal-producing areas have not been subjected to tectonic activity, the coals are generally lignitic.

The greenhouse climate of the Cretaceous was replaced by global cooling near the start of the Cenozoic, when fragments from Gondwana began to approach and collide with the southern margin of Eurasia. One reason for cooling was that the process of accretion interfered with, and ultimately terminated, the equatorial flow of warm ocean water around the earth. A second reason may have been reduction in atmospheric $CO_2$ concentrations by weathering of rapidly uplifted rocks in the Himalayan and nearby mountain chains that began to develop in the Middle Cenozoic (chapter 5; Raymo, 1994; Beck et al., 1998).

Despite this general cooling, major glaciation did not begin until 30–25 Ma. It was caused by separation of South America from the Antarctic Peninsula, thus allowing water to flow from west to east around the entire Antarctic continent (fig. 11.11). This current effectively isolated the Antarctic from any warm water circulating through the rest of the world's oceans and led to eventual covering of almost the entire Antarctic with glacial ice.

Glaciation did not begin in the northern hemisphere until communication between the Caribbean and Pacific Oceans was terminated by closure of the Central American Isthmus (chapter 9). Counterclockwise rotation of blocks in northern Central America and construction of a new volcanic arc in the southern part of the isthmus was completed by ~4–3 Ma, thus preventing effective distribution of warm water in the North Atlantic Ocean. As a result, repeated glacial and interglacial intervals of the Pleistocene began in the northern hemisphere at ~2–1 Ma. These cycles were apparently controlled largely by orbital variations, with the last several cycles consisting of ~100,000 years of slow cooling interspersed with ~10,000 years of rapid warmup (Petit et al., 1999).

## Summary

Variations in the earth's climate have always been controlled partly by processes that operate within the solid earth and partly by processes in and above the atmosphere. The internal processes cause continents to move, thus changing the patterns of ocean currents and the patterns of wind and rainfall. The external processes include long-term increase in solar irradiance and changes in the concentrations of greenhouse gases, which may be linked to continental movements and other tectonic processes.

In the Early Archean, high concentrations of atmospheric greenhouse gases probably made the earth's surface much warmer than it is now. Temperatures apparently decreased significantly by the end of the Archean, and glaciation occurred at several places on the earth during a period of ~200 million years in the Paleoproterozoic. Mesoproterozoic atmospheres seem to have been similar to those at present, although a complete absence of glaciation anywhere on the earth between 2200 Ma and 800 Ma suggests that worldwide warmth was caused by some combination of increase in solar irradiance, elevated $CO_2$ concentrations, and location of continental blocks in temperate and tropical latitudes. Neoproterozoic glaciations have been attributed to high obliquity of the earth's axis of rotation, ice covering to form a snowball earth, formation of glaciers on rifted and compressional margins during the fragmentation of Rodinia, and location of most continental blocks near the South Pole.

During the Paleozoic much of Gondwana passed across or near the South Pole. The South Pole was apparently near Antarctica during the Permo-Carboniferous glaciation that occurred when a nearly assembled Pangea blocked the equatorial flow of ocean water. Rifting of Pangea led to rapid warmup as ocean currents flowed more easily, and global cooling did not begin until near the start of the Cenozoic. East Antarctica remained over the South Pole since the end of the Paleozoic but was not glaciated until 30–25 Ma, when northward movement of South America opened a passage that allowed the Southern Ocean to isolate Antarctica from warm water. Cooling continued throughout the Cenozoic, possibly because of increased tectonic activity, but the Pleistocene glaciation of the northern hemisphere did not begin until construction of the Central American Isthmus prevented exchange of water between the Atlantic and Pacific Oceans.

# 12

# Effects of Continents and Supercontinents on Organic Evolution

The earth's organic life has changed continually for more than 3.5 billion years. This evolution may have resulted partly from environmental stress generated by tectonic activity within the earth and partly from processes independent of the earth's interior. This chapter investigates these different effects in an attempt to determine the role that continents played in the evolution of organisms.

Continents and tectonics associated with them may have influenced organic evolution in both active and passive ways. Active effects include several processes that partly controlled the earth's surface environment. Climate change was caused partly by movements of continents and construction of orogenic belts. Continental rifting increased the area of shallow seas as new continental margins subsided. Changes in volume of ocean ridges and epeiric movements of continents caused marine transgressions and regressions. Temperatures of water in shallow seas increased or decreased as continents moved across latitudes.

The major passive effects of continents and supercontinents result from their influence on diversity of organisms. When continents were broadly dispersed and occupied most latitudes, as on the present earth, this isolation resulted in shallow-water and subaerial families that contained numerous genera, genera with large numbers of species, and species divided among many different varieties. This diversity was clearly smaller at times when continents were aggregated into a few landmasses and particularly low when supercontinents permitted exchange of organisms throughout most of the world's land and shallow seas. During times of major environmental stress, these differences would have restricted extinction of organisms to local species and genera during times of high diversity but might have permitted disappearance of whole orders and classes when diversity was low.

Organic evolution was almost certainly affected by species diversity, but it may have occurred without any active control by tectonic processes. Although evolution probably occurs only when changing environments place stresses on organisms that enhance the competition among them, it is also possible that competition between organisms can cause evolution even without significant environmental change. Furthermore, some environmental change probably resulted from processes that are not related to the tectonics of the solid earth. Such processes include climate change caused by variations in solar irradiance, progressive modification of the concentrations of greenhouse gases, and possibly major variations in the obliquity of the earth's orbit (further discussion in chapter 11). Another external cause of evolution may have been the impacts of meteorites, which are referred to as "bolides" when they explode on impact. Bolides were probably responsible for sudden ("catastrophic") extinction events followed by development of completely new varieties of organisms, and impacts of several small meteorites within a short interval of time may also have led to biotic changes.

We begin this chapter with a section on the ways in which the record of organisms is preserved in the earth. They include recognizable fossils and trace fossils (ichnofossils), preserved organic matter of uncertain source, and variation in $^{12}C/^{13}C$ isotope ratios. The next section discusses whether the sizes and distributions of continents could have influenced various major steps in organic evolution. The final section attempts to answer the question of the relative significance of external and internal (tectonic) processes on evolution.

## Preservation of Organic Activity

The most obvious evidence of former organisms is fossils. Some shells and bones, either whole or fragmented, are preserved in their original or

Figure 12.1. Comparison of sediment not affected by bioturbation (a) and sediment in which original layering has been almost completely destroyed by bioturbation (b). Both of these pictures are of Phanerozoic sediment, but the structures in b were possible only after the evolution of burrowing organisms. (Photos courtesy of Duncan Heron.)

mineralized forms. Some fossils occur either as internal molds or external casts of the organisms. Organisms that are soft bodied or have only chitinous skeletons leave imprints on soft sediments.

In addition to these recognizable forms of fossils, some types of organisms are recognized largely by the traces that they leave of their former activity. These "trace fossils" ("ichnofossils") are produced by animals when they burrow into sediments, walk or crawl across them, or leave some other imprint where no remains of the actual animal can be found. Burrowing and other movements cause "bioturbation" of loose sediments, leaving them with stratification that is more disrupted than in sediments deposited in areas where animals have not been active (fig. 12.1).

Organisms leave not only traces of their physical activity but also chemical compounds, referred to as "molecular fossils," that resulted from their biochemical activities (see Engel and Macko, 1993, for more detailed information). The record consists of organic compounds ranging from simple $CH_4$ to assemblages of complex compounds commonly referred to as "kerogen." The original compounds are altered by numerous processes after the organism that produced them dies, under certain conditions developing to different forms of coal, asphalt, oil, and natural gas. Because of these changes and the complexity of the original biochemical processes, it is extremely difficult to determine the variety and age of the organisms that produced these remains.

Some organic compounds preserved in sediments may not have been left by organisms but, instead, were produced by inorganic ("abiotic") processes. The ability of organic chemicals to be generated from inorganic compounds has been understood since the production of urea by reacting $CO_2$ and $NH_3$ by Friedrich Wohler in 1828. Organic compounds found in carbon-bearing meteorites confirm that inorganic processes in nature are capable of generating complex molecules, including several types of amino acids (Botta and Bada, 2002). The possibility of inorganic generation of organic compounds in ancient rocks has led to several efforts to recognize chemicals that can be generated only by the activity of organisms. Some compounds are sufficiently complex that they are generally regarded as biochemical residues, but all suggestions are controversial, and we refer readers who want additional information to Peters and Moldowan (1993).

With all of the complexities of distinguishing inorganic and organic production of organic molecules, plus the changes caused by post-depositional activity, much of the history of organic activity on the earth has focused on the isotopic effects of biochemical activity (review in Hayes, 2001). Biochemical reactions cause fractionation of several light isotopes, including H, C, N, O, and S, but in this book we only consider changes in the ratios of $^{12}C$ and $^{13}C$, which are commonly referred to in terms of variations in $\delta^{13}C$ (Holser, 1997). Values of $\delta^{13}C$ are calculated as

$$\delta^{13}C = 1000 \times (^{13}C/^{12}C_{sample} - ^{13}C/^{12}C_{standard})/^{13}C/^{12}C_{standard}$$

The commonly used standard is PDB, the ratio of $^{13}C/^{12}C$ in a belemnite from the Pee Dee Formation of South Carolina, United States.

The value of $\delta^{13}C$ can be measured in many different types of material. In bulk limestones it is referred to as $\delta^{13}C_{carb}$, and in bulk kerogen it is $\delta^{13}C_{org}$. Values can be obtained separately for different types of carbonate, including general values for benthic or planktic organisms. Variations also occur among different species of animals and plants and within species depending on other environmental factors (see Des Marais, 2001; Freeman, 2001; and Hayes, 2001 for further information).

Organisms preferentially take $^{12}C$ into their living tissue, causing $\delta^{13}C$ to be low in organic compounds produced by biologic activity. This concentration into organisms leaves the environment in which they lived, such as seawater, enriched in $^{13}C$. Shells and other carbonates that precipitate in seawater consequently show variations in $\delta^{13}C$ that are largely controlled by the amount of organic carbon that is in equilibrium with the surface environment.

During most of geologic time the amount of organic carbon available to the surface environment has been approximately constant as organisms extracted $^{12}C$ into their tissues during life and then released it back into the environment by oxidation when they died. The principal variations from this constancy arose during times of rapid environmental change, resulting in periods of organic evolution and extinction. The variations were both positive (to positive $\delta^{13}C$ values) and negative (to negative values), and they occurred either as long-term trends or as "spikes" that lasted only a few million years or less.

Periods of positive $\delta^{13}C$ in carbonate rocks were generated by a variety of processes. One was removal of $^{12}C$ from the surface during periods of intense global orogeny, which caused burial of organic carbon in foredeeps. Periods of widespread glaciation were also associated with positive $\delta^{13}C$ in carbonate rocks formed by deposition of planktic organisms when anoxic (or dysoxic) bottom water permitted accumulation of organic carbon in the deep oceans.

Periods of negative $\delta^{13}C$ also developed by numerous processes. Negative spikes commonly followed extended periods of glaciation (chapter 11) when sealevels rose and normal patterns of circulation were established throughout the oceans. Negative $\delta^{13}C$ spikes in carbonate rocks also occurred during major extinction events when photosynthetic activity in the upper parts of oceans decreased, thus reducing the amount of organic material that formed and settled to the bottom of the ocean after death. At this same time during some extinction events, however, $\delta^{13}C$ in benthic organisms showed positive variations because of the lack of organic carbon falling to the ocean floors. Some negative spikes may have been caused by widespread burning of land plants or destabilization of methane hydrates on the ocean floor, thus

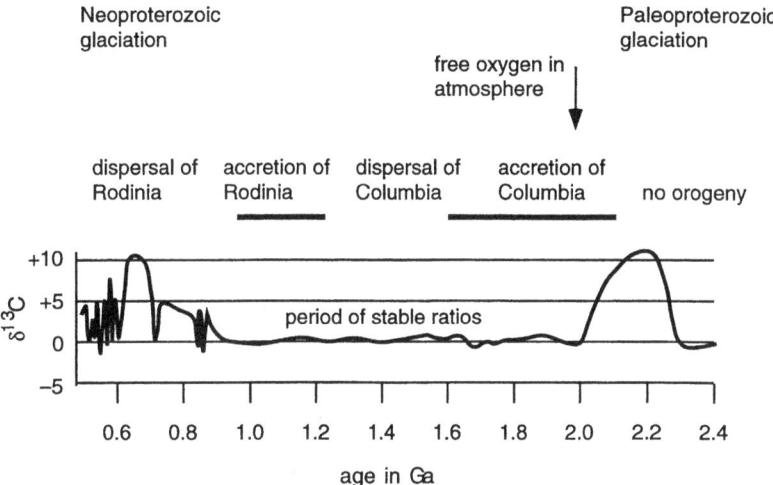

Figure 12.2. Variation in $\delta^{13}C$ during the Proterozoic (modified from Brasier and Lindsay, 1998).

rapidly introducing large amounts of $^{12}C$ into the atmosphere and surface water.

We illustrate long-term variability with fig. 12.2, which shows $\delta^{13}C$ throughout the Proterozoic (Brasier and Lindsay, 1998; Calver and Lindsay, 1998; Lindsay and Brasier, 2000, 2002). The generalized high $\delta^{13}C$ in the Paleoproterozoic began in the latter stages of glaciation and the early stages of the worldwide 2.1–1.8-Ga orogenies (chapter 11 and appendix H). The lack of variation between ~2.1 Ga and 0.8 Ga presumably indicates continued recycling of carbon through organisms and the surface environment without burial of organic carbon. This constancy seems not to have been affected by accretion and rifting of the supercontinent Columbia in the Mesoproterozoic. A more detailed record of $\delta^{13}C$ is available for the Neoproterozoic, when rapid oscillations began after Grenville-age orogenies and presumably were related to a series of glacial intervals during the breakup of Rodinia (chapters 7 and 11).

## Major Episodes of Evolution and their Relationship to Continents and Supercontinents

This section recounts major changes in the types of organisms inhabiting the earth and attempts to relate these changes to the histories of continents and supercontinents. More information about evolutionary changes is provided by numerous books on paleontology and historical geology, and a short summary can be found in Rogers (1993a).

### Prokaryotes

Prokaryotes are distinguished from all other organisms by their lack of a cell nucleus, with their DNA and regulatory organs distributed throughout the cell or attached to the outer cell wall. The first life on earth probably developed by ~3.8 Ga (Schidlowski, 2001) and undoubtedly consisted of prokaryotes, all of which are single-celled organisms commonly referred to as bacteria. Bacteria include an enormous number of forms with different lifestyles (Hagadorn et al., 1999), and their geologic record is difficult to interpret (Nisbet and Sleep, 2001). Two of these types of bacteria, "archaebacteria" (Archaea) and "cyanobacteria" (blue-green algae), are particularly important to our investigation and we discuss them briefly here.

Archaebacteria are different from all other organisms on earth because the fatty acids (lipids) that constitute most of their cells are linked to each other by oxygen atoms (forming ethers). Archaebacteria provide important information about conditions on the earth's surface when life evolved. Modern varieties live mostly in three different environments, including boiling water in hot springs, highly saline lakes, and swamps, where the archaebacteria produce $CH_4$. These characteristics suggest that the surface of the early earth was hot (chapter 11), that at least some bodies of water were rich in dissolved minerals, and that the atmosphere contained $CH_4$.

Cyanobacteria (blue-green algae) are different from all other bacteria because they release oxygen by photosynthesis and can live in the presence of oxygen. Cyanobacteria form sticky mats on the surface of sediments deposited in shallow water, and these mats can trap clastic particles as they wash across the surface. Continued accumulation can develop finely layered rocks known as "stromatolites," which include both limestones and silicic varieties. Stromatolites formed bacterially have been proposed to occur in chert as old as 3.5 Ga, although the evidence is highly controversial (Brasier et al., 2002; Schopf et al., 2002). Microbial mats in siliciclastic rocks have also been proposed for rocks of 2.9-Ga age (Noffke et al., 2003).

The origin of prokaryotes, basically the origin of life, is a problem far beyond the scope of this book, and it is not clear that their evolution was influenced by any of the fundamental processes that we consider. Prokaryotes had evolved well before the development of long-lived continental platforms (cratons; chapter 4), and the only emergent land was probably short-lived areas of sialic crust and possibly platforms similar to modern island arcs. The effect of bolide impacts is also unclear. The Archean was presumably a time of frequent impacts, although the only remaining evidence is tektite layers as old as 3.5 Ga (Byerly et al., 2002).

## Eukaryotes

Eukaryote cells differ from those of prokaryotes by having genetic material inside a nucleus and much of the regulatory material separated into organelles. Eukaryotes also differ from all prokaryotes in requiring oxygen to live, either in water or air, whereas all prokaryotes except cyanobacteria die in an environment that contains oxygen. All prokaryotes and some eukaryotes (such as foraminifera) are single celled, but eukaryotes also have the ability to form multicellular organisms. Recognition of single-celled eukaryotes in the fossil record is controversial. It is generally based on the observation that single-celled eukaryotes tend to be larger and have more complex surfaces than prokaryotes, and it has also been based on discovery of organic molecules in kerogen that are larger and more complex than those synthesized by modern prokaryotes.

The exact time of evolution of eukaryotes is highly uncertain, and estimates vary widely. Some studies of the varieties of organic molecules in kerogens suggest that eukaryotes may have existed in the Archean (Brocks et al., 1999). Because they require oxygen in order to live, however, most workers assume that eukaryotes could not have evolved until at least low concentrations of oxygen developed in the atmosphere and oceans. Fedonkin (2003) proposed that the evolution of eukaryotes was forced by gradually increasing atmospheric oxygen throughout the Proterozoic.

The initial development of free oxygen in the atmosphere at ~2.1 Ga is indicated by several types of evidence, including the deposition of iron formation before that time and red beds afterward (summary in Des Marais, 1997a,b). The age of the oldest observed single-celled eukaryote fossils is consistent with evolution at 2.1 Ga or possibly later in the Proterozoic. Spiral forms of organisms that were presumably eukaryotic have been reported from ~2.1-Ga iron formations (Han and Runnegar, 1992). Shales with an age of 1.5 Ga contain acritarchs that may have been the resting cysts of algae (Javaux et al., 2001).

A time of 2.1 Ga is at the end of, or immediately following, the 200-million-year period of glaciation that occurred in many parts of the earth during the Paleoproterozoic (chapter 11). The glaciation might have led to the development of eukaryotes by creating environmental stresses necessary for organic evolution to occur, and the possibility that glaciation occurred on uplifted flanks of continental rifts (chapter 11) suggests an underlying relationship between tectonics and evolution. Approximately 2.1 Ga is also the beginning of the long period of constant $\delta^{13}C$ that characterized much of the Proterozoic (see above). The times of all tectonic, climatic, and evolutionary events are too uncertain to yield a definitive conclusion, but we can suggest the following possibilities for the influence of tectonics on the evolution of eukaryotes:

- Whether or not caused by tectonic uplift, a prolonged period of glaciation created environmental stresses that led to evolution (we discuss this possibility further in the section on Neoproterozoic evolution below).
- Rifting of continents may have caused the glaciation by exposing more rock to weathering, thereby reducing atmospheric $CO_2$ concentrations (chapter 11).
- The lack of compressive orogenic activity between ~2.5–2.1 Ga (chapter 7) may have permitted expansion of epicontinental seas in areas not undergoing active rifting, a possibility suggested by the constancy of $\delta^{13}C$ ratios from 2.1–0.8 Ga (fig. 12.2). This broad area of warm shallow water may have provided enough different ecological niches to stimulate evolution of new organisms that could occupy them.

Figure 12.3. Ediacaran biota from the Ediacara Formation of South Australia. Upper left—Dickinsonia costata; upper middle—Spriggina; upper right—Tribrachidium; lower left—Parvancorina; lower middle—Aspidella; lower right—Charniodiscus. (Photo courtesy of Yasuo Kondo.)

## Multicellular Organisms

Multicellular eukaryotes include metazoans (animals), metaphytes (plants), and fungi. The time at which multicellular organisms first appeared is questionable because of the difficulty of distinguishing them from other organic remains in the Precambrian. Neoproterozoic organisms clearly included metazoans (see below), and they almost certainly evolved by some time in the Late Mesoproterozoic or possibly earlier (Seilacher et al., 1998; Butterfield, 2000, 2001). The limited information available does not provide a cause for the evolution of multicellular organisms, and whether they were influenced by the accretion of the supercontinent Columbia or other tectonic processes during the Mesoproterozoic (chapter 7) is unknown.

## Ediacaran and Similar Assemblages

Complex soft-bodied assemblages of the Neoproterozoic were first discovered in the Ediacara sediments of southern Australia, and similar assemblages have now been located at numerous other places. Whether these organisms have any relationships to modern animals or plants, however, is uncertain (fig. 12.3). Some have forms similar to jellyfish and others to worms, both of which evolved as recognizable forms in the Cambrian (see below). On the basis of these similarities, some investigators regard some Ediacaran organisms as soft-bodied precursors to animals that developed skeletal parts in the Cambrian (Narbonne, 1998). Other investigators, however, find very little relationship with Cambrian organisms and consider the Ediacaran assemblage to consist of forms that became almost entirely extinct at the beginning of the Cambrian (Seilacher, 1992).

The Ediacaran assemblage flourished from ~565 Ma to ~543 Ma, although some forms may have evolved earlier, and some may have continued into the Paleozoic (Narbonne and Gehling, 2003). Thus, the Ediacaran biota apparently developed during the latest stages of rifting of Rodinia and the accretion of Gondwana (chapters 7 and 8), and Waggoner (1999) showed that slightly different suites of organisms were clustered into three different areas of Rodinia (fig. 12.4). This grouping suggests that the fragmentation of Rodinia exerted some control over the evolution of Ediacaran organisms, possibly because it created such large areas of submerged continental crust that shallow seas provided new environments for Ediacaran organisms to fill.

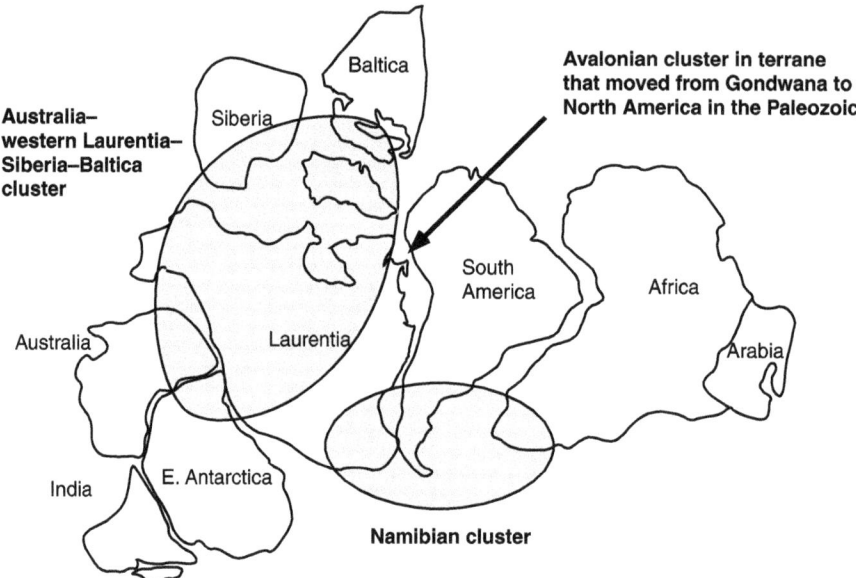

Figure 12.4. Localized areas of different types of Ediacaran biota plotted on configuration of Gondwana proposed by Rogers (1996). Modified from Waggoner (1999).

The development of Ediacaran organisms in the latest part of the Neoproterozoic may have been influenced by any or all of a complex series of events in addition to the rifting of Rodinia. Evolutionary stress may have been provided by Neoproterozoic glaciation, which probably ended at ~580 Ma, although it may have lasted to a younger time. Whether the glaciation was caused by tectonic processes and the location of continents or by external forces is uncertain (chapter 11). A bolide that created the Acraman impact site in Australia at 580 Ma was large enough to cause worldwide modification of atmosphere and oceans, including reduction of light reaching the surface and consequent reduction in the amount of photosynthesis (Grey et al., 2003). The ages of all of these events except the Acraman impact (580 Ma) have broad ranges and uncertainties, and for this reason the cause of the development of Ediacaran organisms remains highly controversial.

*Precambrian–Cambrian Transition*

The transition from the Precambrian to the Cambrian was not instantaneous but apparently occurred in only a few million years between approximately 545 Ma and 540 Ma (Bowring et al., 1993; Landing et al., 1998). During this time the Ediacaran biota became almost completely extinct, and many modern varieties of animals developed skeletal parts. The transition is preserved at numerous localities, and we present a generalized diagram in fig. 12.5 that is based mostly on work in Newfoundland (Landing et al., 1989).

The earliest indication of the development of modern animals is the presence of burrows in soft sediments. Prior to the Cambrian, sediment on the seafloor retained its original bedding, with no indication of the bioturbation that disturbs much Phanerozoic sediment. At a somewhat younger time, the fauna became dominated by a group of organisms known as "small shelly fossils." Some of the tiny shells are recognizable as primitive molluscs, but many have no certain affiliation to modern organisms, and some are probably "platelets" that formed a disorganized skeletal protection around larger organisms (Conway Morris and Peel, 1990). By the time of the small shelly fossils, all of the types of mineralization that characterize modern animals were being produced, including calcite, phosphate, and silica. Recognizable animal skeletons occur higher in transition sections than the small shelly fossils. They represent many of the modern animal phyla, including: trilobites and other arthropods; siliceous and calcareous sponges; gastropods and bivalves of the molluscan phyla; both articulate and inarticulate brachiopods; and primitive forms of echinoderms (Zhuravlev and Riding, 2001).

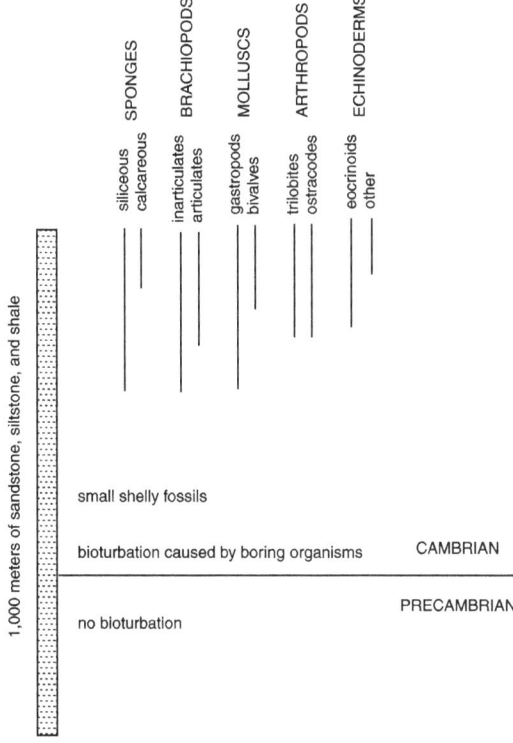

Figure 12.5. Biotic changes across the Precambrian–Cambrian transition. Based largely on the sequence in the Burin Peninsula of Newfoundland (from Landing et al., 1989).

All of these changes occur in stratigraphic sections that contain either no unconformities or only minor ones. This continuity indicates both an absence of tectonic activity and an absence of abrupt changes in sealevel, which is consistent with the lack of any evidence for glaciation during the period 545–540 Ma. Several $\delta^{13}C$ positive and negative spikes occur across the Precambrian–Cambrian boundary, but it has not been possible to correlate any of them with specific biotic changes.

The cause of the Precambrian–Cambrian transition remains unclear. It occurred at a time when the dispersal of Rodinia was complete and the assembly of Gondwana was in its last stages. The South American part of Gondwana was beginning to pass across the South Pole, but other continental blocks were in temperate latitudes, and there is no evidence that significant parts of the earth were glaciated. Sediments that span the transition contain no evidence of bolide impact, and the period of a few million years required by the transition is probably too long to have been the result of impact. It seems likely that understanding the sudden development of mineralized skeletons of animals will require more information than is currently available.

### Middle Cambrian Black Shale Fauna

In 1909, C.D. Walcott discovered a remarkable fauna in Middle Cambrian black shale of the Burgess Formation in the Rocky Mountains of British Columbia (review in Whittington, 1985). Similar assemblages are now known in black shales of Middle Cambrian and slightly older age throughout the world (Conway Morris, 1992). Most of the individual organisms are arthropods, but representatives of almost all of the animal phyla are present. Although some of the fauna seem to be ancestral to younger forms, many of them apparently existed only during the Middle Cambrian and died out after a few million years of existence.

Any proposed origin of these unusual organisms must explain both their development and their existence for only a short period of geologic time. Black shales are common throughout the Phanerozoic without preserving fauna very different from those in other environments, and no fauna of other ages contain so many varieties that did not evolve to other forms. This unique situation suggests that the earth was in some very unusual environmental and/or tectonic condition, but none has been found. The Middle Cambrian was a time with almost no tectonic activity, and no glacial or other unusual climatic conditions seem to have existed. The only explanation currently available seems to be some unrestrained development of many different types of animals that were ultimately noncompetitive.

### The Oldest Land Plants

Evidence of the oldest land plants occurs in Middle Ordovician sediments (Gray, 1993). It consists of spores that presumably indicate the development of bryophytes (mosses), which do not contain any "vascular" or other materials that are capable of being preserved as fossils. This evolution came at a time for which there is no evidence of special environmental or tectonic conditions. It was before Late Ordovician glaciation (see below) and before compressive orogeny began in eastern North America and northern Europe. Thus, there seems to be no special reason for the occupation of the land by plants, and presumably it was simply a continual evolutionary development.

## Extinction at the End of the Ordovician

The transition from the Ordovician to the Silurian over a period of several million years is marked by extinction of almost all graptolites, most conodonts, more than 80% of brachiopod genera, and about 50% of bivalve genera (Hallam and Wignall, 1997). The extinction was mostly toward the end of a period that began with a sharp negative spike in $\delta^{13}C$ followed by a few million years of positive values as marine regression exposed some land areas and developed shallow seas in areas of formerly deeper water. Several short positive and negative spikes in $\delta^{13}C$ occurred as sealevels returned to normal at the end of the extinction event.

During the Late Ordovician compressive orogeny was beginning in eastern North America and northern Europe, and subduction occurred beneath almost the entire outer margin of Gondwana (chapters 5 and 8). Rifting continued along the inner (eastern) margin of Gondwana, and Avalonian–Cadomian terranes were in transit across the Iapetus Ocean. Gondwana had moved across the South Pole so that the South Pole was centered in northwestern Africa, causing widespread glaciation in that region.

Glaciation may have been partly responsible for the Late Ordovician extinction because it caused regression from the generally high sealevels of the earlier part of the Ordovician (chapter 11). This would have resulted in a reduction of the number of ecological niches available to marine organisms and forced greater competition between them. Biotic recovery in the Silurian may have begun when glacial melting was completed near the end of the Ordovician, although the generally low sealevels of the Silurian probably were caused by at least limited continental glaciation as Gondwana passed across the South Pole throughout the period.

## Late Devonian Extinction

Extinction near the end of the Devonian has commonly been referred to as the "Frasnian–Famennian" event because it was originally thought to have been restricted to the last two epochs of the Devonian period. Further work, however, has now demonstrated that it occurred over approximately 15 million years in most of the later half of the Devonian (House, 2002). The episodic nature of extinction and recovery is shown by at least two positive and negative excursions in $\delta^{13}C$ rather than a single event (Joachimski et al., 2002).

The extinction and succession by new organisms was mostly restricted to marine fauna that were affected by widespread development of anoxic/dysoxic water (Hallam and Wignall, 1997). Most types of trilobites became extinct at this time or earlier, with only one group surviving into the Carboniferous. A few types of molluscs, some brachiopods, all cystoids, and all jawless fish (Agnatha) also became extinct. New animals that developed in the Late Devonian or Early Carboniferous include some types of brachiopods and several varieties of cephalopods, with almost all Paleozoic ammonites first appearing at this time. There was no significant extinction among land organisms, with both animals and plants continuing to evolve toward more advanced varieties. The principal development among land animals was the first appearance of amphibians. Vascular plants (Tracheophytes), which had just evolved in the Silurian, continued to develop into the varieties that would flourish in the Carboniferous (Gray, 1993).

No cause of the Late Devonian extinctions is immediately apparent. Some climatic variation is implied by several $\delta^{13}C$ variations and transgression–regression cycles. No direct evidence of glaciation has been found, although waxing and waning of small icecaps within Gondwana may have caused the sealevel variations. The assembly of Asia had begun, and compressive orogeny was active around all margins of North America, the outer margin of Gondwana, and much of southern Europe. Whether any of this climatic and tectonic activity was responsible for the biotic events is unclear, however, and recent discovery of evidence of several small meteorite impacts during the Late Devonian suggests that they may have been responsible for the climatic and environmental changes (McGhee, 2001; Sandberg et al., 2002; Warme et al., 2002).

## Paleozoic–Mesozoic Boundary

Mass extinction at the end of the Permian destroyed ~90% of marine species and ~70% of vertebrate species on land (Erwin, 1993; Erwin et al., 2002). Extinction occurred through a period of a few million years in the Late Permian, and by its end the world's oceans contained no more trilobites, fusulinids, tabulate and rugose corals, blastoids, and stalked crinoids. All of the animal phyla survived the extinction, however, and diversified to new forms in the Triassic. This expansion created new types of corals, caused enormous expansion of ammonites, and began the diversification of reptiles and mammals. Most major groups of plants survived the end of the Permian, but the species that produced coal in the Carboniferous became extinct and were not replaced by new coal-producing types until the Jurassic (chapter 11).

These biotic developments occurred during a time when Pangea was at, or near, its condition of maximum packing (chapter 8). As we discussed in chapter 11, the configuration of Pangea produced extensive glaciation in Gondwana, warm global climates, arid conditions in the center of the supercontinent, and marine regression. During regression the oceans developed lower regions of anoxic water, which may have lasted as long as 20 million years (Knoll et al., 1996; Isozaki, 1997; Wignall and Twitchett, 2002). Environmental stress was also created by the eruption in the Siberian basalt province, which began at ~250 Ma and was so large that it could have caused worldwide atmospheric contamination with ash and poisonous gases (Reichow et al., 2002; chapter 10).

More than one episode of transgression and regression are suggested by extinction of different groups at two or more times, largely in pulses at ~260 Ma and ~250 Ma (Erwin et al., 2002). Whether these pulses were nearly instantaneous or merely part of a continual extinction process is controversial. If the extinction was gradual, then tectonic control by the assembly of Pangea is a likely cause. If the extinctions were rapid, however, then external control by widespread volcanism or bolide impact is more likely. Available evidence cannot demonstrate which, if any, of these causes is more likely.

## Development of Calcareous Nanoplankton and Microplankton

Photosynthetic organisms must have been the base of the oceanic food chain throughout the history of the earth. The oldest known groups that secrete calcite, however, did not develop before the Mesozoic. Because remains of these organisms are easily destroyed after deposition, the stratigraphic record is not certain, but it appears that coccolithophores first appeared in the Triassic, with diatoms and planktic foraminifera in the Cretaceous.

The reasons for evolution of the calcareous plankton are obscure. The highly uncertain dates of their appearance do not seem to correlate with any tectonic or external events. The climatic warmup that began in the Triassic continued without significant interruption, and sealevels rose continually to their peak in the Late Cretaceous (chapter 11). Also, the arrival of calcareous plankton cannot be correlated with any known meteorite impacts or with the development of anoxic/dysoxic layers in the lower part of the oceans at several times in the Jurassic and Cretaceous.

## Late Triassic–Early Jurassic

During a period of about 5 million years in the latest Triassic and earliest Jurassic the world's oceans lost ~10% of their animal families (Palfy and Smith, 2000; Palfy et al., 2002). On the basis of fossil and $\delta^{13}C$ data, the extinction may have occurred in more than one series of transgression and regression. At least one of the regressions coincided with development of anoxia (or dysoxia) in the lower oceans, during which organic matter accumulated widely on ocean floors (Hasselbo et al., 2002). Full recovery of organic diversity did not occur until the Middle Jurassic (Hallam and Wignall, 1997).

The causes of the anoxia and extinction are unclear. General climate warmth and lack of any evidence for glaciation at these times indicate little possibility that the anoxic conditions were caused by glacial regression and terminated when glaciers melted and flushed out the lower oceans. The extinction cannot be correlated with any specific tectonic activity, and there is no credible evidence for meteorite impact at this time. One possibility is based on the age of the massive Karoo basalts of southern Africa and the synchronous Ferrar mafic province of Antarctica (Palfy et al., 2002). Their eruption at the same time as the extinction and anoxic events raises the possibility that volcanism placed enough ash and poisonous gases into the atmosphere to alter climate and ocean composition.

## Cretaceous–Tertiary Boundary

The great faunal diversity of the Cretaceous was dramatically reduced at the end of the period (papers in Koeberl and MacLeod, 2002). Although all phyla survived, several groups of organisms became extinct. They included: dinosaurs and all closely related forms (only crocodiles survived); rudistid and similar reef-forming bivalves; all ammonites and belemnites; all flying and marine reptiles. Even the surviving groups were greatly affected, with extinction of nearly 90% of planktonic micro- and nano-organisms, 50% of macro-invertebrates, 75% of angiosperms, and perhaps 50% of all species. Revival of species diversity did not fully occur until 5–10 million years later.

The extinction took place in a shorter period of time than can be measured at the end of the

Figure 12.6. Cretaceous–Tertiary boundary at a depth of 219.82 m below the seafloor at site 1210A of Ocean Drilling Project Leg 198 (Shatsky Rise, Pacific Ocean). All sediments are calcareous abyssal plain deposits. Lowermost Tertiary (Danian) toward the left is separated by several centimeters of bioturbated sediment from underlying uppermost Cretaceous (Maastrichtian) deposits. The bioturbated zone contains micro- and nanofossils both older and younger than the boundary plus ~0.05-mm microspherules distributed around the earth from the Chicxulub impact site (see text). (Photo courtesy of Harumasa Kano.)

Cretaceous. It may have required a few thousand years, or perhaps only a few months. The extinction was clearly caused by reduced solar radiation for a long enough time to destroy much of the photosynthetic production at the bottom of the food chain. On land, it was also caused by fire, for the first organisms to spread on land at the base of the Tertiary were ferns, preserved almost entirely as pollen, and ferns are well known to colonize areas burned by modern fires. So much organic carbon was burned that $\delta^{13}C$ decreased abruptly at the beginning of the Tertiary.

The extinction occurred so rapidly that it could not possibly have been caused by a tectonic and/or climatic event. The possibility of meteorite impact was first suggested by the discovery of a "boundary clay" at Gubbio, Italy, that separated rocks with Cretaceous fossils from rocks with Tertiary fossils (fig. 12.6; L. Alvarez et al., 1980; W. Alvarez, 1997). Similar clays were soon discovered elsewhere. They are generally a few centimeters thick and contain materials that formed in some type of explosion, including shocked (microlamellar) quartz, a variety of $SiO_2$ (stishovite) that develops only at extremely high pressure, and small tektites that formed by melting of rocks before they were blown into the air. The clays also have a high content of Ir, which is very rare on the earth's surface and could only have come from a meteorite.

Final confirmation of meteorite impact came from the discovery of a crater more than 200 km wide in Cretaceous rocks at Chicxulub, Mexico (Morgan et al., 2002). This crater was apparently caused by a bolide that was large enough to fill the atmosphere with ash worldwide and also to ignite forest fires over most of the earth. Precise dating now shows that this impact occurred at 65 Ma, consistent with ages in the boundary clays and previous estimates of the age of the Cretaceous–Tertiary boundary.

## Paleocene–Eocene Faunal Change

Extinction and evolution rates for many animal species seem to have been higher at the end of the Paleocene and in the Early Eocene than at other times in the Cenozoic (Hallam and Wignall, 1997).

About 50% of benthic foraminifera became extinct, and planktic foraminifera and land mammals evolved rapidly during the "Paleocene–Eocene Thermal Maximum" (PETM). The PETM was a time when all of the earth's seawater became warmer, reducing the longitudinal gradient of decreasing temperature of surface water toward the poles and also the vertical gradient of decreasing temperature from the top to the bottom of the oceans. The origin of the PETM is highly controversial, but because it coincided with a sharp reduction in $\delta^{13}C$, many investigators attribute it to sudden release of methane from seafloor sediments (D. Thomas et al., 2002).

*Late Eocene Faunal Change*

The latter part of the Eocene is characterized by rapid disappearance and evolution of planktic organisms at the species level, although no mass extinction occurred in any major groups. This development occurred during a time of general cooling from high temperatures of the Cretaceous to the present. No cause for the rapid biotic change is immediately apparent, but it may have resulted from the impact of at least three meteorites (all smaller than at Chicxulub) within a few million years (McGhee, 2001).

## Summary of the Effect of Continents and Supercontinents on Organic Evolution

Numerous causes have been proposed for major extinctions and development of new types of organisms. We start this section by reviewing possible causes and their relationship to continental/tectonic processes before investigating the role that continents may have played in each of the stages of biotic development discussed in the preceding section (more complete discussion in Hallam and Wignall, 1997).

- Global warming or cooling caused by variation in concentrations of greenhouse gases, changes in solar intensity, or other factors that affected the earth at all latitudes. Warming might have eliminated organisms that normally live in polar latitudes, and cooling could have caused extinction of tropical species. Temperature increase might also have resulted in anoxia/dysoxia as warming of the oceans reduced their ability to dissolve oxygen. This type of generalized temperature change seems more likely to have been caused by external processes than by movement of continents or assembly and dispersal of supercontinents.
- Change in temperature gradients across latitudes, generally accompanied by development of continental glaciers in high latitudes. These changes would normally have caused organisms to move across latitudes to remain in their normal temperature zones and might only result in extinctions by changing ocean circulation and causing transgression or regression. Most glacial intervals were probably caused by positions of continents and supercontinents, although other possibilities have been proposed (chapter 11).
- Transgression and regression that affected habitats both on shallow shelves and in the deep ocean. Worldwide regression could have eliminated some shallow-water niches and caused extinction of benthic organisms in them, whereas general transgression may have filled isolated basins and created habitats for the evolution of new organisms. Organisms that lived at any depth in the open oceans were probably affected by transgression and regression only if they caused changes in the amounts of dissolved oxygen, particularly resulting in anoxic/dysoxic conditions (see below). Short-term variations in sealevel were almost certainly caused by glaciation and deglaciation, but variations over periods of millions of years resulted mostly from changes in volumes of ocean ridges and partly by epeirogenic movements of continents (chapter 2).
- Anoxic/dysoxic conditions both in epeiric seas and in the deep oceans. Shallow-water black shales developed during both transgressions and regressions wherever basins were sufficiently isolated that surface and deep water could not interchange. Deep-water anoxic/dysoxic events seem to have been independent of sealevel, and numerous proposals have been made for their origin. Some of these deep-water events apparently occurred at the same time as periods of rapid biotic change, but others are apparently unrelated. The only effect of continents and supercontinents on anoxia/dysoxia was probably the provision of shallow basins on continental shelves.
- Eruption of large quantities of ash and volcanic gases that caused global cooling and also

Table 12.1. Fundamental causes of rapid evolution and possible involvement of continents and supercontinents

| Event | Climatic stress | Volcanic stress | Bolide impact | Tectonic involvement |
|---|---|---|---|---|
| Development of prokaryotes | | | | |
| Development of eukaryotes | | | | none unless related to development of atmospheric $O_2$ |
| Development of multicellular organisms | | | | |
| Ediacaran biota | × | | × | possible control of glaciation |
| Precambrian–Cambrian boundary | × | | | possible control of preceding glaciation |
| Middle Cambrian fauna | | | | |
| Development of land plants | | | | |
| Late Ordovician extinction | × | | | possible control of glaciation |
| Late Devonian extinction | | | × | |
| Permiam–Triassic boundary | × | × | | Pangea controlled climate change and disruption of oceanic circulation |
| Development of calcareous plankton | | | | |
| Triassic–Jurassic extinction | | × | | |
| Cretaceous–Tertiary boundary | | | × | |
| Paleocene–Eocene extinction | | | | |
| Late Eocene extinction | | | × | |

Possible controls are marked with an ×. Blank spaces indicate no discernible control. All information based on references cited in text.

made atmosphere and surface water poisonous to many organisms. Some extinction events have been attributed partly to eruption of large regions of plateau basalts, but some plateau basalt provinces developed at times when no significant biotic change occurred. Large areas of subaerial plateau basalts can form only on continents, and they may have formed only during continental rifting and/or passage over a plume (chapter 2).

- Impacts of large bolides apparently could have caused almost instantaneous extinction, and repeated impact of smaller meteorites may have contributed to extinction events that occurred over several million years. The general effect on biotic evolution is unclear, however, because several very large impacts in the Phanerozoic occurred at times when there were no significant biotic overturns.

Table 12.1 summarizes the 15 episodes of biotic change discussed in the previous section and shows our best estimate of the possible involvement of various fundamental causes. The table does not show transgressive/regressive or anoxic/dysoxic episodes that may have been directly responsible for biotic change but only the underlying stresses that led to these direct causes.

Although all of the 15 episodes listed in table 12.1 presumably had one or more fundamental causes, we have found none for seven of them. The earliest prokaryotes and multicellular organisms undoubtedly evolved for fundamental reasons, but none are easily identified. The only apparent reason for the evolution of eukaryotes is some unknown and hypothetical involvement of tectonics/continents in the development of atmospheric oxygen. Similarly we find no fundamental cause for the evolution of the unusual Middle Cambrian fauna, the first land plants, the first calcareous plankton, or the thermal maximum at the Paleocene–Eocene boundary.

Four of the events shown in table 12.1 occurred during or shortly after periods of climatic disruption, mostly associated with glaciation. They include development of Ediacaran biota during the last stages of Neoproterozoic glaciation and the Precambrian–Cambrian transition that occurred several million years after glaciation ended. Whether the Neoproterozoic glaciation was caused by the position of continents near the South Pole is controversial. The Late Ordovician and Permian–Triassic extinctions also occurred at or near the end of glacial intervals, and they clearly formed when different parts of Gondwana passed across the South Pole during the Paleozoic.

The Permian–Triassic extinction was probably caused by the effects of Pangea in addition to Permo-Carboniferous glaciation. The supercontinent affected global climates by cutting the east–

west flow of ocean water and also by the development of mountains along its western edge. This gave Pangea a generally warm climate despite glaciation in Gondwana and created large areas of aridity. Eruption of the Siberian basalt province at the same time as the major extinction undoubtedly contributed to the Permian–Triassic biotic overturn, but it seems likely that the major source of stress was the existence and configuration of Pangea.

In addition to its contribution to the Permian–Triassic extinction, eruption of plateau basalts may have been the direct cause of the biotic overturn at the Triassic–Jurassic boundary. Volcanism, however, seems to have had no effect on other times of major biotic crises.

Bolide impact clearly was the major cause of the Cretaceous–Tertiary extinction. A large bolide may also have been responsible for some of the development of Ediacaran organisms, and groups of smaller bolides may have contributed to extinctions in the Late Devonian and Late Eocene. No other extinction events can be attributed to bolides, and some large impacts occurred at times when there was no significant biotic overturn.

## Summary

The continual evolution of new organisms and extinction of old ones throughout almost all of earth history has passed through several times when totally new types of organisms developed. These times in the Phanerozoic commonly follow episodes of mass extinction, but extinction events are unknown in the Precambrian until near the end of the Proterozoic.

Whether the fundamental causes of these developments among organisms are internal to the earth (tectonic) or external is largely unknown. Movement of continents and the assembly and dispersal of supercontinents seems to have influenced some of the events, particularly the Permian–Triassic extinction, but apparently had no effect on other events. This conclusion implies that organic evolution on the earth's surface was mostly decoupled from processes within the solid earth.

# Appendix A

## Seismic Methods

An elastic material is one that deforms and then returns to its original condition without any permanent distortion. This happens if we stretch a rubber band and then let it go without breaking it, and it also happens whenever the earth is subjected to short-term stresses such as earthquake movements. The earth is elastic to rapid motions, such as seismic waves, and in this appendix we show how seismic waves have been used to locate the earth's two major discontinuities: the Moho and the core–mantle boundary.

The energy that causes an earthquake or other stress is dissipated through the deformed elastic material as a series of seismic waves. In rocks, these waves consist of primary (P) waves that oscillate back and forth like sound waves in air (fig. A.1). A secondary (S) set of waves is also caused by energy release in rocks and travels by a side-to-side (transverse) motion (fig. A.1). Because of their motions, P waves can travel through solids, liquids, and gases, but S waves can travel only through solids. Velocities of the waves are determined by the elastic properties of the media through which they travel according to the equations

$$V_P = [(K + \frac{4}{3}\mu)/\rho]^{1/2}$$

and

$$V_S = (K/\rho)^{1/2}$$

where

$V_P$ is the velocity of the P wave
$V_S$ is the velocity of the S wave
$K$ is the bulk modulus (incompressibility)
$\mu$ is the rigidity modulus (a measure of shear strength)
$\rho$ is the density

Both $K$ and $\mu$ are constants that are used to describe elastic behavior. Because $V_P$ is determined by both $K$ and $\mu$ but $V_S$ depends only on the bulk modulus, the P wave is faster than the S wave, and in almost all earth materials $V_P/V_S$ is typically about 5/3.

The two equations above suggest that velocities of both P and S waves are inversely proportional to the densities of the materials they pass through. In actual rocks, however, the velocities of both waves increase as density increases because increase in $\rho$ also causes rapid increase in both $K$ and $\mu$. In the earth, this means that both $V_P$ and $V_S$ generally increase downward as density increases under increasing pressure of overlying

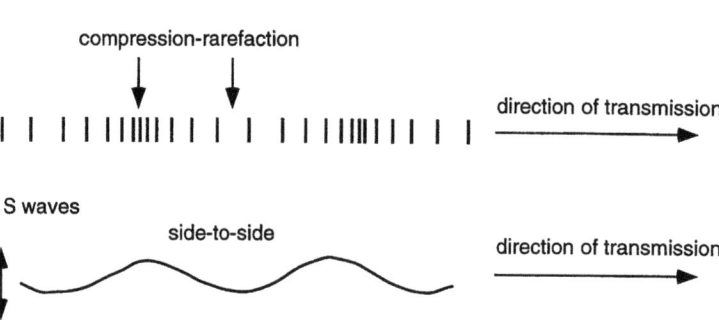

Figure A.1. Methods of transmission of seismic waves.

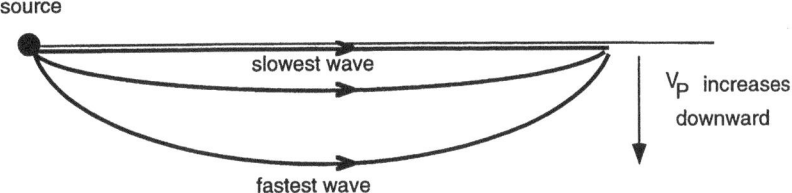

Figure A.2. Relative velocities of seismic waves at different depths.

rock. Increase of velocity downward in the earth causes seismic waves that travel deeper in the earth to arrive earlier than those that remain at shallow levels (fig. A.2).

Discontinuities disrupt simple travel-time curves such as those in fig. A.2. Discontinuities occur where seismic velocities change abruptly across a boundary, because the rock types either have different compositions or consist of different mineral phases (see discussions of various discontinuities in chapters 1 and 4). Seismic waves that pass across discontinuities are refracted so that the angle between the wave and the discontinuity on one side is different from the angle on the other side (fig. A.3). The relationship is referred to as Snell's law

$$(\sin \alpha)/V_1 = (\sin \beta)/V_2$$

The equation shows that waves travel closer to the discontinuity (farther from the vertical) in rocks with higher seismic-wave velocities. This relationship is identical to the one followed by light waves, with the index of refraction used in optics inversely proportional to the velocity of light in the medium of transmission.

Determination of the depth to discontinuities is explained in fig. A.4, which shows how seismic energy can be transmitted as P waves from a source, such as an earthquake, to a set of receiving seismographs along two different paths. A direct wave passes through the upper layer and arrives at the seismographs at times controlled by the distance between each seismograph and the source. The velocity of this wave can be measured from the slope of the time–distance graph. An indirect wave passes down from the energy source through the upper layer, moves laterally along the lower layer, and then must pass upward through the upper layer before it arrives at any of the seismographs. At distances close to the energy source, the indirect wave arrives later than the direct one, but the higher velocity in the lower layer causes the indirect wave to arrive earlier at seismographs at some distance from the source. The distance at which both waves arrive at the same time depends on their velocities in both layers and the depth to the discontinuity, which can be calculated because the wave velocities can be measured from the time–distance graph.

Time–distance graphs such as fig. A.4 are commonly used for general investigations based on earthquake sources and for academic, petroleum, and other commercial work using explosives as energy sources. In chapters 1 and 4 we use the results of these studies largely for discussions of the depth and configuration of the Moho.

The depth of the core–mantle boundary cannot be obtained as easily as the depth of the Moho and other discontinuities that are relatively shallow. The existence of the core and its radius was originally recognized by analysis of the arrival times of waves that had traveled through the entire earth from their earthquake sources (fig. A.5). Measuring these distances as angles from the source, the time–distance curve for P waves shows a smooth increase in travel time with distance up to 103.5° from the source. At 103.5°, however, this curve is interrupted, with delayed and irregular arrivals in the region from 103.5° to 143°, which is referred to as the "shadow zone." Beyond this region, from 143° to 180° the smooth curve of arrival time versus distance is also present but is delayed in comparison with the curve in the region from the source to 103.5°.

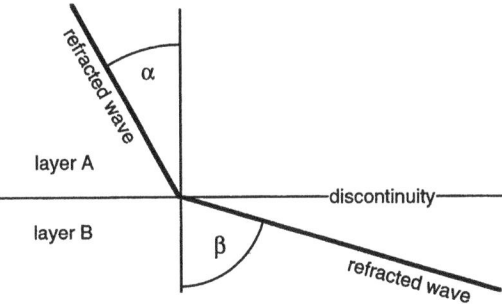

Figure A.3. Refraction of seismic waves across discontinuities. The velocity in layer A is less than the velocity layer B.

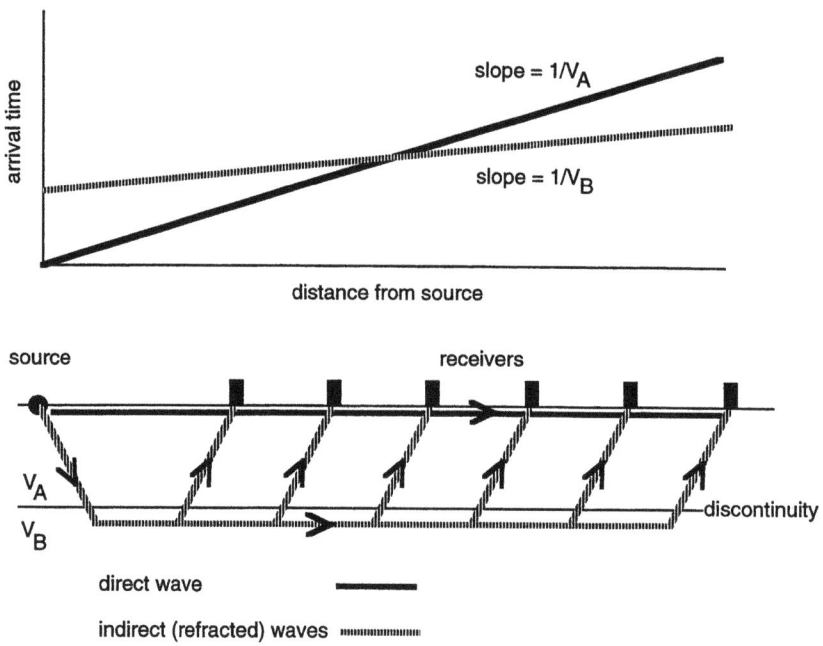

Figure A.4. Determination of depth to seismic discontinuity. $V_A$ is less than $V_B$.

The observations shown in fig. A.5 lead to the concept that the earth contains a core that is significantly different from the overlying mantle. The simplest interpretation of the time–distance curves is that seismic waves deeper than the ones that emerge at 103.5° are refracted downward into a core with slower travel times than in the overlying mantle. This direction of refraction and the lower velocities cannot be the result of decreasing density downward, and the easiest explanation is that at least the outer core is a high-density liquid (although the inner core is probably solid). This liquid nature is further confirmed by the observation that S waves do not travel through the core (see above). We provide a brief description of the core in chapter 1 and in discussions of the earth's heat generation in chapter 2.

Seismic investigations yield further information about the earth by a large variety of techniques that we cannot discuss here. Velocity changes across the Conrad discontinuity are too small to be detected by the method described in this appendix. The location of one or more low-velocity zones in the mantle also requires more sophisticated techniques. In the past decade, simultaneous processing of seismic data from the whole earth has permitted construction of three-dimensional models of large parts of the earth's interior. We refer people who want more information to Lowrie (1993) and Bolt (1999).

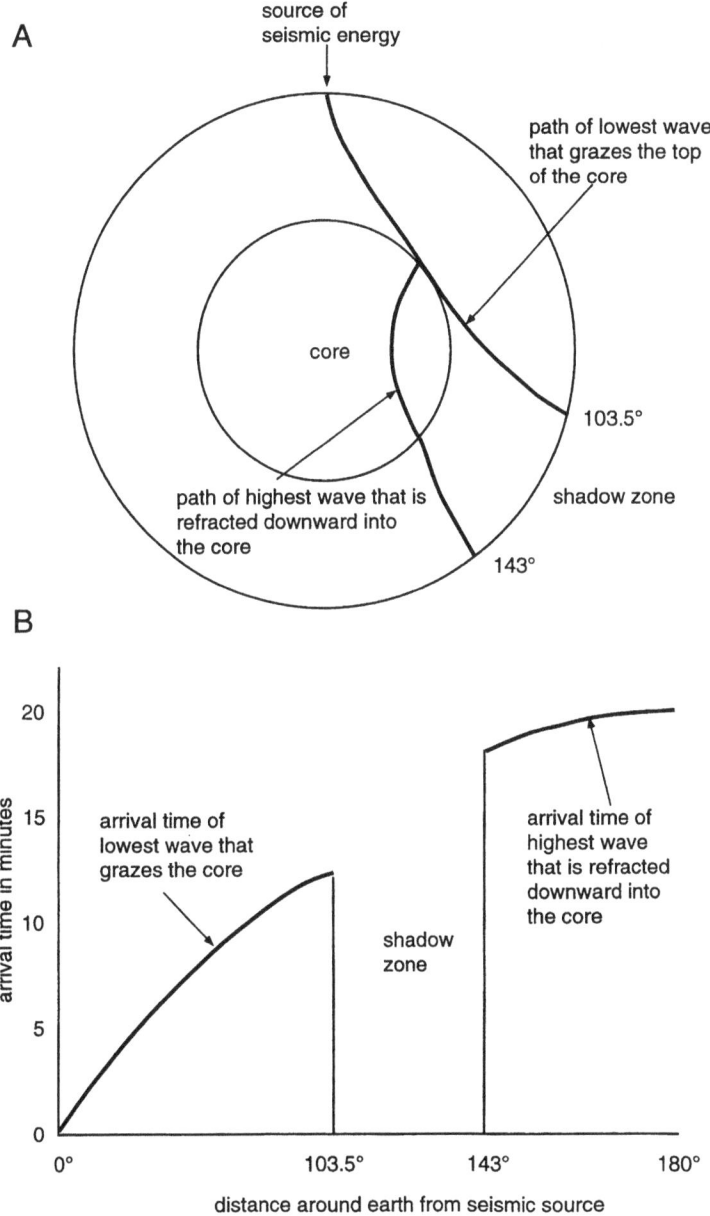

Figure A.5. Recognition of core.

# Appendix B

## Heat Flow and Thermal Gradients

Despite the number of heat sources in the earth and uncertainties over their relative importance in the past and at present, only a limited number of measurements can be made that provide any information about the earth's present thermal condition and its history. In this appendix we describe measurements of the present heat flow, the reason that the earth's mantle is convecting, and inferences that can be drawn about past thermal conditions by determining temperatures and pressures at which rocks came to equilibrium in the earth's interior. This appendix is closely linked to discussions in chapters 1, 2, and 3.

Heat flow is measured by determining temperature gradients near the earth's surface and the thermal conductivity of the rocks that these gradients are established in. Gradients are measured by dropping a set of thermometers down a hole and making simultaneous readings of the temperature at several depths. Holes on land generally must be hundreds or more meters deep in order to obtain temperature measurements that are not disturbed by percolating groundwater, but holes in deep ocean sediments can be as shallow as 10 m. Thermal conductivity is obtained by making laboratory measurements of the ease with which heat flows through rocks in cores from the holes, and heat flow (Q) is calculated by the equation

$$Q = \text{thermal gradient} \times \text{conductivity}$$

Where thermal gradient is in degrees Celsius per kilometer and conductivity is in watts per meter per degree, heat flow is reported in units of milliwatts per square meter ($mW/m^2$). This unit replaces an older one referred to as "heat flow unit" (HFU), which is microcalories per square centimeter per second and is equal to 41.87 $mW/m^2$.

Surface heat flow can be combined with the abundances of radioactive elements in near-surface rocks to obtain a value referred to as "reduced heat flow" ($Q^*$). Measurement of the concentrations of K, Th, and U in cores from the holes where temperature gradients were obtained permits calculation of near-surface radioactive heat production ($A$) in units of microwatts per cubic meter. Combining rates of heat flow and rates of production at different places within a terrane commonly yields a linear relationship of the type shown in fig. B.1.

Extrapolation of the line in fig. B.1 to zero heat production yields $Q^*$, which is the heat flow that would be measured in the terrane if near-surface rocks generated no heat. Because all radioactive elements are also LIL elements (chapter 3), they are concentrated in the upper continental crust and commonly account for more than one half of the heat flow in investigated terranes. This yields reduced heat flows that are very low in most continental regions. Because the rate of decrease in concentration of radioactive elements downward in the upper crust can only be estimated, the exact meaning of the reduced flow is unclear, but it is commonly regarded as heat flowing out of the lower crust and mantle.

The existence of convection in the earth's mantle that we discuss in chapter 1 results from the relationship between actual temperature gradients and the gradients required for stability in a gravitational field. A thick body, such as the mantle, is mechani-

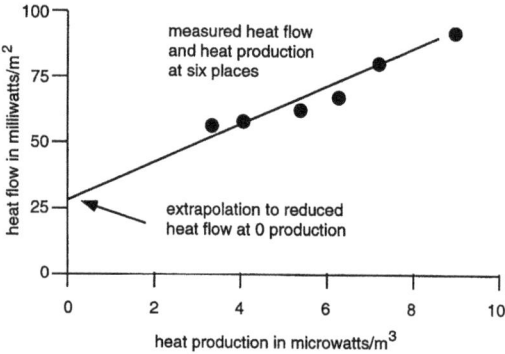

Figure B.1. Calculation of reduced heat flow. Plotted points represent measured heat flow and production at different places within a terrane.

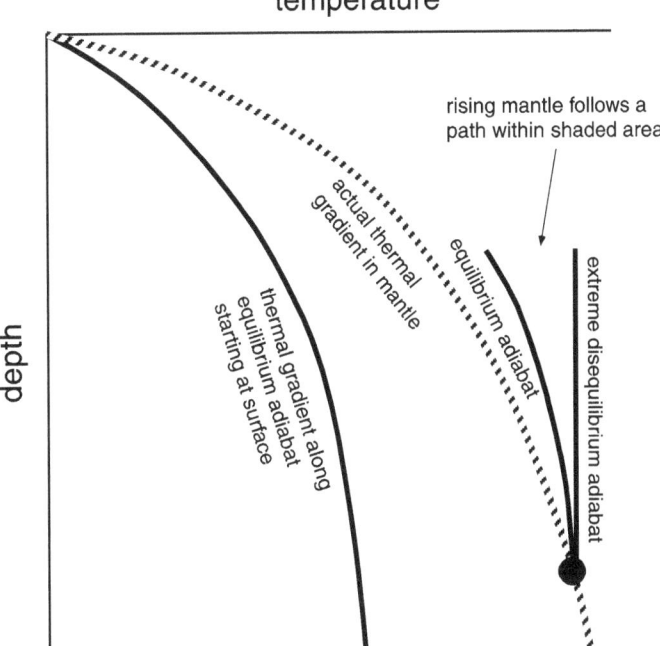

Figure B.2. Cause of convection in the mantle.

cally stable in a gravity field only if the downward increase in temperature to areas of higher density (caused by higher pressure) follows the equation

$$\left(\frac{\delta T}{\delta P}\right)_S = \frac{\alpha T}{\rho c_P}$$

where $T$ and $p$ are temperature and pressure, $S$ refers to constant entropy, $\alpha$ is the coefficient of thermal expansion, $\rho$ is density, and $c_P$ is heat capacity at constant pressure. This equation defines a temperature gradient known as an "equilibrium adiabat," the rate of change of $T$ and $p$ for a body moving vertically without exchanging heat with its surroundings and at a rate that is infinitely slow.

We use fig. B.2 to demonstrate that mantle convection occurs if the mantle is mobile and actual thermal gradients are higher than the gradient along the equilibrium adiabat (further discussion in chapter 1). Because of the time needed for thermal equilibration in the mantle, any spontaneous movement of material will probably be adiabatic. Thus, material moving vertically without gain or loss of heat from any point on the p–T grid follows an adiabatic path that shows its change in $T$ and $p$. Along an adiabatic path, when material rises to lower pressure it expands and converts internal energy into work, thus reducing the remaining energy and decreasing the temperature. Similarly, material moving downward heats up because of compression. Infinitely slow movement forms an equilibrium adiabat, with maximum change in temperature. More rapid movement occurs along any of a series of disequilibrium adiabats, with slopes as high as vertical (no change in temperature).

Because all adiabats are steeper on this diagram than the actual thermal gradient in the mantle, any rising material has a higher temperature than its surroundings, and it expands and becomes less dense than its surroundings. Similarly, falling material becomes denser than the surroundings. Both types of movement cause mechanical instability that results in continual convection in the mantle (fig. B.2). Thus, if the actual temperature gradient in the mantle is higher than the stable gradient along an equilibrium adiabat, the mantle must undergo convection.

Measurement of past thermal gradients requires estimates of temperatures and pressures preserved in ancient rocks and minerals. Many of these estimates can be made from assemblages of metamorphic minerals using a facies diagram such as the one shown in fig. B.3b. A rock containing mineral suites that plot at any point in this diagram presumably formed at the temperature and depth

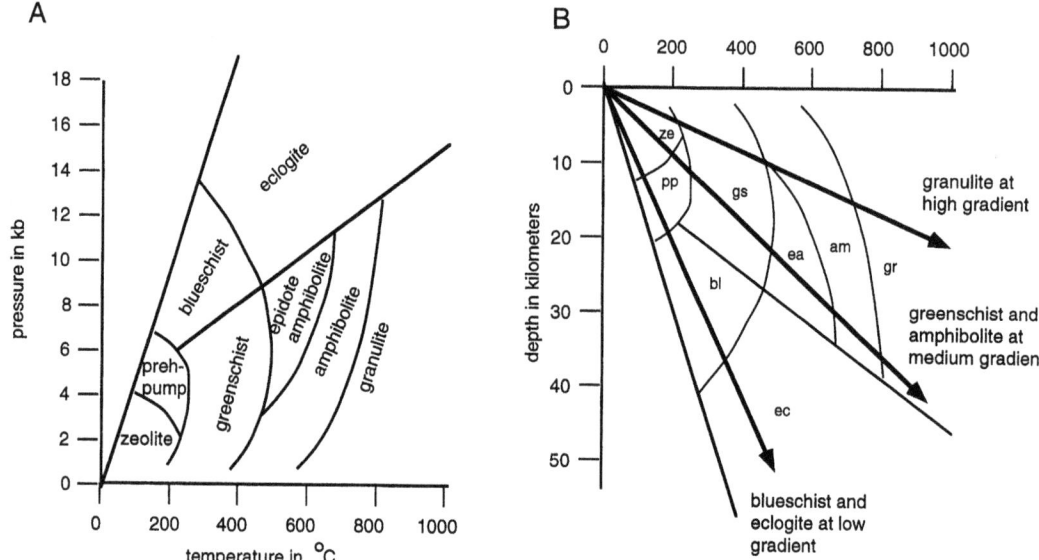

Figure B.3. a. Phase diagram for rocks that occur in continental crust, showing p–T ranges of metamorphic facies. b. Phase diagram inverted to show facies of metamorphic rocks at different depths (using abbreviations of facies shown in fig. B.3a). Three different thermal gradients illustrate development of different metamorphic sequences. The lowest gradient is commonly developed in the seaward part of zones where oceanic lithosphere is subducted and passes through stability fields of blueschist and eclogite. The intermediate gradient passes through greenschist and amphibolite fields and is typical of metamorphism under continental margins and during continental collisions. The highest gradient is commonly formed by contact metamorphism.

shown by that point, thus permitting an estimate of the thermal gradient when the rock formed. Figure B.3b illustrates this method by showing how metamorphism at different thermal gradients develops different sequences of rocks.

Diagrams such as fig. B.3b are less useful for the mantle than the crust because almost all rocks in the mantle are some variety of peridotite dominated by olivine. An approximate depth zonation based on minerals less abundant than olivine shows a zonation from plagioclase-bearing rocks near the top of the mantle to garnet peridotite at greater depths. Spinel peridotites commonly occupy a depth region between those of both the plagioclase- and garnet-bearing varieties, with the spinels consisting of oxides containing some mixture of Fe, Cr, Mg, Al, and Ti.

The approximate estimates that can be obtained from fig. B.3b and the bulk mineralogy of peridotites can be greatly refined by detailed measurement of the temperatures and pressures at which minerals and rocks formed (geothermometry and geobarometry). Numerous techniques are available, and we mention only a few of the major ones here.

Broad estimates of the thermal history of rocks are sometimes inferred from certain index minerals or by identifying key mineral assemblages such as those that characterize different metamorphic facies. With the advent of the electron microprobe, it became possible to analyze the chemistry of very small domains of minerals within rock sections that can be used to precisely compute pressures and temperatures based on thermodynamic application of theoretical, empirical, and experimental mineral phase equilibria. One such technique is the determination of equilibration temperature based on the distribution of Fe and Mg between two ferromagnesian minerals that formed at equilibrium conditions within the same rock. This method is commonly applied to rocks that contain garnet associated with orthopyroxene, clinopyroxene, amphibole, biotite, cordierite, or olivine. The procedure involves the calculation of the ratio:

$$(Fe/Mg)_{\text{mineral A}} / (Fe/Mg)_{\text{mineral B}}$$

Since the "partitioning" of Fe and Mg between coexisting mineral pairs is a function of temperature, the measured Fe and Mg contents in minerals can be used to compute the temperature of equilibration of the mineral pair. The variations in Fe and Mg from core to rim of single grains can also be used to calculate the changing thermal gradients. This technique finds wide application in estimating the temperature of equilibration of metamorphic mineral assemblages at various depths in the earth's crust. It should be noted that the computations are also dependent on the concentrations of various other components in coexisting minerals and the presence or absence of different mineral phases in the rock.

The distribution of Fe and Mg among mineral phases, however, provides little information about the pressure of equilibration. Determination of pressure is mostly based on the ability of Al to occupy sites of tetrahedral coordination and also sites of 6-fold coordination in minerals. In low-pressure minerals such as plagioclase, Al is almost invariably in 4-fold coordination, whereas Al has a 6-fold coordination in denser minerals such as garnet. Precise calculations require equilibrium assemblages with coexisting minerals that have variable Al concentrations. One of the commonly used systems takes into consideration the distribution of Al in the assemblage:

$$\text{Garnet(6-fold Al)} + \text{plagioclase (4-fold Al)} + Al_2SiO_5 + \text{quartz}$$

Many high-grade supracrustal rocks contain the assemblage garnet–aluminosilicate–plagioclase–quartz, and are hence suitable for geobarometry. Dry granulite-facies rocks that characterize a large part of the exposed middle and lower continental crust commonly contain a garnet–orthopyroxene/clinopyroxene–plagioclase–quartz assemblage, which also provides a potential barometer. More details on thermobarometry based on mineral phase equilibria can be found in Ferry (1986).

Estimation of pressure and temperature from minerals and rocks is also done using a variety of other techniques. Out of these, the application of fluid inclusion and stable isotope techniques is now widely employed. Fluid inclusions trapped in minerals preserve various proportions and combinations of gases, liquids, and melts from metamorphic, magmatic, or hydrothermal environments (chapter 4 and appendix F). By observing the temperature of phase changes within the inclusions, the composition and density of the fluids can be characterized. The data are then used to compute isochores (lines of constant volume) in pressure–temperature space. Simultaneous estimation of pressures and temperatures can be derived by intersection of isochores of different categories of fluids trapped at the same time. Where an independent estimate of temperature is available from mineral thermometry, the intersection of the temperature with the isochore determined from fluid inclusions yields the pressure. One of the underlying assumptions of this technique is that fluid inclusions remained as closed systems from the time of trapping, which is verified by finding inclusions trapped at peak pressure and temperature during rock formation still preserved even in rocks that equilibrated in the continental lower crust (Santosh and Tsunogae, 2003).

Temperatures can also be estimated from the partitioning of stable isotopes between coexisting minerals. One of the most widely applied methods is based on carbon-isotope exchange between coexisting calcite and graphite in high-grade metamorphic rocks. The partitioning of $^{13}C$ and $^{12}C$ between coexisting calcite and graphite is a function of the temperature of equilibration of these carbon-bearing phases (Valley, 2001). Experimental and empirical calibrations of the calcite–graphite thermometer have been applied to derive the thermal structure of continental crust, even in the case of extreme crustal metamorphism involving ultrahigh-temperature (>1100°C) metamorphic conditions (Satish-Kumar et al., 2002).

Under suitable conditions, studies of the variation of pressure–temperature conditions in rocks can be expanded to obtain information on the changes of these parameters through time. Thus, the "$p$–$T$–$t$" path of a rock can be defined by care-

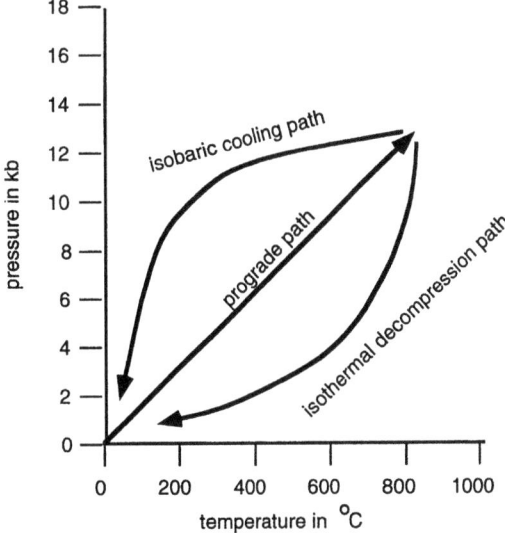

Figure B.4. Idealized isothermal decompression (ITD) and isobaric cooling (IBC) paths.

ful studies of mineral assemblages and reaction textures that developed in a rock at various stages of its formation or exhumation. The $p$–$T$–$t$ path has two components: the "prograde" path in which the rock was taken down to deeper portions of the earth and was subjected to high temperatures and pressures, and the "retrograde" path through which it was brought back (exhumed) to shallower levels after metamorphism.

The nature of prograde and retrograde paths is different in different tectonic environments, such as in zones of collision or extension. The two common types of $p$–$T$–$t$ paths described from metamorphic rocks are "isothermal uplift" or "isothermal decompression" (ITD), where hot rocks are brought up rapidly, and "isobaric cooling" (IBC) where rocks reside at a certain depth for longer time and undergo slow cooling. We show idealized ITD and IBC paths in fig. B.4. A combination of studies from mineral assemblages, textural relations, and chemical zoning preserved in minerals help in reconstructing the pressure–temperature–fluid history and $p$–$T$–$t$ paths and thus aid in defining the thermal gradients within the crust.

# Appendix C

## Paleomagnetism

The earth's magnetic field affects all rocks and minerals at all times, but minerals that contain iron can retain a "memory" ("remanence") of the field that existed at the time they were originally formed. The strongest magnetism is in magnetite, which is attracted by a simple hand magnet, but weaker magnetic fields are retained in all of these "ferromagnetic" minerals and can be measured with laboratory instruments. Because the magnetic properties of naturally occurring minerals have been imposed by the earth's magnetic field, the minerals contain a record of their past positions on the earth and of variations in the earth's magnetic field through time. We use this appendix to show how this information can be used to decipher both the history of individual suites of rocks and the history of tectonic processes.

Magnetic fields are generated by the movement of electrically charged particles. Although the exact causes of the earth's magnetic field are controversial, it is clearly related to rotation of the earth's iron–nickel core, which presumably indicates that the earth has had a magnetic field since the core segregated from the rest of the earth shortly after the earth accreted. This field is enormously complicated, but about 80% can be described as a bar magnet (magnetic dipole) with north and south "geomagnetic poles" that passes through the center of the earth and is tilted about 11.5° away from the earth's axis of rotation (fig. C.1). Torsvik and Van der Voo (2002) estimated that as much as 25% of the magnetic field in the Phanerozoic was not a simple dipole.

Compass needles do not point toward the geomagnetic poles but to the north and south "magnetic poles." If the magnetic field were a simple dipole, then these poles would be at the two ends of the bar magnet and would be identical to the geomagnetic poles. Other influences on the magnetic field, however, place the present north magnetic pole at about latitude 80° N, longitude 109° W and the south magnetic pole at 65° N, 138° E.

The positions of the magnetic poles vary slightly through time. In the past century the north magnetic pole has moved about 10 km/yr, and the south magnetic pole has shown similar motion in the southern hemisphere. Despite this variability and the difference between the magnetic axis and rotational axis of the earth, accumulated paleomagnetic data indicate that the magnetic and rotational axes have been close to each other throughout geologic time and that magnetic pole positions averaged over a period of 3000–10,000 years are identical to geographic poles. The possibility that the magnetic axis has had a very different orientation in the past ("true polar wander") has been suggested, and we discuss it briefly in chapter 11.

Because the magnetic poles are not at the rotational poles, the orientation of the magnetic field at

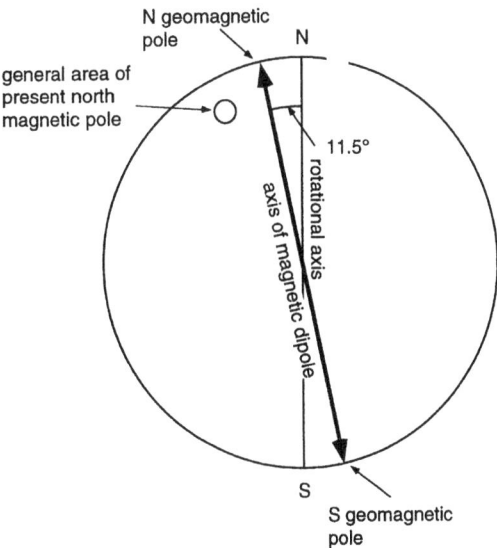

Figure C.1. Comparison of the earth's rotational axis, magnetic dipole axis, and locations of magnetic poles. The south magnetic pole is on the hemisphere not shown here.

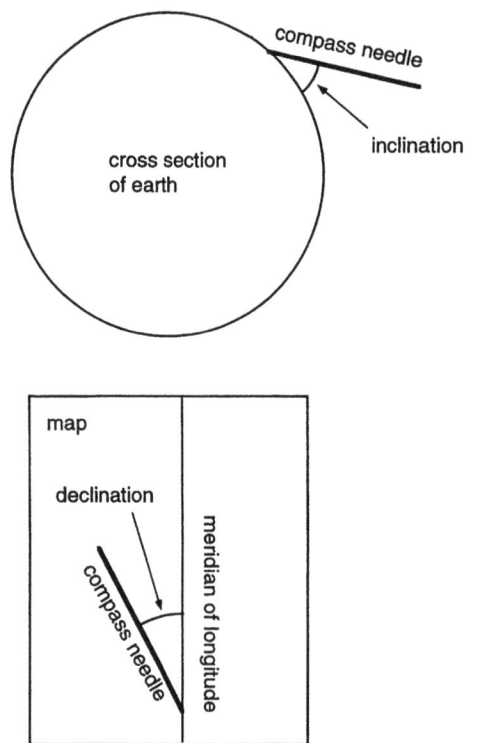

Figure C.2. Definition of inclination and declination.

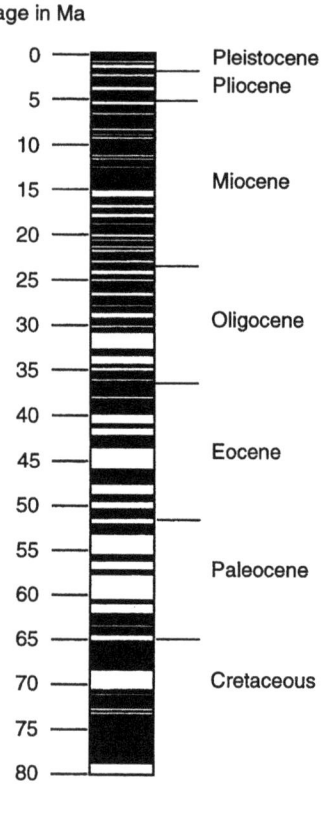

Figure C.3. Magnetic time scale for past 80 million years (based on McElhinny and McFadden, 1999).

any point on the earth's surface must be described by two different measurements (fig. C.2). The "inclination" is the angle that the field makes with the earth's surface at that point. It is 90° at the north and south magnetic poles and at all latitudes is described by the equation tan(inclination) = 2tan(latitude). The "declination" is the angle measured on the surface between true north and south (along a meridian of longitude) and the direction of the compass needle. If a compass needle were mounted in a sphere, the inclination would be approximately the angle between the needle and the earth's surface, and the declination would be the angle between the needle and a meridian of longitude that passes through the earth's geographic poles.

In addition to small movements of the magnetic field, the north and south magnetic poles episodically reverse themselves ("flip-flop"), with magnetic north switching to magnetic south and vice versa. The reason and mechanism for this exchange are unclear, but detailed magnetic measurements of the most recent reversals suggest that the magnetic poles need a few thousand years to rotate around the earth from their previous positions to their new positions. When the reversal is complete, the former north-seeking pole of a compass points to the south, and the south-seeking pole toward the north. The present is defined as a time of "normal" polarity, with magnetic north near the north geographic pole, and "reverse" polarity is a time when magnetic north and south are in opposite positions.

Efforts to find some periodicity to the reversals have failed to detect anything except random variation. In the past ~80 million years, major reversals have occurred approximately every 2 million years, and there have been numerous short-lived reversals, many of which have lasted less than 100,000 years (fig. C.3). A period of normal polarity without reversals occurred between 118 Ma and 83 Ma, and a period of reverse polarity lasted from about 316 Ma to 262 Ma.

The earth's magnetic field imposes magnetic orientations on iron-bearing minerals both when they form and during later events. Initial orientations (remanences) develop when minerals grow in metamorphic rocks and sediments and also during deposition of clastic sediments. In igneous and high-temperature metamorphic rocks, the crystallization of these minerals may be above the "Curie temperature," which prevents minerals from retaining a record of the earth's magnetic field (and hence the rock's position) until they cool down below the Curie temperature. All of these initial orientations may then be further modified when the rocks are altered ("overprinted") by chemical and thermal processes as they move through the earth's magnetic field because of seafloor spreading, continental drift, and other processes. For example, metamorphism at temperatures above the Curie temperatures of the different minerals may destroy ("reset") the original orientations in the rocks.

Because initial orientations are complexly altered by movement of minerals around the earth and by other events, it is necessary for paleomagneticists to investigate the entire magnetic history of materials they are studying by isolating the different directions ("components") of magnetization within a rock sample. In some cases it is possible to recognize a series of discrete events, such as initial crystallization, followed by low-temperature metamorphism, and even recent weathering that modified the original magnetic orientation. For many studies, the principal task is to remove later magnetizations in order to determine the initial one, which is then used to determine where the mineral or rock was on the earth's surface when it was first magnetized.

Paleomagneticists "clean" later magnetizations from minerals and rocks in three ways in their laboratories. One is by chemical solution, which can destroy diagenetic and other late-forming minerals. A second procedure is measuring the orientation of the magnetic field as a mineral or rock is progressively heated. This removes later, and presumably weaker, magnetizations as the temperature rises toward the Curie point. A third method is to subject the mineral to rapidly alternating magnetic fields, which also removes the weaker magnetizations.

Paleomagnetic information has been used in two principal ways. First we discuss the construction of a time scale based on magnetic reversal sequences, which can be correlated with radiometric dates and is then useful for dating suites of rocks for which absolute dates are not available. The reversal sequence is very important in showing the patterns of opening of modern ocean basins (chapter 1) and has also been used to correlate some events on continents. The second use of paleomagnetic data is mapping the movement of continental regions through time by locating apparent pole positions for rocks of different ages on the continents. This yields "apparent polar wandering (APW) curves" that provide information on continental movements and their relationships to each other.

Rocks that are magnetized when they form along mid-ocean spreading centers carry their magnetic orientations with them when they move away from the centers. Rocks that are forming now and during past times of normal polarity have magnetic orientations that enhance the strength of the earth's magnetic field near them, but those that formed when polarity was reversed reduce the strength of the magnetic field. This alternation of enhancement and reduction produces an alternating series of positive and negative deviations ("anomalies") from average magnetic intensities on each side of the spreading ridge as older rocks move farther away. These "stripes" provided some of the best initial support for the concept of seafloor spreading, and we discuss them further in chapter 1 and fig. 1.7.

By combining magnetic patterns in the oceans with absolute dates from oceanic crust, it is possible to construct a magnetic time scale for both large and small reversals during the past ~175 million years. Older rocks in ocean basins have now been completely subducted (chapter 1), and it is necessary to investigate dated rocks on land in order to extend the time scale further back. Figure C.3 shows the complexity of the time scale for the past 80 million years.

Because the magnetic reversals show no repetitive periodicity, the pattern of reversals can be used as "finger prints" to date suites of rocks that cannot otherwise be dated or to confirm dates obtained by other means. This method has been used to date regions of ocean basins that have been isolated from spreading ridges either because the ridges have been subducted or because of plate reorganizations within the basins. Some suites of rocks on land have also been dated magnetically, although these dates are commonly accepted only when confirmed by other methods.

Apparent polar wandering curves are constructed by measuring the magnetic orientations of rocks of different ages within one coherent region, such as a drifting continent. In each rock, the magnetic inclination shows the latitude (paleolatitude) at which the rock formed, but it does not show whether the indicated pole was north or south because the rock may have formed during a period of either normal or reversed polarity. The declina-

202  Continents and Supercontinents

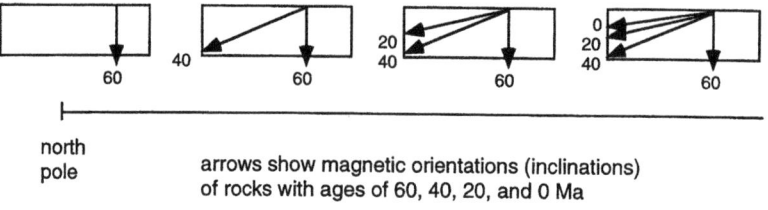

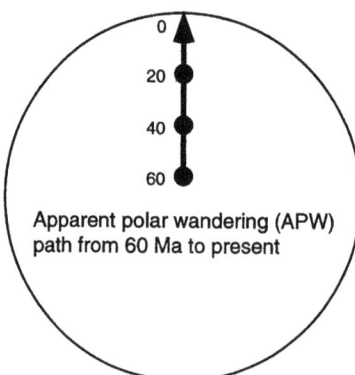

Figure C.4. Idealized APW path for continent moving south from North Pole in past 60 million years.

tion indicates the amount of rotation undergone by the rock since it formed.

We begin to illustrate the process of constructing APW paths with the simple example of a block of continental crust moving along a meridian of longitude away from the North Pole between 60 Ma and the present (fig. C.4). The oldest rocks carry magnetic orientations that are nearly vertical because they formed near the pole, and progressively younger rocks show increasing distances from the pole. This sequence of poles yields an APW path that appears to proceed in the opposite direction from the actual movement of the rocks measured. In constructing fig. C.4, the uncertainty between normal and reversed positions of poles has been resolved by placing apparent poles along a smooth path. For rock suites formed during the past few hundred million years, it is generally also possible to determine whether the path shows a north or south pole, but in older rocks this distinction is commonly difficult.

The paleolongitude at which a rock formed cannot be determined purely from magnetic orientation because the rock's magnet points toward the north or south magnetic poles wherever the rock is on the earth's surface. Consequently, paleomagnetic data

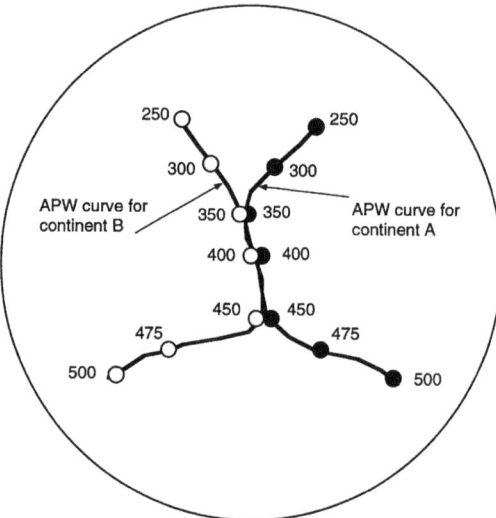

Figure C.5. Idealized APW paths for continents A and B (from J. Meert, personal communication). Ages of apparent pole positions are shown for both curves.

can only support other evidence, such as patterns of seafloor spreading, to indicate longitudinal movements of rocks. Figure 1.5 shows how early paleomagnetic data helped to confirm the separation of North America and Europe by opening of the Atlantic Ocean.

Locations of apparent pole positions for two different areas, such as two continents, can be used to show their relative movements. Figure C.5 shows the apparent pole positions for two continents that were separated before 450 Ma, joined at 450 Ma, moved together until 350 Ma, and then split apart again. These APW data would permit determination of absolute movements of both continents across latitudes and indicate that there was no longitudinal separation between them from 450–350 Ma. They would, however, provide no information on absolute longitudes of either continent.

This review has covered only a few aspects of the study of paleomagnetism, and we refer readers who wish more information to Van der Voo (1993) and McElhinny and McFadden (1999).

# Appendix D

## Isotopic Systems

Isotopic studies of the evolution of the earth inevitably start with the concept of a "primitive" earth that had not undergone any compositional fractionation. If we knew the concentrations of parent and daughter isotopes in this original earth, then we could use the half lives of the various parent isotopes to calculate the concentrations of daughter isotopes in the whole earth at any time in its 4.55-billion-year history. Because the earth has undergone compositional fractionation, however, it is impossible to measure these original concentrations directly, and they must be inferred by some indirect method.

Most methods for estimating original isotope ratios either depend on, or can be checked by, studies of meteorites. One variety of meteorites, known as chondrites, is used for most isotopes because it is hypothesized to represent the unfractionated composition of the earth. Because the meteorites are unfractionated, we can use present concentrations of parent and daughter isotopes, plus the half lives of the parents, to calculate back to the ratios that the meteorites had at any time since they were formed 4.55 billion years ago. Applying this method to the earth is commonly referred to as viewing the earth as a "chondritic uniform reservoir" (CHUR), and isotopic ratios calculated from this concept are designated with the subscript CHUR. The concept of CHUR is similar to the one used for Sr isotopes, but they are compared to meteorites classified as basaltic achondrites, which provide an initial Sr isotope ratio known as BABI ("basaltic achondrite best initial").

Five isotopic systems are important in our studies of the evolution of the earth: Rb–Sr, Sm–Nd, Lu–Hf, Re–Os, and the more complex U–Th–Pb system. We discuss each of them briefly below.

### Rb–Sr

Decay of $^{87}$Rb by beta emission with a half life of $48.9 \times 10^9$ years produces $^{87}$Sr, which is not radioactive. Concentrations of both isotopes are commonly reported as ratios with the nonradiogenic $^{86}$Sr, allowing construction of graphs such as fig. D.1. This diagram shows present isotope ratios measured in several different phases that formed at the same time from a common source, such as a suite of comagmatic igneous rocks or individual minerals in the same rock. If these phases have different proportions of Rb and Sr, then a plot of $^{87}$Sr/$^{86}$Sr versus $^{87}$Rb/$^{86}$Sr forms a straight line that extrapolates back to an "initial ratio" (Sr$_i$) at $^{87}$Rb/$^{86}$Sr = 0, which is the $^{87}$Sr/$^{86}$Sr ratio of the source at the time the suite of rocks or minerals originally formed. The caption for figure D.1 also shows how the age of the suite can be measured from the plotted data.

The Sr$_i$ of the primitive earth (BABI) is estimated to be approximately 0.699 based on studies of basaltic achondrites. If the earth had remained unfractionated, its present $^{87}$Sr/$^{86}$Sr would have been 0.702–0.703, but both Rb and Sr are fractionated upward into the upper mantle and crust. Furthermore, Rb is a LIL element (chapter 2) and is highly concentrated into continental crust, yielding very high Rb/Sr ratios that produced high $^{87}$Sr/$^{86}$Sr ratios. These fractionations have continued through the entire history of the earth, forming present $^{87}$Sr/$^{86}$Sr ratios that are high in continental crust and much lower in the depleted mantle that melts to form oceanic tholeiites along mid-ocean ridges. Figure D.2 shows the growth curve for depleted mantle (asthenosphere) and also the general ranges for isotopes in continental crust and lithospheric mantle beneath continents.

An extraordinary amount of information on the sources of magmatic rocks can be obtained by comparing their Sr$_i$ values at the time of magma generation against the growth curves in fig. D.2. For example, modern lavas with Sr$_i$ of 0.703 have probably been derived from a mantle source, whereas volcanic rocks with Sr$_i$ of 0.710 presumably were generated by partial melting of continental crust. Similarly, magmas that are 1 billion years old should have comparatively low Sr$_i$ if derived from

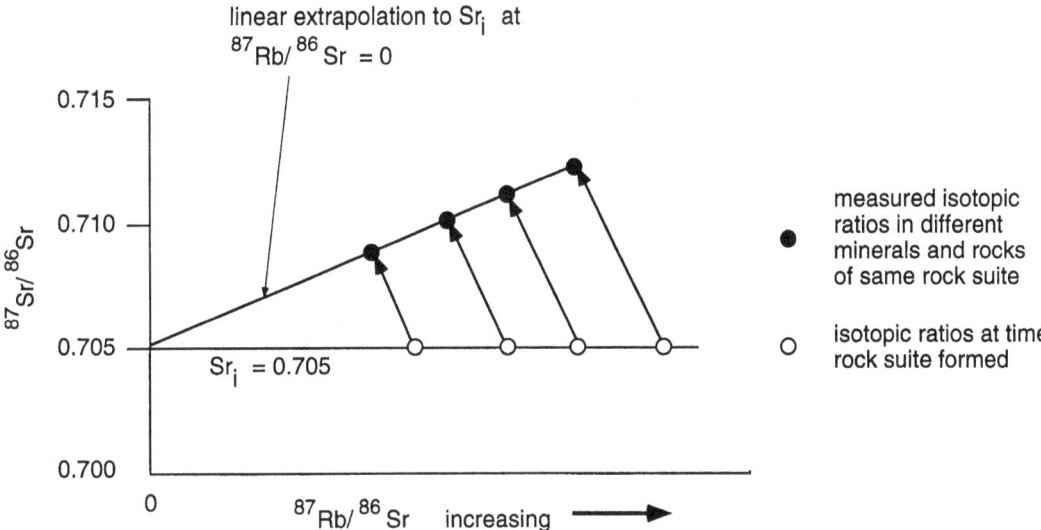

Figure D.1. Rb–Sr isochron for a comagmatic suite of rocks and/or minerals. The intercept at $^{87}$Rb/$^{86}$Sr = 0 is the initial ratio (Sr$_i$) for this suite of rocks. All rocks and minerals in this suite start with this ratio, and rocks with higher $^{87}$Rb/$^{86}$Sr ratios generate more Sr through time as the suite evolves. This yields a sloped line of present $^{87}$Rb/$^{86}$Sr versus $^{87}$Sr/$^{86}$Sr values, with the slope of the line proportional to the time since formation of the suite.

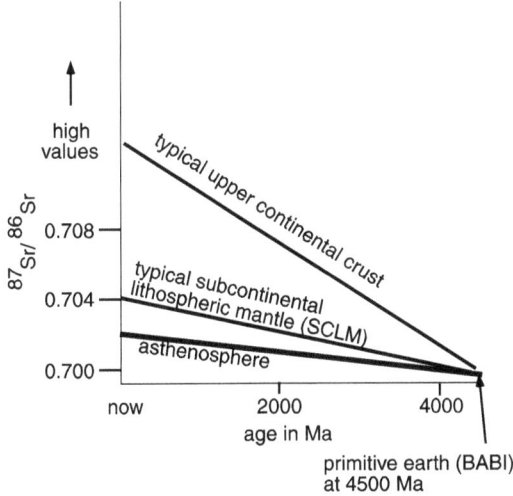

Figure D.2. Growth of $^{87}$Rb/$^{86}$Sr in the bulk earth and continental crust starting from the initial $^{87}$Rb/$^{86}$Sr of the unfractionated earth.

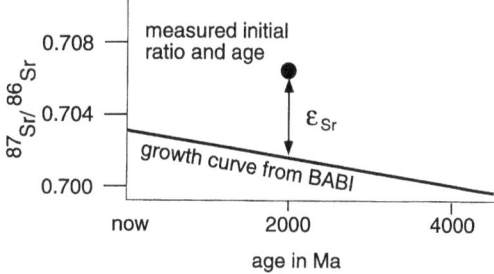

Figure D.3. Determination of $\varepsilon_{Sr}$.

the mantle and higher Sr$_i$ if from the continental crust.

The curves shown in fig. D.2 are difficult to interpret and it is customary to report Sr$_i$ values for individual suites in terms of their $\varepsilon_{Sr}$ values (fig. D.3), which are their deviations from mantle values at the age of the suite.

## Sm–Nd

In most ways, the Sm–Nd isotopic system is similar to the Rb–Sr system and can be analyzed in a similar fashion. Variations in Nd isotopes are caused by the radioactivity of $^{147}$Sm, which decays to $^{143}$Nd by emission of an alpha particle and has a half life of $106 \times 10^9$ years. This causes the concentration of $^{143}$Nd in the whole earth to increase with time and also increases the ratio of $^{143}$Nd to the nonradiogenic $^{144}$Nd. Individual rocks and rock suites can be dated by measuring their $^{143}$Nd/$^{144}$Nd ratios and their $^{147}$Sm/$^{144}$Nd ratios and determining ages and initial ratios using isochrons of the same type that we used above for the Rb–Sr system. For our purposes, however, it is more useful to calculate

$^{143}$Nd/$^{144}$Nd ratios in the mantle at the times when different volumes of juvenile crust were formed.

Just as in the Rb–Sr system, decay of $^{147}$Sm causes the $^{143}$Nd/$^{144}$Nd ratio to increase with time in the whole earth, but the major difference is that the Rb parent is the LIL element in the Rb–Sr system, whereas the Nd daughter is the LIL element in the SmNd system. Both Sm and Nd are rare earth elements with +3 valences, but the lighter Nd$^{+3}$ is a larger ion than Sm$^{+3}$ and fractionates more readily into melts than Sm. This causes the mantle to become depleted in Nd relative to the crust, and the $^{143}$Nd/$^{144}$Nd ratio increases much more rapidly in the mantle than in either oceanic crust or continental crust. The numerical differences between $^{143}$Nd/$^{144}$Nd ratios in the crust and mantle are smaller than the differences between the $^{87}$Sr/$^{86}$Sr ratios because the fractionation between Rb and Sr is much greater than the fractionation between Sm and Nd.

Figure D.4 shows variations in the $^{143}$Nd/$^{144}$Nd ratio throughout the history of the earth, starting from the initial ratios for the primitive earth using the CHUR model (initial $^{143}$Nd/$^{144}$Nd = 0.512; initial $^{147}$Sm/$^{144}$Nd = 0.197). The $^{143}$Nd/$^{144}$Nd ratio in an unfractionated earth grows slowly along the line labeled CHUR in the diagram. Because the mantle has selectively fractionated Nd upward into the crust at all times, the $^{143}$Nd/$^{144}$Nd ratio in the mantle has become higher than the $^{143}$Nd/$^{144}$Nd ratio in the bulk earth. Figure D.4 shows this change as a "depleted" (asthenospheric) mantle curve, labeled DM.

Instead of discussing $^{143}$Nd/$^{144}$Nd ratios as shown in fig. D.4, it is easier to use a diagram in which the growth curve for the entire earth (CHUR)

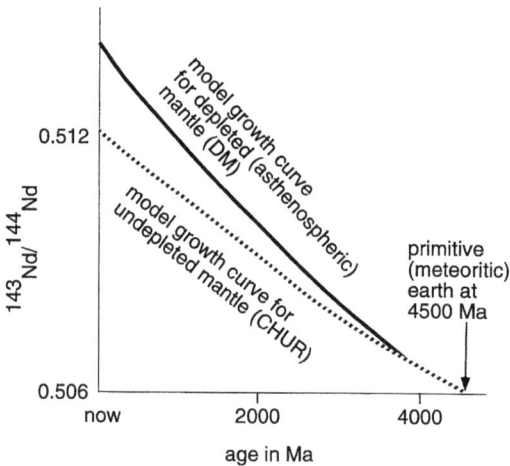

Figure D.4. Evolution of $^{143}$Nd/$^{144}$Nd in undepleted mantle (CHUR) and depleted (asthenospheric) mantle (DM) starting from the initial $^{143}$Nd/$^{144}$Nd of the unfractionated earth.

is "rotated" to horizontal (fig. D.5). This permits calculation of $^{143}$Nd/$^{144}$Nd ratios in other parts of the earth in terms of their deviation ($\varepsilon_{Nd}$) from the values in the bulk earth or depleted mantle at different times.

Because selective loss of Nd from the asthenosphere steadily increases the Sm/Nd ratio in the lithosphere, all asthenospheric $\varepsilon_{Nd}$ values are positive. Similarly, all $\varepsilon_{Nd}$ values in rocks that have been in the lithosphere for significant periods of time are negative, and those in crustal rocks are lowest. This provides another method for distinguishing between mantle and crustal sources of

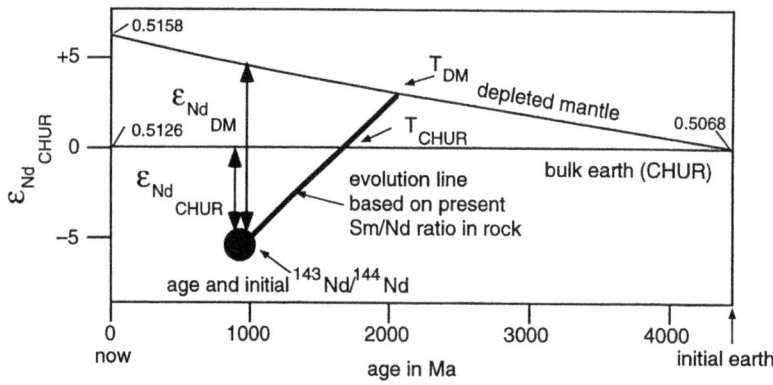

Figure D.5. Determination of $T_{DM}$, $T_{CHUR}$, and $\varepsilon_{Nd}$. The diagram shows the curve for bulk-earth evolution in fig. D.4 as a horizontal line with a value of zero in epsilon units. Numbers show the values of $^{143}$Nd/$^{144}$Nd at three points on the diagram. Explanation in text.

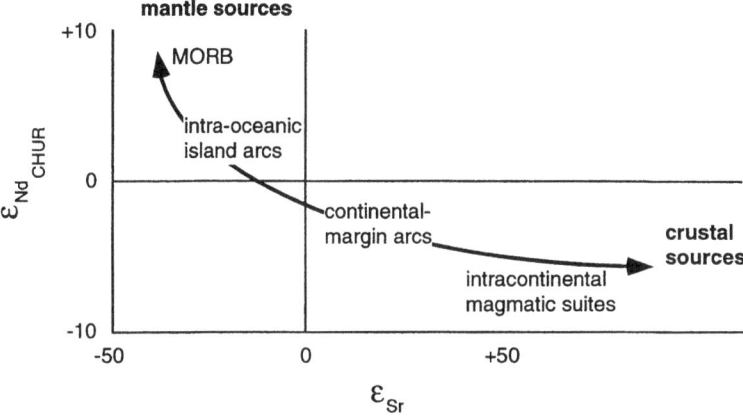

Figure D.6. Typical relationship between $\varepsilon_{Nd}$ and $\varepsilon_{Sr}$.

modern igneous rocks. It also enables us to obtain information from the negative relationship between $\varepsilon_{Sr}$ and $\varepsilon_{Nd}$, where rocks from island arcs and continental margins plot between the fields for MORB and suites formed within continents (fig. D.6).

We can use fig. D.5 to demonstrate graphically one method to calculate the time when a juvenile rock in the crust separated from its original mantle source. The calculation assumes that Sm and Nd are chemically similar enough that they did not fractionate in the rock during metamorphism or other post-magmatic processes and requires measuring the age of formation of the rock, its present $^{143}$Nd/$^{144}$Nd ratio, and its present $^{147}$Sm/$^{144}$Nd ratio. With this information, the present $^{143}$Nd/$^{144}$Nd ratio can be extrapolated back either to a time of separation from a primitive earth, referred to as "$T_{CHUR}$," or to a time of separation from a depleted mantle, referred to as "$T_{DM}$." These values are referred to as "model ages" and do not necessarily correspond to specific geologic events.

## Lu–Hf

Variation in Hf isotopes is produced by radioactivity of $^{176}$Lu, which undergoes beta decay with a half life of $35 \times 10^9$ years. The $^{176}$Lu isotope produces $^{176}$Hf, which is compared with the nonradiogenic $^{177}$Hf to yield variable $^{176}$Hf/$^{177}$Hf ratios. Estimated concentrations in a primitive earth (CHUR) yield an initial $^{176}$Hf/$^{177}$Hf $= 0.278$ and initial $^{176}$Lu/$^{177}$Hf $= 0.33$. The Lu–Hf system is analyzed in the same way as the Rb–Sr and Sm–Nd systems by calculating $\varepsilon_{Hf}$ in ways similar to those used for Sr and Nd isotopes.

Isotopic data for Lu and Hf can be collected from both whole rocks and minerals, and some of the most useful information comes from zircons. Because zircons contain significant concentrations of Hf but virtually no Lu, the $^{176}$Hf/$^{177}$Hf ratios acquired by zircons during their crystallization undergo almost no change during their post-crystallization history. Consequently, $^{176}$Hf/$^{177}$Hf ratios in zircons are identical to the ratios in the magmas from which they crystallized.

During petrologic processes, the Lu and Hf behave very similarly to Sm and Nd. Both Hf and Nd daughters are more fractionated into melts than Lu and Sm parents, causing increase in both Lu/Hf and Sm/Nd ratios in the asthenosphere and corresponding decrease in the lithosphere through time. This yields a linear relationship in which $\varepsilon_{Hf}$ is approximately $(1.5-2) \times \varepsilon_{Nd}$ in many studied rock suites and shows that $\varepsilon_{Hf}$ can be used in the same way as $\varepsilon_{Nd}$ to determine source regions of igneous rocks. Rocks in which $\varepsilon_{Nd}$ and $\varepsilon_{Hf}$ are not proportional presumably have been affected by fractionation of Sm, Nd, Lu, and/or Hf during metamorphism or in some other process, all of which lead to controversial interpretations of the history of the rocks being investigated.

## Re–Os

The Re–Os system has many similarities to the Lu–Hf system. $^{187}$Re decays to $^{187}$Os by beta decay with a half life of $42 \times 10^9$ years, and the concentration of $^{187}$Os is reported as a ratio with nonradiogenic $^{186}$Os or with $^{188}$Os, which has shown virtually no change in abundance since the formation of the earth. During production of magmas from the mantle, Re is preferentially fractionated into melts, and Os remains in the solid residue. Analyses of $^{187}$Os/$^{186}$Os or $^{187}$Os/$^{188}$Os ratios are commonly made of whole-rock mantle xenoliths and also of chromite separates.

Limited evidence suggests that the initial $^{187}$Os/$^{186}$Os ratio for the bulk earth (CHUR) was about 0.81, and the initial $^{187}$Re/$^{186}$Os was approximately 3. Depletion of Re causes this ratio in the mantle to increase more slowly than in the whole earth, and the present $^{187}$Os/$^{186}$Os ratios in the mantle are ~1.06. This variation permits measured $^{187}$Os/$^{186}$Os ratios in mantle-derived rocks to provide an estimate of the time when regions of the upper mantle lost Re and apparently became stable. Calculation of $\varepsilon_{Os}$ is done by similar methods as are used for Sr and Nd isotopes.

Another parameter is $T_{RD}$, the time at which depletion of the mantle in Re leaves rocks with Os isotope ratios that remain unchanged during the remaining history of the earth. This time is determined from the linear change of $^{187}$Os/$^{186}$Os from 0.81 at 4.5 Ga to 1.06 now.

Investigations of Os isotopes that use the ratio $^{187}$Os/$^{188}$Os instead of $^{187}$Os/$^{186}$Os yield the same interpretations but with different numerical values because the ratio of $^{186}$Os/$^{188}$Os is 0.12043.

## U–Th–Pb

Two long-lived radioactive isotopes occur in natural U and one in Th. All three of them decay through complex schemes of intermediate isotopes until they reach stable isotopes of Pb. Summary equations for these decays are:

$$^{238}U \text{ yields } ^{206}Pb + 8\alpha + 6\beta$$
$$T_{1/2} = 4.47 \times 10^9 \text{ years}$$

$$^{235}U \text{ yields } ^{207}Pb + 7\alpha + 4\beta$$
$$T_{1/2} = 0.704 \times 10^9 \text{ years}$$

$$^{232}Th \text{ yields } ^{208}Pb + 6\alpha + 4\beta$$
$$T_{1/2} = 14.0 \times 10^9 \text{ years}$$

It is convenient to express many of the parent and daughter isotope concentrations as ratios with the nonradiogenic $^{204}$Pb.

An enormous amount of information can be obtained from the U–Th–Pb system, but we discuss only two of the numerous uses that can be made of it: measurement of the ages of zircons and detection of changes in U/Pb ratios in different regions of the earth at different times.

When zircons crystallize they readily include U in their lattices but exclude Pb. This means that any Pb now measured in a zircon was formed by radioactive decay of U in the zircon, and the U/Pb ratio is an indication of the age of the zircon. Determinations are complicated, however, by the tendency of Pb to diffuse out of zircons as soon as it is produced. Figure D.7 shows the problem and method of determining the age of a zircon diagrammatically.

The "concordia" curve in fig. D.7 shows the $^{206}$Pb/$^{238}$U and $^{207}$Pb/$^{235}$U ratios that zircons crystallized at different ages from 4.55 Ga to 0 Ga (present) would have now if they had not lost any Pb after crystallization. It starts from the point 0,0 because zircons crystallizing now contain no Pb, and proceeds to higher concentrations of $^{206}$Pb and $^{207}$Pb as their radioactive parents decay. The shape of the curve results from the much shorter half life of $^{235}$U than of $^{238}$U, which has caused the $^{235}$U/$^{238}$U ratio to decrease from a high value in the original earth to 1/137.8 today. Analyzed zircons whose $^{206}$Pb/$^{238}$U and $^{207}$Pb/$^{235}$U ratios plot on concordia at any specific age are regarded as "concordant" and have clearly crystallized at that age.

If the zircons lose Pb, either continually or in a single event such as metamorphic reheating, however, then their various U/Pb ratios fall below the concordia (fig. D.7). Loss of Pb in one event commonly produces a suite of zircons that lie along a "discordia," which forms a straight line between the concordia at the time of original zircon crystallization and the date of Pb loss. Connecting this discordia line to the upper intercept with concordia shows the time of initial zircon crystallization, and connecting it to the lower intercept reveals the time of the event that caused Pb loss.

The evolution of Pb isotopes in the earth can be shown diagrammatically in fig. D.8, in which $^{207}$Pb/$^{204}$Pb and $^{206}$Pb/$^{204}$Pb are plotted against each other starting from initial values in the initial earth of $^{206}$Pb/$^{204}$Pb = 9.31, $^{207}$Pb/$^{204}$Pb = 10.3, and U/Pb = 7.2. The "conformable" curve is a primary growth curve that shows how these radiogenic isotopes of Pb would have grown relative to the nonradiogenic $^{204}$Pb in an unfractionated earth from 4.55 Ga to the present (0 Ga). The short half life of $^{235}$U relative to $^{238}$U causes the curve to rise rapidly at first and more slowly later. Minerals such as galena, which contains no U, are referred to as "conformable" if they have the Pb isotope ratios

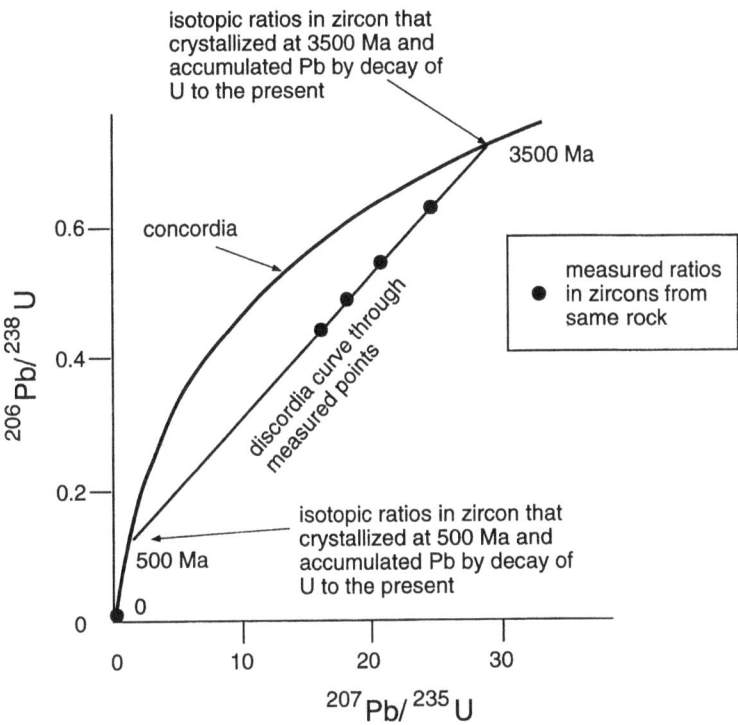

Figure D.7. Calculation of ages of concordant and discordant zircons. This diagram amplifies the discussion in the text with the specific example of a zircon that crystallized at 3500 Ma and lost Pb in a single event at 500 Ma.

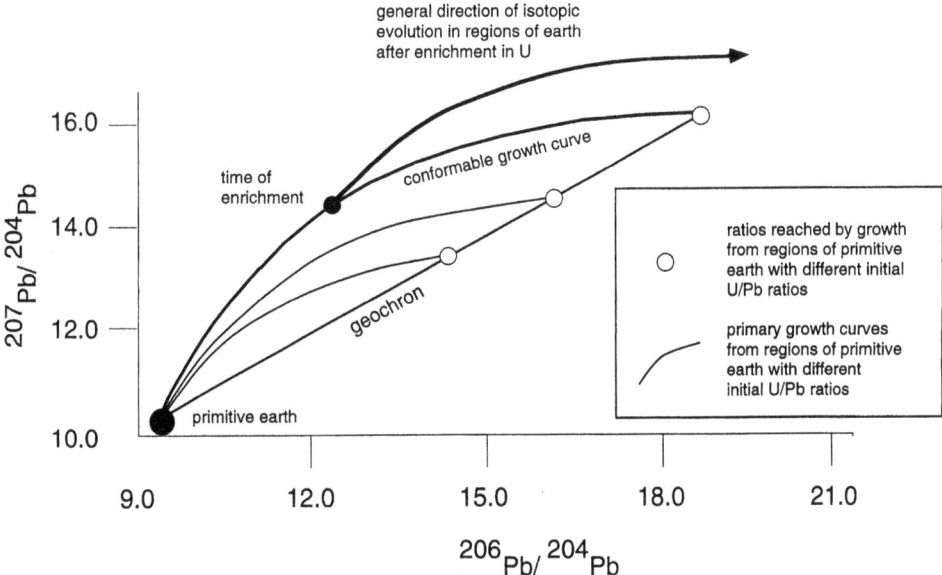

Figure D.8. Evolution of $^{207}$Pb/$^{204}$Pb and $^{206}$Pb/$^{204}$Pb in earth history. The conformable curve starts at $^{206}$Pb/$^{204}$Pb = 9.31, $^{207}$Pb/$^{204}$Pb = 10.3, and a value of $^{235}$U/$^{238}$U that evolved to the present $^{235}$U/$^{238}$U ratio of 1/137.8. All primary curves terminate on a straight line referred to as the "geochron." Primary evolution is interrupted by events that change U/Pb ratios, which cause isotopic ratios on this diagram to diverge from the conformable and other primary curves.

that the whole earth had at the time the minerals formed.

Some Pb isotopes apparently evolved in parts of the initial earth that had different U/Pb ratios than the average for the bulk earth. They formed other "primary" growth curves leading to present $^{207}Pb/^{204}Pb$ and $^{206}Pb/^{204}Pb$ ratios that lie along a straight line known as a "geochron."

Many $^{207}Pb/^{204}Pb$ and $^{206}Pb/^{204}Pb$ ratios do not lie along primary growth curves and are referred to as "disconformable." Disconformable minerals result from the ease of fractionation of U from Pb by various earth processes, particularly because U is an LIL element and Pb is not. Thus, when U rises in the mantle and then becomes further concentrated in the upper levels of the continental crust, the U/Pb ratio also increases in these regions The result is that many of the Pb isotope systems that are measured in near-surface rocks show one or more times of U enrichment, and we show the general result in fig. D.8.

Further information on isotope geology can be found in Faure (1986) and Dickin (1995).

# Appendix E

## Cratons

These descriptions of cratons amplify the discussions in chapter 4, and the Barberton Mountain region is also referred to in the discussion of destruction of continental crust in chapter 3.

### Barberton Mountain Region of Southern Africa (Stabilized at 3.1–3.0 Ga)

The Barberton Mountain region displays the history of the eastern part of the Kaapvaal craton of southern Africa (fig. 4.12). The history of the craton is best known from this area because the central Kaapvaal craton is mostly covered by Late Archean and Early Proterozoic supracrustal suites, and the basement in the west is poorly exposed. Limited information on these outcrops, however, suggests a similar history throughout the entire Kaapvaal craton (Poujol et al., 2002).

We summarize problems in understanding the destruction of ancient continental crust by discussing the evolution of the well-studied Barberton Mountain region of Swaziland and South Africa. Lowe (1999) divides the region into five major blocks separated by numerous faults, but we show only a generalized map in fig. E.1. The following discussion is based on Kroner and Tegtmeyer (1994) and Lowe (1999).

The known history of the Barberton Mountain region begins at ~3.6 Ga with eruption of silicic volcanic and volcaniclastic rocks and emplacement

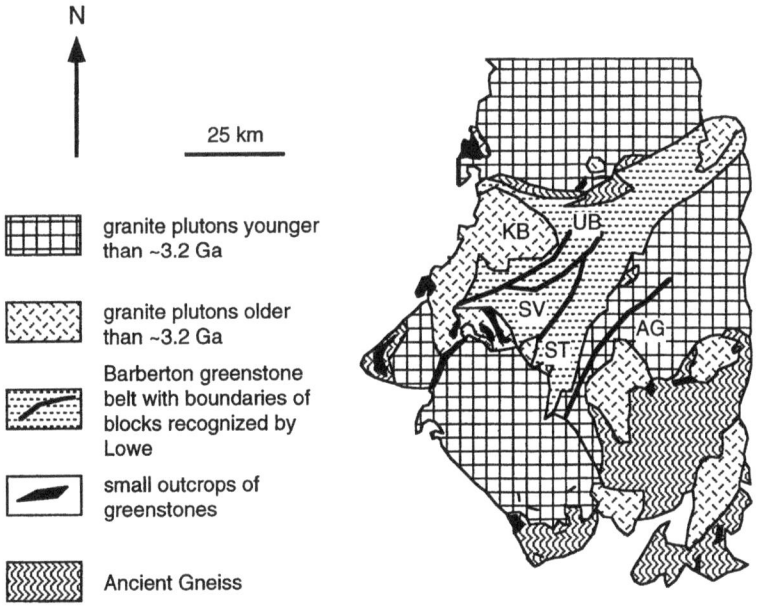

Figure E.1. Map of Barberton Mountain region (adapted from Anhaeusser, 1983; Kroner and Tegtmeyer, 1994; and Lowe, 1999). Blocks recognized by Lowe are KB, Kaap Valley; UB, Umudhua; SV, Songimvalo; ST, Steynsdorp; and AG, Archean gneiss complex.

of TTG plutons that were magmatically related to the siliceous volcanics and are now converted to gneisses. A very small amount of metagraywacke that was deposited at this time contains some debris eroded from the gneisses.

Intrusion of TTG gneisses at ~3.6 Ga was the first episode in a series of similar intrusions that were emplaced, deformed, and metamorphosed from ~3.6 Ga to ~3.2 Ga. All of the gneisses show positive $\varepsilon_{Nd}$, negative $\varepsilon_{Sr}$, and $T_{DM}$ ages a few hundred million years older than their emplacement ages, indicating continuous extraction from the mantle over a period of at least 500 million years. The entire suite of gneisses has been widely referred to as the "Ancient Gneiss Complex," and some workers have proposed that the successive phases formed at least a partial basement for deposition of overlying volcanic and sedimentary rocks.

From 3.50 to 3.43 Ga, these oldest rocks were followed by a basal series of mafic/ultramafic volcanic and intrusive rocks (including komatiites) and overlying silicic flows and volcaniclastic rocks. They were intruded by precursors of TTG gneisses at 3.46–3.45 Ga. Some debris from the gneisses occurs in minor graywackes in the supracrustal section, but most of the sedimentary rocks obtained their components solely from the silicic volcaniclastics (some geologists refer to these rocks as "epiclastic").

From 3.43 Ga to about 3.2 Ga, more komatiites and mafic rocks were erupted and overlain by younger sequences of silicic volcanics and associated sediments. Thick sequences of cherts were also deposited during this time interval. They were formed by syn- or post-depositional silicification of volcaniclastic rocks, sediments containing debris from gneisses, and both biochemically and chemically deposited sediments. Additional members of the Ancient Gneiss Complex were also intruded during this period.

The Barberton Mountain region became a stable craton at about 3.1 Ga (Moser et al., 2001) with the emplacement of undeformed, diapiric granites/granodiorites between 3.2 and 3.1 Ga (we discuss cratonic stabilization in chapter 3). Immediately after this stabilization, graywackes of the Moodies Formation were widely distributed over the new platform and contain debris from all of the underlying rock types.

The information summarized above yields three significant observations that must be explained in any effort to determine rates of crust production and destruction in the Barberton Mountain region. (1) For approximately 500 million years, the Barberton Mountain region was an area of synchronous and repeated production of mantle-derived TTG, mafic/ultramafic lavas, silicic lavas, and silicic pyroclastic and epiclastic rocks. (2) Continuous extraction of TTG precursors from the mantle suggests that a large volume of continental crust was generated during this period, but it did not produce a stable basement for deposition of supracrustal rocks. (3) Only a small amount of sedimentary debris was produced by erosion of TTG until intrusion of post-orogenic granites and rapid production of crust developed a stable craton at 3.2–3.1 Ga.

Observations in the Barberton region illustrate the difficulty of determining rates of production and destruction of crust in the past. Both rates could have been low in the Barberton region if the scarcity of preserved ancient crust was caused simply by the inability of the ancient mantle to produce new crust quickly. This seems unlikely because of the large volume of silicic volcanism and also because of isotopic data (see below) that suggest worldwide separation of large amounts of continental crust from the mantle early in earth history.

For these reasons it seems more probable that both growth and destruction in the Barberton region were rapid instead of slow. Also, the lack of debris produced by erosion of basement TTG shows that this rapid destruction could not have been caused by erosion of significant areas of continental crust above sealevel. For these reasons, we conclude that the destruction could only have been accomplished by direct subduction of thin, newly formed, and unstable crust.

## Guiana (Guyana) Craton (Stabilized at 2.1–2.0 Ga)

The Guiana craton is the part of the Amazonian shield north of the Amazon River (fig. E.2; Gibbs and Barron, 1993; Sidder et al., 1995). With the exception of an exotic 3.0–2.5-Ga suite at the northern end of the craton, all of the rocks in the craton are younger than 2.5 Ga (Gaudette et al., 1996). Stabilization occurred during the Transamazonian event between 2.1 Ga and 2.0 Ga.

More than 90% of the Guiana craton consists of TTG gneisses. Limited dating and other isotopic information indicates that they evolved from the mantle between about 2.5 Ga and 2.1 Ga. Intense deformation apparently occurred throughout their formation, culminating in the Transamazonian event.

Two suites of supracrustal rocks occur in the Guiana craton. Most are typical greenstone assem-

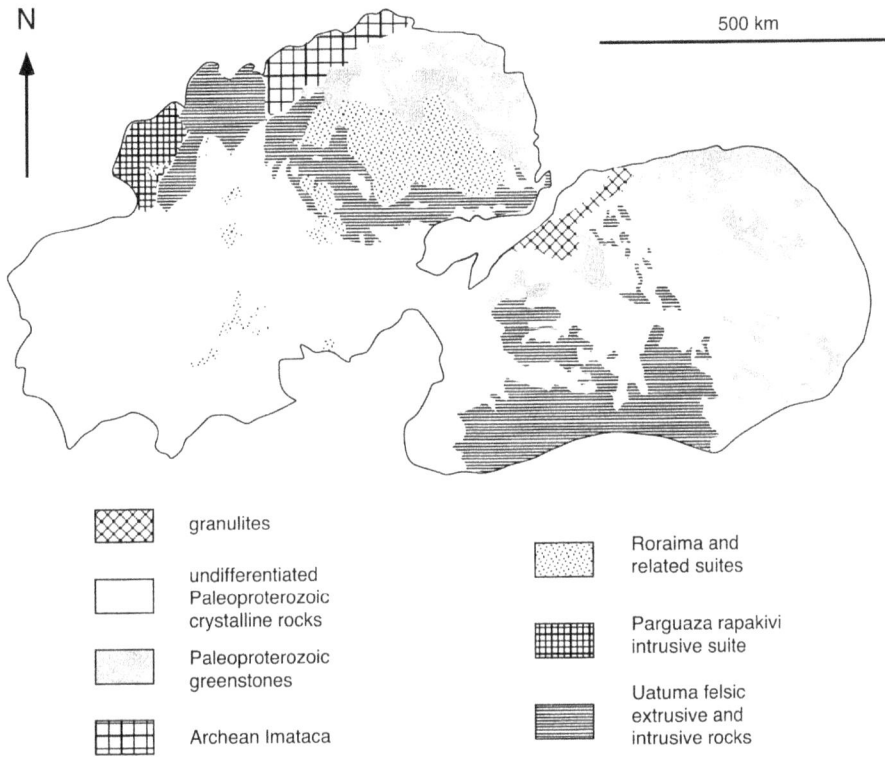

Figure E.2. Map of Guiana craton (compiled from Gibbs and Barron, 1983; Gibbs and Barron, 1993; and Sidder et al., 1995).

blages that consist mainly of metavolcanic rocks, with basalts dominant near the base of the sequence and silicic rocks increasing in abundance upward (Vanderhaeghe et al., 1998). Komatiites have not been reported, and a complete basalt–andesite–dacite sequence is present instead of the bimodal basalt–rhyolite assemblages of greenstone belts in many cratons. Sediments include cherts and iron formation in addition to graywackes derived by erosion of metavolcanic rocks within the greenstone belts.

A second group of pre-stabilization supracrustal rocks consists of shales and shaly sandstones without associated volcanic rocks. They apparently were deposited in comparatively stable shallow-water environments, but they were all intensely deformed and intruded during the Transamazonian event.

A large variety of silicic igneous rocks were intruded during the Transamazonian orogeny in a few tens of millions of years centered around 2.1 Ga. The older suites are trondhjemitic (sodic), and magmas emplaced during the later stages of tectonism are more granitic (potassic). This syntectonic emplacement of the older magmas produced deformational structures similar to those in the older TTG gneisses, and the various suites have not everywhere been distinguished in the field. The younger intrusives are diapiric granites whose lack of deformation demonstrates the end of compressive tectonism in the Guiana craton.

Almost immediately after stabilization of the Guiana craton, large parts of it were covered by flow and pyroclastic rocks of the Uatuma Group (Brito Neves, 2002). The eruptives and associated shallow intrusions were probably formed over an age range of about 100 million years between ~2.0 Ga and ~1.8 Ga. Rock types are mostly rhyolitic and dacitic, but they range from andesites to mildly alkalic trachytes.

Extension of the Guiana craton began in Uatuma time, and it clearly affected deposition of the overlying Roraima Group, a series of sandstones with stratigraphic thicknesses of at least 1800 m. The suite occurs in one large outcrop and several smaller ones scattered over the craton, and it is not known whether they represent deposits in different basins or erosional remnants of one basin. Most of the rocks are fluvio-deltaic or aeolian sandstones, with minor conglomerates, siltstones, and shales. The sediments are extremely mature, and the preponderance of quartz and lack of unstable minerals shows that the Guiana shield had undergone a long period of chemical weathering when the Roraima suite was deposited.

## Nubian–Arabian Craton (Nubian–Arabian Shield; Stabilized at 0.6 Ga to 0.5 Ga)

When the Red Sea began opening at about 30 Ma, it split the Nubian (African) side of the Nubian–Arabian craton away from the Arabian side, and fig. E.3 shows the two sides after palinspastic closure. The reconstructed craton is almost completely surrounded by undeformed Phanerozoic sediments, which overlap an unknown area of cratonic basement on both the east and west. The only information available on the western side of the craton is that a few Mesoproterozoic rocks crop out in the deserts of Egypt and Sudan, but no suture is exposed or inferred from geophysical studies. Consequently the structures and rock suites of the exposed shield may extend farther west than is shown in fig. E.3.

Magnetic studies on the eastern side of the Nubian–Arabian shield show that it extends under Phanerozoic cover on its eastern edge until they terminate against a different terrane west of the Arabian Gulf (P. Johnson and Stewart, 1995). This covered area contains a continental-margin subduction zone where oceanic lithosphere to the east of the shield was subducted under the Afif terrane (Al Saleh and Boyle, 2001).

The exposed part of the shield west of the Afif terrane consists of assembled terranes and intervening suture zones with a NE–SW trend (Vail, 1985; Abdelsalam and Stern, 1996). Each terrane is dominated by basaltic and andesitic lavas and pyroclastics with some volcanogenic (epiclastic) sediments. These supracrustal suites are extensively intruded by dioritic to minor granitic plutons, which probably formed at the same time as metamorphism of the supracrustals to greenschist and low-amphibolite facies. The general lithologies and compositions of the igneous rocks show that most of this part of the shield consists of island arcs and intervening oceanic sediments. Suture zones consist largely of dismembered ophiolite complexes and other oceanic rock suites. Individual rock types include ultramafic suites, serpentinites, gabbros, pillow lavas, and rare sheeted dikes. Locally these rock suites are mixed with rocks from the island-arc terranes to form extensive melanges.

Isotopically determined ages and $T_{DM}$ ranging from 1 Ga to 0.6 Ga all show that the island arcs formed between about 900 Ma to 550 Ma. Much of this material is probably juvenile growth from the mantle, but incorporation of some amount of old continental crust or upper mantle is indicated by several types of isotopic information (summary in Abdelsalam and Stern, 1996; Stern, 2002). This includes old cores in zircons, which may have come from old felsic rocks within the present area of the shield or may be detrital zircons washed into the area from surrounding older terranes. Some isotopic systems also indicate inheritance of older ages around both the eastern and western margins of the shield.

Numerous undeformed granite plutons with diameters of several kilometers invade both arc terranes and suture zones (fig. E.4; Greenberg, 1981). Their ages closely center around 550 Ma, apparently coinciding with the final suturing of the island-arc terranes. Initial $^{87}Sr/^{86}Sr$ ratios of ~0.702 and positive $\varepsilon_{Nd}$ values show that the source of the granite magmas was a mantle-derived crustal underplate without any components from older sialic crust (Moghazi, 1999).

The Arabian side of the shield is partly covered by fluvio-deltaic suites that contain Cambrian trace fossils (fig. E.5; Dabbagh and Rogers, 1993). They lie primarily on the arc terranes and also locally on the 550-Ma granites. The Cambrian age of these sediments can be combined with the ages of the youngest arc suites and the diapiric granites to show that suturing of the craton, invasion by granite, and uplift and erosion to form a platform covered by shallow water all took place within a few tens of millions of years.

Although the exposed part of the craton is largely in greenschist facies, xenoliths in Tertiary volcanic rocks in Arabia show that deeper levels in the crust range from amphibolite to granulite (McGuire and Stern, 1993). The granulite xenoliths are almost entirely basaltic in composition. Some amphibolite-facies xenoliths are from higher levels in the crust and have an average andesitic composition very similar to rocks exposed at the surface. Isotopic properties of the xenoliths are all virtually identical to those of exposed arc rocks, indicating that the basaltic granulites may be the residue from movement of more silicic magmas upward to present levels of exposure.

Cratons 215

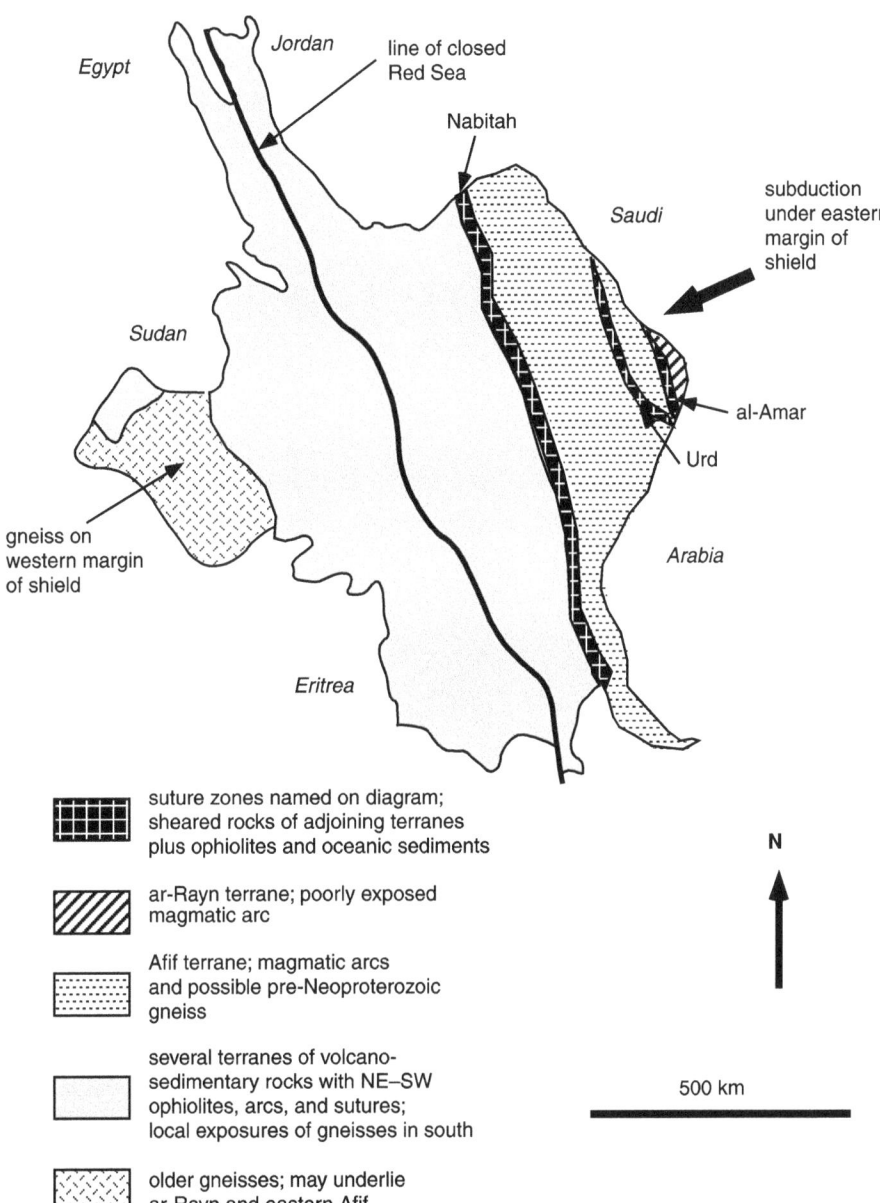

Figure E.3. Map of Nubian–Arabian shield (compiled from Vail, 1985, and Al Saleh and Boyle, 2001). Named suture zones are Nabitah, Urd, and al-Amar. The gray pattern that occupies most of the shield includes numerous terranes with small lithologic differences and various names used by different investigators.

Figure E.4. Younger granite pluton intruding Neoproterozoic wall rocks in eastern Egypt (photo by J.J.W.R.).

Figure E.5. Wajid Formation of Saudi Arabia (top of peak), showing flat-lying Lower Cambrian sediments resting unconformably on deformed Neoproterozoic basement (photo by J.J.W.R.).

# Appendix F

## Anorogenic Magmatic Suites

Descriptions of these anorogenic magmatic suites amplify discussions in chapter 4.

### Arsikere Granite

The 2.5-Ga Arsikere Granite is a roughly circular pluton with an outcrop area of ~75 km$^2$ in the Western Dharwar craton (WDC) of India (fig. 4.2; Rogers, 1988). A slight indication of ring structure is shown along part of the margin by a narrow zone of granite that apparently fractionated from the main part of the pluton. A very discordant age of euhedral zircons and a whole-rock Rb–Sr age show original crystallization at ~2.6 Ga (J. Miller et al., 1996), the same time as granulite-facies metamorphism at the southern margin of the WDC (see below). An initial $^{87}Sr/^{86}Sr$ value of 0.702 suggests derivation of the Arsikere magma from a lower crust that had been depleted in Rb at some time older than 2.5 Ga, presumably when the WDC was stabilized at 3.0 Ga.

Major minerals are typical quartz, K feldspar, and sodic plagioclase, but many of the minor and trace minerals are very unusual. Euhedral epidote was formed by crystallization in the magma rather than by alteration of primary Ca-bearing minerals. Euhedral sphene is abundant, and high oxygen fugacities are shown by incorporation of all of the Ti in sphene, leaving Ti-free magnetite as the only opaque mineral. High concentrations of HF in the magmatic fluids are demonstrated by the presence of magmatic fluorite and high fluorine contents of both biotite and muscovite.

The mineralogy, composition, and isotopic properties of the Arsikere Granite suggest two properties of the stabilized WDC. One is that the Arsikere magma was produced from a lower crust that had a low water content and was depleted in LIL elements. The second is that generation of the magma may have been possible only because of the presence of nonhydrous fluids such as HF.

### Alno Island

The alkaline–carbonatite complex of Alno Island and related suites in Sweden were emplaced between 610 Ma and 530 Ma into ~2-Ga gneisses of the Baltic shield (Morogan and Lindblom, 1995). The principal Alno complex has a diameter of ~4 km and consists of at least one ring dike and two small intrusions. Associated rocks include volcanic vents containing breccia and tuff, carbonatite plugs, and dikes consisting of silicate rocks, carbonatite, and kimberlite. Principal intrusive rocks are ijolites, accompanied by pyroxenite, other alkalic mafic rocks, nepheline syenite, and siliceous carbonatite. Much of the intrusion is surrounded by a zone of fenitization, where $CO_2$-rich fluids permeated the surrounding wall rocks. Rock compositions show that the silicate magmas contain some components from the mantle, but whether some are from the lower crust is unclear.

Fluid inclusions in the Alno suite have been intensely studied. Compositions of different types of fluids in various minerals and rocks include: a three-phase suite consisting of liquid $H_2O$ + liquid $CO_2$ + vapor; liquid $H_2O$ + solids that commonly include NaCl; liquid $H_2O$ + vapor containing $H_2O$ and $CO_2$; and a single phase of miscible $H_2O$ and $CO_2$. Some of the differences in fluids were controlled by immiscibility between silicate and carbonate magmas at depth. Fluids containing as much as 40% NaCl developed during crystallization of silicates as magmas rose in the complex. Many of the fluids in crystals that formed near the surface seem to have been diluted with surface water.

The Alno suite demonstrates that magmatism can occur in areas that appear to have been dormant for 1.5 billion years provided sufficient fluids are available to mobilize the melts. At Alno, those fluids contained high concentrations of NaCl as well as $H_2O$ and $CO_2$.

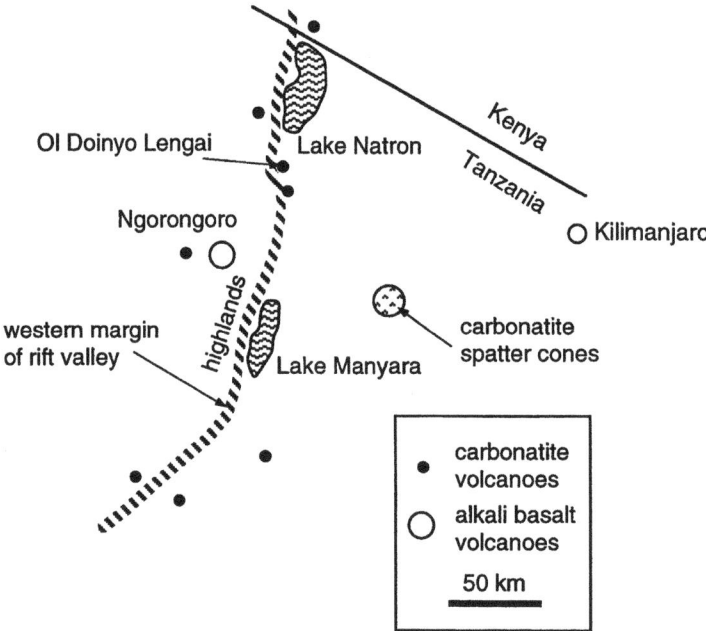

Figure F.1. Map showing volcanism and rifting in Ol Doinyo Lengai area (adapted from J. Dawson, 1989).

## Abu Khrug

The ~90-Ma ring complex at Abu Khrug, Egypt, is one of numerous ring complexes erupted through the Nubian–Arabian craton throughout the Phanerozoic (Landoll et al., 1994). Rock types include rhyolitic and trachytic volcanics and shallow intrusive alkali syenites, quartz syenites, nepheline syenites, and minor dikes that range from gabbro to trachyte. Sodic minerals such as aegerine and riebeckite are common, along with feldspathoids in the undersaturated rocks. The only evidence of fluid activity at Abu Khrug is widespread hydrothermal alteration, but Sn mineralization in a similar complex in Saudi Arabia (Silsilah) shows the importance of HF in its formation (DuBray, 1986).

Isotopic evidence indicates that gabbros that are regarded as parental magmas at Abu Khrug were derived almost completely from a basaltic upper mantle. The $\varepsilon_{Nd}$ value of the gabbros is +4.4, significantly higher than could have been derived from any sialic rocks of the crust. Similar positive $\varepsilon_{Nd}$ values also occur in alkali syenites and nepheline syenites, but quartz syenites have lower $^{143}Nd/^{144}Nd$ ratios that indicate crustal assimilation.

Magmatism at Abu Khrug and elsewhere in the Nubian–Arabian craton demonstrates that many areas of the underlying mantle were fertile (capable of producing magmas) for several hundred million years after cratonic stabilization. There is no indication that fluids caused the magmatism at Abu Khrug, but HF and probably other species were important in some of the other anorogenic complexes in the Nubian–Arabian shield.

## Ol Doinyo Lengai

The Eastern Branch of the East African Rift Valley was the site of eruption of voluminous alkali-olivine basalt, both as valley filling and as large cones (chapter 10). In the southern part of the rift, this magmatism produced comparatively recent volcanoes such as Kilimanjaro and Ngorongoro in Tanzania, but about 1 million years ago renewed faulting developed a sharp western edge of the rift valley and caused a sudden change in volcanic activity (fig. F.1). At the same time, the typical basaltic volcanism ceased and was replaced by eruption of magmas that were highly alkaline and rich in carbonates. This magmatism built major volcanoes, both within and just outside of the rift, plus areas of carbonate-rich spatter cones.

The alkaline magmatism produced mostly phonolite and some rocks so undersaturated that they

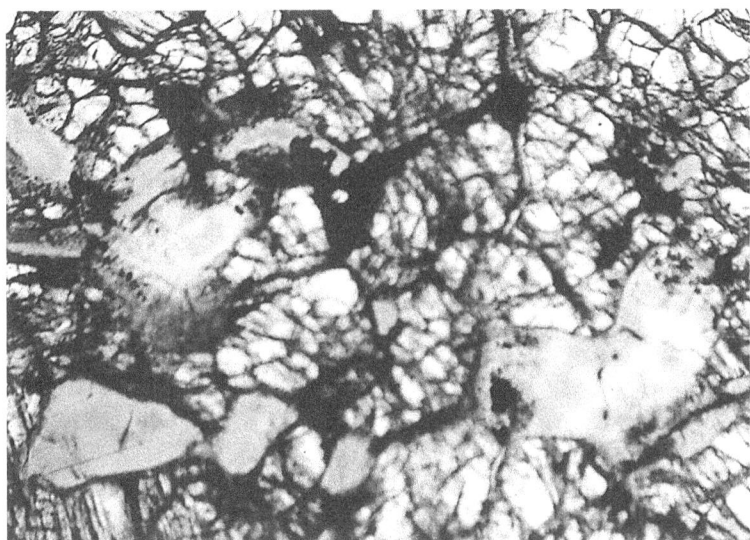

Figure F.2. Photomicrograph of metasomatic vein of mica, amphibole, rutile, and ilmenite (MARID) formed by infiltration of melt into ancient lithospheric mantle less than 100 million years before eruption in a kimberlite (courtesy of Graham Pearson).

are referred to nephelinite. These silicate rocks plus minor carbonatite built the modern cones as stratovolcanoes. The most recent volcanic activity is in Ol Doinyo Lengai, which is the only active volcano in the world to erupt lavas that consist of natrocarbonatite, mostly sodium carbonate plus small amounts of other carbonates (J. Dawson, 1989). The lavas and fragments erupt at temperatures of ~600°C, and the liquids are so fluid that they have viscosities only slightly higher than water. Because they are completely soluble, the natrocarbonatites remain only briefly on the surface before they are dissolved and washed away, partly to form the alkalic Lake Natron.

The eruption of nepheline-rich magmas in this part of the East African rift system clearly was related to the increase in extension that caused faulting at ~1 Ma. This relaxation provided the new volcanoes access to some magma reservoir with very low $SiO_2$ concentrations, high concentrations of $CO_2$ and $H_2O$, and apparently a source of Na. Similar fluids containing $CO_2$, $H_2O$, and Na-rich brine have been observed in other alkali–carbonate complexes, but the concentrations are generally lower than at Ol Doinyo Lengai.

## Matsoku Kimberlite

Kimberlites are volatile-rich eruptions from the upper mantle that commonly reach the surface as "pipes." Many of them consist mostly of breccia and other pyroclastic material, some of which has been blown through the air. Minor liquid phases crystallize as dikes, usually at greater depth than the breccias. Components are all from the upper mantle and include crystals of mantle minerals (xenocrysts), xenoliths of peridotite and rocks that had been altered by upper-mantle processes, and hydrous alteration products of almost everything (fig. F.2). Some kimberlites contain diamonds, showing their derivation from a depth of more than 100 km.

Kimberlites have been formed at virtually all ages from the Middle Proterozoic to the Tertiary. They are all continental and are most common in exposed shields. They also occur, however, in cratonic areas covered by sediments and volcanics, and some were erupted through comparatively young orogenic belts.

We use the pipe at Matsoku, Lesotho, as an example of kimberlites. It contains an assemblage of components that provides a large amount of information about the origin of kimberlite and its significance for studies of cratons and their underlying mantle (Olive et al., 1997). It was intruded at 90 Ma into basalts of the supracrustal Karoo basin, which lies on the margin between the Kaapvaal craton and Late Proterozoic orogenic belts to the south. The kimberlite contains a large variety of xenoliths from the lithospheric upper mantle, including peridotite that is apparently unaltered and peridotite that had been intruded by metasomatic veins and dikes.

Osmium isotopic studies of the various xenoliths show different times when Re was almost completely removed from them and their $^{187}Os/^{188}Os$ ratios were established. For unaltered peridotites and wall rocks around dikes these ages range from about 2200 Ma to 1200 Ma, whereas they vary from ~300 Ma to ~150 Ma for veins, dikes, and other metasomatized fragments. All of these ages are much younger than the 3100-Ma age of the Kaapvaal craton and suggest that the entire underlying upper mantle has been modified since cratonic evolution was complete. The latest and most recognizable metasomatism probably occurred a few hundred million years before eruption of the kimberlite.

The Matsoku kimberlite demonstrates the extent to which upper mantle is modified as long as 3 billion years after stabilization of the overlying craton. It also shows that this modification probably could not have occurred without the introduction of new fluids from elsewhere in the earth.

# Appendix G

## Orogenic Belts of Grenville Age

Belts of Grenville age are shown in fig. 7.1, which uses the classification into interior magmatic and exterior thrust belts identified in the discussion of the Grenville belt of eastern North America (see below). We begin with an extended discussion of the type Grenville belt of eastern North America and continue with belts of similar age elsewhere in the world.

### Grenville Belt of Eastern Canada

The Grenville "front" cuts abruptly across the southeastern margin of the Superior craton of the Canadian shield (fig. G.1). The front separates NE–SW-trending granulite-facies rocks of the Grenville province from rocks of various orientations in the Superior province. Early work found that some rock suites of the Superior province could be correlated across the front into the Grenville belt, showing that at least part of the Grenville belt was reworked crust of continental North America. Further studies, aided by an intense program of reflection-seismic work, now demonstrate a complex history of the Grenville belt that began at about 1.8 Ga and continued until final consolidation followed by rifting at some time younger than 1 Ga (reviews by Rivers, 1997, and Davidson, 1998).

Figure G.1 illustrates the basic structure of the Grenville belt in eastern Canada. It is dominated by northwest-vergent thrusts that bring a series of different terranes on top of each other (Louden and Fan, 1998; Carr et al., 2000; Ludden and Hynes, 2000; Mereu, 2000; D.J. White et al., 2000). The thrusts generally root into some type of decollement at depths of ~25 km. The rock underlying the decollement is generally unknown, and some workers have speculated that it might be an extension of the Superior craton. This implies overthrusting of the craton by several hundred kilometers, which has been demonstrated in numerous younger orogens (chapter 5).

On the northwestern margin of the Grenville belt, rocks of the Superior province are covered by a "parautochthon" of granulite-facies rocks that consist at least partly of metamorphosed Superior suites. The rocks are mostly felsic with some possible metamorphosed greenstone assemblages. Because of the intense metamorphic overprinting, ages of protoliths are difficult to determine, but at least some rocks younger than the Superior craton are apparently interspersed.

The next belt to the southeast is the Central Gneiss belt (also referred to as Laurentia + Laurentian margin), which consists of para- and orthogneisses generally in granulite to high-amphibolite facies (Ketchum and Davidson, 2000). The boundary between the Central Gneiss belt and the parautochthon is clearly one or more of the northwest-vergent thrusts, but the exact location is uncertain because some of the suites regarded as parautochthonous may be allochthonous. Most lithologies in the gneiss belt are felsic, including metasediments originally deposited along the margin of the Superior craton. Ages and $T_{DM}$ values are 1.7 Ga and slightly younger, indicating that most components of the belt were newly created in the Proterozoic (Guo and Dickin, 1996; Dickin, 1998, 2000). Efforts to find a suture zone within the belt have been complicated by the scarcity of mafic rocks that might represent oceanic lithosphere, such as metagabbro and eclogite, but some outcrops suggest ocean closure within the gneiss belt at ~1.2 Ga.

All components of the Composite Arc belt are clearly allochthonous to the Superior craton. Rocks in the belt consist of a variety of lithologies commonly formed in island arcs, including both calcalkaline and bimodal volcanic suites, clastic sediments, and minor carbonate rocks. Some sediments probably formed in backarc basins and contain debris from both the developing arcs and the North American margin. Ages and $T_{DM}$ values indicate juvenile formation at ~1.3–1.2 Ga. The arc belt contains numerous individual terranes,

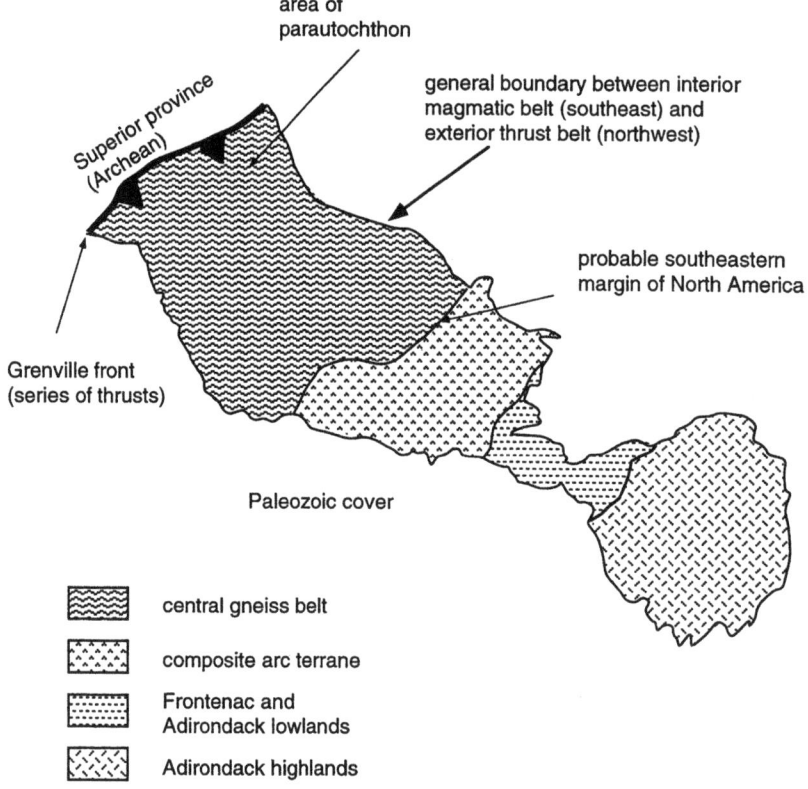

Figure G.1. Grenville orogen (adapted from Rivers, 1997; Davidson, 1998; Louden and Fan, 1998; Carr et al., 2000; Ludden and Hynes, 2000; Mereu, 2000; D. White et al., 2000).

and the entire assemblage was clearly swept together by the early part of the Grenville orogeny.

The Frontenac–Adirondack belt lies between the Composite Arc belt and Phanerozoic rocks that cover the southeastern margin of the Grenville orogen. Much of the belt contains quartzofeldspathic gneiss and marble in low-granulite to high-amphibolite facies. The Adirondack highlands, however, consist largely of AMCG massifs, dominated by anorthosite and charnockite. These intrusive suites were emplaced mostly between ~1.2–1.0 Ga into metasediments that probably had been deposited at ~1.3–1.2 Ga and metamorphosed at approximately the same time as igneous intrusion.

Assembly of most of the terranes in the Grenville belt may have been complete by ~1.4 Ga, with metamorphism and compression of the terranes occurring later (Martignole and Friedman, 1998; Hanmer et al., 2000). Very little activity occurred from ~1.4–1.3 Ga, consistent with data from the Sveconorwegian belt (see below), and possibly related to the breakup of the supercontinent Columbia (see below). The earliest phases of the Grenville compression and metamorphism occurred at ~1.3–1.2 Ga, regarded by some workers as a separate Elzivirian orogeny. Thrusting and metamorphism continued from ~1.2–1.0 Ga throughout the Grenville belt, with brief interruptions when local crustal extension accompanied the intrusion of AMCG complexes. The final stages of the Grenville orogeny recorded in present exposures took place slightly younger than 1.0 Ga, when post-orogenic granites were intruded. Extension began at ~0.9 Ga (Streepey et al., 2000).

In addition to the classification of terranes used above, Gower et al. (1990) separated the Grenville orogen into an "exterior thrust belt" and an "interior magmatic belt." The thrust belt includes all of the parautochthon and the northwestern part of the Central Gneiss terrane, both of which exhibit only very minor magmatism. This absence may strengthen arguments that the inner part of the gneiss belt contains significant components of crust from the Superior craton. All areas southeast of the thrust belt contain various types of granitic rock suites, presumably associated with the consolidation of juvenile terranes.

The interior magmatic belt is defined largely by the presence of granites that were intruded in the general age range of 1.2–0.9 Ga. The granites commonly occur as individual plutons and are not associated with other igneous rocks of comparable ages. These granites are apparently post-orogenic varieties characteristic of intercontinental belts that result from collision of two continental land masses (chapter 5).

## Grenville Rocks in Areas of North America Outside the Type Grenville Belt

The Grenville front has been traced beneath Phanerozoic cover through much of the eastern and southern United States. Grenville outcrops occur in scattered locations in the thrust complexes of New England, the Blue Ridge Mountains, and parts of West Texas (Rankin et al., 1993; Faill, 1997). Rocks of Grenville age are very sparsely exposed in northern Mexico, but it is not clear whether they are a southern continuation of older Proterozoic accretionary belts in the southwestern United States or are exotic (Sanchez-Zavala et al., 1999). Rocks of the Oaxaquia terrane in southern Mexico are clearly exotic to North America, and we discuss them below.

Grenville exposures in the Blue Ridge Mountains consist mostly of ortho- and paragneisses with ages of parent rocks in the general range of 1.2–1.0 Ga (Rankin et al., 1993). Metamorphism to low-granulite or high-amphibolite facies occurred at ~1 Ga, consistent with ages in parts of the Grenville orogen that are better exposed. Many of the outcrops show no evidence of intrusion at this time and presumably represent the exterior thrust belt of the Grenville orogen, but some of the gneissic suites were intruded by granites and similar rocks at ~1.0 Ga, placing them as part of the interior magmatic belt. All of the outcrops are in thrust complexes and can be regarded either as parautochthonous or as allochthonous blocks that originated very close to the North American continental margin.

Grenville outcrops in Texas clearly are parts of the interior magmatic belt. Metasedimentary and metavolcanic rocks of the basement in the Llano area apparently evolved at some time before 1.1 Ga, when they were compressed and metamorphosed to high-amphibolite facies (Rougvie et al., 1999). This basement was extensively intruded by diapiric granites at ~1 Ga, and this emplacement stabilized the crust against further deformation and metamorphism (D. Smith et al., 1997). The Franklin Mountains at the western tip of Texas contain a suite of sedimentary rocks metamorphosed at ~1.3 Ga and intruded by granites and related igneous rocks at ~1 Ga (Rankin et al., 1993; D.R. Smith et al., 1997).

## Sveconorwegian Belt

The Sveconorwegian belt along the southern margin of the Baltic shield is almost identical to the outcropping part of the Grenville belt (Starmer, 1993, 1996; Cosca et al., 1998). The Sveconorwegian frontal deformation zone is a north-vergent thrust complex that is equivalent to the Grenville front (Juhlin et al., 2000). It separates the Baltic shield from rocks deformed and intruded mostly during the period 1.1–1.0 Ga (Andersson et al., 1999; Bingen et al., 2001; Johansson et al., 2001). The belt consists mostly of supracrustal rocks metamorphosed to low-granulite or high-amphibolite facies and separated by terrane boundaries indicating accretion of individual arcs. Some AMCG complexes within the belt developed at ~1.1 Ga, roughly synchronous with similar suites in the Grenville terrane. The metamorphic rocks were intruded by granite between ~1.0–0.9 Ga, indicating that the Sveconorwegian belt is mostly within the internal magmatic part of Grenville orogens.

## Oaxaquia

Much of eastern Mexico appears to be underlain by a basement of Grenville age, and Ortega-Gutierrez et al. (1995) proposed the name Oaxaquia for this very large terrane. The major exposure is the Oaxacan complex in southern Mexico (Keppie and Ortega-Cutierrez, 1999; Keppie et al., 2001), but numerous smaller suites of Proterozoic rocks are exposed through Phanerozoic cover farther north (Lawlor et al., 1999; Weber and Koehler, 1999; Lopez et al., 2001). Principal rock types include paragneisses formed from clastic rocks and calcsilicates, felsic and basaltic orthogneisses, anorthosites and charnockites of AMCG complexes, and calcalkaline granite. Northward and westward thrusting and metamorphism to low-granulite and high-amphibolite facies occurred at

~1 Ga and slightly younger. This assemblage indicates a geologic history consisting of formation of an island arc or continental-margin arc terminated by metamorphism, deformation, and intrusion. As in other Grenville belts, a brief period of relaxation of crustal stress is shown by the presence of the AMCG suites.

Many geologists have proposed that the Grenville rocks in Mexico are a southern extension of the sparsely exposed Grenville belt in the eastern and southern United States (see above). In some reconstructions of Rodinia, this extension would provide a link with Grenville terranes in East Antarctica, thus forming a continuous belt around most of North America. This connection seems unlikely, however, for two reasons. One is isotopic correlation of Oaxaquia with Colombia instead of North America (Ruiz et al., 1999). The second reason is that supracrustal rocks lying on the Oaxaquia basement contain Lower Paleozoic fossils more similar to those of the western margin of Amazonia than to fauna in North America (Stewart et al., 1999). Both of these types of evidence demonstrate that Oaxaquia is essentially the southernmost of a series of Avalonian terranes in eastern North America derived from Gondwana (chapter 5).

## Sunsas Belt

The Sunsas belt extends along the western margin of the Amazonian shield and westward through Bolivia until it is buried by sediments of the Andean orogeny. We summarize the history of this belt and related suites from Litherland et al. (1989), Tassinari and Macambira (1999), Santos et al. (2000), Almeida et al. (2000), and Geraldes et al. (2001).

Beginning at ~1.3 Ga, thick sequences of flysch-type sediments were deposited in the central zone, exposed mostly in Columbia. At approximately the same time, thinner sequences of orthoquartzites and related suites began to accumulate on the margin of the Amazonian shield to the east. Orogenic activity began at some time before 1.1 Ga, accompanied by eruption of minor basalts and larger volumes of syntectonic granites. The sediments were metamorphosed mostly to amphibolite facies, although granulites exposed in tectonic windows in the Colombian Andes may be correlative with the Sunsas belt and indicate a higher grade of metamorphism characteristic of Grenville belts elsewhere. The syntectonic granites have ages of ~1.1 Ga, presumably the peak of metamorphism, and post-orogenic suites slightly younger than 1 Ga indicate the termination of orogeny at that time. The orogenic activity developed numerous thrust sequences that carried metamorphosed platform sediments farther east over the Amazonian shield and contributed to the intense deformation of the thick metasediments in the central part of the orogen.

The Sunsas belt clearly shows many similarities to the Grenville and Sveconorwegian belts. They include peak metamorphism at ~1.1 Ga, transport of metasediments on numerous shear zones from the axis of the orogen toward adjoining shields, and termination of orogeny with intrusion of post-orogenic granites just after 1.0 Ga. The Sunsas belt also shows the same division into interior magmatic zone and exterior thrust zone as in the Grenville province.

Much of present South America west of the Sunsas belt is underlain by rocks of Grenville age (discussion of Andes in chapter 6). Because the belt presumably marks the western margin of the Amazonian shield in Grenville time, rocks to the west must have accumulated against the Sunsas margin at some time younger than 1 Ga. At present their origin is unknown, partly because they have been discovered recently and partly because so many of them are exposed only in small windows. One explanation is that they are fragments of the Grenville terrane in North America, transported toward South America at the same time as South American blocks with Grenville basements accreted to North America as Oaxaquia and related Avalonian terranes (see above).

## Southern Africa and Dronning Maud Land

The southern part of the Mozambique belt (Lurio belt) bounds the eastern margin of the Kaapvaal craton, and the Natal belt forms the southern margin The Mozambique belt was active primarily during Pan-African (~0.5-Ga) time, but the Lurio belt contains metasediments and metavolcanics that were metamorphosed and thrust northwestward over the craton in a series of Grenville-age shear zones (Sacchi et al., 2000; Kroner et al., 2001). Metamorphism is generally in amphibolite facies in the south, grading to granulite facies toward the north. Granites include both highly deformed syntectonic varieties and undeformed post-tectonic plutons.

The Natal belt has a Grenville history partly superimposed on a series of rock suites that evolved much earlier in the Proterozoic (R. Thomas et al.,

1993). The northern zone of the belt consists largely of volcanic rocks in amphibolite facies, and the southern zone is mostly granulite-facies metasediments plus syn- to post-tectonic granites. Metamorphism of all of these rocks was near its peak between ~1.1–1.0 Ga.

The Maudheim province of adjacent East Antarctica contains a fragment of the Kaapvaal craton (Grunehogna) plus metamorphosed supracrustal rocks. They include calcalkaline volcanic rocks and sediments consisting of graywackes, shales, and minor quartz–carbonate sequences (Groenewald et al., 1991). Metamorphism at 1.1–1.0 Ga converted most of these rocks to amphibolite facies, with some granulites thrust over them. Synchronous deformation caused tight folding and displacement of all rocks to the north in a series of thrust zones.

The similarity of geologic history in southern Africa and Maudheim suggests that they developed as one unit during Grenville time (Moyes et al., 1993; Golynski and Jacobs, 2001). In particular, rocks in all provinces were thrust toward the Kaapvaal craton, suggesting that a Grenville-age ocean was somewhere to the south of Maudheim, presumably now covered by the Antarctic icecap. On the basis of isotopic mapping, Wareham et al. (1998) suggested that southern Africa and Maudheim were also connected to parts of West Antarctica and the Falkland Islands in Grenville time.

## Eastern Ghats and Rayner Belts

The Eastern Ghats belt forms the eastern margin of the Archean shield of eastern India (fig. 7.1). It consists almost entirely of felsic charnockites and khondalites plus minor granulitic mafic rocks metamorphosed two (perhaps three) times during the Proterozoic. The Terrane Boundary Shear Zone (TBSZ) is similar to the Grenville front and transported granulite-facies rocks westward over the dominantly amphibolite-facies suites of the Archean cratons (Biswal et al., 2002). Retrogression along the TBSZ converted some rocks from granulitic to high-amphibolite facies, indicating that metamorphic activity continued during thrusting.

Crustal separation (including $T_{DM}$) ages of the protoliths of virtually all rocks are older than 2 Ga (Shaw et al., 1997; Rickers et al., 2001). This resetting of older rocks and the absence of ophiolites or other evidence of ocean closure suggests that the entire Eastern Ghats belt formed by metamorphism and deformation of an eastern extension of the Archean crust of India. The oldest recognizable metamorphism was at ~1.6 Ga, but it was overprinted by the major tectonometamorphic event at ~1 Ga, which established almost all of the structural patterns in the belt. Very minor overprinting occurred locally at ~0.5 Ga.

In Pangea, the Eastern Ghats was separated from the Rayner belt of Antarctica by the Archean Napier complex (Biswal et al., 2002). The Rayner belt is similar to the Eastern Ghats, showing granulite-facies metamorphism at ~1 Ga and northward (present orientation) thrusting over the Napier complex (L. Black et al., 1987). In both belts, granulite-facies metamorphism occurred primarily during Grenville time, both belts were thrust away from central East Antarctica over adjoining Archean cratons, and neither belt exhibits evidence of ocean closure within the belt. Presumably the closure that created both orogenies was somewhere within the ice-covered part of East Antarctica. The Eastern Ghats appears to represent an exterior thrust belt, and the Rayner suite the interior magmatic belt, of a typical Grenville orogen.

## Belts in Australia and Adjoined East Antarctica

The Darling belt follows the western margin of the Yilgarn craton of southwestern Australia, and the Albany–Fraser belt wraps around its southern and southeastern cratonic margins (Harris, 1995; Myers et al., 1996). The orogenic belt of the Musgrave Range connects to the eastern end of the Fraser belt and extends eastward until covered by younger rocks. The Albany–Fraser belt also connects with coastal East Antarctica, and we discuss all of these belts in this section.

The Darling belt consists of three basement inliers surrounded by Phanerozoic sediments that have been only slightly deformed (Harris, 1995; Wilde, 1999). Exposed or drilled basement shows a few ages slightly greater than 2 Ga, but most rocks of the orogen are significantly younger. Most of the rocks consist of para- and orthogneisses and some late-tectonic granites with ages ranging from ~1.1–1.0 Ga. Deformation was mostly transpressional, with tectonic transport toward the Yilgarn craton.

Rocks in the Albany belt are largely reworked suites of the Yilgarn craton, but they include juvenile magmas and also sediments formed by erosion of both the Yilgarn craton and the juvenile rocks. Metamorphic grades range from high amphibolite to low granulite, with tectonic transport toward the

Yilgarn craton during one or more periods of dextral transpression. The major tectonism was at ~1.3 Ga, with significant overprint at ~1.2 Ga (Clark et al., 2000). Reworked rocks of the Yilgarn craton form at least the northern part of the Albany belt and may form the basement throughout the belt (Black et al., 1992). Ages of inherited zircons in the metasediments and $T_{DM}$ ages of magmatic protoliths range up to 3 Ga. The northern part of the orogen appears to be an exterior thrust belt, but a large granitic intrusion in the south places that part in the interior magmatic zone. Harris (1993) correlated the Albany orogen with the Central Indian Tectonic Zone (chapter 6).

The Fraser belt differs from the Albany belt mainly by the presence of the large Fraser mafic igneous complex, which has been deformed and interleaved with granulitic metasediments. In addition, the belt contains reworked rocks of the Yilgarn craton, ortho- and paragneisses, migmatites, and widespread granites. Some intrusive activity began by ~1.7 Ga, but most of the tectonism was during a period of dextral transpression from 1.3–1.1 Ga.

The Musgrave belt forms the northern margin of the Gawler craton, possibly joining most of Proterozoic Australia to the Gawler craton and attached parts of East Antarctica (R. White et al., 1999). Deformation and metamorphism in the Musgrave belt occurred mostly from 1.3–1.2 Ga. During this time pre-existing felsic magmatic rocks and syntectonic orthogneisses were metamorphosed to granulite and high-amphibolite facies. Zircons in these older rocks have ~1.2-Ga overgrowths, approximately the same age as the emplacement of post-tectonic granites.

Several areas of coastal East Antarctica have almost identical histories to the Albany–Fraser orogen (Harris, 1995). The Windmill Islands, Bunger Hills, and Mirnvy region all show high-grade metamorphism at ~1.2–1.0 Ga of protoliths that range in age from Archean to Middle Proterozoic. These regions also exhibit the same northward (present orientation) tectonic transport away from an apparent suture zone under the Antarctic icecap. Boger et al. (2000) correlate this belt westward around much of coastal East Antarctica, with the exception that orogenic activity with complex vergences occurred in one suite of the Prince Charles Hills from 1.0–0.9 Ga.

## Transantarctic Mountains

The Transantarctic Mountains are mostly the site of the Ross orogeny, a tectonometamorphic event that occurred from the latest Proterozoic to the beginning of the Phanerozoic. Early investigations suggested the presence of a Grenville-age orogeny, but the evidence is questionable. By contrast, recent work indicates the possibility of an orogeny at ~1.7 Ga (Nimrod orogeny). At this time overgrowths were formed on inherited Archean zircons, conformable zircons crystallized in intrusive rocks, and zircons crystallized in eclogites (Goodge et al., 2001).

Present data show no evidence of a Grenville event in the Transantarctic Mountains. Consequently, any collision between this margin of East Antarctica and any other continental block presumably occurred either before the creation of Rodinia or in Pan-African time.

## South China (Yangtze) Craton

The oldest rocks in the South China craton have ages greater than 3.0 Ga, and they are followed by an extremely complex series of events (Qiu and Gao, 2000). The entire craton apparently did not develop as a coherent block until terranes in the southeast (present orientation) collided with older areas in the north at some time in the Middle to Late Proterozoic (Chen and Jahn, 1998). The last stages of development seem to have occurred in Grenville time, and the craton was sufficiently established that some models show its location in Rodinia (Li et al., 1995).

## Kibaran Event of Central Africa

The type Kibaran of central Africa is an orogenic belt in Congo on the eastern margin of the Congo craton, and its name is widely used to refer to orogenies in the general age range of 1.4–1.0 Ga. Areas along the western flank of the Tanzania craton in Burundi and Tanzania, however, are better known than the type Kibaran (Rumvegeri, 1991; Meert et al., 1994; Deblond et al., 2001). This latter area shows deposition of thin sediments on the margin of the craton, an orogenically active zone to the west during the period of 1.4–1.2 Ga, with thrusting toward the Tanzanian craton interior and intrusion of granitic plutons at ~1.2 Ga. This history is consistent with collision of the Congo and Tanzanian cratons between ~1.4–1.2 Ga, and it is possible that the Kibaran orogeny in its type area terminated before orogenies in areas commonly regarded as Grenville.

# Appendix H

## Orogenic Belts of 2.1–1.3-Ga Age

The description in this appendix of orogenic belts developed from 2.1–1.3 Ga amplifies the discussion of Columbia in chapter 7 and also parts of chapter 6 (locations in fig. 7.6). The most intense activity was from 1.9–1.8 Ga, but we include a general discussion of orogenic activity from the later part of the Paleoproterozoic until the earliest stages of Grenville-age deformation (appendix G). We divide this appendix into five geographic areas: North America/Greenland (fig. H.1); Baltica and Russian platform (fig. H.2); eastern Asia (fig. H.3); South America and western Africa (fig. H.4); and southern Africa, India, Australia, and East Antarctica (fig. H.5). More complete discussions of 2.1–1.8-Ga belts can be found in Zhao et al. (2002a).

### North America/Greenland

The orogenic belts shown in North America and Greenland were mostly active between about 1.9–1.8 Ga (fig. H.1). Their significance is interpreted in different ways by different geologists (chapter 6), but nearly all investigators agree that most of North America became a stable block by ~1.8 Ga and has remained in this condition since that time.

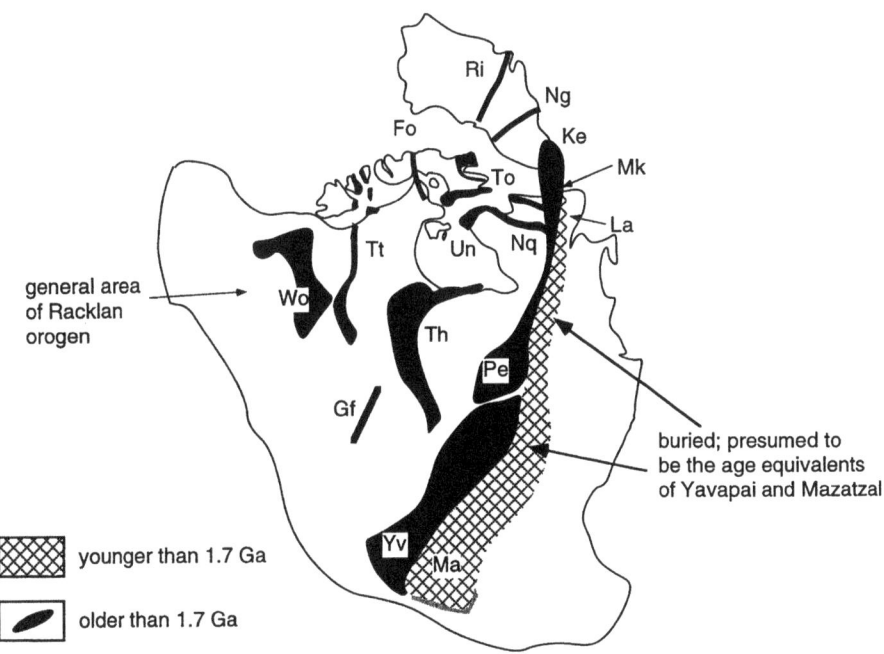

Figure H.1. Locations of orogenic belts with ages from 2.1–1.3 Ga in North America and Greenland. Abbreviations are: Fo, Foxe; Wo, Wopmay; Tt, Taltson–Thelon; Ri, Rinkian; Ng, Nagssugtoqidian; To, Torngat; Un, Ungava; Nq, New Quebec; Ke, Ketilidian; Mk, Makkovikian; La, Labradorian; Th, Trans-Hudson; Gf, Great Falls; Pe, Penokean; Yv, Yavapai; Ma, Mazatzal.

## Wopmay (Bear Province)

The Wopmay orogen represents the earliest phase of the growth of North America westward from the Slave craton (chapter 6). The eastern part of the orogen is largely a highly metamorphosed wedge of continental-margin sediments thrust eastward toward the craton between ~1.9–1.84 Ga. Compression may have been caused by collision of a complex terrane, possibly continental fragment, that now forms the western part of the orogen (Gandhi et al., 2001). Eastward subduction beneath the continental margin formed the Great Bear and Fort Simpson batholithic suites from ~1.9–1.8 Ga (Ross, 1991).

## Racklan

To the west of the Wopmay orogen, the Racklan belt is exposed primarily as the Wernecke Group, which underwent deformation both before and after 1.3 Ga (Cook, 1992), and the Coppermine homocline, where sedimentation both preceded and followed eruption of the subduction-related Narakay volcanic suite at ~1.7 Ga (Bowring and Ross, 1985).

## Great Falls

The Great Falls tectonic zone is one of the major right-lateral shears that cut the western part of the Canadian shield (chapter 6). Mueller et al. (2002) proposed that volcanic and sedimentary rocks characteristic of island arcs were attached to the Wyoming craton at ~1.85 Ga, when the craton was sutured to the Rae province (cratons shown in fig. 6.7).

## Trans-Hudson and Taltson–Thelon

The Trans-Hudson and Taltson–Thelon orogens cut the Canadian shield into three parts and are critical to an understanding of its history. We discuss in chapter 6 how different investigators have interpreted these orogens as the result of continental collisions following the closure of wide ocean basins and, conversely, as intracratonic orogens that developed largely within a Canadian shield that had accreted by the end of the Archean.

## Foxe, New Quebec, Torngat, and Ungava/Cape Smith

Several belts are concentrated in eastern Canada (Foxe, New Quebec, Torngat, Ungava/Cape Smith). All of these belts were active at ~2.0–1.8 Ga, and if they were not internal (intracratonic or confined), they contributed to a continental assembly that became coherent by no later than 1.8 Ga. They all contain abundant Archean rock that apparently consists of reworked material from adjacent cratons, but they also contain rock suites that presumably formed in island arcs and may represent ophiolitic material from closing ocean basins (papers in Brewer, 1996; Wardle et al., 2002). The Foxe orogen may be an eastward continuation of the Snowbird zone, which separates the Rae and Hearne provinces. The Snowbird zone is clearly a right-lateral shear, but whether it represents a collisional belt or simply an intracontinental shear is very controversial (Kopf, 2002).

## Penokean

The Penokean orogen developed from 1.9–1.8 Ga along the southern margin of the Superior craton. It consists mostly of a wedge of continental-margin sediments plus reworked parts of the Archean craton and a batholithic suite formed during northward subduction beneath the Superior craton (Attoh et al., 1997). The southern margin of the orogen is covered by platform sediments of interior North America, but apparently the southward growth of North America continued throughout most of the Proterozoic (see discussion of Yavapai and Mazatzal below).

## Nagssugtoqidian

The Nagssugtoqidian orogen apparently crosses Greenland from the west to east coasts although much of it is covered by ice. The northern and southern parts consist largely of reworked rocks from adjacent Archean cratons, but the central part consists of highly deformed Paleoproterozoic metasediments intruded at ~1.9 Ga by batholithic suites and at ~1.8 Ga by pink granites that may represent the final stages of orogenic deformation (Kalsbeek and Nutman, 1996; Whitehouse et al., 1998).

## Rinkian

The Rinkian orogen of Greenland is almost completely covered by ice, but it may be a continuation of the Foxe orogen of Labrador (Kalsbeek et al., 1998).

## Makkovikian and Ketilidian

The Ketilidian province of Greenland and Makkovik province of Labrador consist of felsic gneisses among voluminous batholithic suites. Most of the intrusions formed between approximately 1.8–1.7 Ga and post-date thrusting toward the continental interiors of Greenland and Canada (Kerr et al., 1996). In Labrador these suites are abruptly terminated southward by the Labradorian batholithic suite, which has ages in the range of 1.7–1.6 Ga.

The history of eastern North America is unclear between the time of the Makkovikian and Grenville orogenies, a period of several hundred million years. Gower et al. (1990) and Rivers (1997) described the pre-Grenville history of eastern Canada as a period of continental-margin outgrowth beginning at ~1.8 Ga followed by backarc rifting and stability from approximately 1.5–1.3 Ga. This stability was terminated when the early stages of the Grenville orogeny began shortly after 1.3 Ga. Farther south in eastern North America, rocks older than Grenville crop out only in a few isolated patches where they have been uncovered by erosion of Paleozoic thrust complexes in the Blue Ridge province. They are all granulites and exhibit poorly defined ages in the range of 1.8–1.6 Ga (Fullagar, 2002).

## Labradorian

The Labradorian episode is primarily a time of intrusion of batholithic magmas along the margin of North America in Labrador (Rivers, 1997).

## Yavapai and Mazatzal

The Yavapai and Mazatzal orogens both contain a series of accreted arcs and continental-margin supracrustal rocks intruded by batholithic suites (Karlstrom and Humphreys, 1998). The Yavapai province apparently completed its accretion in the period from 1.8–1.7 Ga, and the isotopically distinct Mojave area has the same history as the rest of the Yavapai province (Barth et al., 2000; D. Coleman et al., 2002). The Mazatzal province followed from about 1.7–1.5 Ga and continued the process of outgrowth of the North American craton. Mesoproterozoic growth was followed by the early stages of the Grenville collisional orogeny (1.3–1.2 Ga), which consolidated this portion of Rodinia (Condie, 1992; Van Schmus et al., 1996).

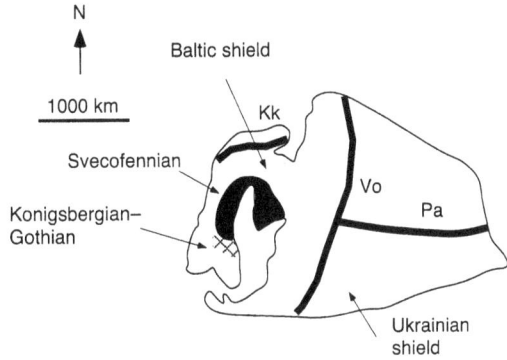

Figure H.2. Locations of orogenic belts with ages from 2.1–1.3 Ga in Baltica and Russia. Abbreviations are: Kk, Kola–Karelian; Vo, Volhyn; Pa, Pachemel.

Midcontinent North America is almost entirely covered by undeformed sediments that began to accumulate when the underlying basement was stabilized, but it is apparently underlain by extensions of the Yavapai and Mazatzal orogens. The oldest rocks in this area belong to the Central Plains Orogen and consist of metaigneous and metasedimentary rocks of amphibolite grade or lower (Van Schmus et al., 1993). Ages commonly range from 1.8–1.6 Ga, consistent with those of the Yavapai and Mazatzal orogens to the southwest. The granite–rhyolite terrane south of the Central Plains Orogen is identified by widespread magmatism centered around 1.4 Ga, but the age of the underlying basement is unclear.

## Baltica and Russia

### Baltic Shield and East European Platform

Precambrian basement underlies a broad area known as the East European platform (discussion of Urals in chapter 5) and is exposed only in the Baltic and Ukrainian cratons (fig. H.2). The covered region may consist of three Late Archean and Paleoproterozoic blocks sutured at ~1.9 Ga along the completely covered Volhyn and Pachemel (Pachelma) belts (Claesson et al., 2001). An ~1.9-Ga suture is exposed in the northern part of the Baltic shield, where the Kola–Karelia belt brings together two Late Archean blocks, apparently by subduction beneath the northern block (Gorbatschev and Bogdanova, 1993).

## Svecofennide and Younger Belts

Belts dominated by batholithic rocks intrusive into metasupracrustal assemblages are all parts of a long-continued outgrowth of the Baltic shield that began in the Paleoproterozoic (fig. H.2; Lindh and Persson, 1990). The Svecofennian orogen underwent a peak of metamorphism during the approximate period of 1.9–1.8 Ga and the granite-porphyry belt from about 1.7–1.65 Ga. The Western Gneiss region is a suite of migmatitic leucogneisses with ages of ~1.7–1.6 Ga (Tucker et al., 1990). The Southwest Scandinavian gneiss complex represents a broad suite of rocks with age ranges from 1.7–0.8 Ga. The Konigsbergian–Gothian is a magmatic suite with ages of ~1.6–1.5 Ga (Ahall et al., 2000). A period centered around 1.4 Ga seems to have been a time of crustal stability and/or rifting when active compression did not occur throughout the region. This southward growth generally continued into Grenville time, forming the Sveconorwegian belt in the Baltic area (appendix G).

## Eastern Asia

Eastern Asia contains two principal sutures, both of ~1.85-Ga age (fig. H.3). The Trans-North China orogen formed by collision of Archean/Paleoproterozoic blocks on both the east and west sides of the deformed belt (Zhao et al., 2000; Wilde et al., 2002). The orogen contains reworked older rocks from both sides plus juvenile sedimentary and igneous suites. Some mafic and ultramafic zones have been proposed to represent remnants of oceanic lithosphere that was destroyed by subduction beneath the eastern block. The suturing at 1.85 Ga established the present North China craton and was followed by widespread rifting (Lu et al., 2002).

The Akitkan orogen crops out in the area of Lake Baikal but is known only from geophysical data along most of its extent because of covering by younger rocks. The orogen apparently forms a suture between the Aldan craton, to the southeast, and the Anabar–Angara craton to the north. Both cratons contain gneiss and greenstone suites that were assembled in the Late Archean, with extensive overprinting at ~2 Ga in the Aldan block. Late syntectonic granites of 1.84-Ga age in the exposed part of the Akitkan belt suggest suturing of the entire Siberian region at this time (Rosen et al., 1994). Condie and Rosen (1994) correlated the Akitkan belt with the Taltson–Thelon orogen of the Canadian shield (chapter 6), and Condie (2002) also suggested that Siberia and Canada

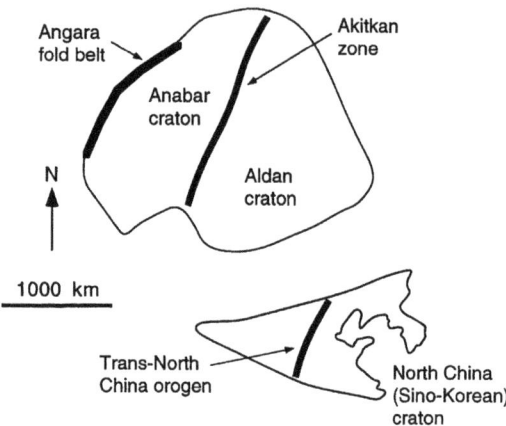

Figure H.3. Locations of orogenic belts with ages from 2.1–1.3 Ga in Siberia and North China.

may have been oriented so that the Wopmay and Angara orogens were correlated.

## South America and Western Africa
### Transamazonian and Eburnian

The dominant orogenic event in eastern South America and western Africa is referred to as Transamazonian in South America and Eburnian in Africa (fig. H.4; Feybesse and Milesi, 1994; Feybesse et al., 1998; summary in Zhao et al., 2002b). The orogens are essentially parallel to modern continental margins, with supracrustal suites thrust eastward in Africa and westward in South America. Most of the deformation occurred during largely sinistral transpression from ~2.1–2.0 Ga, with late magmatic stages from 2.0–1.9 Ga. Orogenic activity at this time apparently was not restricted to the two belts but affected large areas at 2.1–2.0 Ga. This led to the stabilization of numerous cratons, and we discuss them in chapter 6 in connection with the accretion of the very large continental assembly Atlantica.

### Westward Growth of Amazonia

The western part of the Amazon craton underwent marginal growth at about the same time as eastern North America and the Baltic craton until it was incorporated into Rodinia during Grenville time (Sadowski and Bettencourt, 1996; Brito Neves, 2002). The two major belts are Rio Negro–Juruena and Rondonian, both of which consist of supracrustal and intrusive rocks formed during sub-

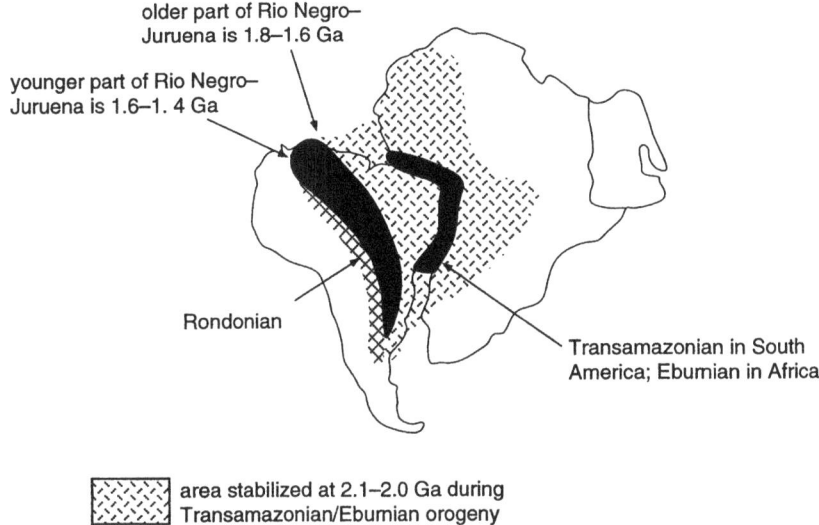

Figure H.4. Locations of orogenic belts with ages from 2.1–1.3 Ga in eastern South America and western Africa.

duction under the western margin of Amazonia. The Rio Negro–Juruena suite developed approximately during the period 1.8–1.5 Ga and lies on a basement consisting of the 2.0-Ga and older rocks of the Amazonian shield. Rondonian growth and metamorphism west of the Rio Negro–Juruena suite apparently continued until ~1.3 Ga, when it was accompanied by crustal relaxation associated with emplacement of the Rondonian intrusive complex (Bettencourt et al., 1999).

## Southern Africa, India, Australia, and East Antarctica

Figure H.5 shows areas of ~2.1–1.8-Ga activity in southern Africa, India, Australia, and coastal East Antarctica (see discussion of Ur in chapter 5). Many of these areas were overprinted during Grenville and/or Pan-African time, and we discuss them more completely in appendices G and I.

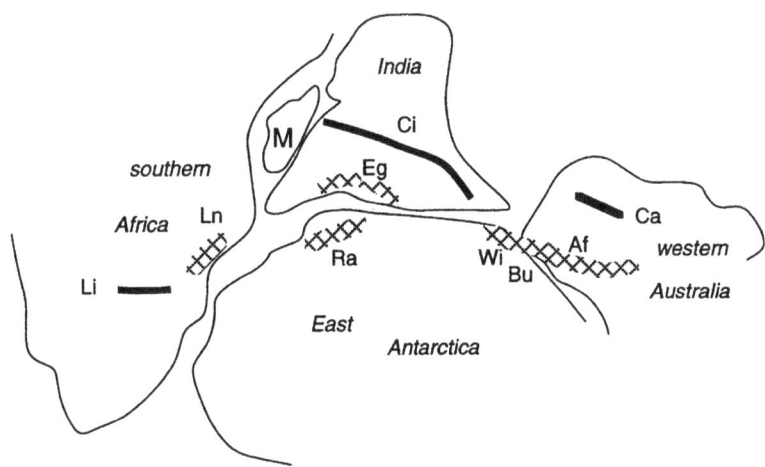

Figure H.5. Locations of orogenic belts with ages from 2.1–1.3 Ga in southern Africa, India, Australia, and part of coastal East Antarctica. Abbreviations are: Li, Limpopo; Ln, Lurio–Namama; Ci, Central Indian Tectonic Zone; Eg, Eastern Ghats; Ra, Rayner; Wi, Windmill Islands; Bu, Bunger Hills; Af, Albany–Fraser; Ca, Capricorn. M is Madagascar.

## Southern Africa

The Zimbabwe craton was attached to the Kaapvaal craton by collision along the Limpopo belt (Rollinson and Whitehouse, 2001). Most of the belt consists of Archean rocks that were deformed and metamorphosed between 2.0–1.9 Ga (Holzer et al., 1999). The northern edge of the Zimbabwe craton seems to have undergone rifting at ~1.4 Ga, the age of the Chewore ophiolite in the Zambezi belt (S. Johnson and Oliver, 2000).

## India

Events with ages of 1.9 Ga to 1.4 Ga are widespread throughout India. The Delhi orogen of northwestern India showed continued sedimentation and deformation from the Paleoproterozoic into the Mesoproterozoic, with ages of magmatic rocks from 1.9 Ga to 1.5 Ga (Chaudhary et al., 1984). Gneisses in the Meghalaya area of northeastern India have a Rb–Sr age of ~1.7 Ga, and inclusions in a younger pluton have a Rb–Sr age of ~1.5 Ga (Ghosh et al., 1994). Closure of northern and southern India occurred along the Central Indian Tectonic Zone at ~1.8 Ga (Acharyya and Roy, 2000), and L. Harris (1993) correlated it with the Albany belt of Australia.

The southernmost part of India consists largely of rocks originally formed in the Paleoproterozoic and then overprinted during the Pan-African (Santosh et al., 2003). The southern margin of the Archean cratons of central India is a transition zone to granulites (chapter 4), and the terrane directly south of the transition zone consists of Archean rocks overprinted by younger events. The southern border of this region is a dextral fault zone, which separates the overprinted Archean suites from gneisses, granulites, and supracrustal rocks that formed originally at ~1.9–1.8 Ga and were overprinted in Pan-African time.

The Eastern Ghats granulite terrane of eastern India is mostly a suite of silicic metaigneous and metasedimentary rocks that show abundant isotopic evidence of Mesoproterozoic events prior to metamorphism during the Grenville (~1 Ga) and Pan-African (~0.5 Ga) events (Bhattacharya, 1996; Arima et al., 2000; Dobmeier and Simmat, 2002). Ultrahigh temperature metamorphism of metapelites occurred at ~1.6 Ga (Dasgupta et al., 1999). Cooling ages of allanite and monazite from a pegmatite are from 1.7–1.6 Ga (Mezger and Cosca, 1999). C. Shaw et al. (1997) reported ages up to 1.5 Ga for a broad range of rock types and also show that both felsic and mafic rocks have Sm–Nd whole-rock ages of ~1.4 Ga, with some indication of older ages based mostly on U–Pb systems.

The Aravalli–Delhi orogens of northwestern India began to develop at ~1.9 Ga and continued until later in the Proterozoic.

## Australia

Eastward growth of Australia occurred from the Early Proterozoic to the Mesozoic. It began during and following the suturing of the Pilbara and Yilgarn cratons along the Capricorn orogen (Van Kranendonk and Collins, 1998), and possibly the complete fusion of all of the Precambrian of western Australia by ~1.8 Ga (G. Dawson et al., 2002). The Capricorn orogen contains oceanic lithosphere, continental-margin sediments, banded iron formation, and carbonates, all of which suggest development in a backarc or foreland-basin environment. Deformation apparently occurred when the Pilbara and Yilgarn cratons collided at ~2.0–1.9 Ga.

The Albany and Fraser orogens of Australia extend into the Bunger Hills and Windmill Islands of coastal East Antarctica, which were attached throughout most of the Proterozoic (L. Harris, 1995). Together they form the southern and southeastern margins of the western Australian craton. The oldest well-known Mesoproterozoic activity in the Fraser belt consists of intrusion of subduction-related batholiths between approximately 1.7–1.6 Ga (Nelson et al., 1995), and the next well-dated event was dextral transpression and intrusion of the Albany enderbite at ~1.3 Ga. This history is remarkably similar to that of the Windmill Islands and Bunger Hills of Antarctica (L. Harris, 1995). The Windmill Islands contain gneisses at least as old as 1.5 Ga, and orthogneisses in the Bunger Hills range from 1.7–1.5 Ga. These Meso-proterozoic ages have not been found in the Albany belt, which shows the oldest orogenic activity at ~1.3 Ga, partly reworking Archean and Paleoproterozoic rocks in the southern margin of the Yilgarn shield (western Australian craton).

## East Antarctica

Areas of Proterozoic orogeny in coastal East Antarctica correlate well with orogens in South Africa, India, and Australia. The Windmill Islands and Bunger Hills are so similar to the Albany and Fraser orogens of Australia that they presumably developed together both in Grenville and pre-Grenville time (see above). Similarly the Rayner orogen (Kelly et al., 2002) shares a common history with the Eastern Ghats of India, including northward vergence of thrusts (present orientation in Antarctica and pre-drift orientation of India in Gondwana). The Transantarctic (Ross) orogen was primarily active during the Neoproterozoic and Early Paleozoic (chapter 8; appendix I), and older events are controversial.

# Appendix I

## Orogenic Belts of Pan-African–Brasiliano Age

The names and locations of the major Pan-African–Brasiliano orogenic belts and the stable areas that they separate are shown in fig. I.1. It also shows four important regions that we discuss in more detail below. We begin with a discussion of individual orogenic belts and then proceed to these areas of greater detail.

The Central African fold belt formed along the continental margin between the northern part of the Congo craton and the Pan-African region that occupied much of the area from West Africa to central Arabia (Feybesse et al., 1998; Toteu et al., 2001). Development that began in the Archean is shown by Archean xenocrysts and isotopic inheritance in high-grade metamorphic rocks formed during the Eburnian (~2.1 Ga) orogeny. The area was unaffected by events of Grenville age, and at some time before the Pan-African orogeny it became the site of an ocean basin that extended northward from the Congo craton. In a period centered around 600 Ma, oceanic volcanic rocks and sediments, continental-margin sediments, and Eburnian-age metamorphic rocks were thrust onto the Congo craton, accompanied by syntectonic magmatism and right-lateral thrusting. This Pan-African orogeny may have been caused by collision with an unknown block to the north, or it may have been an Andean-style orogeny along a continental margin.

Although the Central African fold belt appears to extend into the Borborema province, the history of Borborema is very different (da Silva Filho and de Lima, 1995). Borborema is dominated by intense right-lateral shears that extend into it from the Central African fold belt and also from orogenic belts on the east side of the West African craton. The shearing is contemporaneous with, and may be partly responsible for, emplacement of voluminous high-K magmatic rocks with a range of $SiO_2$ contents from intermediate to felsic. Derivation of these magmas from an enriched lithospheric mantle is demonstrated by high concentrations of LIL elements and several isotopic studies, including $T_{DM}$ values of 2–1.8 Ga for magmas emplaced at 600–500 Ma (Ferreira et al., 1998; Guimares et al., 1998; Neves et al., 2000). This evidence clearly indicates that Pan-African oceanic lithosphere did not extend into Borborema, and the orogen is intracontinental.

The Aracuai belt, including West Congo, is the type example of a confined orogen (chapter 5; Pedrosa-Soares et al., 2001). It represents closure of an ocean basin that extended briefly into the Sao Francisco and Congo cratons, which had been fused at some time much earlier than the approximately 600-Ma age of orogenic activity.

The Kaoko and Gariep belts of Africa and the Dom Feliciano belt of South America developed on opposite sides of a closing ocean basin. The eastern part of the Gariep belt consists of a parautochthonous sequence of unmetamorphosed shelf sediments thrust over a basement of Grenville age on the western margin of the Kalahari block (fig. I.1; Frimmel and Frank, 1998). The western part of the Gariep belt formed on oceanic lithosphere, and it contains oceanic basalts and sediments and bimodal basalt–rhyolite suites, all of which are highly deformed. Deformation, thrusting, and emplacement of post-orogenic granites all occurred during ~600–500 Ma.

The Dom Feliciano belt also consists of predominantly oceanic and continental terranes (fig. I.1; Leite et al., 1998; Frantz and Botelho, 2000). The western part of the belt contains parautochthonous rocks of the Rio de la Plata craton that were metamorphosed to high-amphibolite facies and thrust toward the craton at various times culminating at ~580 Ma. East of this sequence is a terrane dominated by ophiolites and oceanic sediments that was highly deformed, invaded by syntectonic batholithic rocks, and also thrust westward at ~580 Ma. The occurrence of batholithic suites in the Dom Feliciano belt and their absence in the Gariep and Kaoko belts suggest that the Pan-African ocean closed by subduction westward under the Rio de la Plata craton.

Orogenic Belts of Pan-African–Brasiliano    235

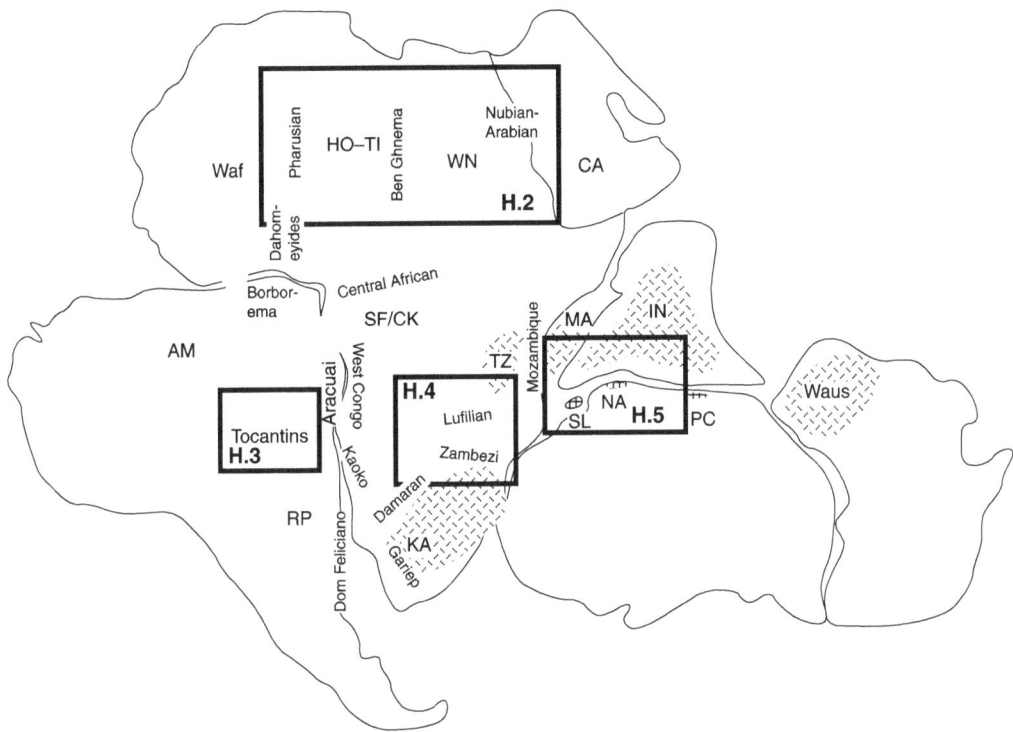

Figure I.1. Locations of major belts of Pan-African age plus four areas described in detail in this appendix. Numbers in each area are the figure numbers of maps in this appendix. Abbreviations are: Waf, West Africa; HO–TI, Hoggar–Tibesti; WN, West Nile; CA, Central Arabia; Am, Amazonia; SF/CK, Sao Francisco/Congo–Kasai; MA, Madagascar; IN, India; RP, Rio de la Plata; KA, Kalahari; SL, Sri Lanka; NA, Napier; PC, Prince Charles; Waus, western Australia.

The Mozambique belt extends southward from the Nubian–Arabian shield along almost the entire east coast of Africa and is commonly referred to as the "East African Orogen" (reviews by Stern, 1994, 2002). As we discussed in chapter 3, the Nubian–Arabian shield clearly resulted from the accumulation of volcanic arcs during a few hundred million years before ~600 Ma, with intrusion of post-tectonic granites and stabilization of the terrane between 600–500 Ma. This region of juvenile crust extends to the northern margin of the Ethiopian flood basalts and also appears in scattered outcrops just south of the basalts (Teklay et al., 2001; Yibas et al., 2001), but the Mozambique belt farther south apparently had a different history.

The Mozambique belt through Kenya, Tanzania, and Mozambique is the eastern margin of the Tanzanian and Kalahari blocks. The margin is best exposed in Tanzania, where granulites of the Mozambique belt are thrust over the Archean craton in a pattern similar to the Grenville front and other granulite belts (Muhongo, 1999; chapter 7). Some rocks from the craton have been intermixed in the thrust sequences, suggesting that western parts of the belt are parautochthonous. Rocks throughout the belt consist almost entirely of para- and orthogneisses and migmatites in high-amphibolite to granulite facies. Mafic rocks occur locally, but their ages are unclear, and no unquestionable ophiolite or ophiolite fragments have been found.

The Mozambique belt exhibits a complex series of orogenic events showing almost continuous activity from about 1100–600 Ma, and all of them show vergences toward the west. The oldest compressive deformation was probably in the Lurio belt of northern Mozambique, which may have been associated with suturing of the Maudheim and Kalahari blocks (chapter 7; Pinna et al., 1993; Manhica et al., 2001). This suturing is consistent with proposals that formation of juvenile crust terminated about 1000 Ma, and 600-Ma Pan-African orogenesis merely overprinted older

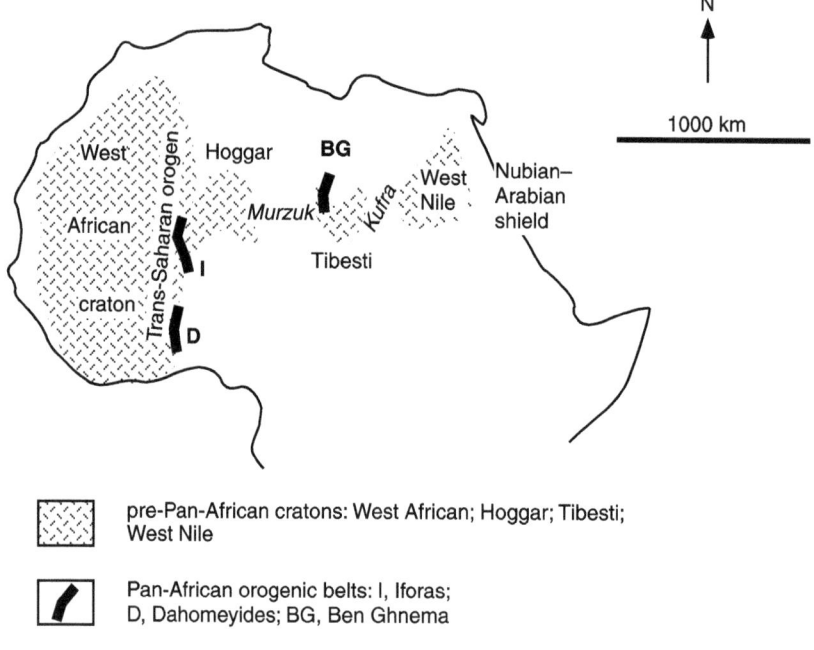

Figure I.2. General geology of area between Nubian–Arabian shield and West African craton.

rocks (Kroner et al., 2001; Muhongo et al., 2001a, b). Other workers, however, suggest that the final closure of the Lurio belt occurred during Pan-African time (Appel et al., 1998; Sacchi et al., 2000).

Isotopic systems north of the Lurio belt show numerous events between about 900–600 Ma, with cooling ages generally in the range of 500–400 Ma (Moeller et al., 1998; Muhongo et al., 2001a). The amount of juvenile crust formed in these events is controversial, but apparently most of the activity consisted of overprinting of older rocks. This question is an integral part of the effort to locate a suture between East and West Gondwana, and we discuss it more completely below.

Pan-African activity along coastal East Antarctica consists primarily of reworking of rocks deformed during Grenville time. Few juvenile magmatic rocks were developed during the Pan-African, and vergences of thrusts are all away from interior Antarctica, just as they were during the Grenville orogeny. Major Pan-African activity is confined to the area from the Rayner belt westward to Maudheim (Yoshida, 1995).

Most of Australia shows no evidence of compressive activity or thermal resetting of isotopic systems in Pan-African time. The Darling belt on the western side of Australia was a strike-slip zone that extended into Antarctica, but the principal Pan-African activity in Australia was rifting along its eastern margin and extension of shallow seas into most of the continental interior (Preiss, 2000). Rifting to form the Adelaidian trough formed a subsiding shelf that permitted the accumulation of thick Neoproterozoic sediments (they contain the Ediacaran soft-bodied biota, and we discuss them further in chapter 12). The Adelaide trough was continuous with the Neoproterozoic subsiding margin of the Transantarctic Mountains (Goodge and Fanning, 1999), and sediments deposited there were deformed beginning in the Early Paleozoic (perhaps starting in the latest Neoproterozoic). This deformation developed the Delamerian orogeny in Australia and Ross orogeny in Antarctica, both of which occurred on the outer margin of Gondwana through much of the Paleozoic.

## North Africa and Arabian Peninsula

The area shown in fig. I.2 is an arid region where rocks of Pan-African and older age are almost completely covered by Phanerozoic sediments and desert sands. The lack of exposure throughout the region has led some geologists to refer to much of the area from the Nubian–Arabian shield to the West African craton as a "Ghost Craton." Abdelsalam et al. (2002) recently termed it the "Saharan Metacraton" because many of the exposed rocks seem to have been formed by reworking of older continental crust, possibly following delamination of continental lithosphere shortly before the Pan-African event. We discuss the area from east to west.

The West Nile craton forms the western margin of the Nubian–Arabian shield, which was consolidated in Pan-African time after a few hundred million years of accretion of island arcs and oceanic volcanosedimentary suites (chapter 4; appendix E). Exposures in southern Egypt and Sudan show Sr and Nd isotopic values that indicate Pan-African metamorphism of older rocks and also generation of granitic magmas from the same sources (Schandelmeier et al., 1983).

The West Nile craton is terminated westward by the Kufra basin, which is particularly deep along a zone of weakness separating the West Nile craton from the Tibesti massif. The eastern part of the Tibestis consists of Pan-African intrusive rocks emplaced into a suite of presumed, but not well dated, Proterozoic metasediments. The basement of the western Tibestis is apparently similar to the basement farther east, but the Pan-African intrusive suite is very different. The western Tibestis are dominated by the Ben Ghnema batholith, with an age of 550 Ma (Ghuma and Rogers, 1978). This batholith is lithologically very similar to such typical continental-margin suites as the Sierra Nevadas, particularly in showing a gradation from rocks of intermediate composition on the east to highly felsic granites to the west. This gradation and general lithology strongly indicate that the Ben Ghnema batholith formed by subduction of oceanic lithosphere westward under a continental block that crops out only in the western Tibesti massif.

A buried basement west of the Tibestis is now covered by the extensive Murzuk basin, which appears to consist entirely of shallow-water Phanerozoic sediments. The western margin of this basin is the Hoggar massif, which may represent an outcrop of the crystalline rocks that underlies the Murzuk basin or may be a different massif structurally separated from the Murzuk block.

The Hoggar massif contains a complex sequence of fault slices apparently assembled by right-lateral transpression during Pan-African time (Boullier, 1991). The eastern part of the massif consists mostly of Mesoproterozoic rocks metamorphosed in the Pan-African and also partly melted to yield Pan-African granites (Liegeois and Black, 1993). The western part of Hoggar consists of oceanic materials formed in the Pharusian ocean basin that may have developed during the rifting of Rodinia (chapter 7). These materials consist of oceanic basalts, arc volcanic and sedimentary rocks, and sediments shed into the ocean from both the West African craton and the eastern Hoggar crystalline rocks.

Toward the end of Pan-African time, the Pharusian ocean closed by transpression between the eastern part of the Hoggar massif and the West African craton. Arc and oceanic rocks, including some ophiolites, were metamorphosed and thrust westward onto the West African craton to form the Dahomeyide and Pharusian (Iforas) orogenic belts. Subduction of oceanic lithosphere beneath West Africa at this time is demonstrated by synchronous development of calcalkaline batholithic suites in the orogens and in some rocks of the West African craton (Dostal et al., 1994; Attoh et al., 1997).

Roughly synchronous closure of the Pharusian ocean and an ocean in the Tibesti region probably completed the suturing of a mixture of juvenile and older Proterozoic terranes. By the end of Pan-African time, the entire area shown in fig. I.2 had been converted from a region of oceanic lithosphere containing an unknown number of continental fragments into a coherent continental block. Suturing of this block to the West African craton and Congo craton accompanied the final assembly of Gondwana by ~500 Ma.

## Tocantins Region

Tocantins is a region of Pan-African instability flanked by blocks that have been stable since Transamazonian time (fig. I.3; reviews by Pimentel et al., 1997, and Strieder and Freitas Suita, 1999; Brueckner et al., 2000). The Araguaia belt is a series of sediments and orthogneisses deformed and metamorphosed in Brasiliano (Pan-African) time between the Amazonian and Sao Francisco cratons. Similar rocks and structures extend into the Paraguay belt, which is thrust onto the southern margin of the Amazonian craton. Some ultramafic rocks that might be remnants of oceanic lithosphere have been found in the Araguaia belt but not in the Paraguay belt.

The Brasilia belt contains abundant evidence of closure of an ocean basin. Sediments deposited

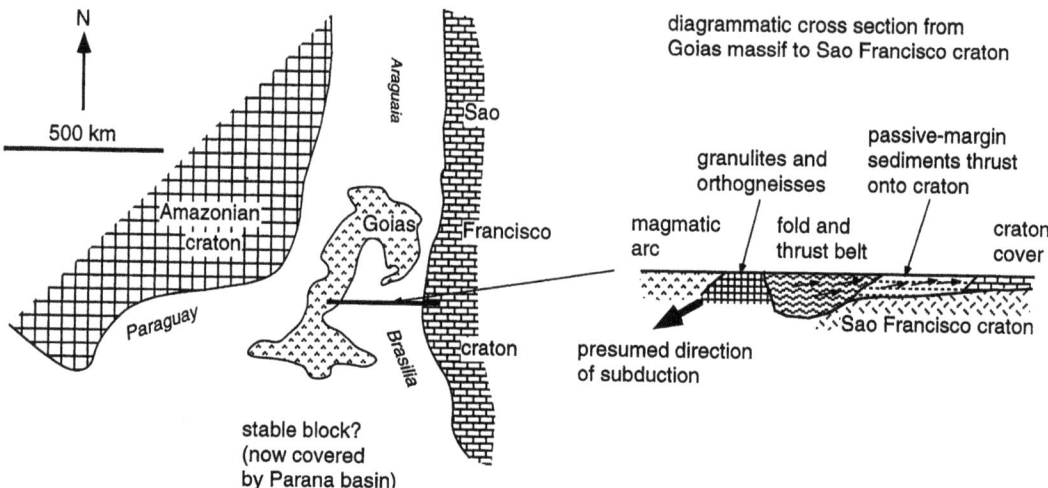

Figure I.3. General map of Tocantins region.

along the margin of the Sao Francisco craton were derived largely from the ~2-Ga rocks of the craton, with increasing mixtures of a source with $T_{DM}$ values of ~0.9 Ga upward in the section (Pimentel et al., 2001). These sediments were metamorphosed from low-amphibolite to granulite facies and thrust toward the craton between ~800–600 Ma. Several suites of ultramafic rocks interpreted to be fragments of ophiolites were incorporated in the metasedimentary suite during thrusting.

The Goias magmatic suite crops out west of the Brasilia metasediments. It contains some gneisses with ages greater than 2 Ga but is dominated by metamorphosed intrusive and volcanic rocks with intermediate compositions formed between ~900–600 Ma (Pimentel et al., 1997). These younger rocks have Nd and Sr isotopic ratios that indicate derivation mostly from oceanic lithosphere, and they are clearly the source for sediments derived from a provenance with $T_{DM}$ ages of ~0.9 Ga.

The end of orogenic activity in the Tocantins region is marked by intrusion of granites that are concentrated mostly in the southern part of the Goias arc (Pimentel et al., 1996). Suites older than 550 Ma are calcalkaline, with typical post-tectonic varieties somewhat younger (discussion of types of granites in chapter 3). The two suites have similar isotopic values, with initial $^{87}Sr/^{86}Sr$ from 0.703–0.710, and $\varepsilon_{Nd}$ from −4 to +3. These values indicate derivation partly from a mantle source and partly from continental lithosphere.

The southern extension of the Tocantins region is covered by the Parana basin, a series of Phanerozoic sediments partly topped by the Parana flood basalts that formed during the Cretaceous opening of the South Atlantic Ocean. This cover prevents correlation of the Tocantins region with the Rio de la Plata craton, which crops out only in its southern part and may form the basement for the Parana sequence.

Neoproterozoic rocks that separate older cratons in the Tocantins region clearly contain some juvenile magmatic suites and sediments derived from them, but the size of the ocean basin in the area is unclear. Because the Amazonian and Sao Francisco cratons may have been joined by Transamazonian time (chapter 5), it seems likely that the Araguaia belt was an intracratonic or a confined orogen (chapter 5). The Brasilia belt and Goias terrane, however, clearly contain oceanic materials, and were either a confined orogen or a zone of collision between the Sao Francisco craton and a block of uncertain age now covered by the Parana basin.

## Damaran–Zambezi–Lufilian Region

The Damaran–Zambezi system of belts forms a zone of Pan-African activity between stable blocks in southern and central Africa (fig. I.4; Porada and Berhorst, 2000). The Kalahari block, to the south, includes the Kaapvaal and Zimbabwe craton, which were joined by ~2.5 Ga (chapter 5), and central Africa consists of the Tanzanian and southeastern (Kasai) part of the Congo craton, which were fused along the type Kibaran belt at ~1.4–

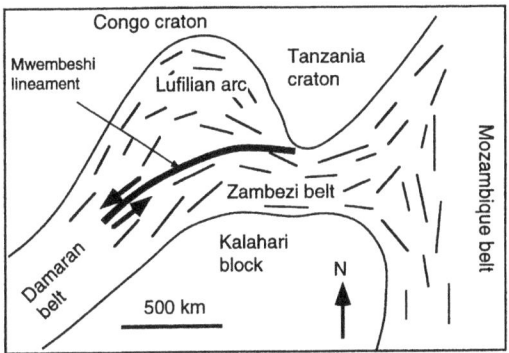

Figure I.4. General map of Damaran–Zambezi area.

1.2 Ga (chapter 7). The Zambezi belt connects eastward with the Lurio belt, which was deformed primarily in Grenville time and has an intense Pan-African overprint (see above). Left-lateral shearing of Pan-African age in both the Zambezi and Damaran belts may also continue into the Lurio belt, signifying that all of southern Africa moved eastward (present orientation) past central Africa.

The Damaran belt consists largely of quartzofeldspathic and carbonate–evaporite sediments metamorphosed to amphibolite facies and deformed during an age range centered around 600 Ma (reviews in R. Miller, 1983). Some of the sediments are turbidites that contain both quartzose and carbonate debris. Gneissic foliation is mostly parallel to the orogen, indicating compression across the belt. Migmatization is common, and syntectonic and post-tectonic granites are locally abundant. All granites were formed at ~500 Ma, with post-tectonic varieties slightly younger than syntectonic ones, and derivation of both types by crustal anatexis is shown by $^{87}Sr/^{86}Sr$ ratios greater than 0.716 and negative $^{143}Nd/^{144}Nd$ ratios (McDermott et al., 1996; Jung et al., 1997, 2000, 2001). Only a few thin zones of amphibolite are present, and they have been variously regarded as fragments of ocean floor, parts of a magmatic arc, or intracontinental volcanic suites (papers in R. Miller, 1983).

Derivation of granites by crustal anatexis and the absence of unquestioned ophiolites has led many investigators to conclude that the Damaran orogen did not form by closure of an ocean basin and should be regarded as intracontinental. Some workers, however, regard the presence of thick basins of accumulation and local occurrence of turbidites as evidence of a history of rifting followed by collision (Kukla and Stanistreet, 1991; Stanistreet et al., 1991). An example of the controversy is provided by Prave (1996), Duerr et al. (1997), and Vinyu et al. (1999).

The Zambezi belt contains mostly a sequence of shallow-water sediments and an extensive suite of bimodal (basalt–rhyolite) volcanic rocks lying on a continental basement of uncertain age (T. Wilson et al., 1993, 1997; Hanson et al., 1994). Compositions of the volcanic rocks indicate their formation in intracratonic rift basins rather than on oceanic lithosphere (Munyanwiya et al., 1997; Kampunzu et al., 2000). All units were metamorphosed to amphibolite facies or higher at ~800 Ma, with widespread resetting of isotopic systems at ~600 Ma (Hanson et al., 1994; Vinyu et al., 1999). At some time between ~800–600 Ma, rocks in the southern part of the belt were thrust onto the Kalahari block, and rocks to the north were thrust over stable areas of central Africa as the Lufilian arc (Kampunzu and Cailteux, 1999). The left-lateral Mwembeshi shear zone separates the northern and southern parts of the Zambezi belt and extends at least partly along the Damaran belt.

Because igneous rocks in the Damaran belt are anatectic and those in the Zambezi belt formed in intracontinental basins, it is possible that neither the Damaran nor Zambezi belts retains evidence of ocean closure in Pan-African time. This conclusion implies that a large area extending from the southern part of the Kalahari block through much of central Africa was a coherent continental block before Pan-African closure of other areas in Gondwana. We discuss below the difficulty this poses in discovering a suture between East and West Gondwana.

## Southern India, Madagascar, Sri Lanka, and Attached East Antarctica

Southern India, Madagascar, Sri Lanka, and attached East Antarctica occupied the central part of Gondwana and formed the core of a broad region of high-grade metamorphism, deformation, magmatism, and mineralization associated with the assembly of the supercontinent (fig. I.5). Almost identical metamorphic, magmatic, and mineralization patterns in all of these terranes indicate that they were a coherent assembly at the time of collision of East and West Gondwana. Because of their location and importance, we describe them in somewhat more detail than other areas.

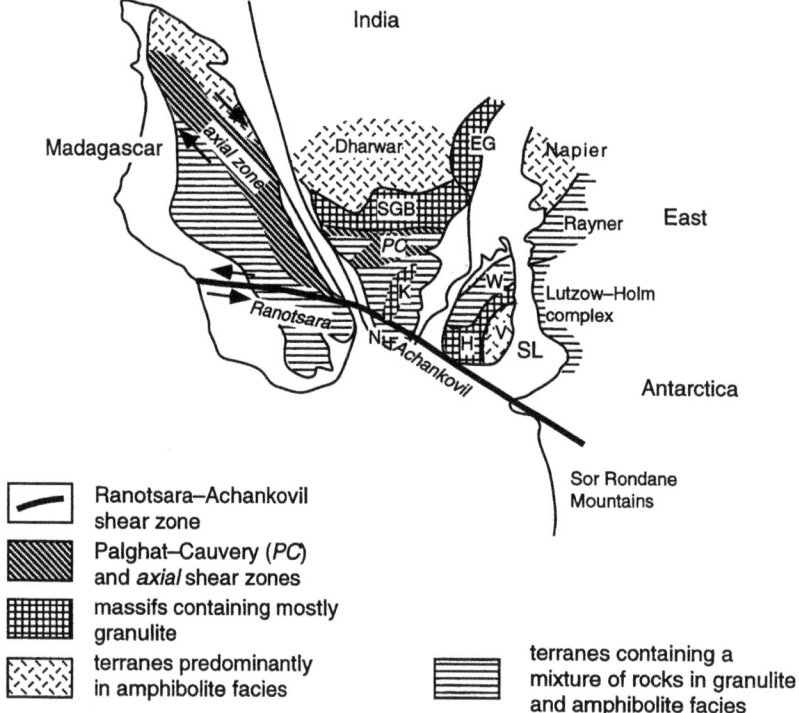

Figure I.5. General map of southern India, Madagascar, Sri Lanka, and adjoining East Antarctica. EG, Eastern Ghats; SGB, Southern Granulite Belt; PC, Palghat-Cauvery shear zone; k, Palni-Kodaikanal-Cardamom massif; N, Nagercoil massif; W, Wanni complex; H, highlands complex; V, Vijayan complex; SL, Sri Lanka

*Southern India*

Southern India comprises two distinct domains separated by the right-lateral Palghat–Cauvery and related shear zones. The Archean Dharwar craton to the north was stabilized at 3.0 Ga (chapter 4). It has a southern margin fringed by high-pressure granulite massifs with $T_{DM}$ ages ranging from 3.4–2.6 Ga and metamorphism at ~2.5 Ga (Peucat et al., 1993; N. Harris et al., 1994; discussion of transition zone in chapter 3). The craton and marginal massifs are largely unaffected by Pan-African orogeny.

By contrast, terranes south of the Palghat–Cauvery shear zone show extensive Pan-African reworking. This region contains the Palni-Kodaikanal–Cardamom and the Nagercoil massifs, which consist of charnockites and slivers of granulite-grade metasediments. Between and around the massifs is a vast supracrustal terrane composed of leptynites, khondalites, calcsilicates, and charnockites. Pan-African granulite-facies metamorphism and extensive crustal rejuvenation at ~550 Ma characterize all of these blocks (Bartlett et al., 1995, 1998; Santosh et al., 2003). Most of them, however, preserve Paleoproterozoic $T_{DM}$ ages (2.4–2.2 Ga), showing that the Pan-African event was one of crustal reworking and not of crustal growth. Mineral reaction textures and exhumation paths of these granulites indicate rapid and virtually isothermal uplift of various levels of the continental crust along a clockwise $p–T$ path during late Pan-African time.

The northern and southern parts of this high-pressure terrane are separated by the left-lateral Achankovil Shear Zone. A belt of cordierite gneisses, charnockites, and calcsilicate rocks ~50 km wide along the shear zone shows $T_{DM}$ ages in the range of 1.5–1.2 Ga (Bartlett et al., 1995; Brandon and Meen, 1995), suggesting that the sediment was derived from both Paleoproterozoic (~2.0 Ga) and Neoproterozoic (~1.0–0.5 Ga) sources. Several felsic, mafic, and ultramafic intrusives occur along the shear zone, and both the high-grade metamorphic rocks and the intrusives show Pan-African ages. It is possible that the Achankovil metasedimentary belt represents an internal or confined orogen (chapter 5), but the possibility remains to be investigated.

The Pan-African orogeny in southern India was accompanied by widespread felsic magmatism represented by a suite of alkali granite and syenite plutons that puncture the granulite-facies basement (U–Pb zircon ages of ~580–555 Ma; J. Miller et al., 1996). Most of the plutons occur along or proximal to regional fault lineaments, suggesting that the magma tectonics and emplacement occurred in an extensional tectonic regime (Santosh and Drury, 1988). The magmatism might represent high temperatures above a rising plume or a zone of rifting related to the extensional collapse following the collision of East and West Gondwana.

The final phase of magmatism is marked by the emplacement of pegmatites and veins that carry numerous types of mineralization. The timing of mineralization has been established from a Re–Os age of 525 Ma for molybdenites associated with alkali granite (Santosh et al., 1994) and a U–Pb age of 512 Ma for gem-quality zircons associated with pegmatite (J. Miller et al., 1996). Gemstone mineralizations in southern India, Sri Lanka, Madagascar, and parts of East Antarctica show comparable ages, similar tectonic settings, and remarkably identical mineral associations, suggesting a Pan-African gemstone province within East Gondwana (Menon and Santosh, 1995).

The late Pan-African in southern India and other parts of East Gondwana was also a period of extensive infiltration of $CO_2$-rich fluids from mantle sources. A variety of mineralogic and isotopic alterations in high-grade metamorphic rocks and the presence of abundant $CO_2$-rich fluid inclusions in metamorphic, magmatic, and pegmatitic assemblages provide evidence for the transfer of $CO_2$ through deep-rooted shear zones or magmatic conduits. Rich graphite deposits also formed within veins and pegmatites in southern India, Sri Lanka, and Madagascar by precipitation from $CO_2 \pm H_2O$ fluids (Santosh and Wada, 1993a,b; Santosh and Yoshida, 2001).

## East Antarctica

Major Pan-African orogenic belts crop out in Lutzow–Holm Bay and the Rayner complex (fig. I.5) and also in Prydz Bay and the southern Prince Charles Mountains farther east. Lutzow–Holm Bay, perhaps the locus of Pan-African extension into Antarctica, shows granulite-facies metamorphism at 550–520 Ma based on SHRIMP U–Pb ages of zircons in high-grade metamorphic rocks (Shiraishi et al., 1994). This area may be an extension of the Highland and Southwestern complexes of Sri Lanka (see below). The Rayner belt, separated from the Archean Napier complex by a poorly exposed fault, has $T_{DM}$ ages of 2.3–1.3 Ga and may correlate with the Wanni complex in Sri Lanka, which also shows a metamorphic event at 1.1 Ga (Shiraishi et al., 1994). All of the Antarctic belts shown in fig. 8.7 exhibit extensive Pan-African overprints, and ages from Prydz Bay and the southern Prince Charles Mountains also indicate prominent late Pan-African orogeny between 550–490 Ma (Boger et al., 2000, 2001, 2002). Significant volumes of felsic plutons were emplaced throughout the area during Pan-African time, correlating with identical late Pan-African magmatism in southern India, Sri Lanka, and Madagascar.

The lithologic units, metamorphic style, and deformation patterns in all of these areas of East Antarctica are closely comparable with those in southern India, Sri Lanka, and Madagascar. All of these areas also have Mesoproterozoic $T_{DM}$ ages, with the youngest ages (1.1–0.9 Ga) from the Sor Rondane Mountains of East Antarctica and the Vijayan complex of Sri Lanka. The assembly thus represents a Mesoproterozoic–Neoproterozoic terrane that underwent extensive crustal reworking but limited crustal growth during the Pan-African event.

## Madagascar

In Madagascar, the Ranotsara shear zone separates the Precambrian terrane into the southern and northern blocks and probably correlates with the Achankovil shear zone in southern India (Windley and Razakamanana, 1996). Both shear zones contain abundant granulite-facies supracrustal rocks and charnockites that yield ages in the range of ~570–520 Ma, late Pan-African alkali granites, and mineralized pegmatites.

South of the Ranotsara shear zone are N–S trending granulites and upper-amphibolite-facies paragneisses that contain conformable layers of marble, quartzite, diopsidite, and amphibolite. Prominent Neoproterozoic ductile shear zones divide the southern block into different units, indicating an accelerated exhumation accompanied by extensional tectonics between 520–490 Ma (de Wit et al., 2001). The southern block also carries evidence for widespread Pan-African overprinting ranging in age from 570–523 Ma (Paquette et al., 1994), representing reworking of early

Precambrian protoliths. Massif-type anorthosites yield crystallization ages of ~660 Ma and metamorphic ages of 559 Ma (Ashwal et al. 1998), and numerous stratoid sheets of alkali granites were emplaced at 585 Ma (Nicollet, 1990). Late Pan-African pegmatites contain ruby, beryl, columbite, niobotantalites, and uranium-bearing minerals. Both magmatism and pegmatite formation in Madagascar are identical to those in southern India with respect to timing of formation, tectonic environment, and mineralization (Menon and Santosh, 1995). All of this magmatism is correlated with an extensional setting and high geothermal gradients accompanying collapse of the Pan-African orogen that may have sutured East and West Gondwana.

Central and northern Madagascar consists of Precambrian rocks that form three distinct N–S belts (Windley and Razakamanana, 1996; A. Collins and Windley, 2002). The central (axial) belt exhibits dextral shearing and comprises granulite-facies rocks and their partly retrogressed equivalents of amphibolite-facies gneisses: paragneisses and graphitic gneisses (Nedelec et al., 2000). The region also contains a number of granite sheets; lenses of mafic–ultramafic rocks with ages of 787 Ma that include harzburgite, dunite, pyroxenite, and gabbro that contain nickel mineralization; and some lenses of low-pressure granulite. Ages in this terrane range from 726–527 Ma. Structurally above this central belt, toward the north of the island, are several synclinal belts which contain granulite and amphibolite-facies hornblende gneisses with ages of ~2700 Ma. Magnetite quartzite, hornblende-rich gabbro, metavolcanic amphibolite layers, basic granulites, and folded stratiform complexes of norite and pyroxenite also occur within these belts.

The axial belt of northern Madagascar shares many similarities with the Palghat–Cauvery shear zone in southern India, with both zones marking the boundary between southern terranes characterized by crustal reworking and high-grade metamorphism during the Pan-African and a northern Archean craton where a Pan-African overprint is less recognizable.

### Sri Lanka

The Sri Lankan fragment of Gondwana has been divided into three distinct provinces based on $T_{DM}$ ages (Milisenda et al., 1994). The oldest is the Highland/Southwestern complex (HSWC), which is dominated by high-grade supracrustal rocks and granitoids with $T_{DM}$ ages of 2.9–1.9 Ga that compare closely with the $T_{DM}$ ages from southern India. The western margin of the HSWC is a younger province known as the Wanni complex (WC), characterized by lower-grade (amphibolite-facies) rocks with $T_{DM}$ ages of 1.9 1.2 Ga. A similar range of $T_{DM}$ ages has been obtained from the Achankovil metasediments in southern India (see above). The third and youngest age province is the Vijayan complex (VC), comprising mostly amphibolite-facies rocks exposed on the eastern edge of the HSWC and clearly separated from it by a major thrust. The Vijayan complex has $T_{DM}$ ages from 1.5–0.8 Ga, indicative of a younger period of crustal growth.

In common with other East Gondwana terranes, Sri Lankan terranes also show upper-amphibolite to granulite-grade metamorphism and felsic magmatism during the Pan-African orogeny. The regional high-grade metamorphism, charnockite formation, and granite emplacement at about 550 Ma are broadly coeval and occurred within a relatively short time span of 100 Ma (Kroner et al., 1994). Distinct Pan-African thrust-nappe structures developed along both the eastern and western boundaries of the HSWC (Kriegsman, 1995). Sri Lanka is also one of nature's museums for gems, and the gemstone mineralization here is closely comparable to that in southern India.

# References

Abbott, D.H. (1996) Plumes and hotspots as sources of greenstone belts. Lithos, v. 37, 113–127.

Abdelsalam, M.G., and Stern, R.J. (1996) Sutures and shear zones in the Arabian–Nubian shield. J. African Earth Sci., v. 23, 289–310.

Abdelsalam, M.G., Liegeois, J.-P., and Stern, R.J. (2002). The Saharan metacraton. J. African Earth Sci., v. 34, 289–310.

Acharyya, S.K., and Roy, A. (2000) Tectonothermal history of the Central Indian Tectonic Zone and reactivation of major faults/shear zones. J. Geol. Soc. India, v. 55, 239–256.

Adiyaman, O., Chorowicz, J., Arnaud, O.N., Gundogdu, M.N., and Gourgaud, A. (2001) Late Cenozoic tectonics and volcanism along the North Anatolian fault; new structural and geochemical data. Tectonophysics, v. 338, 135–165.

Ahall, K.-I., Connelly, J.N., and Brewer, T.S. (2000) Episodic rapakivi magmatism due to distal orogenesis?: correlation of 1.69–1.50 Ga orogenic and inboard "anorogenic" events in the Baltic shield. Geology, v. 28, 823–826.

Al Saleh, A.M., and Boyle, A.P. (2001) Neoproterozoic ensialic back-arc spreading in the eastern Arabian shield: geochemical evidence from the Halaban ophiolite. J. African Earth Sci., v. 33, 1–15.

Alavi, M. (1994) Tectonics of the Zagros orogenic belt of Iran: new data and interpretations. Tectonophysics, v. 229, 211–238.

Aleinikov, A.L., Belavin, O.V., Bulashevich, Y.P., Tavrin, I.F., Maksimov, E.M., Rudkevich, M.Y., Nalivkin, V.D., Shablinskaya, N.V., and Surkov, V.S. (1980) Dynamics of the Russian and West Siberian platforms. In Dynamics of plate interiors (A.W. Bally, Ed.). Amer. Geophys. Union Geodynamics Ser. 1, 53–71.

Algeo, T.J., and Wilkinson, B.H. (1991) Modern and ancient continental hypsometries. Geol. Soc. London J., v. 148, 643–653.

Alkmim, F.F., Marshak, S., and Fonseca, M.A. (2001) Assembling West Gondwana in the Neoproterozoic; clues from the Sao Francisco Craton region, Brazil. Geology, v. 29, 319–322.

Almeida, F.F.M. de, Brito Neves, B.B. de, and Dal Re Carneiro, C. (2000) The origin and evolution of the South American platform. Earth-Sci. Rev., v. 50, 77–111.

Alvarez, L.W., Alvarez, W., Asaro, F., and Michel, H.V. (1980) Extraterrestrial cause for the Cretaceous–Tertiary extinction. Science, v. 208, 1095–1098.

Alvarez, W. (1997) T. Rex and the crater of doom. Princeton, N.J. Princeton University Press.

Ampferer, O., and Hammer, W. (1911) Geologischer querschnitt durch die Ostalpen von Allgau zum Gardarsee, Jahrbuch Geologischen Reichsanstalt, v. 61, 531–710.

Amri, I., Benoit, M., and Ceuleneer, G. (1996) Tectonic setting for the genesis of oceanic plagiogranites: evidence from a paleo-spreading structure in the Oman ophiolite. Earth Planet. Sci. Lett., v. 139, 177–194.

Anderson, D.L. (2002) How many plates? Geology, v. 30, 411–414.

Anderson, J.B. (1999) Antarctic marine geology. New York, Cambridge University Press.

Anderson, J.L., and Cullers, R.L. (1999) Paleo- and Mesoproterozoic granite plutonism of Colorado and Wyoming. Rocky Mountain Geol., v. 34, 149–164.

Anderson, J.L., and Morrison, J. (1992) The role of anorogenic granites in the Proterozoic crustal development of North America. In Proterozoic crustal evolution (K.C. Condie, Ed.). New York, Elsevier, 263–299.

Andersson, J., Soderlund, U., Cornell, D., Johansson, L., and Moller, C. (1999) Sveconorwegian (Grenvillian) deformation, metamorphism and leucosome formation in SW Sweden, SW Baltic Shield; constraints from a Mesoproterozoic granite intrusion. Precamb. Res., v. 98, 151–171.

Anhaeusser, C.R. (Ed.) (1983) Contributions to the geology of the Barberton Mountain Land. Geol. Soc. South Africa Spec. Publ. 9.

Appel, P., Moeller, A., and Schenk, V. (1998) High-pressure granulite-facies metamorphism in the Pan-African Belt of eastern Tanzania; $P$–$T$–$t$ evidence against granulite formation by continent collision. J. Metam. Geol., v. 16, 491–509.

Arima, M., Takono, N., and Saradhi, P. (2000) Tectonothermal evolution of Eastern Ghats belt, India: implications for the East Gondwana–Rodinia amalgamation. Tokyo, 20th Symposium on Antarctic Geosciences, 32–33. Tokyo, National Institute of Polar Research.

Ashwal, L.D., Hamilton, M.A., Morel, V.P.I., and Rambeloson, R.A. (1998) Geology, petrology and isotope geochemistry of massif-type anorthosites from southwest Madagascar. Contrib. Mineral. Petrol., v. 133, 389–401.

Aspler, L.B., and Chiarenzelli, J.R. (1998) Two Neoarchean supercontinents? Evidence from the Paleoproterozoic. Sedimentary Geol., v. 120, 75–104.

Astini, R.A., and Thomas, W.A. (1999) Origin and evolution of the Precordillera terrane of western Argentina; a drifted Laurentian orphan. In Laurentia–Gondwana connections before Pangea (J.D. Keppie, Ed.). Geol. Soc. Amer. Spec. Paper 336, 1–20.

Attoh, K., Dallmeyer, R.D., and Affaton, P. (1997) Chronology of nappe assembly in the Pan-African Dahomeyide orogen, West Africa: evidence from $^{40}Ar/^{39}Ar$ mineral ages. Precamb. Res., v. 82, 153–171.

Atwater, T. (1989) Plate tectonic history of the Northeast Pacific and western North America. In The eastern Pacific Ocean and Hawaii (E.L. Winterer, D.M. Hussong, and R.W. Decker, Eds.). The Geology of North America, V.N., Geol. Soc. Amer., 21–72.

Atwater, T., and Stock, J. (1998) Pacific–North America plate tectonics of the Neogene southwestern United States: an update. Internat. Geol. Rev., v. 40, 375–402.

Austrheim, H. (1986) Eclogitization of lower crustal granulites by fluid migration through shear zones. Earth Planet. Sci. Lett., v. 81, 221–232.

Axen, G. Jr., Lam, P.S., Grove, M., Stockli, D.F., and Hassanzadeh, J. (2001) Exhumation of the west-central Alborz Mountains, Iran, Caspian subsidence, and collision-related tectonics. Geology, v. 29, 559–562.

Ballance, P.F. (1999) Simplification of the southwest Pacific Neogene arcs; inherited complexity and control by a retreating pole of rotation. In Continental tectonics (C. Mac Niocaill and P.D. Ryan, Eds.). Geol. Soc. London Spec. Publ. 164, 7–19.

Barley, M.E. (1993) Volcanic, sedimentary and tectonostratigraphic environments of the approximately 3.46 Ga Warrawoona Megasequence; a review. Precamb. Res., v. 60, 47–67.

Barth, A.P., Wooden, J.L., Coleman, D.S., and Fanning, C.M. (2000) Geochronology of the Proterozoic basement of southwesternmost North America, and the origin and evolution of the Mojave crustal province. Tectonics, v. 19, 616–629.

Bartlett, J.M., Dougherty-Page, J.S., Harris, N.B.W., Hawkesworth, C.J., and Santosh, M. (1998) The application of single zircon evaporation and model Nd ages to the interpretation of polymetamorphic terrains: an example from the Proterozoic mobile belt of south India. Contrib. Mineral. Petrol., v. 131, 181–195.

Bartlett, J.M., Harris, N.B.W., Hawkesworth, C.J., and Santosh, M. (1995) New isotopic constraints on the crustal evolution of South India and Pan-African granulite metamorphism. Geol. Soc. India Memoir 34, 391–397.

Beck, R.A., Sinha, A., Burbank, D.W., Sercombe, W.J., and Khan, A.M. (1998) Climatic, oceanographic, and isotopic consequences of the Paleocene India–Asia collision. In Late Paleocene–Early Eocene climatic and biotic events in the marine and terrestrial records (M.-P. Aubrey, S.G. Lucas, and W.A. Berggren, Eds.). New York, Columbia University Press, 103–117.

Behrendt, J.C. (1999) Crustal and lithospheric structure of the West Antarctic Rift system from geophysical investigations; a review. In Lithosphere dynamics and environmental change of the Cenozoic West Antarctic Rift system (F.M. van der Wateren and S.A.P.L. Cloetingh, Eds.). Global and Planetary Change, v. 23, 25–44.

Bekker, A., Kaufman, A.J., Karhu, J.A., Beukes, N.J., Swart, Q.D., Coetzee, L.L., and Eriksson, K.A. (2001) Chemostratigraphy of the Paleoproterozoic Duitschland Formation, South Africa: implications for coupled climatic change and carbon cycling. Amer. J. Sci., v. 301, 261–285.

Bell, K., and Blenkinsop, J. (1987) Archean depleted mantle: evidence from Nd and Sr initial isotopic ratios of carbonatites. Geochim. Cosmochim. Acta, v. 51, 291–298.

Benioff, H. (1954) Orogenesis and deep crustal structure—additional evidence from seismology. Geol. Soc. Amer. Bull., v. 65, 385–400.

Berry, E.W. (1928) Comments on the Wegener hypothesis. In Theory of continental drift (W.A.J.M. van Waterschoot van der Gracht, Ed.). Amer. Assoc. Petrol. Geol., 194–196.

Besse, J., and Courtillot, V. (1988) Palaeogeographic maps of the continents bordering the Indian Ocean since the early Jurassic. J. Geophys. Res., v. 93, 11,791–11,808.

Besse, J., Torcq, F., Gallet, Y., Ricou, L.E., Krystyn, L., and Saidi, A. (1998) Late Permian to late Triassic palaeomagnetic data from Iran; constraints on the migration of the Iranian block through the Tethyan Ocean and initial destruction of Pangaea. Geophys. J. Internat., v. 135, 77–92.

Bettencourt, J.S., Tosdal, R.M., Leite, W.B., and Payolla, B.L. (1999) Mesoproterozoic rapakivi granites of the Rondonian tin province, southwestern border of the Amazon craton, Brazil; I. reconnaissance U–Pb geochronology and regional implications. In Rapakivi granites and related rocks (I. Haapala and O.T. Ramo, Eds.). Precamb. Res., v. 95, 41–67.

Bhattacharya, S. (1996) Eastern Ghats granulite terrain of India: an overview. In Precambrian India within East Gondwana (M. Yoshida, M. Santosh, and M. Arima, Eds.). J. Southeast Asian Earth Sci., v. 14, 165–174.

Bhushan, S.K. (2000) Malani rhyolites; a review. Gondwana Res., v. 3, 65–77.

Bickford, M.E., and Anderson, A.L. (1993) Middle Proterozoic magmatism. In Precambrian: conterminous U.S. (J.C. Reed, Jr., M.E. Bickford, R.S. Houston, P.K. Link, D.W. Rankin, P.K. Sims, and W.R. Van Schmus, Eds.). The Geology of North America, v. C-2, Geol. Soc. Amer., 281–292.

Bickford, M.E., Collerson, K.D., and Lewry, J.F. (1994) Crustal history of the Rae and Hearne provinces, southwestern Canadian shield, Saskatchewan; constraints from geochronologic and isotopic data. Precamb. Res., v. 68, 1–21.

Bingen, B., Davis, W.J., and Austrheim, H. (2001) Zircon U–Pb geochronology in the Bergen Arc eclogites and their Proterozoic protoliths, and implications for the pre-Scandian evolution of the Caledonides in western Norway. Geol. Soc. Amer. Bull., v. 113, 640–649.

Biswal, T.K., Biswal, B., Mitra, S., and Maitry Roy, M. (2002) Deformation pattern of the NW terrane boundary of the Eastern Ghats Mobile Belt, India; a tectonic model and correlation with Antarctica. Gondwana Res., v. 5, 45–52.

Black, L.P., Harley, S.L., Sun, S.S., and McCulloch, M.T. (1987) The Rayner Complex of East Antarctica: complex isotopic systematics within a Proterozoic mobile belt. J. Metam. Geol., v. 5, 126.

Black, L.P., Harris, L.B., and Delor, C.P. (1992) Reworking of Archaean and early Proterozoic components during a progressive, middle Proterozoic tectonothermal event in the Albany mobile belt, Western Australia. Precamb. Res., v. 59, 95–123.

Blewett, R.S. (2002) Archaean tectonic processes: a case for horizontal shortening in the North Pilbara granite–greenstone terrane, Western Australia. Precamb. Res., v. 113, 87–120.

Boger, S.D., Carson, C.J., Fanning, C.M., Hergt, J.M., Wilson, C.J.L., and Woodhead, J.D. (2002) Pan-African intraplate deformation in the nothern Prince Charles Mountains, east Antarctica. Earth Planet. Sci. Lett., v. 195, 195–210.

Boger, S.D., Carson, C.J., Wilson, C.J.L., and Fanning, C.M. (2000) Neoproterozoic deformation in the Radok Lake region of the northern Prince Charles Mountains, east Antarctica; evidence for a single protracted orogenic event. Precamb. Res., v. 104, 1–24.

Boger, S.D., Wilson, C.J.L., and Fanning, C.M. (2001) Early Paleozoic tectonism within the East Antarctic craton: the final suture between east and west Gondwana? Geology, v. 29, 463–466.

Boher, M., Abouchami, W., Michard, A., Albarede, F., and Arndt, N.T. (1992) Crustal growth in West Africa at 2.1 Ga. J. Geophys. Res., v. 97, 345–369.

Bohlen, S.R., and Mezger, K. (1989) Origin of granulite terranes and the formation of the lowermost continental crust. Science, v. 244, 326–329.

Bolt, B.A. (1999) Earthquakes (4th ed.). New York, W.H. Freeman Co.

Bond, G.C., Nickerson, P.A., and Kominz, M.A. (1984) Breakup of a supercontinent between 625 Ma and 555 Ma: new evidence and implications for continental histories. Earth Planet. Sci. Lett., v. 70, 325–345.

Borg, G., and Shackleton, R.M. (1997) The Tanzania and NE-Zaire cratons, In Greenstone belts (M.J. DeWit and L.D. Ashwal, Eds.). Oxford Monographs on Geology and Geophysics 35, 608–619.

Botta, G., and Bada, J.L. (2002) Extraterrestrial organic compounds in meteorites. Surv. in Geophys., v. 23, 411–467.

Boullier, A.-M. (1991) The Pan-African trans-Saharan belt in the Hoggar shield (Algeria, Mali, Niger); a review. In The West African orogens and circum-Atlantic correlatives (R.D. Dallmeyer and J.P. Lecorche, Eds.). Berlin, Springer-Verlag, 85–105.

Bowring, S.A., and Grotzinger, J.P. (1992) Implications of new chronostratigraphy for tectonic evolution of Wopmay orogen, northwest Canadian shield. Amer. J. Sci., v. 292, 1–20.

Bowring, S.A., and Housh, T. (1995) The earth's early evolution. Science, v. 269, 1535–1540.

Bowring, S.A., and Ross, G.M. (1985) Geochronology of the Narakay Volcanic Complex: implications for the age of the Coppermine Homocline and Mackenzie events. Canadian J. Earth Sci., v. 22, 774–781.

Bowring, S.A., Grotzinger, J.P., Isachsen, C.E., Knoll, A.H., Pelechaty, S.M., and Kolosov, P. (1993) Calculating rates of Early Cambrian evolution. Science, v. 261, 1293–1298.

Brandon, A.D., and Meen, J.K. (1995) Nd isotopic evidence for the position of southernmost Indian terrains within East Gondwana. Precamb. Res., v. 70, 269–280.

Brasier, M.D., and 7 others (2002). Questioning the evidence for Earth's oldest fossils. Nature, v. 416, 76–81.

Brasier, M.D., and Lindsay, J.F. (1998) A billion years of environmental stability and the emergence of eukaryotes; new data from northern Australia. Geology, v. 26, 555–558.

Braun, J., and Shaw, R. (2001) A thin-plate model of Palaeozoic deformation of the Australian lithosphere; implications for understanding the dynamics of intracratonic deformation, In Continental reactivation and reworking (J.A. Miller, R.E. Holdsworth, I.S. Buick, and M. Hand, Eds.). Geol. Soc. London Spec. Pub. 184, 165–193.

Brewer. T.S. (Ed.) (1996) Precambrian crustal evolution in the North Atlantic region. Geol. Soc. London Spec. Pub. 112.

Brito Neves, B.B. de (2002) Main stages of the development of the sedimentary basins of South America and their relationship with the tectonics of supercontinents. Gondwana Res., v. 5, 175–196.

Brocks, J.J., Logan, G.A., Buick, R., and Summons, R.E. (1999) Archean molecular fossils and the early rise of eukaryotes. Science, v. 285, 1033–1036.

Brookfield, M.E. (1993) Neoproterozoic Laurentia–Australia fit. Geology, v. 21, 683–686.

Brown, D., and Spadea, P. (1999) Processes of forearc and accretionary complex formation during arc-continent collision in the southern Ural Mountains. Geology, v. 27, 649–652.

Brueckner, H.K., Cunningham, D., Alkmim, F.F., and Marshak, S. (2000) Tectonic implications of Precambrian Sm–Nd dates from the southern Sao Francisco craton and adjacent Aracuai and Riberira belts, Brazil. Precamb. Res., v. 99, 255–269.

Bullen, M.E., Burbank, D.W., Garver, J.I., and Abdrakhmatov, K.Ye. (2001) Late Cenozoic tectonic evolution of the northwestern Tien Shan: new age estimates for the initiation of mountain building. Geol. Soc. Amer. Bull., v. 113, 1544–1559.

Burchfiel, B.C., Cowan, D.S., and Davis, G.A. (1992) Tectonic overview of the Cordilleran orogen in the western United States. In The Cordilleran orogen: conterminous U.S. (B.C. Burchfiel, P.W. Lipman, and M.M. Zoback, Eds.). The Geology of North America, v. G-3, Geol. Soc. Amer., 407–479.

Burrett, C., and Berry, R. (2000) Proterozoic Australia–western United States (AUSWUS) fit between Laurentia and Australia. Geology, v. 28, 103–106.

Burrett, C., and Berry, R. (2001). Unpublished data.

Burrett, C., and Berry, R. (2002) A statistical approach to defining Proterozoic crustal provinces and testing continental reconstructions of Australia and Laurentia—SWEAT or AUSWUS? Gondwana Res., v. 5, 109–122.

Burton, K.W., Schiano, P., Birck, J.-L., Allegre, C.J., Rehkamper, M., Halliday, A.N., and Dawson, J.B. (2000) The distribution and behaviour of rhenium and osmium amongst mantle minerals and the age of the lithospheric mantle beneath Tanzania. Earth Planet. Sci. Lett., v. 183, 93–106.

Butterfield, N.J. (2000) *Bangiomorpha Pubescens* n. gen., n. sp.: implications for the evolution of sex, multicellularity, and the Mesoproterozoic/Neoproterozoic radiation of eukaryotes. Paleobiology, v. 26, 386–404.

Butterfield, N.J. (2001) Paleobiology of the late Mesoproterozoic (ca. 1200 Ma) Hunting Formation, Somerset Island, Arctic Canada. Precamb. Res., v. 111, 235–256.

Byerly, G.R., Lowe, D.R., Wooden, J.L., and Xie, X. (2002) An Archean impact layer from the Pilbara and Kaapvaal cratons. Science, v. 297, 1325–1327.

Calver, C.R., and Lindsay, J.F. (1998) Ediacaran sequence and isotope stratigraphy of the Officer basin, South Australia. Australian J. Earth Sci, v. 45, 513–532.

Card, K.D. (1990) A review of the Superior province of the Canadian shield. Precamb. Res., v. 48, 99–156.

Carignan, J., Ludden, J., and Francis, D. (1996) On the recent enrichment of subcontinental lithosphere: a detailed U–Pb study of spinel lherzolite xenoliths, Yukon, Canada. Geochim. Cosmochim. Acta, v. 60, 4241–4252.

Carlson, R.W., and 16 others (2000) Continental growth, preservation, and modification in southern Africa. GSA Today, v. 10, no. 2, 1–7.

Carr, S.D., Easton, R.M., Jamieson, R.A., and Culshaw, N.G. (2000) Geologic transect across the Grenville orogen of Ontario and New York. Canadian J. Earth Sci., v. 37, 193–216.

Castle, J.W. (2001) Foreland-basin sequence response to collisional tectonism. Geol. Soc. Amer. Bull., v. 113, 801–812.

CD-ROM Working Group (2002) Structure and evolution of the lithosphere beneath the Rocky Mountains: initial results from the CD-ROM experiment. GSA Today, v. 12, no. 3, 4–10.

Chacko, T., De, S.K., Creaser, R.A., and Muehlenbachs, K. (2000) Tectonic setting of the Taltson magmatic zone at 1.9–2.0 Ga: a granitoid-based perspective. Canadian J. Earth Sci., v. 37, 1597–1609.

Chacko, T., Ravindra Kumar, G.R., Meen, J.K., and Rogers, J.J.W. (1992) Geochemistry of high-grade supracrustal rocks from the Kerala Khondalite Belt and adjacent massif charnockites, South India. Precamb, Res., v. 55, 469–489.

Chadwick, B., Vasudev, V.N., and Hegde, G.V. (2000) The Dharwar Craton, southern India, interpreted as the result of late Archaean oblique convergence. Precamb. Res., v. 99, 91–111.

Chamberlain, R.T. (1928) Some of the objections to Wegener's theory. In Theory of continental drift (W.A.J.M. van Waterschoot van der Gracht, Ed.). Amer. Assoc. Petrol. Geol., 83–87.

Chappell, B.W., White, A.J.R., and Hine, R. (1988) Granite provinces and basement terranes in the Lachlan fold belt, southeastern Australia. Australian J. Earth Sci., v. 35, 505–521.

Chaudhary, A.K., Gopalan, K., and Sastry, C.A. (1984) Present status of the geochronology of the Precambrian rocks of Rajasthan. Tectonophysics, v. 105, 131–140.

Chaudhuri, A.K., Saha, D., Deb, G.K., Patranabis Deb, S., Mukherjee, M.K., and Ghosh, G. (2002) The Purana basins of southern province of India—a case for Mesoproterozoic fossil rifts. Gondwana Res., v. 5, 23–33.

Chen, J., and Jahn, B.-M. (1998) Crustal evolution of southeastern China: Nd and Sr isotopic evidence. Tectonophysics, v. 284, 101–133.

Chiarenzelli, J.R., Aspler, L.B., Villeneuve, M., and Lewry, J.F. (1998) Early Proterozoic evolution of the Saskatchewan Craton and its allochthonous cover, Trans-Hudson Orogen. J. Geol., v. 106, 247–267.

Claesson, S., Bogdanova, S.B., Bibikova, E.V., and Gorbatschev, R. (2001) Isotopic evidence for the Palaeoproterozoic accretion in the basement of the East European craton. Tectonophysics, v. 339, 1–18.

Clark, D.J., Hensen, B.J., and Kinny, P.D. (2000) Geochronological constraints for a two-stage history of the Albany–Fraser Orogen, Western Australia. Precamb. Res., v. 102, 155–183.

Clark, P.O., Webb., R.S., and Keigwin, L.D. (Eds.) (1999) Mechanisms of global climate change at millennial time scales. Amer. Geophys. Union Monograph 112.

Cloos, M. (1982) Flow melanges: numerical modeling and geologic constraints on their origin in the Franciscan subduction complex, California. Geol. Soc. Amer. Bull., v. 93, 330–345.

Cocks, L.R.M. (2001) Ordovician and Silurian global geography. J. Geol. Soc. London, v. 158, 197–210.

Cocks, L.R.M., and Torsvik, T.H. (2002) Earth geography from 500 to 400 million years ago: a faunal and paleomagnetic review. J. Geol. Soc. Lond., v. 159, 631–644.

Coleman, D.S., Barth, A.P., and Wooden, J.L. (2002) Early to Middle Proterozoic reconstruction of the Mojave Province, southwestern United States. Gondwana Res., v. 5, 75–78.

Coleman, R.G., and Irwin, W.P. (1974) Ophiolites and ancient continental margins, In The geology of continental margins (C.A. Burk and C.L. Drake, Eds.). New York, Springer Verlag, 921–931.

Collins, A.S., and Windley, B.F. (2002) The tectonic evolution of central and northern Madagascar and its place in the final assembly of Gondwana. J. Geol., v. 110, 325–329.

Collins, W.J. (1993) Melting of Archaean sialic crust under high alpha$_{H_2O}$ conditions; genesis of 3300 Ma Na-rich granitoids in the Mount Edgar batholith, western Australia. Precamb. Res., v. 60, 151–174.

Collinson, J.W., Isbell, J.L., Elliot, D.H., Miller, M.F., Miller, J.M.G., and Veevers, J.J. (1994) Permian–Triassic Antarctic basins. In Permian–Triassic Pangean basins and foldbelts along the Panthalassan margin of Gondwanaland (J.J. Veevers and C.McA. Powell, Eds.). Geol. Soc. Amer. Memoir 184, 173–222.

Condie, K.C. (1992) Proterozoic terrane and continental accretion in southwestern North America. In Proterozoic continental evolution (K.C. Condie, Ed.). Amsterdam, Elsevier, 447–480.

Condie, K.C. (1999) Mafic crustal xenoliths and the origin of the lower continental crust. In Oceanic plateaus and hotspot islands; identification and role in continental growth (K.C. Condie and D.H. Abbott, Eds.). Lithos, v. 46, 95–101.

Condie, K.C. (2001) Mantle plumes and their record in earth history. Cambridge, Cambridge University Press.

Condie, K.C. (2002) Breakup of a Paleoproterozoic supercontinent. Gondwana Res., v. 5, 41–43.

Condie, K.C., and Rosen, O.M. (1994) Laurentia–Siberia connection revisited. Geology, v. 22, 168–170.

Condie, K.C., Allen, P., and Narayana, B.L. (1982) Geochemistry of the Archean low- to high-grade transition zone, southern India. Contrib. Mineral. Petrol., v. 81, 157–167.

Conway Morris, S. (1992) Burgess Shale-type faunas in the context of the "Cambrian explosion": a review. J. Geol. Soc. London, v. 149, 631–636.

Conway Morris, S., and Peel, J.S. (1990) Articulated halkieriids from the Lower Cambrian of north Greenland. Nature, v. 345, 802–805.

Cook, F.A. (1992) Racklan orogen. Canadian J. Earth Sci., v. 29, 2490–2496.

Corfu, F., Davis, D.W., Stone, D., and Moore, M.L. (1998) Chronostratigraphic constraints on the genesis of Archean greenstone belts, northwestern Superior Province, Ontario, Canada. Precamb. Res., v. 92, 277–295.

Correa-Gomes, L.C., and Oliveira, L.P. (2000) Radiating 1.0 Ga mafic dyke swarms of eastern Brazil and western Africa; evidence of postassembly extension in the Rodinia supercontinent? Gondwana Res., v. 3, 325–332.

Cosca, M.A., Mezger, K., and Essene, E.J. (1998) The Baltica-Laurentia connection; Sveconorwegian (Grenvillian) metamorphism, cooling, and unroofing in the Bamble Sector, Norway. J. Geol., v. 106, 539–552.

Courtillot, V., Jaupart, C., Manighetti, I., Tapponnier, P., and Besse, J. (1999) On causal links between flood basalts and continental breakup. Earth Planet. Sci. Lett., v. 166, 177–195.

Cox, R., Gutmann, E.D., and Hines, P.G. (2002) Diagenetic origin for quartz-pebble conglomerates. Geology, v. 30, 323–326.

Cronin, T.M. (1999) Principles of paleoclimatology. New York, Columbia University Press.

Crowley, T.J. (1994) Pangean climates, In Pangea; paleoclimate, tectonics, and sedimentation during accretion, zenith, and breakup of a supercontinent (G.D. Klein, Ed.). Geol. Soc. Amer. Spec. Paper 288, 25–39.

Cunningham, W.D. (2001) Cenozoic normal faulting and regional doming in the southern Hangay region, central Mongolia: implications for the origin of the Baikal rift province. Tectonophysics, v. 331, 389–411.

D'Lemos, R.S., Strachan, R.A., and Topley, C.G. (Eds.) (1990) The Cadomian orogeny. Geol. Soc. London Spec. Paper 51.

da Silva Filho, A.F., and Lima, E.S. de (Eds.) (1995) Geology of the Borborema Province, northeast Brazil. Spec. Issue of J. South Amer. Earth Sci., v. 8, 233–424.

Dabbagh, M.E., and Rogers, J.J.W. (1983) Depositional environments and tectonic significance of the Wajid Sandstone of southern Saudi Arabia. J. African Earth Sci., v. l, 47–57.

Dallmeyer, R.D. (1990) The West African orogens and Circum-Atlantic correlatives. In Avalonian and Cadomian geology of the North Atlantic (R.A. Strachan and G.K. Taylor, Eds.). Glasgow, Blackie, 134–165.

Dalziel, I.W.D. (1991) Pacific margins of Laurentia and East Antarctica-Australia as a conjugate rift pair: evidence and implications for an Eocambrian supercontinent. Geology, v. 19, 598–601.

Dalziel, I.W.D. (1997) Overview: Neoproterozoic-Palaeozoic geography and tectonics. Review, hypothesis, environmental speculations. Geol. Soc. Amer. Bull., v. 109, 16–42.

Dalziel, I.W.D., Lawver, L.A., and Murphy, J.B. (2000a) Plumes, orogenesis, and supercontinental fragmentation. Earth Planet. Sci. Lett., v. 178, 1–11.

Dalziel, I.W.D., Mosher, S., and Gahagan, L.M. (2000b) Laurentia-Kalahari collision and the assembly of Rodinia. J. Geol., v. 108, 499–513.

Dasgupta, P., Sen, J., Dasgupta, M., Raith, M., Bhui, U.K., and Ehl, J. (1999) Ultra-high temperature metamorphism of metapelitic granulites from Kondapalle, Eastern Ghats belt: implications for the Indo–Antarctic correlation. J. Petrol., v. 40, 1065–1087.

Davidson, A. (1998) An overview of Grenville Province geology, Canadian Shield. In Geology of the Precambrian Superior and Grenville provinces and Precambrian fossils in North America (S.B. Lucas and M.R. St. Onge, Eds.). Geol. Soc. Amer., 207–270.

Davidson, J.P., Harmon, R.S., and Gerhard, W. (1991) The source of central Andean magmas; some considerations. In Andean magmatism and its tectonic setting (R.S. Harmon and C.W. Rapela, Eds.). Geol. Soc. Amer. Spec. Paper 265, 171–334.

Davies, H.L., Sun, S.S., Frey, F.A., Gautier, I., McCulloch, M.T., Price, R.C., Bassias, Y., Klootwijk, C.T., and Leclaire, L. (1989) Basalt basement from the Kerguelen plateau and the trail of a Dupal plume. Contrib. Mineral. Petrol., v. 103, 457–469.

Davis, W.J. (1997) U–Pb zircon and rutile ages from granulite xenoliths in the Slave Province; evidence for mafic magmatism in the lower crust coincident with Proterozoic dike swarms. Geology, v. 25, 343–346.

Davis, W.J., Lacroix, S., Gariepy, C., and Machado, N. (2000) Geochronology and radiogenic isotope geochemistry of plutonic rocks from the central Abitibi subprovince: significance to the internal subdivision and plutono–tectonic evolution of the Abitibi belt. Canadian J. Earth Sci., v. 37, 117–133.

Dawson, G.C., Krapez, B., Fletcher, I.R., McNaughton, N.J., and Rasmussen, B. (2002) Did late Paleoproterozoic assembly of proto-Australia involve collision between the Pilbara, Yilgarn, and Gawler cratons? Geochronological evidence from the Mount Barren Group in the Albany–Fraser orogen of Western Australia. Precamb. Res., v. 118, 195–220.

Dawson, J.B. (1989) Sodium carbonatite extrusions from Oldoinyo Lengai, Tanzania: implications for carbonatite complex genesis. In Carbonatites—genesis and evolution (K. Bell, Ed.). London, Unwin Hyman, 255–277.

De, S.K., Chacko, T., and Creaser, R.A. (2000) Geochemical and Nd-Pb-O isotope systematics of granites from the Taltson magmatic zone, NE Alberta; implications for early Proterozoic tectonics in western Laurentia. Precamb. Res., v. 102, 221–249.

de Wit, M.J., Bowring, S.A., Ashwal, L.D., Randrianasolo, L.G., Morel, V.P.I., and Rambeloson, R.A. (2001) Age and tectonic evolution of Neoproterozoic ductile shear zones in southwestern Madagascar, with implications for Gondwana studies. Tectonics, v. 20, 1–45.

Deblond, A., Punzalan, L.E., Boven, A., and Tack, L. (2001) The Malagarazi Supergroup of southeast Burundi and its correlative Bukoba Supergroup of northwest Tanzania: Neo- and Mesoproterozoic chronostratigraphic constraints from Ar–Ar ages on mafic intrusive rocks. J. African Earth Sci., v. 32, 435–449.

Debon, F., LeFort, P., Sheppard, S.M.F., and Sonet, J. (1986) The four plutonic belts of the Transhimalaya–Himalaya: a chemical, mineralogical, isotopic, and chronological synthesis along a Tibet–Nepal section. J. Petrol., v. 27, 219–250.

Dehler, C.M., Elrick, M., Karlstrom, K.E., Smith, G.A., Cressey, L.J., and Timmons, J.M. (2001) Neoproterozoic Chuar Group (approximately 800–742 Ma), Grand Canyon; a record of cyclic marine deposition during global cooling and supercontinent rifting. Sedimentary Geol., v. 141–142, 465–499.

Delvaux, D., Moeys, R., Stapel, G., Petit, C., Levi, K., Miroschnichenko, A., Ruzhich, V., and San'Kov, V. (1997) Paleostress reconstructions and geodynamics of the Baikal region, central Asia, Part 2: Cenozoic rifting. Tectonophysics, v. 282, 1–38.

Des Marais, D.J. (1997a) Isotopic evolution of biogeochemical carbon cycle during the Proterozoic Eon. Organic Geochem., v. 27, 185–193.

Des Marais, D.J. (1997b) Long-term evolution of the biogeochemical carbon cycle; interactions between microbes and minerals. Geomicrobiology v. 35, 429–448.

Des Marais, D. J. (2001) Isotopic evolution of the biogeochemical carbon cycle during the Precambrian. In Stable isotope geochemistry (J.W. Valley and D.R. Cole, Eds.). Rev. Mineralogy and Geochem., v. 43, Mineralogical Soc. Amer., 555–578.

Dewey, J.F., and Bird, J.M. (1970) Mountain belts and the new global tectonics. J. Geophys. Res., v. 75, 2626–2647.

Dickin, A.P. (1995) Radioactive isotope geology. Cambridge, Cambridge University Press.

Dickin, A.P. (1998) Nd isotope mapping of a cryptic continental suture, Grenville Province of Ontario. Precamb. Res., v. 91, 433–444.

Dickin, A.P. (2000) Crustal formation in the Grenville Province: Nd-isotope evidence. Canadian J. Earth Sci., v. 37, 165–181.

Dietz, R.S. (1961) Continent and ocean basin evolution by spreading of the sea floor. Nature, v. 190, 854–857.

Dinter, D.A. (1998) Late Cenozoic extension of the Alpine collisional orogen, northeastern Greece; origin of the North Aegean Basin. Geol. Soc. Amer. Bull., v. 110, 1208–1226.

Dobmeier, C., and Simmat, R. (2002) Post-Grenvillean transpression in the Chilka Lake area, Eastern Ghats Belt—implications for the geological evolution of peninsular India. Precamb. Res., v. 113, 243–268.

Donaldson, J.A., and de Kemp, E.A. (1998) Archaean quartz arenites in the Canadian shield: examples from the Superior and Churchill Provinces. Sedimentary Geol., v. 120, 153–176.

Donnelly, T.W. (1989) Geologic history of the Caribbean and Central America. In The Geology of North America, v. A (A.W. Bally and A.R. Palmer, Eds.). Geol. Soc. Amer., 299–321.

Donnelly, T.W., and Rogers, J.J.W. (1980) Igneous series in island arcs—the northeastern Caribbean compared with worldwide island-arc assemblages. Bull. Volcanologique, v. 43, 347–382.

Dostal, J., Dupuy, C., and Caby, R. (1994) Geochemistry of the Neoproterozoic Tilemsi belt of Iforas (Mali, Sahara): a crustal accretion of an oceanic island arc. Precamb. Res., v. 65, 55–69.

DuBray, E.A. (1986) Jabal Silsilah tin prospect, Najd region, Kingdom of Saudi Arabia. J. African Earth Sci., v. 4, 237–247.

Dueker, K., Yaun, H., and Zurek, B. (2001) Thick-structured Proterozoic lithosphere of the Rocky Mountain region. GSA Today, v. 11, no. 12, 1–9.

Duerr, S.B., Dingeldey, D.P., and Prave, A.R. (1997) Tale of three cratons; tectonostratigraphic anatomy of the Damara Orogen in northwestern Namibia and the assembly of Gondwana; discussion and reply. Geology, v. 25, 1149–1151.

Ebinger, C.J., and Casey, M. (2001) Continental breakup in magmatic provinces; an Ethiopian example. Geology, v. 29, 527–530.

Ebinger, C.J., and Sleep, N.H. (1998) Cenozoic magmatism throughout East Africa resulting from impact of a single plume. Nature, v. 395, 788–791.

Einsele, G., Ratschbacher, L., and Wetzel, A. (1996) The Himalaya–Bengal fan denudation–accumulation system during the past 20 Ma. J. Geol., v. 104, 163–184.

Emery, K.O., and Uchupi, E. (1984) The geology of the Atlantic Ocean. New York, Springer Verlag.

Emslie, R.F., and Hunt, P.A. (1990) Ages and petrogenetic significance of igneous mangerite–charnockite suites associated with massif anorthosites, Grenville province. J. Geol., v. 98, 213–231.

Encarnacion, J.P., and Grunow, A.M. (1996) Changing magmatic and tectonic cycles along the palaeo-Pacific margin of Gondwana and the

onset of early Paleozoic magmatism in Antarctica. Tectonics, v. 15, 1325–1341.
Engel, M.H., and Macko, S.A. (Eds.) (1993) Organic geochemistry: principles and applications. New York, Plenum.
Ernesto, M., Marques, L.S., Piccirillo, E.M., Molina, E.C., Usami, N., Comin-Chiaramonti, P., and Bellieni, G. (2002) Parana magmatic province–Tristan da Cunha plume system; fixed versus mobile plume, petrogenetic considerations and alternative heat sources. J. Vol. Geotherm. Res., v. 118, 15–36.
Ernst, R.E., and Buchan, K.L. (1997) Giant radiating dyke swarms: their use in identifying pre-Mesozoic large igneous provinces and mantle plumes. In Large igneous provinces—continental, oceanic, and planetary flood volcanism (J.J. Mahoney and M.F. Coffin, Eds.). Amer. Geophys. Union Geophys. Monograph 100, 297–333.
Ernst, W.G. (1970) Tectonic contact between the Franciscan melange and the Great Valley sequence—crustal expression of a Late Mesozoic Benioff zone. J. Geophys. Res., v. 75, 886–902.
Erwin, D.H. (1993) The great Paleozoic crisis: life and death in the Permian. New York, Columbia University Press.
Erwin, D.H., Bowring, S.A., and Yugan, J. (2002) End-Permian mass extinctions: a review. In Catastrophic events and mass extinctions: impacts and beyond (C. Koeberl and K.G. MacLeod, Eds.). Geol. Soc. Amer. Spec. Paper 356, 363–383.
Evans, D.A., Beukes, N.J., and Kirschvink, J.L. (1997) Low latitude glaciation in the Paleoproterozoic era. Nature, v. 386, 262–266.
Evans, D.A.D. (2000) Stratigraphic, geochronological, and paleomagnetic constraints upon the Neoproterozoic climatic paradox. Amer. J. Sci., v. 300, 347–433.
Eyles, N. (1993) Earth's glacial record and its tectonic setting. Earth-Sci. Rev., v. 35, 1–248.
Eyles, N., and Januszczak, N. (2003) Interpreting the Neoproterozoic glacial record; the importance of tectonics. Amer. Geophys. Union Monograph (in press).
Faill, R.T. (1997) A geologic history of the north-central Appalachians. 1. Orogenesis from the Mesoproterozoic through the Taconic orogeny. Amer. J. Sci., v. 297, 551–619.
Faure, G. (1986) Principles of isotope geology (2nd. ed.). New York, John Wiley and Sons.
Fedo, C.M. (2000) Setting and origin for problematic rocks from the > 3.7 Ga Isua greenstone belt, southern West Greenland; Earth's oldest coarse clastic sediments. Precamb. Res., v. 101, 69–78.
Fedo, C.M., Young, G.M., Nesbitt, H.W., and Hanchar, J.M. (1997) Potassic and sodic metasomatism in the Southern Province of the Canadian Shield: evidence from the Palaeproterozoic Serpent Formation, Huronian Supergroup, Canada. Precamb. Res., v. 84, 17–36.
Fedonkin, M.A. (2003) The origin of Metazoa in the light of the Proterozoic fossil record. Paleontol. Res., v. 7, 9–41.
Fernandez-Alonso, M., and Theunissen, K. (1998) Airborne geophysics and geochemistry provide new insights in the intracontinental evolution of the Mesoproterozoic Kibaran belt (Central Africa). Geol. Mag., v. 135, 203–216.
Ferreira, V.P., Sial, A.N., and Jardim de Sa, E.F. (1998) Geochemical and isotopic signatures of Proterozoic granitoids in terranes of the Borborema structural province, northeastern Brazil. J. South Amer. Earth Sci., v. 11, 439–455.
Ferry, J.M. (Ed.) (1986) Characterization of metamorphism through mineral equilibria. Min. Soc. Amer., Rev. Mineralogy, v. 10, 397 pp.
Fershtater, G.B., Montero, P., Borodina, N.S., Pushkarev, E.V., Smirnov, V.N., and Bea, P. (1997) Uralian magmatism: an overview. Tectonophysics, v. 276, 87–102.
Feybesse, J.L., and Milesi, J.P. (1994) The Archean/Proterozoic contact zone in West Africa: a mountain belt of decollement, thrusting and folding on a continental margin related to the 2.1 Ga convergence of Archean cratons? Precamb. Res., v. 69, 199–227.
Feybesse, J.L., Johan, V., Triboulet, C., Guerrot, C., Mayaga-Mikolo, F., Bouchot, V., and Eko N'Dong, J. (1998) The West Central African Belt: a model of 2.5–2.0 Ga accretion and two-phase orogenic evolution. Precamb. Res., v. 87, 161–216.
Fielding, E.J. (1996) Tibet uplift and erosion. Tectonophysics, v. 260, 55–84.
Fitzsimons, I.C.W. (2000) Grenville-age basement provinces in East Antarctica: evidence for three separate collisional orogens. Geology, v. 28, 879–882.
Flagler, P.A., and Spray, J.G. (1991) Generation of plagiogranite by amphibolite anatexis in oceanic shear zones. Geology, v. 19, 70–73.
Foster, D.A., and Gray, D.R. (2000) Evolution and structure of the Lachlan fold belt (orogen) of eastern Australia. Ann. Rev. Earth Planet. Sci., v. 28, 47–80.
Fountain, D.M., Arculus, R., and Kay, R.W. (Eds.) (1992) Continental lower crust. Amsterdam, Elsevier.
Frantz, J.C., and Botelho, N.F. (2000) Neoproterozoic granitic magmatism and evolution of the eastern Dom Feliciano Belt in southernmost Brazil; a tectonic model. Gondwana Res., v. 3, 7–19.
Freeman, K.H. (2001) Isotopic biogeochemistry of marine organic carbon. In Stable isotope geochemistry (J.W. Valley and D.R. Cole, Eds.).

Rev. Mineralogy and Geochem., v. 43, Mineralogical Soc. Amer., 579–605.
Frey, M., Desmons, J., and Neubauer, F. (1999) The new metamorphic map of the Alps. Schweiz. Mineral. Petrog. Mitteilungen, v. 79, 230 pp.
Friberg, M., and Petrov, G.A. (1998) Structure of the Middle Urals, east of the main Uralian fault. Geological J., v. 33, 37–48.
Frimmel, H.E., and Frank, W. (1998) Neoproterozoic tectono-thermal evolution of the Gariep Belt and its basement, Namibia and South Africa. Precamb. Res., v. 90, 1–28.
Frost, C.D., Schellekens, J.H., and Smith, A.L. (1998) Nd, Sr, and Pb isotopic characterization of Cretaceous and Paleogene volcanic and plutonic island arc rocks from Puerto Rico. In Tectonics and geochemistry of the northeastern Caribbean (E.G. Lidiak and D.K. Larue, Eds.). Geol. Soc. Amer. Spec. Paper 322, 123–132.
Fullagar, P.D. (2002) Evidence for early Mesoproterozoic (and earlier?) crust in the southern and central Appalachians of North America. Gondwana Res., v. 5, 197–204.
Gale, A.S., Hardenbol, J., Hathway, B., Kennedy, W.J., Young, J.R., and Phansalkar, V. (2002) Global correlation of Cenomanian (Upper Cretaceous) sequences: evidence for Milankovitch control on sea level. Geology, v. 30, 291–294.
Gamble, J.A., Christie, R.H.K., Wright, I.C., and Wysoczanski, R.J. (1997) Primitive K-rich magmas from Clark volcano, southern Kermadec arc: a paradox in the K–depth relationship. Canadian Mineral., v. 35, 275–290.
Gandhi, S.S., Mortensen, J.K., Prasad, N., and van Breemen, O. (2001) Magmatic evolution of the southern Great Bear continental arc, northwestern Canadian shield: geochronological constraints. Canadian J. Earth Sci., v. 38, 767–785.
Garfunkel, Z. (1998) Constraints on the origin and history of the eastern Mediterranean basin. In Collision-related processes in the Mediterranean region (A.H.F. Robertson and M.C. Comas, Eds.). Tectonophysics Spec. Issue, v. 298, 5–32.
Gaudette, H., Olszewski, W.J., Santos, J., and Orestes, S. (1996) Geochronology of Precambrian rocks from the northern part of the Guiana shield, state of Roraima, Brazil. J. South Amer. Earth Sci., v. 9, 183–195.
Geraldes, M.C., Van Schmus, W.R., Condie, K.C., Bell, S., Texeira, W., and Babinski, M. (2001) Proterozoic geologic evolution of the SW part of the Amazonian craton in Mato Grosso State, Brazil. Precamb. Res., v. 111, 91–128.
Ghosh, S., Chakraborty, S., Paul, D.K., Bhalla, J.K., Bishui, P.K., and Gupta, S.N. (1994) New Rb–Sr isotopic ages and geochemistry of granitoids from Meghalaya and their significance in middle- to late-Proterozoic crustal evolution. Indian Minerals, v. 48, 33–44.
Ghuma, M.A., and Rogers, J.J.W. (1978) Geology, geochemistry, and tectonic setting of the Ben Ghnema batholith, Tibesti massif, southern Libya. Geol. Soc. Amer. Bull., v. 89, 1351–1358.
Gibbs, A.K., and Barron, C.N. (1983) The Guiana shield reviewed. Episodes, v. 1983, no. 2, 7–14.
Gibbs, A.K., and Barron, C.N. (1993) The geology of the Guiana Shield. New York, Oxford University Press.
Glazner, A.F., and Brandon, M.T. (2002) Origin of enriched geochemical signatures in continental arc rocks by antithetic subduction of continental lithosphere (abstract). Eos (Trans. Amer. Geophys. Union), v. 83, no. 47, Abstract V12C–10.
Golynsky, A., and Jacobs, J. (2001) Grenville-age versus Pan-African magnetic anomaly imprints in western Dronning Maud Land, East Antarctica. J. Geol., v. 109, 136–142.
Goodge, J.W., and Fanning, C.M. (1999) 2.5 b.y. of punctuated earth history as recorded in a single rock. Geology, v. 27, 1007–1010.
Goodge, J.W., Fanning, C.M., and Bennett, V.C. (2001) U–Pb evidence of ~1.7-Ga crustal tectonism during the Nimrod Orogeny in the Transantarctic Mountains, Antarctica: implications for Proterozoic plate reconstructions. Precamb. Res., v. 12, 261–288.
Gorbatschev, R., and Bogdanova, S. (1993) Frontiers in the Baltic shield. Precamb. Res., v. 64, 3–21.
Gower, C.F., Ryan, A.B., and Rivers, T. (1990) Mid-Proterozoic Laurentia–Baltica; an overview of its geological evolution and a summary of the contributions made by this volume. In Mid-Proterozoic Laurentia–Baltica (C.F. Gower, T. Rivers, and A.B. Ryan, Eds.). Geol. Assoc. Canada Spec. Paper 38, 1–20.
Grantz, A., Clark, D.L., Phillips, R.L., and Srivastava, S.P. (1998) Phanerozoic stratigraphy of Northwind Ridge, magnetic anomalies in the Canada basin, and the geometry and timing of rifting in the Amerasia basin, Arctic Ocean. Geol. Soc. Amer. Bull., v. 110, 801–820.
Gray, J. (1993) Major Paleozoic land plant evolutionary bio-events. Palaeogeog., Palaeoclim., Palaeontol., v. 104, 153–169.
Greenberg, J.K. (1981) Characteristics and origin of Egyptian Younger Granites. Geol. Soc. Amer. Bull., v. 89, Pt. 1, pp. 224–232; Pt. 2, pp. 749–840.
Gregory, J.W. (1921) The rift valleys and geology of East Africa; an account of the origin and history of the rift valleys of East Africa and their relation to the contemporary earth move-

ments which transformed the geography of the world. London, Seeley, Service and Co.

Grey, K., Walter, M.R., and Calver, C.R. (2003) Neoproterozoic biotic diversification: snowball earth or aftermath of the Acraman impact? Geology, v. 31, 459–462.

Griffin, W.L., O'Reilly, S.Y., and Ryan, C.G. (1999) The composition and origin of subcontinental lithospheric mantle. In Mantle petrology: field observations and high-pressure experimentation: a tribute to Francis R. (Joe) Boyd (Y. Fei, C.M. Bertka, and B.O. Mysen, Eds.). Geochemical Soc. Spec. Pub. 6, 13–43.

Griffin, W.L., O'Reilly, S.Y., Ryan, C.G., Gaul, O., and Ionov, D. (1998) Secular variation in the composition of the subcontinental lithospheric mantle. In Structure and evolution of the Australian continent (J. Braun et al., Eds.). Amer. Geophys. Union Geodynamics Series, v. 26, 1–26.

Groenewald, P.B., Grantham, G.H., and Watkeys, M.K. (1991) Geological evidence for a Proterozoic to Mesozoic link between southeastern Africa and Dronning Maud Land, Antarctica. J. Geol. Soc. London, v. 148, 1115–1123.

Grow, J.A., and Sheridan, R.E. (1988) U.S. Atlantic continental margin; a typical Atlantic-type or passive continental margin. In The Atlantic continental margin (J.A. Grow and R.E. Sheridan, Eds.). Geol. Soc. Amer., Geology of North America v. I-2, 1–7.

Gruau, G., Rosing, M., Bridgwater, D., and Gill, R.C.O. (1996) Resetting of Sm–Nd systematics during metamorphism of >3.7-Ga rocks; implications for isotopic models of early Earth differentiation. Chem. Geol., v. 133, 225–240.

Gueguen, E., Doglioni, C., and Fernandez, M. (1998) On the post-25 Ma evolution of the western Mediterranean. In Collision-related processes in the Mediterranean region (A.H.F. Robertson and M.C. Comas, Eds.). Tectonophysics Spec. Issue, v. 298, 259–270.

Guimares, I. deP., da Silva Filho, A.F., and Adejardo, F. (1998) Nd- and Sr- isotopic and U–Pb geochronologic constraints for evolution of the shoshonitic Brasiliano Bom Jardin and Toritama complexes; evidence for a Transamazonian enriched mantle under Borborema tectonic province, Brazil. Internat. Geol. Rev., v. 40, 500–527.

Guiraud, R., and Bosworth, W. (1999) Phanerozoic geodynamic evolution of northeastern Africa and the northwestern Arabian Platform. Tectonophysics, v. 315, 73–108.

Guo, A., and Dickin, A.P. (1996) The southern limit of Archean crust and significance of rocks with Paleoproterozoic model ages; Nd model age mapping in the Grenville Province of western Quebec. Precamb. Res., v. 77, 231–241.

Gurnis, M. (1988) Large-scale mantle convection and the aggregation and dispersal of supercontinents. Nature, v. 232, 695–699.

Gutenberg, B., and Richter, C.F. (1934, 1935, 1936) On seismic waves. Gerlands Beitrage Geophysik, v. 43, pp. 56–133; v. 45, pp. 280360; v. 47, pp. 73–131.

Haapala, I., and Ramo, O.T. (1990) Petrogenesis of the Proterozoic rapakivi granites of Finland. In Ore-bearing granite systems: petrogenesis and mineralizing processes (H.J. Stein and J. Hannah, Eds.). Geol. Soc. Amer. Spec. Paper 246, 275–286.

Hagadorn, J.W., Pflueger, F., and Bottjer, D. (Eds.) (1999) Unexplored microbial worlds. Palaios, v. 14, 93 pp.

Hall, R., and Spakman, W. (2002) Subducted slabs beneath the eastern Indonesia–Tonga region; insights from tomography. Earth Planet. Sci. Lett., v. 201, 321–336.

Hallam, A., and Wignall, P.B. (1997) Mass extinctions and their aftermath. Oxford, Oxford University Press.

Hamilton, W. (1979) Tectonics of the Indonesian region. U.S. Geol. Surv. Spec. Pub. 1078.

Han, T.-M., and Runnegar, B. (1992) Megascopic eukaryotic algae from the 2.1-billion-year-old Negaunee iron-formation, Michigan. Science, v. 257, 232–235.

Hanmer, S., Corrigan, D., Pehrsson, S., and Nadeau, L. (2000) SW Grenville Province, Canada; the case against post-1.4 Ga accretionary tectonics. Tectonophysics, v. 319, 33–51.

Hansen, E.C., Newton, R.C., Janardhan, A.S., and Lindenberg, S. (1995) Differentiation of late Archean crust in the eastern Dharwar craton, Krishnagiri–Salem area, South India. J. Geol., v. 103, 629–651.

Hansen, V.L. (1992) Backflow and margin-parallel shear within an ancient subduction complex. Geology, v. 20, 71–74.

Hanson, R.E., Wilson, T.J., and Munyanwiya, H. (1994) Geologic evolution of the Neoproterozoic Zambezi Orogenic Belt in Zambia. J. African Earth Sci., v. 18, 135–150.

Haq, B.U., Hardenboll, J., and Vail, P.R. (1987) Chronology of fluctuating sea levels since the Triassic. Science, v. 238, 1237–1242.

Hargraves, R.B. (1976) Precambrian geologic history. Science, v. 193, 363–371.

Harmon, R.S. and Rapela, C.W. (Eds.) (1991) Andean magmatism and its tectonic setting. Geol. Soc. Amer. Spec. Paper 265. 309 pp.

Harris, L.B. (1993) Correlations of tectonothermal events between the Central Indian Tectonic Zone and the Albany Mobile Belt of Western Australia. In Gondwana Eight: assembly, evolution and dispersal (R.H. Findlay, R. Unrug, M.R. Banks, and J.J.

Veevers, Eds.). Rotterdam, A.A. Balkema, 165–180.
Harris, L.B. (1995) Correlation between the Albany, Fraser, and Darling mobile belts of western Australia and Mirnvy to Windmill Islands in the East Antarctic shield: implications for Proterozoic Gondwanaland reconstructions. In India and Antarctica during the Precambrian (M. Yoshida and M. Santosh, Eds.). Geol. Soc. India Mem. 34, 47–71.
Harris, N.B.W., Santosh, M., and Taylor, P.N. (1994) Crustal evolution in South India: constraints from Nd isotopes. J. Geol., v. 102, 139–150.
Harrison, T.M., Copeland, P., Kidd, W.S.F., and An Yin (1992) Raising Tibet. Science, v. 255, 1663–1670.
Hartmann, L.A. (2002) The Mesoproterozoic supercontinent Atlantica in the Brazilian shield—review of geological and U–Pb zircon and Sm–Nd isotopic evidence. Gondwana Res., v. 5, 157–163.
Hartz, E.B., and Torsvik, T.H. (2002) Baltica upside down: a new plate tectonic model for Rodinia and the Iapetus Ocean. Geology, v. 30, 255–258.
Hashimoto, M. (Ed.) (1991) Geology of Japan. Tokyo, Terra Scientific Publishing Co., and Dordrecht, Kluwer Academic Publishers.
Hasselbo, S.P., Robinson, S.A., Surlyk, F., and Piasecki, S. (2002) Terrestrial and marine extinction at the Triassic–Jurassic boundary synchronized with major carbon-cycle perturbation: a link to initiation of massive volcanism? Geology, v. 30, 251–254.
Hatcher, R.D., Jr. (2002). Alleghanian (Appalachian) orogeny, a product of zipper tectonics; rotational transpressive continent–continent collision and closing of ancient oceans along irregular margins. In Variscan–Appalachian dynamics; the building of the late Paleozoic basement (J.R. Martinez Catalan, R.D. Hatcher, Jr., R. Arenas, and F. Diaz Garcia, Eds.). Geol. Soc. Amer. Spec. Paper 364, 1199–1208.
Hauck, M.L., Nelson, K.D., Brown, L.D., Zhao, Wenjin, and Ross, A.R. (1998) Crustal structure of the Himalayan orogen at ~90° east longitude from Project INDEPTH deep reflection profiles. Tectonics, v. 17, 481–500.
Hawkesworth, C.J., Kelley, S., Turner, S., Le Roex, A., and Storey, B. (1999) Mantle processes during Gondwana break-up and dispersal. J. African Earth Sci., v. 28, 239–261.
Hayes, J.M. (2001) Fractionation of carbon and hydrogen isotopes in biosynthetic processes. In Stable isotope geochemistry (J.W. Valley and D.R. Cole, Eds.). Rev. Mineralogy and Geochem., v. 43, Mineralogical Soc. Amer., 225–277.

Hendrix, M.S., and Davis, G.A. (Eds.) (2001) Paleozoic and Mesozoic tectonic evolution of central Asia: from continental assembly to intracontinental deformation. Geol. Soc. Amer. Mem. 194.
Henry, P., Stevenson, R.K., Larbi, Y., and Gariepy, C. (2000) Nd isotopic evidence for Early to Late Archean (3.4–2.7 Ga) crustal growth in the Western Superior Province (Ontario, Canada). Tectonophysics, v. 322, 135–151.
Hess, H.H. (1962) History of ocean basins. In Petrologic studies—a volume in honor of A.F. Buddington (A.E.J. Engel, H.L. James, and B.F. Leonard, Eds.). Geol. Soc. Amer., New York, 599–620.
Heubeck, C. (2001) Assembly of central Asia during the middle and late Paleozoic. In Paleozoic and Mesozoic tectonic evolution of central Asia: from continental assembly to intracontinental deformation (M.S. Hendrix and G.A. Davis, Eds.). Geol. Soc. Amer. Mem. 194, 1–22.
Hill, K.C., and Raza, A. (1999) Arc-continent collision in Papua Guinea; constraints from fission track thermochronology. Tectonics, v. 18, 950–966.
Hodges, K.V. (2000) Tectonics of the Himalaya and southern Tibet from two perspectives. Geol. Soc. Amer. Bull., v. 112, 324–350.
Hoffman, P.F. (1991) Did the breakout of Laurentia turn Gondwanaland inside out? Science, v. 252, 1409–1412.
Hoffman, P.F. (1992) Rodinia, Gondwanaland, Pangea, and Amasia; alternating kinematic scenarios of supercontinental fusion (abstract). Eos (Trans. Amer. Geophys. Union), v. 73, no. 14 supplement, 282.
Hoffman, P.F., and Schrag, D.P. (2002) The snowball Earth hypothesis: testing the limits of global change. Terra Nova, v. 14, 129–155.
Hoffman, P.F., Kaufman, A.J., Halverson, G.P., and Schrag, D.P. (1998) The Neoproterozoic snowball earth. Science, v. 281, 1342–1346.
Holmes, A. (1928) Radioactivity and continental drift. Geol. Mag., v. 65, 236–238.
Holser, W.T. (1997) Geochemical events documented in inorganic carbon isotopes. Palaeogeog., Palaeoclim., Palaeontol., v. 132, 173–182.
Holtta, P., Huhma, H., Manttari, I., Peltonen, P., and Juhanoja, J. (2000) Petrology and geochemistry of mafic granulite xenoliths from the Lahtojoki kimberlite pipe, eastern Finland. Lithos, v. 51, 109–133.
Holzer, L., Barton, J.M., Paya, B.K., and Kramers, J.D. (1999) Tectonothermal history of the western part of the Limpopo Belt; tectonic models and new perspectives. J. African Earth Sci., v. 28, 383–402.
Homewood, P., Allen, P.A., and Williams, G.D. (1986) Dynamics of the molasse basin of wes-

tern Switzerland. *In* Foreland basins (P.A. Allen and P. Homewood, Eds.). Internat. Assoc. Sedimentol. Spec. Pub. 8, 199–217.

Hooper, P.R. (1997) The Columbia River flood basalt province: current status. *In* Large igneous provinces: continental, oceanic, and planetary flood volcanism (J.J. Mahoney and M.F. Coffin, Eds.). Geophysical Monograph 100, Amer. Geophys. Union, 1–27.

Hooper, P.R., and Hawkesworth, C.J. (1993) Isotopic and geochemical constraints on the origin and evolution of the Columbia River Basalt. J. Petrol., v. 34, 1203–1246.

House, M.R. (2002) Strength, timing, setting and cause of mid-Paleozoic extinctions. Palaeogeog., Palaeoclim., Palaeontol., v. 181, 5–25.

Hurich, C.A. (1996) Kinematic evolution of the lower plate during intracontinental subduction: an example from the Scandinavian Caledonides. Tectonics, v. 15, 1248–1263.

Ichikawa, K. (1990) Pre-Cretaceous terranes of Japan. *In* Pre-Jurassic evolution of eastern Asia (K. Ichikawa, S. Mizutani, I. Hara, S. Hada, and A. Yao, Eds.). Osaka, IGCP–IUGS Project 224 Publication, 1–12.

Ionov, D. (2002) Mantle structure and rifting processes in the Baikal–Mongolia region: geophysical data and evidence from xenoliths in volcanic rocks. Tectonophysics, v. 351, 41–60.

Isozaki, Y. (1997) Permo-Triassic boundary superanoxia and stratified superocean: records from lost deep sea. Science, v. 276, 235–238.

Jackson, S.L., Fyon, J.A., and Corfu, F. (1994) Review of Archean supracrustal assemblages of the southern Abitibi greenstone belts in Ontario, Canada; products of microplate interaction within a large-scale plate-tectonic setting. Precamb. Res., v. 65, 183–205.

Jacobs, J. (1999) Neoproterozoic/lower Paleozoic events in central Dronning Maud Land (East Antarctica). Gondwana Res., v. 2, 473–480.

Jacobs, J., Fanning, C.M., Henjes-Kunst, F., Olesch, M., and Paech, H.J. (1998) Continuation of the Mozambique belt into East Antarctica: Grenville-age metamorphism and polyphase Pan-African high-grade events in central Dronning Maud Land. J. Geol., v. 106, 385–406.

Jacobshagen, V., Muller, J., Wemmer, K., Ahrendt, H., and Manutsoglu, E. (2002) Hercynian deformation and metamorphism in the Cordillera Oriental of southern Bolivia, central Andes. Tectonophysics, v. 345, 119–130.

Jaupart, C., and Mareschal, J.C. (1999) The thermal structure and thickness of continental roots. Lithos, v. 48, 93–114.

Javaux, E.J., Knoll, A.H., and Walter, M.R. (2001) Morphological and ecological complexity in early eukaryotic ecosystems. Nature, v. 412, 66–69.

Jayananda, M., Moyen, J.F., Martin, H., Peucat, J.J., Auvray. B., and Mahabaleswar, B. (2000) Late Archaean (2550–2520 Ma) juvenile magmatism in the eastern Dharwar craton, southern India; constraints from geochronology, Nd–Sr isotopes and whole rock geochemistry. Precamb. Res., v. 99, 225–254.

Joachimski, M.M., Pancost, R.D., Freeman, K.H., Ostertag-Henning, C., and Buggisch, W. (2002) Carbon isotope geochemistry of the Frasnian–Fammenian transition. Palaeogeog., Palaeoclim., Palaeontol., v. 181, 91–109.

Johansson, L., Moller, C., and Soderlund, U. (2001) Geochronology of eclogite facies metamorphism in the Sveconorwegian Province of SW Sweden. Precamb. Res., v. 106, 261–275.

Johnson, M.R.W. (1994) Volume balance of erosional loss and sediment deposition related to Himalayan uplifts. J. Geol. Soc. London, v. 151, 217–220.

Johnson, P.R., and Stewart, I.C.F. (1995) Magnetically inferred basement structure in central Saudi Arabia. Tectonophysics, v. 245, 37–52.

Johnson, S.P., and Oliver, G.J.H. (2000) Mesoproterozoic oceanic subduction, island-arc formation and the initiation of back-arc spreading in the Kibaran belt of central, southern Africa: evidence from the ophiolite terrane, Chewore inliers, northern Zimbabwe. Precamb. Res., v. 103, 125–146.

Johnston, S.T., and Thorkelson, D.J. (2000) Continental flood basalts; episodic magmatism above long-lived hotspots. Earth Planet. Sci. Lett., v. 175, 247–256.

Jolivet, M., Brunel, M., Seward, D., Xu, A., Yang, J., Roger, F., Tapponier, P., Malavielle, J., and Arnaud, N. (2001) Mesozoic and Cenozoic tectonics of the northern edge of the Tibetan Plateau; fission-track constraints. Tectonophysics, v. 343, 111–134.

Joly, J. (1925) The surface history of the earth. Oxford, Clarendon Press.

Jordan, T.H. (1975) The continental Tectosphere. Nature, v. 257, 745–750.

Juhlin, C., Wahlgren, C.-H., and Stephens, M.B. (2000) Seismic imaging in the frontal part of the Sveconorwegian Orogen, south-western Sweden. Precamb. Res., v. 102, 135–154.

Jung, S., Hoernes, S., and Mezger, K. (2000) Geochronology and petrogenesis of Pan-African syn-tectonic, S-type and post-tectonic A-type granite (Namibia); products of melting of crustal sources, fractional crystallization and wall rock entrainment. Lithos, v. 50, 259–287.

Jung, S., Mezger, K., and Hoernes, S. (1997) Petrology and geochemistry of syn- to post-collisional metaluminous A-type granites; a major and trace element and Nd–Sr–Pb–O isotope study from the Proterozoic Damara Belt, Namibia. Lithos, v. 45, 147–175.

Jung, S., Mezger, K., and Hoernes, S. (2001) Trace element and isotopic (Sr, Nd, Pb, O) arguments for a mid-crustal origin of Pan-African garnet-bearing S-type granites from the Damara orogen (Namibia). Precamb. Res., v. 110, 325–355.

Kalsbeek, F., and Nutman, A.P. (1996) Anatomy of the early Proterozoic Nagssugtoqidian Orogen, West Greenland, explored by reconnaissance SHRIMP U–Pb zircon dating. Geology, v. 24, 515–518.

Kalsbeek, F., Pulvertaft, T.C.R., and Nutman, A.P. (1998) Geochemistry, age, and origin of metagraywackes from the Palaeoproterozoic Karrat Group, Rinkian belt, West Greenland. Precamb. Res., v. 91, 383–399.

Kalsbeek, F., Thrane, K., Nutman, A.P., and Jepsen, H.F. (2000) Late Mesoproterozoic to early Neoproterozoic history of the East Greenland Caledonides: evidence for Grenvillian orogenesis? J. Geol. Soc. London, v. 157, 1215–1225.

Kampunzu, A.B., and Cailteux, J. (1999) Tectonic evolution of the Lufilian Arc (Central Africa copper belt) during Neoproterozoic Pan-African orogenesis. Gondwana Res., v. 2, 401–421.

Kampunzu, A.B., Tembo, F.M., Kapenda, D., and Hunstman-Mapila, P. (2000) Geochemistry and tectonic setting of mafic igneous units in the Neoproterozoic Katangan Basin, central Africa; implications for Rodinia breakup. Gondwana Res., v. 3, 125–153.

Kano, T. (2001) Major geologic units of south-west Japan. ISRGA Field Guide Book. Gondwana Res. Group Misc. Publ. No. 11, Gondwana Institute of Geology and Environment (Japan).

Karlstrom, K.E., and Humphreys, E.D. (1998) Persistent influence of Proterozoic accretionary boundaries in the tectonic evolution of southwestern North America: interaction of cratonic grain and mantle modification events. Rocky Mountain Geol., v. 33, 161–179.

Karlstrom, K.E., Harlan, S.S., Williams, M.L., McLelland, J., Geissman, J.W., and Ahall, A.-I. (1999) Refining Rodinia: geologic evidence for the Australia–western U.S. connection. GSA Today, v. 9, no. 10, 1–7.

Karson, J.A. (1999) Geological investigation of a lineated massif at the Kane transformation fault; implications for oceanic core complexes. In Response of the earth's lithosphere to extension (R.S. White, R.F.P. Hardman, A.B. Watts, and R.B. Whitmarsh, Eds.). Phil. Trans. Roy. Soc. Mathematical, Physical and Engineering Sciences, v. 357, 713–740.

Katili, J.A., and Reinemund, J.A. (1982) Southeast Asia: tectonic framework, earth resources and regional geological programs. Internat. Union Geol. Sci. Pub. 13.

Kay, S.M., Mpodozis, C., Ramos, V.A., and Munizaga, F. (1991) Magma source variations for mid-late Tertiary magmatic rocks associated with a shallowing subduction zone and a thickening crust in the central Andes (28° to 33°S). In Andean magmatism and its tectonic setting (R.S. Harmon and C.W. Rapela, Eds.). Geol. Soc. Amer. Spec. Paper 265, 113–137.

Kearey, P. (1996) Global tectonics (2nd ed.). Oxford, Blackwell Science.

Kelly, N.M., Clarke, G.L., and Fanning, C.M. (2002) A two-stage evolution of the Neoproterozoic Rayner structural episode: new U–Pb sensitive high resolution ion microprobe constraints from the Oygarden Group, Kemp Land, East Antarctica. Precamb. Res., v. 116, 307–330.

Kempton, P.D., Downes, H., and Embey-Isztin, A. (1997) Mafic granulite xenoliths in Neogene alkali basalts from the western Pannonian Basin; insights into the lower crust of a collapsed orogen. J. Petrol., v. 38, 941–970.

Kempton, P.D., Fitton, J.G., Saunders, A.D., Nowell, G.M., Taylor, R.N., Hardarson, B.S., and Pearson, G. (2000) The iceland plume in space and time: a Sr–Nd–Pb–Hf study of the North Atlantic rifted margin. Earth Planet. Sci. Lett., v. 177, 255–271.

Kennedy, W.Q. (1964) The structural deformation of Africa in the Pan-African (±500 m.y.) tectonic episode. Leeds Univ. Res. Inst. African Geol., Annual Rept., v. 8, 48–49.

Kent, R.W., and Fitton, J.G. (2000) Mantle sources and melting dynamics in the British Palaeogene igneous province. J. Petrology, v. 41, 1023–1040.

Keppie, J.D., and Ortega-Gutierrez, F. (1999) Middle American Precambrian basement: a missing piece of the reconstructed 1-Ga orogen. Geol. Soc. Amer. Spec. Paper 336, 199–210.

Keppie, J.D., Dostal, J., Ortega-Gutierrez, F., and Lopez, R. (2001) A Grenvillian arc on the margin of Amazonia; evidence from the southern Oaxacan Complex, southern Mexico. Precamb. Res., v. 112, 165–181.

Kerr, A., Ryan, B., Gower, C.F., and Wardle, R.J. (1996) The Makkovik province: extension of the Ketilidian mobile belt in mainland North America. In Precambrian crustal evolution in the North Atlantic region (T.S. Brewer, Ed.). Geol. Soc. London Spec. Publ. 112, 155–177.

Kerrich, R., and Ludden, J.N. (2000) The role of fluids during formation and evolution of the southern Superior Province lithosphere; an overview. Canadian J. Earth Sci., v. 37, 135–164.

Ketchum, J.W.F., and Davidson, A. (2000) Crustal architecture and tectonic assembly of the Central Gneiss Belt, southwestern Grenville Province, Canada: a new interpretation. Canadian J. Earth Sci., v. 37, 217–234.

Kirschvink, J.L. (1992) Late Proterozoic low-latitude global glaciation: the snowball Earth. In The Proterozoic biosphere: a multidisciplinary study (J.W. Schopf and C. Klein, Eds.). Cambridge, Cambridge Univ. Press, 51–52.

Kirstein, L.A., Kelley, S., Hawkesworth, C., Turner, S., Mantovani, M., and Wijbrans, J. (2001) Protracted felsic magmatic activity associated with the opening of the South Atlantic. J. Geol. Soc. London, v. 158, 583–592.

Kloppenberg, A., White, S.H., and Zegers, T.E. (2001) Structural evolution of the Warrawoona Greenstone Belt and adjoining granitoid complexes, Pilbara craton, Australia: implications for Archean tectonic processes. Precamb. Res., v. 112, 107–147.

Knauth, L.K. (1998) Salinity history of the earth's early ocean. Nature, v. 395, 554–555.

Knauth, P.L., and Lowe, D.R. (2003) High Archean climatic temperature inferred from oxygen isotope geochemistry of cherts in the 3.5 Ga Swaziland Supergroup, South Africa. Geol. Soc. Amer. Bull, v. 115, 566–580.

Knoll, A.H., Bambach, R.K., Canfield, D.E., and Grotzinger, J.P. (1996) Comparative earth history and Late Permian mass extinction. Science, v. 273, 452–457.

Koeberl, C., and MacLeod, K.G. (Eds.) (2002) Catastrophic events and mass extinctions: impacts and beyond. Geol. Soc. Amer. Spec. Paper 356.

Kogbe, C.A., and Burollet, P.F. (1990). A review of continental sediments in Africa. In Major African continental Phanerozoic complexes and dynamics of sedimentation (C.A. Kogbe and J. Lang, Eds.). J. African Earth Sci., v. 10, 1–25.

Kopf, C.F. (2002) Archean and early Proterozoic events along the Snowbird tectonic zone in northern Saskatchewan, Canada. Gondwana Res., v. 5, 79–83.

Kriegsman, L.M. (1995) The Pan-African event in East Antarctica; a view from Sri Lanka and the Mozambique belt. Precamb. Res., v. 75, 263–277.

Kristofferson, Y. (1990) Eurasia basin. In The Arctic Ocean region (A. Grantz, L. Johnson, and J.F. Sweeney, Eds.). The geology of North America, v. L, Geol. Soc. Amer., 365–378.

Krogh, T.H., and Moser, D.E. (1994) U-Pb zircon and monazite ages from the Kapuskasing uplift; age constraints on deformation within the Ivanhoe Lake fault zone. Canadian J. Earth Sci., v. 7, 1096–1103.

Kroner, A., and Tegtmeyer, A. (1994) Gneiss-greenstone relationships in the Ancient Gneiss complex of southwestern Swaziland, southern Africa, and implications for early crustal evolution. Precamb. Res., v. 67, 109–139.

Kroner, A., Jaeckel, P., and Williams, I.S. (1994) Pb-loss patterns in zircons from a high-grade metamorphic terrain as revealed by different dating methods: U–Pb and Pb–Pb ages for igneous and metamorphic zircons from northern Sri Lanka. Precamb. Res., v. 66, 151–181.

Kroner, A., Willner, A.P., Hegner, E., Jaeckel, P., and Nemchin, A. (2001) Single zircon ages, P-T evolution and Nd isotopic systematics of high-grade gneisses in southern Malawi and their bearing on the evolution of the Mozambique belt in southeastern Africa. Precamb. Res., v, 109, 257–291.

Kuester, D., and Liegeois, J.-P. (2001) Sr, Nd isotopes and geochemistry of the Bayuda Desert high-grade metamorphic basement (Sudan); an early Pan-African oceanic convergent margin, not the edge of the East Saharan ghost craton. Precamb. Res., v. 109, 1–23.

Kuhn, W.R., Walker, J.C.G., and Marshall, H.G. (1989) The effect on earth's surface temperature from variations in rotation rate, continent formation, solar luminosity, and carbon dioxide. J. Geophys. Res., v. 94, 11,129–11,136.

Kukla, P.A., and Stanistreet, I.G. (1991) Record of the Damaran Khomas Hochland accretionary prism in central Namibia; refutation of an "ensialic" origin of a late Proterozoic orogenic belt. Geology, v. 19, 473–476.

Laird, M.G. (1991) The late Proterozoic–middle Palaeozoic rocks of Antarctica. In The geology of Antarctica (R.J. Tingey, Ed.). Oxford Monographs on Geology and Geophysics 17, 74–119.

Lamb, S., Hoke, L., Lorcan, K., and Dewey, J. (1997) Cenozoic evolution of the central Andes in Bolivia and northern Chile. In Orogeny through time (J.-P. Burg and M. Ford, Eds.). Geol. Soc. London Spec. Publ. 121, 237–264.

Lambiase, J.J. (1989) The framework of African rifting during the Phanerozoic. J. African Earth Sci., v. 8, 183–190.

Lan, C.Y., Chung, S.L., Lo, C.-H., Lee, Tung-Yi, Wang, P.L., Li, H., and Dinh, V.T. (2001) First evidence for Archean continental crust in northern Vietnam and its implications for crustal and tectonic evolution in Southeast Asia. Geology, v. 29, 219–222.

Landenberger, B., and Collins, W.J. (1996) Derivation of A-type granites from a dehydrated charnockitic lower crust; evidence from the Chaelundi Complex, eastern Australia. J. Petrol., v. 37, 145–170.

Landing, E., Bowring, S.A., Davidek, K.L., Westrop, S.R., Geyer, G., and Heldmaier, W. (1998) Duration of the Early Cambrian: U–Pb ages of volcanic ashes from Avalon and Gondwana. Canadian J. Earth Sci., v. 35, 329–338.

Landing, E., Myrow, P., Benus, A., and Narbonne, G.M. (1989) The Placentian Series: appearance of the oldest skeletized fossils in southeastern Newfoundland. J. Paleontol., v. 63, 739–769.

Landoll, J.D., Foland, K.A., and Henderson, C.M.B. (1994) Nd isotopes demonstrate the role of contamination in the formation of coexisting quartz and nepheline syenites at the Abu Khrug complex, Egypt. Contrib. Mineral. Petrol., v. 117, 305–329.

Larson, R.L. (1991) Latest pulse of the earth: evidence for a mid-Cretaceous superplume. Geology, v. 19, 547–550.

Laubscher, H. (2001) Plate interactions at the southern end of the Rhine Graben. Tectonophysics, v. 343, 1–19.

Lawlor, P.J., Ortega-Gutierrez, F., Cameron, K.L., Ochoa-Camarillo, H., Lopez, R., and Sampson, D.E. (1999) U–Pb geochronology, geochemistry, and provenance of the Grenvillian Huiznopala Gneiss of eastern Mexico. Precamb. Res., v. 94, 73–99.

LeCheminant, A.N., and Heaman, L.M. (1989) Mackenzie igneous events, Canada: Middle Proterozoic hotspot magmatism associated with ocean opening. Earth Planet. Sci. Lett., v. 96, 38–48.

Ledru, P., Johan, V., Millesi, J.P., and Tegyey, M. (1994) Markers of the last stages of the Palaeoproterozoic collision: evidence for a 2 Ga continent involving circum-Atlantic provinces. Precamb. Res., v. 69, 169–191.

Lee, C.-T., Yin, Q., Rudnick, R.L., and Jacobsen, S.B. (2001) Preservation of ancient and fertile lithospheric mantle beneath the southwestern United States. Nature, v. 411, 69–73.

LeGrand, H.E. (1988) Drifting continents and shifting theories. Cambridge, Cambridge University Press.

Leitch, E.C., Fergusson, C.L., Henderson, R.A., and Morand, V.J. (1994) Late Palaeozoic arc flank and fore-arc basin sequence of the New England fold belt in the Stanage Bay region, central Queensland. Australian J. Earth Sci., v. 41, 301–310.

Leite, J.A.D., Hartman, L.A., McNaughton, N.J., and Chemale, F., Jr. (1998) SHRIMP U/Pb zircon geochronology of Neoproterozoic juvenile and crustal-reworked terranes in southernmost Brazil. Internat. Geol. Rev., v. 40, 688–705.

Lenat, J.-F., Gibert-Malengreau, B., and Galdeano, A. (2001) A new model for the evolution of the volcanic island of Reunion (Indian Ocean). J. Geophys. Res., v. 106, 8645–8663.

Lewry, J.F., and 8 others (1994) Structure of a Paleoproterozoic continent–continent collision zone: a LITHOPROBE seismic reflection profile across the Trans-Hudson orogen, Canada. Tectonophysics, v. 232, 143–160.

Li, Z.-X. (1998) Tectonic history of the major East Asian lithospheric blocks since the mid-Proterozoic; a synthesis. In Mantle dynamics and plate interactions in East Asia (F.J. Martin, S.-L. Chung, C.-H. Lo, and T.-Y. Lee, Eds.). Amer. Geophys. Union Geodynamics Series, v. 27, 221–243.

Li, Z.-X., Zhang, L., and Powell, C.McA. (1995) South China in Rodinia: part of the missing link between Australia–East Antarctica and Laurentia? Geology, v. 23, 407–410.

Li, Z.-X., Zhang, L., and Powell, C.McA. (1996) Positions of the East Asian cratons in the Neoproterozoic supercontinent Rodinia. Australian J. Earth Sci., v. 43, 593–604.

Lidiak, E.G., and Larue, D.K. (Eds.) (1998) Tectonics and geochemistry of the northeastern Caribbean. Geol. Soc. Amer. Spec. Paper 322.

Liegeois, J.P., and Black, R. (1993) Cratons, mobile belts, alkaline rocks and continental lithospheric mantle; the Pan-African testimony. J. Geol. Soc. London, v. 150, 89–98.

Lindh, A., and Persson, P.-O. (1990) Proterozoic granitoid rocks of the Baltic shield—trends of development. In Mid-Proterozoic Laurentia–Baltica (C.F. Gower, T. Rivers, and B. Ryan, Eds.). Geol. Assoc. Canada Spec. Paper 38, 23–40.

Lindsay, J.F., and Brasier, M.D. (2000) A carbon isotope reference curve for ca. 1700–1575 Ma, McArthur and Mount Isa basins, Northern Australia. Precamb. Res., v. 99, 271–308.

Lindsay, J.F., and Brasier, M.D. (2002) Did global tectonics drive early biosphere evolution? Carbon isotope record from 2.6 to 1.9 Ga carbonates of Western Australia basins. Precamb. Res., v. 114, 1–34.

Link, P.K., and 12 others (1993) Middle and Later Proterozoic stratified rocks of the western U.S. Cordillera, Colorado Plateau, and Basin and Range province. In Precambrian: conterminous U.S. (J.C. Reed, Jr., M.E. Bickford, R.S. Houston, P.K. Link, D.W. Rankin, P.K. Sims, and W.R. Van Schmus, Eds.). The Geology of North America, v. C-2. Geol. Soc. Amer., 463–595.

Litherland, M., and 10 others (1989) The Proterozoic of eastern Bolivia and its relationship to the Andean mobile belt. Precamb. Res., v. 43, 157–174.

Liu, J., Han, J., and Fyfe, W.S. (2001) Cenozoic episodic volcanism and continental rifting in Northeast China and possible link to Japan Sea development as revealed from K–Ar geochronology. Tectonophysics, v. 339, 385–401.

Lopez, R., Cameron, K.L., and Jones, N.W. (2001) Evidence for Paleoproterozoic, Grenvillian, and Pan-African age Gondwanan crust beneath northeastern Mexico. Precamb. Res., v. 107, 195–214.

Louden, K.E., and Fan, J. (1998) Crustal structures of Grenville, Makkovik, and southern Nain provinces along the Lithoprobe ECSOOT Transect; regional seismic refraction and gravity models and their tectonic implications. Canadian J. Earth Sci., v. 35, 583–601.

Lowe, D.R. (1999) Geologic evolution of the Barberton Greenstone Belt and vicinity. In Geologic evolution of the Barberton Greenstone Belt, South Africa (D.R. Lowe and B.R. Byerly, Eds.). Geol. Soc. Amer. Spec. Paper 329, 287–312.

Lowrie, W. (1993) Fundamentals of geophysics. Cambridge, Cambridge Univ. Press.

Lu, S., Yang, C., Li, H., and Li, H. (2002) A group of rifting events in the North China craton. Gondwana. Res., v. 5, 123–131.

Lucas, S.B., Green, A., Hajnal, Z., White, D., Lewry, J., Ashton, K., Weber, W., and Clowes, R. (1993) Deep seismic profile across a Proterozoic collision zone: surprises at depth. Nature, v. 363, 339–342.

Ludden, J.N., and Hynes, A. (2000) The Lithoprobe Abitibi-Grenville transect; two billion years of crust formation and recycling in the Precambrian Shield of Canada. Canadian J. Earth Sci., v. 37, 459–476.

Luepke, J.J., and Lyons, T.W. (2001) Pre-Rodinian (Mesoproterozoic) supercontinental rifting along the western margin of Laurentia: geochemical evidence from the Belt–Purcell supergroup. Precamb. Res., v. 111, 79–90.

Macdonald, R., and Upton, B.G.C. (1993) The Proterozoic Gardar rift zone, South Greenland: comparisons with the East African rift system. In Magmatic processes and plate tectonics (H.M. Prichard, T. Alabaster, N.B.W. Harris, and C.R. Neary, Eds.). Geol. Soc. London Spec. Publ. 76, 427–442.

Macdonald, R., Rogers, N.W., Fitton, J.G., Black, S., and Smith, M. (2001) Plume-lithosphere interactions in the generation of the basalts of the Kenya Rift, East Africa. J. Petrol., v. 42, 877–900.

Mallard, L.D., and Rogers, J.J.W. (1997) Relationship of Avalonian and Cadomian terranes to Grenville and Pan-African events. In Assembly and dispersal of supercontinents (N. Rast and J.J.W. Rogers, Eds.). J. Geodynamics, v. 23, 197–221.

Manhica, A.D.S.T., Grantham, G.H., Armstrong, R.A., Guise, P.G., and Kruger, F.J. (2001) Polyphase deformation and metamorphism at the Kalahari craton–Mozambique Belt boundary. In Continental reactivation and reworking (J.A. Miller, J.A. Holdsworth, and I.S. Buick, Eds.). Geol. Soc. London Spec. Publ. 184, 303–322.

Markwick, A.J.W., and Downes, H. (2000) Lower crustal granulite xenoliths from the Archangelsk kimberlite pipes; petrological, geochemical and geophysical results. Lithos, v. 51, 135–151.

Marsh, J.S., Ewart, A., Milner, S.C., Duncan, A.R., and Miller, R. McG. (2001) The Entendeka igneous province; magma types and their stratigraphic distribution with implications for the evolution of the Parana–Entendeka flood basalt province. Bull. Volcanologique, v. 62, 464–486.

Martignole, J., and Friedman, R. (1998) Geochronological constraints on the last stages of terrane assembly in the central part of the Grenville Province. Precamb. Res., v. 92, 145–164.

Martin, D.M. (1999) Depositional setting and implications of Paleoproterozoic glaciomarine sedimentation in the Hamersley Province, Western Australia. Geol. Soc. Amer. Bull., v. 111, 189–203.

Martin, H. (1994) Archean grey gneisses and the genesis of the continental crust. In Archean crustal evolution (K.C. Condie, Ed.). Amsterdam, Elsevier, 205–260.

Marvin, U.B. (1973). Continental drift: the evolution of a concept. Washington, Smithsonian Inst. Press.

Master, S. (1990) "Archaean," "Proterozoic," and the Archaean–Proterozoic boundary; semantic minefield in a Precambrian no-man's land. South African J. Geol., v. 93, 417–419.

Mazumder, R., Bosse, P.K., and Sarkar, S.A. (2000) A commentary on the tectono-sedimentary record of pre-2.0 Ga continental growth of India vis-a-vis a possible pre-Gondwana Afro-Indian supercontinent. J. African Earth Sci., v. 30, 201–217.

McDermott, F., Harris, N.B.W., and Hawkesworth, C.J. (1996) Geochemical constraints on crustal anatexis; a case study from the Pan-African Damara granitoids of Namibia. Contrib. Mineral. Petrol., v. 123, 406–423.

McElhinny, M.W., and McFadden, P.L. (1999) Paleomagnetism—continents and oceans. International Geophysics Series, v. 75, San Diego, Academic Press.

McGhee, G.R., Jr. (2001) The "multiple impacts hypothesis" for mass extinction: a comparison of the Late Devonian and late Eocene. Palaeogeog., Palaeoclim., Palaeontol., v. 176, 47–58.

McGuire, A.V., and Stern, R.J. (1993) Granulite xenoliths from western Saudi Arabia; the lower crust of the Precambrian Arabian–Nubian shield. Contrib. Mineral. Petrol., v. 114, 395–408.

McMenamin, M.A.S., and McMenamin, D.L.S. (1990) The emergence of animals: the Cambrian breakthrough. New York, Columbia University Press.

McNamara, A.K., Mac Niocaill, C., van der Pluijm, B.A., and Van der Voo, R. (2001) West African proximity of the Avalon terrane in the latest Precambrian. Geol. Soc. Amer. Bull., v. 113, 1161–1170.

Mechie, J., Egorkin, A.V., Solidilov, L., Fuchs, K., Lorenz, F., and Wenzel, F. (1997) Major features of the mantle velocity structure beneath northern Eurasia from long-range seismic recordings of peaceful nuclear explosions. In Upper mantle heterogeneities from active and passive seismology (K. Fuchs, Ed.). Dordrecht, Kluwer Academic Publ., 33–50.

Meen, J.K., Rogers, J.J.W., and Fullagar, P.D. (1992) Pb isotopic compositions of the Western Dharwar Craton, southern India: evidence for distinct middle Archean terranes in a late Archean craton. Geochim. Cosmochim. Acta, v. 56, 2455–2470.

Meert, J.G. (2001) Growing Gondwana and refining Rodinia: a paleomagnetic perspective. Gondwana Res., v. 4, 279–288.

Meert, J.G. (2002) Paleomagnetic evidence for a Paleo–Mesoproterozoic supercontinent Columbia. Gondwana Res., v. 5, 207–215.

Meert, J.G. (2003) A synopsis of events related to the assembly of eastern Gondwana. Tectonophysics, v. 362, 1–40.

Meert, J.G., and van der Voo, R. (1994) The Neoproterozoic (1000–540 Ma) glacial intervals: no more snowball earth? Earth Planet. Sci. Lett., v. 123, 1–13.

Meert, J.G., Hargraves, R.B., Van der Voo, R., Hall, C.M., and Halliday, A.N. (1994) Paleomagnetic and $^{40}Ar/^{39}Ar$ studies of Late Kibaran intrusives in Burundi, East Africa: implications for late Proterozoic supercontinents. J. Geol., v. 102, 621–637.

Meisel, T., Walker, R.J., Irving, A.J., and Lorand, J.-P. (2001) Osmium isotope compositions of mantle xenoliths: a global perspective. Geochim. Cosmochim. Acta, v. 65, 1311–1323.

Meng, Q.-R., and Zhang, G.-W. (2000) Geologic framework and tectonic evolution of the Qinling orogen, central China. Tectonophysics, v. 323, 183–196.

Menon, R.D., and Santosh, M. (1995) The Pan-African gemstone province of East Gondwana. Geol. Soc. India Memoir 34, 357–371.

Menzies, M., Baker, J., and Chazot, G. (2001) Cenozoic plume evolution and flood basalts in Yemen; a key to understanding older examples. In Mantle plumes; their identification through time (R.E. Ernst and K.L. Buchan, Eds.). Geol. Soc. Amer. Spec. Paper 352, 23–36.

Meredith, D.J., and Egan, S.S. (2002) The geological and geodynamic evolution of the eastern Black Sea basin: insights from 2-D and 3-D tectonic modelling. Tectonophysics, v. 350, 157–179.

Mereu, R.F. (2000) The complexity of the crust and Moho under the southeastern Superior and Grenville provinces of the Canadian Shield from seismic refraction-wide-angle reflection data. Canadian J. Earth Sci., v. 37, 439–458.

Metcalfe, I. (1996) Gondwanaland dispersion, Asian accretion and evolution of eastern Tethys. In Breakup of Rodinia and Gondwanaland and assembly of Asia (Z.X. Li, I. Metcalfe, and C.McA. Powell, Eds.). Australian J. Earth Sci., v. 43, 605–623.

Meyer, M.T., Bickford, M.E., and Lewry, J.F. (1992) The Wathaman batholith: an Early Proterozoic continental arc in the Trans-Hudson orogenic belt, Canada. Geol. Soc. Amer. Bull., v. 104, 1073–1085.

Meyerhoff, A.A., and Meyerhoff, H.A. (1974) Tests of plate tectonics. In Plate tectonics: assessments and reassessments (C.F. Kahle, Ed.). Amer. Assoc. Petrol. Geol. Memoir 23, 43–145.

Meyers, J.B., Rosendahl, B.R., Harrison, C.G.A., and Ding, Z.D. (1998) Deep-imaging seismic and gravity results from the offshore Cameroon volcanic line and speculation of African hotlines. Tectonophysics, v. 284, 31–63.

Mezger, K., and Cosca, M.A. (1999) The thermal history of the Eastern Ghats belt (India) as revealed by U–Pb and $^{40}Ar/^{39}Ar$ dating of metamorphic and magmatic minerals: implications for the SWEAT correlation. Precamb. Res., v. 94, 251–271.

Milisenda, C.C., Liew, T.C., Hofmann, A.W., and Kohler, H. (1994) Nd isotopic mapping of the Sri Lanka basement: update and additional constraints from Sr isotopes. Precamb. Res., v. 66, 95–110.

Millar, I.L., Willan, R.C.R., Wareham, C.D., and Boyce, A.J. (2001) The role of crustal and mantle sources in the genesis of the granitoids of the Antarctic Peninsula and adjacent crustal blocks. J. Geol. Soc. London, v. 158, 855–867.

Miller, B.V., Samson, S.D., and D'lemos, R.S. (2001) U–Pb geochronological constraints on the timing of plutonism, volcanism, and sedimentation, Jersey, Channel Islands, UK. J. Geol. Soc. London, v. 158, 243–252.

Miller, J.S., Santosh, M., Pressley, R.A., Clements, A.S., and Rogers, J.J.W. (1996) A Pan-African thermal event in southern India. In Precambrian India within East Gondwana (M. Yoshida, M. Santosh, and M. Arima, Eds.). Special issue of J. Southeast Asian Earth Sci., v. 14, 127–136.

Miller, R. McG. (Ed.) (1983) Evolution of the Damara Orogen of South West Africa/Namibia. Geol. Soc. South Africa Spec. Pub. 11.

Miyashiro, A. (1961) Evolution of metamorphic belts. J. Petrol., v. 2, 277–311.

Miyashiro, A. (1986) Hot regions and the origin of marginal basins in the western Pacific. Tectonophysics, v. 122, 195–216.

Modie, B.N. (2002) A glacigenic interpretation of a Neoarchean (~2.78 Ga) volcanogenic sedimentary sequence in the Nnywane formation, Sikwane, Southeast Botswana. J. African Earth Sci., v. 35, 163–175.

Moeller, A., Mezger, K., and Schenk, V. (1998) Crustal age domains and the evolution of the continental crust in the Mozambique Belt of Tanzania; combined Sm–Nd, Rb–Sr, and Pb–Pb isotopic evidence. J. Petrol., v. 39, 749–783.

Moghazi, A.-K. M. (1999) Magma source and evolution of late Neoproterozoic granitoids in the Gabal El-Urf area, Eastern Desert, Egypt; geochemical and Sr–Nd constraints. Geol. Mag., v. 136, 285–300.

Moorbath, S., Whitehouse, M.J., and Kamber, B.S. (1997) Extreme Nd-isotope heterogeneity in the early Archaean—fact or fiction? Case histories from northern Canada and West Greenland. Chem. Geol., v. 135, 213–231.

Moores, E.M. (1991) Southwest U.S.–East Antarctica (SWEAT) connection: a hypothesis. Geology, v. 19, 425–428.

Moores, E.M., Kellogg, L.H., and Dilek, Y. (2000) Tethyan ophiolites, mantle convection, and "historical contingency"; a resolution of the ophiolite conundrum. In Ophiolites and oceanic crust (Y. Dilek, E. Moores, D. Elthon, and A. Nicolas, Eds.). Geol. Soc. Amer. Spec. Paper 349, 3–12.

Morgan, J., Warner, M., and Grieve, R. (2002) Geophysical constraints on the size and structure of the Chicxulub impact crater. In Catastrophic events and mass extinctions: impacts and beyond (C. Koeberl and K.G. MacLeod, Eds.). Geol. Soc. Amer. Spec. Paper 356, 39–46.

Morgan, W.J. (1972) Plate motions and deep mantle convection. In Studies in earth and space sciences (R. Shagam, R.B. Hargraves, W.J. Morgan, F.B. van Houten, C.A. Burk, H.D. Holland, and L.C. Hollister, Eds.). Geol. Soc. Amer. Memoir, 132, 7–22.

Morley, C.K. (2002) A tectonic model for the Tertiary evolution of strike-slip faults and rift basins in SE Asia. Tectonophysics, v. 347, 189–215.

Morley, L.W., and Larochelle, A. (1964) Paleomagnetism as a means of dating geological events. In Geochronology in Canada (F.F. Osborne, Ed.). Univ. Toronto Press, Royal. Soc. Canada Spec. Pub. 8, 39–51.

Morogan, V., and Lindblom, S. (1995) Volatiles associated with the alkaline–carbonatite magmatism at Alno, Sweden: a study of fluid and solid inclusions in minerals from the Langarsholmen ring complex. Contrib. Mineral. Petrol., v. 122, 262–274.

Morozova, E.A., Morozov, I.B., Smithson, S.B., and Solodilov, L.N. (1999) Heterogeneity of the uppermost mantle beneath Russian Eurasia from the ultra-long-range profile QUARTZ. J. Geophys. Res., v. 104, 20,329–20,348.

Moser, D.E., and Heaman, L.M. (1997) Proterozoic zircon growth in Archean lower crust xenoliths, southern Superior Craton; a consequence of Matachewan Ocean opening. Contrib. Mineral. Petrol., v. 128, 164–175.

Moser, D.E., Flowers, R.M., and Hart, R.J. (2001) Birth of the Kaapvaal tectosphere 3.08 billion years ago. Science, v. 291, 465–468.

Moyen, J.F., Martin, H., and Jayananda, M. (2001) Multi-element geochemical modeling of crust–mantle interactions during late-Archaean crustal growth: the Closepet granite (South India). Precamb. Res., v. 112, 87–105.

Moyes, A.B., Groenewald, P.B., and Brown, R.W. (1993) Isotopic constraints on the age and origin of the Brattskarvet intrusive suite, Dronning Maud-Land, Antarctica. Chem. Geol., v. 106, 453–466.

Mueller, P.A., Heatherington, A.L., Kelly, D.M., Wooden, J.L., and Mogk, D.W. (2002) Paleoproterozoic crust within the Great Falls tectonic zone: implications for the assembly of southern Laurentia. Geology, v. 30, 127–130.

Muenker, C., and Crawford, A. (2000) Cambrian arc evolution along the SE Gondwana active margin; a synthesis from Tasmania–New Zealand–Australia–Antarctica correlations. Tectonics, v. 19, 415–432.

Muhongo, S. (1999) Anatomy of the Mozambique Belt of eastern and southern Africa; evidence from Tanzania. Gondwana Res., v. 2, 369–375.

Muhongo, S., Kroner, A., and Nemchin, A.A. (2001a) Single zircon evaporation and SHRIMP ages for granulite-facies rocks in the Mozambique Belt of Tanzania. J. Geol., v. 109, 171–189.

Muhongo, S., Kroner, A., Wallbrecher, E., Tuisku, P., Hauzenberger, C., and Sommer, H. (2001b) Cross-section through the Mozambique Belt of East Africa and implications for Gondwana assembly. Gondwana Res., v. 4, 709–710.

Muir, R.J., Ireland, T.R., Weaver, S.D., and Bradshaw, J.D. (1996) Ion microprobe dating of Paleozoic granitoids: Devonian magmatism in New Zealand and correlations with Australia and Antarctica. Chem. Geol. (Isotope Geosci. Sect.), v. 127, 191–210.

Mukherjee, A., and Das, S. (2002) Anorthosites, granulites and the supercontinent cycle. Gondwana Res., v. 5, 147–156.

Mukhopadhyay, D. (2001) The Archaean nucleus of Singhbhum: the present state of knowledge. Gondwana Res., v. 4, 307–318.

Munyanwiya, H., Hanson, R.E., Blenkinsop, T.G., and Treloar, P.J. (1997) Geochemistry of amphibolites and quartzofeldspathic gneisses in the Pan-African Zambezi Belt, Northwest Zimbabwe; evidence for bimodal magmatism in a continental rift setting. Precamb. Res., v. 81, 179–196.

Myers, J.S. (1993) Precambrian history of the west Australian craton and adjacent orogens. Ann. Rev. Earth Planet. Sci., v. 21, 453–485.

Myers, J.S., Shaw, R.D., and Tyler, I.M. (1996) Tectonic evolution of Proterozoic Australia. Tectonics, v. 15, 1431–1446.

Nagler, Th.F., Kramers, J.D., Kamber, B.S., Frei, R., and Prendergast, M.D.A. (1997) Growth of subcontinental lithospheric mantle beneath Zimbabwe started at or before 3.8 Ga: Re-Os study on chromites. Geology, v. 25, 983–986.

Nagy, R.M., Ghuma, M.A., and Rogers, J.J.W. (1976) A crustal suture and lineament in North Africa. Tectonophysics Letter Section, v. 31, T67–T72.

Nance, R.D., and Thompson, H.D. (Eds.) (1996) Avalonian and related peri-Gondwana terranes of the circum-North Atlantic. Geol. Soc. Amer. Spec. Paper 304.

Naqvi, S.M., Condie, K.C., and Allen, P. (1983) Geochemistry of some unusual early Archaean sediments from Dharwar craton, India. Precamb. Res., v. 22, 125–147.

Naqvi, S.M., Divakara Rao, V., and Narain, H. (1974) The protocontinental growth of the Indian shield and the antiquity of its rift valleys. Precamb. Res., v. 1, 345–398.

Naqvi, S.M., Manikyamba, C., Gnaneshwar Rao, T., Subba Rao, D.V., Ram Mohan, M., and Srinivasa Sarma, D. (2002a) Geochemical and isotopic constraints of Neoarchaean fossil plume for evolution of volcanic rocks of Sandur greenstone belt, India. J. Geol. Soc. India, v. 60, 27–56.

Naqvi, S.M., Uday Raj, B., Subba Rao, D.V., Manikyamba, C., Nirmal Charan, S., Balaram, V., and Srinivasa Sarma, D. (2002b) Geology and geochemistry of arenite–quartzwacke from the late Archaean Sandur schist belt—implications for provenance and accretion processes. Precamb. Res., v. 114, 177–197.

Narbonne, G.M. (1998) The Ediacaran biota: a terminal Neoproterozoic experiment in the evolution of life. GSA Today, v, 8, no. 2, 1–6.

Narbonne, G.M., and Gehling, J.G. (2003) Life after snowball: the oldest complex Ediacaran fossils. Geology, v. 31, 27–30.

Neal, C.R., Mahoney, J.W., Kroenke, L.W., Duncan, R.A., and Petterson, M.G. (1997) The Ontong Java plateau. *In* Large igneous provinces; continental, oceanic, and planetary flood volcanism (J.J. Mahoney and M.F. Coffin, Eds.). Amer. Geophys. Union Geophysical Monograph 100, 183–216.

Nedelec, A., Ralison, B., Bouchez, J.-L., and Gregoire, V. (2000) Structure and metamorphism of the granitic basement around Anantanarivo: a key to the Pan-African history of central Madagascar and its Gondwana connections. Tectonics, v. 19, 997–1020.

Nelson, D.R., Myers, J.S., and Nutman, A.P. (1995) Chronology and evolution of the middle Proterozoic Albany–Fraser orogen, Western Australia. Australian J. Earth Sci., v. 42, 481–495.

Nelson, D.R., Trendall, A.F., and Altermann, W. (1999) Chronological correlations between the Pilbara and Kaapvaal cratons. Precamb. Res., v. 97, 165–189.

Nelson, K.D., and 27 others (1996) Partially molten middle crust beneath southern Tibet: Synthesis of project INDEPTH results. Science, v. 274, 1684–1688.

Neves, S.P., Mariano, G., Guimares, I.P., da Silva Filho, A.F., and Melo, S.C. (2000) Intralithospheric differentiation and crustal growth; evidence from the Borborema Province, northeastren Brazil. Geology, v. 28, 519–522.

Nicollet, C. (1990) Crustal evolution of the granulites of Madagascar. *In* Granulites and crustal evolution (D. Veilzeuf and P. Vidal, Eds.). NATO ASI Ser C, v. 311. Dordrecht, Kluwer, 291–310.

Nikishin, A.M., and 14 others (1996) Late Precambrian to Triassic history of the East European Craton: dynamics of sedimentary basin evolution. Tectonophysics, v. 268, 23–63.

Nisbet, E.G. (1987) The young earth—an introduction to Archaean geology. Boston, Allen and Unwin.

Nisbet, E.G., and Sleep, N.H. (2001). The habitat and nature of early life. Nature, v. 409, 1083–1091.

Nixon, G.T., Johnston, A.D., and Martin, R.F. (Eds.) (1997) Nature and origin of primitive magmas at subduction zones, Canadian Mineral., v. 35, 253–569.

Noffke, N., Hazen, R., and Nhleko, N. (2003) Earth's earliest microbial mats in a siliciclastic marine environment (2.9 Ga Mozaan Group, South Africa). Geology, v. 31, 673–676.

Norton, I.O., and Sclater, J.G. (1979) A model for the evolution of the Indian Ocean and the breakup of Gondwanaland. J. Geophys. Res., v. 84, 6803–6830.

Nutman, A.P., and Collerson, K.D. (1991) Very early crustal-accretion complexes preserved in the North Atlantic craton. Geology, v. 19, 791–794.

Nutman, A.P., Bennett, V.C., Friend, C.R.L., and Norman, M.D. (1999) Meta-igneous (nongneissic) tonalites and quartzdiorites from an extensive ca. 3800 Ma terrain south of the Isua supracrustal belt, southern West Greenland; constraints on early crust formation. Contrib. Mineral. Petrol., v. 137, 364–388.

Nyblade, A.A., and Pollack, H.N. (1993) A global analysis of heat flow from Precambrian terrains: implications for the thermal structure of Archean and Proterozoic lithospheres. J. Geophys. Res., v. 98, 12,207–12,218.

O'Reilly, S.Y., Griffin, W.L., Poudjom Domani, Y.H., and Morgan, P. (2001) Are lithospheres forever? Tracking changes in subcontinental lithospheric mantle through time. GSA Today, v. 11, 4–10.

Olive, V., Ellam, R.M., and Harte, B. (1997) A Re–Os isotope study of ultramafic xenoliths from the Matsuko kimberlite. Earth Planet. Sci. Lett., v. 150, 129–140.

Oliver, G.J.H., Johnson, S.P., Williams, I.S., and Herd, D.A. (1998) Relict 1.4 Ga oceanic crust in the Zambezi valley, northern Zimbabwe: evidence for Mesoproterozoic supercontinental fragmentation. Geology, v. 26, 571–573.

Oreskes, N. (1999) The rejection of continental drift. New York, Oxford University Press.

Oreskes, N. (2002) Plate tectonics: an insider's history of the modern theory of the earth. Boulder, Colo, Westview Press.

Orrell, S.E., Bickford, M.E., and Lewry, J.F. (1999) Crustal evolution and age of thermotectonic reworking in the western hinterland of the Trans-Hudson orogen, northern Saskatchewan. Precamb. Res., v. 95, 187–223.

Ortega-Gutierrez, F., Ruiz, J., and Centeno-Garcia, E. (1995) Oaxaquia, a Proterozoic microcontinent accreted to North America during the late Paleozoic. Geology, v. 23, 1127–1130.

Palfy, J., and Smith, P.L. (2000) Synchrony between Early Jurassic extinction, oceanic anoxic event, and the Karoo-Ferrar flood basalt volcanism. Geology, v. 28, 747–750.

Palfy, J., Smith, P.L., and Mortensen, J.K. (2002) Dating the end-Triassic and Early Jurassic mass extinctions, correlative large igneous provinces, and isotopic events. In Catastrophic events and mass extinctions: impacts and beyond (C. Koeberl and K.G. MacLeod, Eds.). Geol. Soc. Amer. Spec. Paper 356, 523–532.

Pandit, M.K., Ashwal, L.D., Tucker, R.D., Carter, L.M., Van Lente, B., Torsvik, T.H., Jamtveit, B., and Bhushan, S.K. (2001) Proterozoic acid magmatism in the Northwest Indian shield and its significance for Rodinia construction. Gondwana Res., v. 4, 726–728.

Pandit, M.K., Sial, A.N., Sukuraman, G.B., Pimentel, M.M., Ramaswamy, A.K., and Ferreira, V.P. (2002). Depleted and enriched mantle sources for Paleo- and Neoproterozoic carbonatites of southern India: Sr, Nd, C–O isotopic and geochemical constraints. Chem. Geol., v. 189, 69–89.

Paquette, J.L., Nedelec, A., Moine, B., and Rakotondrazafy, M. (1994) U–Pb, single zircon Pb evaporation, and Sm–Nd isotopic study of a granulite domain in SE Madagascar. J. Geol., v. 102, 523–538.

Park, J.K., Buchan, K.L., and Harlan, S. (1995) A proposed giant radiating dyke swarm fragmented by the separation of Laurentia and Australia based on paleomagnetism of ca. 780 Ma mafic intrusions in western North America. Earth Planet. Sci. Lett., v. 132, 129–139.

Parrish, J.T. (1993) Climate of the supercontinent Pangea. J. Geol., v. 101, 215–233.

Pearson, D.G., Carlson, R.W., Shirey, S.B., Boyd, F.R., and Nixon, P.H. (1995a) Stabilisation of Archaean lithospheric mantle: a Re–Os isotope study of peridotite xenoliths from the Kaapvaal craton. Earth Planet. Sci. Lett., v. 134, 341–357.

Pearson, D.G., Snyder, G.A., Shirey, S.B., Taylor, L.A., Carlson, R.W., and Sobolev, N.V. (1995b) Archaean Re–Os age for Siberian eclogites and constraints on Archaean tectonics. Nature, v. 374, 711–713.

Pedrosa-Soares, A.C., Noce, C.M., Wiedemann, C.M., and Pinto, C.P. (2001) The Aracuai-West-Congo Orogen in Brazil: an overview of a confined orogen formed during Gondwanaland assembly. Precamb. Res., v. 110, 307–323.

Pelechaty, S.M. (1996) Stratigraphic evidence for the Siberia–Laurentia connection and Early Cambrian rifting. Geology, v. 24, 719–722.

Peresson, H., and Decker, K. (1997) Far-field effects of late Miocene subduction in the Eastern Carpathians; E–W compression and inversion of structures in the Alpine-Carpathian-Pannonian region. Tectonics, v.16, 38–56.

Peters, K.E., and Moldowan, J.M. (1993) The biomarker guides: interpreting molecular fossils in petroleum and ancient sediments. Englewood Cliffs, N.J., Prentice Hall.

Petford, N., Atherton, M.P., and Halliday, A.N. (1996) Rapid magma production rates, underplating and remelting in the Andes: isotopic evidence from northern-central Peru (9°–11°S). J. South Amer. Earth Sci., v. 9, 69–78.

Petit, J.R., and 18 others (1999) Climate and atmospheric history of the past 420,000 years from the Vostok ice core. Nature, v. 399, 429–436.

Peucat, J.J., Bouhallier, H., Fanning, C.M., and Jayananda, M. (1995) Age of the Holenarasipur greenstone belt, relationships with surrounding gneisses (Karnataka, South India). J. Geol., v. 103, 701–710.

Peucat, J.J., Mahabaleswar, B., and Jayananda, M. (1993) Age of younger tonalitic magmatism and granulitic metamorphism in the South Indian transition zone (Krishnagiri area); comparison with older Peninsular gneisses from the Gorur–Hassan area. J. Metam. Petrol., v. 11, 879–888.

Peucat, J.J., Meno, R.P., Monnier, O., and Fanning, C.M. (1999) The Terre Adelie basement in the East-Antarctica shield: geological and isotopic evidence for a major 1.7 Ga thermal event: comparison with the Gawler craton in South Australia. Precamb. Res., v. 94, 205–224.

Pichamuthu, C.S. (1960) Charnockite in the making. Nature, v. 188, 135–136.

Pilger, R.H., Jr. (1978) A closed Gulf of Mexico, pre-Atlantic ocean plate reconstruction and the early rift history of the Gulf and North Atlantic. Trans. Gulf Coast Assoc. Geol. Soc., v. 28, 385–393.

Pimentel, M.M., Dardenne, M.A., Fuck, R.A., Viana, M.G., Junges, S.L., Fischel, D.P., Seer, H.J., and Dantas, E.L. (2001) Nd isotopes and the provenance of detrital sediments of the Neoproterozoic Brasilia Belt, central Brazil. J. South Amer. Earth Sci., v. 14, 571–585.

Pimentel, M.M., Fuck, R.A., and Souza de Alvarenga, C.J. (1996) Post-Brasiliano (Pan-African) high-K granitic magmatism in central Brazil; the role of late Precambrian–early Paleozoic extension. Precamb. Res., v. 80, 217–238.

Pimentel, M.M., Whitehouse, M.J., Viana, M. das G., Fuck, R.A., and Machado, N. (1997) The Mara Rosa arc in the Tocantins Province; further evidence for Neoproterozoic crustal accretion in central Brazil. Precamb. Res., v. 81, 299–310.

Pinna, P., Jourde, G., Calvez, J.Y., Mroz, J.P., and Marques, J.M. (1993) The Mozambique Belt in northern Mozambique; Neoproterozoic (1100–850 Ma) crustal growth and tectogenesis, and superimposed PanAfrican (800–550 Ma) tectonism. Precamb. Res., v. 62, 1–59.

Piper, J.D.A. (1982) The Precambrian Palaeomagnetic record: the case for the Proterozoic supercontinent. Earth Planet. Sci. Lett., v. 59, 61–89.

Piper, J.D.A. (2001) The Neoproterozoic supercontinent: Rodinia or Palaeopangaea? Earth Planet. Sci. Lett., v. 176, 131–146.

Piper, J.D.A. (2003) Consolidation of continental crust in late Archean–early Proterozoic times. Gondwana Res., v. 6, 435–448.

Polat, A., and Kerrich, R. (2001) Geodynamic processes, continental growth and mantle evolution recorded in late Archean greenstone belts of the southern Superior Province, Canada. Precamb. Res., v. 112, 5–25.

Porada, H., and Berhorst, V. (2000) Towards a new understanding of the Neoproterozoic–early Paleozoic Lufilian and northern Zambezi belts in Zambia and the Democratic Republic of Congo. J. African Earth Sci., v. 30, 727–771.

Poudjom Domani, Y.H., O'Reilly, S.Y., Griffin, W.L., and Morgan, P. (2001) The density structure of subcontinental lithosphere through time. Earth Planet Sci. Lett., v. 184, 605–621.

Poujol, M., Anhaeusser, C.R., and Armstrong, R.A. (2002) Episodic granite emplacement in the Amalia–Kraaipan terrane, South Africa: confirmation from single zircon U–Pb geochronology. J. African Earth Sci., v. 35, 147–161.

Powell, C.McA., and Pisarevsky, S.A. (2002) Late Neoprotrozoic assembly of East Gondwana. Geology, v. 30, 3–6.

Powell, C.McA., Li, Z.-X., McElhinny, M.W., Meert, J.G., and Park, J.K. (1993) Paleomagnetic constraints on timing of the Neoproterozoic breakup of Rodinia and the Cambrian formation of Gondwana. Geology, v. 21, 889–892.

Prave, A.R. (1996) Tale of three cratons; tectonostratigraphic anatomy of the Damara Orogen in northwestern Namibia and the assembly of Gondwana. Geology, v. 24, 1115–1118.

Preiss, W.V. (2000) The Adelaide geosyncline of South Australia and its significance in Neoproterozoic continental reconstruction. Precamb. Res., v. 100, 21–63.

Prodehl, C., and Aichroth, B. (1992) Seismic investigations along the European Geotraverse and its surroundings in central Europe. Terra Nova, v. 4, 14–24.

Puchkov, V.N. (1997) Structure and geodynamics of the Uralian orogen. In Orogeny through time (J.-P. Burg and M. Ford, Eds.). Geol. Soc. London Spec. Pub. 121, 201–236.

Puchtel, I.S., Hofmann, A.W., Mezger, K., Jochum, K.P., Shchipansky, A.A., and Samsonov, A.V. (1998) Oceanic plateau model for continental crustal growth in the Archaean: a case study from the Kostomuksha greenstone belt, NW Baltic shield. Earth Planet. Sci. Lett., v. 155, 57–74.

Puura, V., and Floden, T. (1999) Rapakivi–granite–anorthosite magmatism: a way of thinning and stabilization of the Svecofennian crust, Baltic Sea basin. In Tectonics of continental interiors (S. Marshak, B.A. van der Pluijm, and M. Hamburger, Eds.). Tectonophysics, v. 305, 75–92.

Qiu, Y.M.M., and Gao, S. (2000) First evidence of >3.2 Ga continental crust in the Yangtze craton of South China and its implications for Archean crustal evolution and Phanerozoic tectonics. Geology, v. 28, 11–14.

Rankin, D.W., and 11 others (1993) Proterozoic rocks east and southeast of the Grenville front. *In* Precambrian: conterminous U.S. (J.C. Reed, Jr., M.E. Bickford, R.S. Houston, P.K. Link, D.W. Rankin, P.K. Sims, and W.R. Van Schmus, Eds.). The geology of North America, v. C-2, Geol. Soc. Amer., 335–461.

Rast, N., and Skehan, J.W. (1983) The evolution of the Avalon plate. Tectonophysics, v. 100, 257–286.

Raymo, M.E. (1994) The Himalayas, organic carbon burial, and climate in the Miocene. Paleoceanography, v. 3, 399–404.

Rayner, R.J. (1995) The paleoclimate of the Karoo: evidence from plant fossils. Palaeogeog., Palaeoclimat., Palaeoecol., v. 119, 385–394.

Reichow, M.K., Saunders, A.D., White, R.V., Pringle, M.A., Al'Mukhamedov, A.I., Medvedev, A. I., and Kirda, N.P. (2002) $^{40}Ar/^{39}Ar$ dates from the West Siberian basin: Siberian flood basalt province doubled. Science, v. 296, 1846–1849.

Ren, J., Tamaki, K., Li, S., and Zhang, J. (2002) Late Mesozoic and Cenozoic rifting and its dynamic setting in eastern China and adjacent seas. Tectonophysics, v. 344, 175–205.

Restrepo-Pace, P.A., Ruiz, J., Gehrels, G.E., and Cosca, M. (1997) Geochronology and Nd isotopic data of Grenville-age rocks in the Colombian Andes; new constraints for late Proterozoic–early Paleozoic paleocontinental reconstructions of the Americas. Earth Planet. Sci. Lett., v. 150, 427–441.

Retallack, G.J., and Alonso-Zarza, A.M. (1998) Middle Triassic paleosols and paleoclimate of Antarctica. J. Sedimentary Res., v. 68, 169–194.

Rickers, K., Mezger, K., and Raith, M.M. (2001) Evolution of the continental crust in the Proterozoic Eastern Ghats belt, India and new constraints for Rodinia reconstruction: implications from Sm–Nd, Rb–Sr and Pb–Pb isotopes. Precamb. Res., v. 112, 183–210.

Ritsema, J., and van Heijst, H.-J. (2000) New seismic model of the upper mantle beneath Africa. Geology, v. 28, 63–66.

Rivers, T. (1997) Lithotectonic elements of the Grenville Province: review and tectonic implications. Precamb. Res., v. 86, 117–154.

Robertson, A.H.F., and Comas, M.C. (1998) Collision-related processes in the Mediterranean region; introduction. *In* Collision-related processes in the Mediterranean region (A.H.F. Robertson and M.C. Comas, Eds.). Tectonophysics, v. 298, 1–4.

Rogers, J.J.W. (1986) The Dharwar craton and the assembly of peninsular India. J. Geol., v. 94, 129–144.

Rogers, J.J.W. (1988) The Arsikere Granite of southern India: magmatism and metamorphism in a previously depleted crust. Chem. Geol., v. 67, 155–163.

Rogers, J.J.W. (1993a) A history of the Earth. Cambridge, Cambridge Univ. Press.

Rogers, J.J.W. (1993b) India and Ur. J. Geol. Soc. India, v. 42, 217–222.

Rogers, J.J.W. (1996) A history of continents in the past three billion years. J. Geol., v. 104, 91–107.

Rogers, J.J.W., and Greenberg, J.K. (1990) Late-orogenic, post-orogenic, and anorogenic granites: distinction by major-element and trace-element chemistry, and possible origins. J. Geol., v. 98, 291–309.

Rogers, J.J.W., and Mauldin, L.C. (1994) A review of the terranes of southern India. *In* Volcanism (K.V. Subbarao, Ed.). New Delhi, Wiley Eastern Ltd., 157–171.

Rogers, J.J.W., and Ragland, P.C. (1980) Trace elements in continental-margin magmatism— Part I. Trace elements in the Clarno Formation of central Oregon and the nature of the continental margin on which eruption occurred. Geol. Soc. Amer. Bull., v. 91, Part I, pp. 196–198; Part II, Card 3, pp. 1217–1292.

Rogers, J.J.W., and Santosh, M. (2002) Configuration of Columbia, a Mesoproterozoic supercontinent. Gondwana Res., v. 5, 5–22.

Rogers, J.J.W., Callahan, E.J., Dennen, K.O., Fullagar, P.D., Stroh, P.T., and Wood, L.F. (1986) Chemical evolution of Peninsular Gneiss in the Western Dharwar craton, southern India. J. Geol., v. 94, 233–246.

Rollinson, H.R., and Whitehouse, M. (Eds.) (2001) Archaean crustal evolution. Precamb. Res., v. 112, 163 pp.

Rosen, O.M., Condie, K.C., Natapov, L.M., and Nozhkin, A.D. (1994) Archean and early Proterozoic evolution of the Siberian craton: a preliminary assessment. *In* Archean crustal evolution (K.C. Condie, Ed.). Developments in Precamb. Geol. 11, Amsterdam, Elsevier, 411–459.

Rosing, M.T., Rose, N.M., Bridgwater, D., and Thomsen, H. (1996) Earliest part of earth's stratigraphic record; a reappraisal of the >3.7 Ga Isua (Greenland) supracrustal sequence. Geology, v. 24, 43–46.

Ross, G.M. (1991) Precambrian basement in the Canadian cordillera: an introduction. Canadian J. Earth Sci., v. 28, 1133–1139.

Ross, G.M., Parrish, R.R., and Winston, D. (1992) Provenance and U–Pb geochronology of the Mesoproterozoic Belt Supergroup (northwestern United States): implications for age of deposition and pre-Panthalassa plate reconstructions. Earth Planet. Sci. Lett., v. 113, 57–76.

Rougvie, J.R., Carlson, W.D., Copeland, P., and Connelly, J.N. (1999) Late thermal evolution of Proterozoic rocks in the northeastern Llano Uplift, central Texas. Precamb. Res., v. 94, 49–72.

Roy, A.B. (2001) Neoproterozoic crustal evolution of northwestern Indian shield: implications on breakup and assembly of supercontinents. Gondwana Res., v. 4, 289–306.

Rudkevich, M.Ya., and Maksimov, Ye.M. (1988) Cyclicity in the geologic development of the West Siberian platform in the cratonal stage. Petroleum. Geol., v. 24, 265–269.

Rudnick, R.L. (1990) Nd and Sr isotopic compositions of lower-crustal xenoliths from North Queensland, Australia: implications for Nd model ages and crustal growth processes. Chem. Geol., v. 83, 195–208.

Rudnick, R.L., and Taylor, S.R. (1991) Petrology and geochemistry of lower crustal xenoliths from northern Queensland and inferences on lower crustal composition. In The Australian lithosphere (B. Drummond, Ed.). Geol. Soc. Australia Spec. Pub. 17, 189–208.

Rudnick, R.L., McDonough, W.F., and O'Connell, R.J. (1998) Thermal structure, thickness and composition of continental lithosphere. Chem. Geol., v. 145, 395–411.

Ruiz, J., Tosdal, R.M., Restrepo, P.A., and Murillo-Muneton, G. (1999) Pb isotope evidence for Colombia–southern Mexico connections in the Proterozoic. In Laurentia–Gondwana connections before Pangea (V.A. Ramos and J.D. Keppie, Eds.). Geol. Soc. Amer. Spec. Paper 336, 183–197.

Rumvegeri, B.T. (1991) Tectonic significance of Kibaran structures in Central and Eastern Africa. J. African Earth Sci., v. 13, 267–276.

Runcorn, S.K. (1956) Palaeomagnetic comparisons between Europe and North America. Geol. Assoc. Canada Proc., v. 8, 77–85.

Runcorn, S.K. (1962) Palaeomagnetic evidence for continental drift and its geophysical cause. In Continental drift. Academic Press Internat. Geophys. Series 3, Chapter 1, 1–40.

Sacchi, R., Cadoppi, P., and Costa, M. (2000) Pan-African reactivation of the Lurio segment of the Kibaran Belt system; a reappraisal from recent age determinations in northern Mozambique. J. African Earth Sci., v. 30, 629–639.

Sadowski, G.R., and Bettencourt, J.S. (1996) Mesoproterozoic tectonic correlations between eastern Laurentia and the western border of the Amazon craton. Precamb. Res., v. 76, 213–227.

Sanchez-Zavala, J.L., Centeno-Garcia, E., and Ortega-Gutierrez, F. (1999) Review of Paleozoic stratigraphy of Mexico and its role in the Gondwana–Laurentia connections. In Laurentia–Gondwana connections before Pangea (V.A. Ramos and J.D. Keppie, Eds). Geol. Soc. Amer. Spec. Paper 336, 211–226.

Sandberg, C.A., Morrow, J.R., and Ziegler, W. (2002) Late Devonian sea-level changes, catastrophic events, and mass extinctions. In Catastrophic events and mass extinctions: impacts and beyond (C. Koeberl and K.G. MacLeod, Eds.). Geol. Soc. Amer. Spec. Paper 356, pp. 473–487.

Santos, J.O.S., Hartmann, L.A., Gaudette, H.E., Groves, D.I., McNaughton, N.J., and Fletcher, I.R. (2000) A new understanding of the provinces of the Amazon craton based on integration of field mapping and U–Pb and Sm–Nd geochronology. Gondwana Res., v. 3, 453–488.

Santosh, M. (2000) Palaeoproterozoic accretion, Pan-African reworking and fluid-driven processes in the continental deep crust of southern India. In Deep crustal studies on granulite terranes from southern India (C. Srikantappa, Ed.). Indian Mineralogist, v. 34, 22–28.

Santosh, M., and Drury, S.A. (1988) Alkali granites with Pan-African affinities from Kerala, S. India, J. Geology, v. 96, 616–626.

Santosh, M., and Tsunogae, T. (2003) Extremely high density pure $CO_2$ fluid inclusions in a garnet granulite from southern India. J. Geol., v. 111, 1–16.

Santosh, M., and Wada, H. (1993a) A carbon isotope study of graphites from the Kerala Khondalite Belt: evidence for $CO_2$ infiltration in granulites. J. Geol., v. 101, 643–651.

Santosh, M., and Wada, H. (1993b) Microscale isotopic zonation in graphite crystals: evidence for channeled $CO_2$ influx in granulites. Earth Planet. Sci. Lett., v. 119, 19–26.

Santosh, M., and Yoshida, M. (2001) Pan-African extensional collapse along the Gondwana suture. Gondwana Res., v. 4, 188–191.

Santosh, M., and Yoshikura, S. (2001) Charnockite magmatism and charnockitic metasomatism in East Gondwana and Asia. Gondwana Res., v. 4, 768–771.

Santosh, M., Harris, N.B.W., Jackson, D.H., and Mattey, D.P. (1990) Dehydration and incipient charnockite formation: a phase equilibria and fluid inclusion study from South India. J. Geol., v. 98, 915–926.

Santosh, M., Suzuki, K., and Masuda, A. (1994) Re–Os dating of molybdenites from southern India; implication for Pan-African metallogeny. J. Geol. Soc. India, v. 43, 585–590.

Santosh, M., Yokoyama, K., Biju-Sekhar, S., and Rogers, J.J.W. (2003) Multiple tectonothermal events in the granulite blocks of southern India revealed from EPMA dating: implications on the history of supercontinents. Gondwana Res., v. 6, 29–64.

Santosh, M., Yoshida, M., and Yoshikura, S. (2001) Fluids in the Gondwana crust. Gondwana Res., v. 4, 766–768.

Satish-Kumar, M., and Wada, H. (1997) Meteoric water infiltration in Skallen marbles, East Antarctica; oxygen isotope evidence, *In* Proceedings of the 16th NIPR Symposium on Antarctic Geosciences (M. Funaki, Ed.), Tokyo, v. 10, 111–119. Tokyo, National Institute for Polar Research.

Satish-Kumar, M., Wada, H., and Santosh, M. (2002) Constraints on the application of carbon isotope thermometry in high- and ultrahigh-temperature metamorphic terranes. J. Metam. Geol., v. 20, 335–350.

Sato, K., and Siga, O., Jr. (2002) Rapid growth of continental crust between 2.2 to 1.8 Ga in the South American platform: integrated Australian, European, North American, and SW USA crustal evolution study. Gondwana Res., v. 5, 165–173.

Schandelmeier, H., Richter, A.E., and Franz, G. (1983) Outline of the geology of magmatic and metamorphic units between Gebel Uweinat and Bir Safsaf (SW Egypt/NW Sudan). J. African Earth Sci., v. 1, 275–283.

Schandelmeier, H., Wipfler, E., Kuster, D., Sultan, M., Becker, R., Stern, R.J., and Abedelsalam, M.G. (1994) Atmur–Delgo suture: a Neoproterozoic oceanic basin extending into the interior of northeast Africa. Geology, v. 22, 563–566.

Schidlowski, M. (2001) Carbon isotopes as biogeochemical recorders of life over 3.8 Ga of earth history: evolution of a concept. Precamb. Res., v. 106, 117–134.

Schilling, J.G., Zajac, M., Evans, R., Johnston, T., White, W., Devine, J.D., and Kingsley, R. (1983) Petrologic and geochemical variations along the mid-Atlantic ridge from 29°N to 79°N. Amer. J. Sci., v. 283, 510–586.

Schmid, S.M., and Kissling, E. (2000) The arc of the western Alps in the light of geophysical data on deep crustal structure. Tectonics, v. 19, 62–85.

Schmitz, M.D., and Bowring, S.A. (2001) The significance of U–Pb zircon dates in lower crustal xenoliths from the southwestern margin of the Kaapvaal Craton, southern Africa. Chem. Geol., v. 172, 59–76.

Schopf, J.W., Krudytsev, A.B., Agrestl, D.E., Wdowiak, T.J., and Czaja, A.D. (2002) Laser Raman imagery of earth's earliest fossils. Nature, v. 416, 73–76.

Schumacher, M.E. (2002) Upper Rhine Graben; role of preexisting structures during rift evolution. Tectonics, v. 21, 1–17.

Schwartz, M.O., Rajah, S.S., Askury, A.K., Putthapiban, P., and Djaswadi, S. (1995) The Southeast Asian tin belt. Earth-Sci. Rev., v. 38, 95–293.

Scotese, C.R., and McKerrow, W.S. (1990) Revised world maps and introduction. *In* Palaeozoic geography and biogeography (W.S. McKerrow and C.R. Scotese, Eds.). Geol. Soc. London Memoir 12, 1–21.

Scrimgeour, I., and Close, D. (1999) Regional high-pressure metamorphism during intracratonic deformation; the Petermann orogeny, central Australia. J. Metam. Geol., v. 17, 557–572.

Scrimgeour, I., and Raith, J.G. (2001) High-grade reworking of Proterozoic granulites during Ordovician intraplate transpression, eastern Arunta inlier, central Australia. *In* Continental reactivation and reworking (J.A. Miller, R.E. Holdsworth, I.S. Buick, and M. Hand, Eds.). Geol. Soc. London Spec. Pub. 184, 261–287.

Sears, J.W., and Price, R.A. (2000) New look at the Siberian connection: no SWEAT. Geology, v. 28, 423–426.

Sears, J.W., and Price, R.A. (2002) The hypothetical Mesoproterozoic supercontinent Columbia: implications of the Siberian–west Laurentian connection. Gondwana Res., v. 5, 35–39.

Sears, J.W., Chamberlain, K.R., and Buckley, S.N. (1998) Structural and U–Pb geochronological evidence for 1.47 Ga rifting in the Belt basin, western Montana. Canadian J. Earth Sci., v. 35, 467–475.

Seilacher, A. (1992) Vendobionta and Psammocorallia: lost constructions of Precambrian evolution. J. Geol. Soc. London, v. 149, 607–613.

Seilacher, A., Bose, P.K., and Pflueger, F. (1998) Triploblastic animals more than 1 billion years ago: trace fossil evidence from India. Science, v. 281, 80–83.

Selverstone, J., Pun, A., and Condie, K.C. (1999) Xenolithic evidence for Proterozoic crustal evolution beneath the Colorado plateau. Geol. Soc. Amer. Bull., v. 111, 590–606.

Sengor, A.M.C. (1985). East Asian tectonic collage. Nature, v. 318, 16–17.

Sengor, A.M.C. (1987) Tectonics of the Tethysides: orogenic collage development in a collisional setting. Ann. Rev. Earth Planet. Sci., v. 15, 213–244.

Sengor, A.M.C., Altiner, D., Cin, A., Ustaomer, T., and Hsu, K.J. (1988) Origin and assembly of the Tethyside orogenic collage at the expense of Gondwanaland. *In* Gondwana and Tethys (A. Hallam, Ed.). Geol. Soc. London Spec. Pub. 37, 119–181.

Sengor, A.M.C., Natal'in, B.A., and Burtman, V.S. (1993). Evolution of the Altaid tectonic collage and Palaeozoic crustal growth in Eurasia. Nature, v. 364, 299–307.

Shackleton, R.M. (1996) The final collision zone between East and West Gondwana: where is it? J. African Earth Sci., v. 23, 271–287.

Shaw, C.A., Arima, M., Kagami, H., Fanning, C.M., Shiraishi, K., and Motoyoshi, Y. (1997) Proterozoic events in the Eastern Ghats granulite belt, India: evidence from

Rb–Sr, Sm–Nd systematics, and SHRIMP dating. J. Geol., v. 105, 645–656.

Shaw, D.M., Dickin, A.P., Li, H., McNutt, R.H., Schwarcz, H.P., and Truscott, M.G. (1994) Crustal geochemistry in the Wawa–Foleyet region, Ontario. Canadian J. Earth Sci., v. 31, 1104–1121.

Shiraishi, K., Ellis, D.J., Hiroi, Y., Fanning, C.M., Motoyoshi, Y., and Nakai, Y. (1994) Cambrian orogenic belt in East Antarctica and Sri Lanka: implications for Gondwana assembly. J. Geol., v. 102, 47–65.

Shore, M., and Fowler, A.D. (1999) The origin of spiniflex texture in komatiites. Nature, v. 397, 691–694.

Sidder, G.B., Garcia, A.E., and Stoeser, J.W. (Eds.) (1995) Geology and mineral deposits of the Venezuelan Guyana Shield. U.S. Geol. Surv. Bull. 2124.

Simons, F.J., Zielhuis, A., and van der Hilst, R.D. (1999) The deep structure of the Australian continent from surface wave tomography. Lithos, v. 48, 17–43.

Sleep, N.H., and Windley, B.F. (1982) Archean plate tectonics: constraints and inferences. J. Geol., v. 90, 363–379.

Smith, A.L., Schellekens, J.H., and Muriel Diaz, A.-L. (1998) Batholiths as markers of tectonic change in the northeastern Caribbean. In Tectonics and geochemistry of the Northeastern Caribbean (E.G. Lidiak and D.K. Larue, Eds.). Geol. Soc. Amer. Spec. Paper 322, 99–122.

Smith, D. (2000) Insights into the evolution of the uppermost continental mantle from xenolith localities on and near the Colorado plateau and regional comparisons. J. Geophys. Res., v. 105, 16,769–16,781.

Smith, D.R., Barnes, C., Shannon, W., Roback, R., and James, E. (1997) Petrogenesis of Mid-Proterozoic granitic magmas; examples from central and West Texas. Precamb. Res., v. 85, 53–79.

Smith, I.E.M., Worthington, T.J., Price, R.C., and Gamble, J.A. (1997) Primitive magmas in arc-type volcanic associations: examples from the southwest Pacific. In Nature and origin of primitive magmas at subduction zones (G.T. Nixon, A.D. Johnston, and R.F. Martin, Eds.). Canadian Mineral., v. 35, 257–273.

Stampfli, G.M. (2000) Tectonics and magmatism in Turkey and the surrounding area. In Tethyan oceans (E. Bozkurt, J.A. Winchester, and J.D.A. Piper, Eds.). Geol. Soc. London Spec. Pub. 173, 1–23.

Stanistreet, I.G., Kukla, P.A., and Henry, G. (1991) Sedimentary basinal responses to a late Precambrian Wilson cycle; the Damara Orogen and Nama Foreland, Namibia. J. African Earth Sci., v. 13, 141–156.

Starmer, I.C. (1993) The Sveconorwegian Orogeny in Southern Norway, relative to deep crustal structures and events in the North Atlantic Proterozoic supercontinent. Norsk Geologisk Tidsskrift, v. 73, 109–132.

Starmer, I.C. (1996) Accretion, rifting, rotation and collision in the North Atlantic supercontinent, 1700–950 Ma. In Precambrian crustal evolution in the North Atlantic region (T.S. Brewer, Ed.) Geol. Soc. London Spec. Pub. 112, 219–248.

Stein, H.J., and Crock, J.G. (1990) Late Cretaceous–Tertiary magmatism in the Colorado Mineral Belt; Rare earth element and samarium–neodymium isotopic studies. In The nature and origin of Cordilleran magmatism (L.A. Anderson, Ed.). Geol. Soc. Amer. Memoir 174, 195–223.

Steinmann, G. (1927) Die ophiolitischen zone in den Mediterranean Kettengebirgen. Congres Geologique International, XIV Session, Madrid, Spain, 637–667.

Stern, R.A., and Hanson, G.N. (1991) Archean high-Mg granodiorite: a derivative of light rare earth element-enriched monzodiorite of mantle origin. J. Petrol., v. 32, 201–238.

Stern, R.J. (1994) Arc assembly and continental collision in the Neoproterozoic East African orogen: implications for the consolidation of Gondwanaland. Ann. Rev. Earth Planet. Sci., v. 22, 319–351.

Stern, R.J. (2002) Crustal evolution in the East African orogen: a neodymium isotopic perspective. J. African Earth Sci., v. 34, 109–117.

Stewart, A.J. (1979) A barred-basin marine evaporite in the Upper Proterozoic of the Amadeus Basin, central Australia. Sedimentology, v. 26, 33–62.

Stewart, J.H. (1978) Basin–range structure in western North America: a review. In Cenozoic tectonics and regional geophysics of the western cordillera (R.B. Smith and G.P. Eaton, Eds.). Geol. Soc. Amer. Memoir 152, 1–32.

Stewart, J.H., Blodgett, R.B., Boucot, A.J., Carter, J.L., and Lopez, R. (1999) Exotic Paleozoic strata of Gondwanan provenance near Ciudad Victoria, Tamaulipas, Mexico. In Laurentia–Gondwana connections before Pangea (V.A. Ramos and J.D. Keppie, Eds.). Geol. Soc. Amer. Spec. Paper 336, 227–252.

Storey, B.C., and Kyle, P.R. (1997) An active mantle mechanism for Gondwana breakup. South African J. Geol., v. 100, 283–290.

Storey, B.C., Leat, P.T., and Ferris, J.K. (2001) The location of mantle plume centers during the initial stages of Gondwana breakup. In Mantle plumes; their identification through time (R.E. Ernst and K. Buchan, Eds.). Geol. Soc. Amer. Spec. Paper 352, 71–80.

Storey, B.C., Macdonald, D.L.M., Dalziel, I.W.D., Isbell, J.L., and Millar, I.L. (1996) Early Paleozoic sedimentation, magmatism, and deformation in the Pensacola Mountains, Antarctica: the significance of the Ross orogeny. Geol. Soc. Amer. Bull., v. 108, 685–707.

Stott, C.M. (1997) The Superior Province, Canada. In Greenstone belts (M. J. De Wit and L.D. Ashwal, Eds.). Oxford Monographs on Geology and Geophysics 35, 480–507.

Streepey, M.M., van der Pluijm, B.A., Essene, E.J., Hall, C.M., and Macloughlin, J.F. (2000) Late Proterozoic (ca. 930 Ma) extension in eastern Laurentia. Geol. Soc. Amer. Bull., v. 112, 1522–1530.

Strieder, A.J., and Freitas Suita, M.T. de (1999) Neoproterozoic geotectonic evolution of Tocantins structural province, central Brazil. J. Geodynamics, v. 28, 267–289.

Stump, E. (1995) The Ross orogen of the Transantarctic Mountains. Cambridge, Cambridge Univ. Press.

Suayah, I.B., and Rogers, J.J.W. (1991) Petrology of the lower Tertiary Clarno Formation in north-central Oregon: the importance of magma mixing. J. Geophys. Res., v. 96, 13,357–13,371.

Suess, E. (1904–1909) English editions of multivolume work titled "Das Antlitz der Erde." Oxford, Clarendon Press.

Sun, Y., Chen, Z., Lio, Y., Wang, T., and Zhang, Z. (2000) Junction and evolution of the Qinling, Qilian and Kunlun orogenic belts. Acta Geologica Sinica (English edition), v. 74, 223–228.

Sykes, L.R. (1967) Mechanism of earthquakes and nature of faulting on mid-oceanic ridges. J. Geophys. Res., v. 72, 2131–2153.

Tait, J., Schatz, M., Bachtadse, V., and Soffel, H. (2000). Palaeomagnetism and Palaeozoic palaeogeography of Gondwana and European terranes. In Orogenic processes: quantification and modeling in the Variscan belt (W. Franke, V. Haak, O. Oncken, and D. Tanner, Eds.). Geol. Soc. London Spec. Pub. 179, 21–34.

Tarduno, J.A., and Cottrell, R.D. (1997) Paleomagnetic evidence for motion of the Hawaiian hotspot during formation of the Emperor seamounts. Earth Planet. Sci. Lett., v. 153, 171–180.

Tassinari, C.C.G., and Macambira, M.J.B. (1999) Geochronological provinces of the Amazonian craton. Episodes, v. 22, 174–182.

Taylor, R.N., Thirlwall, M.F., Murton, B.J., Hilton, D.R., and Gee, M.A.M. (1997) Isotopic constraints on the influence of the Icelandic mantle plume. Earth Planet. Sci. Lett., v. 148, E1–E8.

Taylor, S.R., and McLennan, S.M. (1985) The continental crust: its composition and evolution. Oxford, Blackwell Scientific.

Teklay, M., Haile, T., Kroner, A., Asmerom, Y., and Watson, J. (2001) A back-arc palaeotectonic setting for the Neoproterozoic magmatic rocks of western Eritea: inferences from geochemistry, geochronology and isotope geology (abstract). Gondwana Res., v. 4, 800.

ten Brink, U.S., Hackney, R.I., Bannister, S., Stern, T.A., and Makovsky, Y. (1997) Uplift of the Transantarctic Mountains and the bedrock beneath the East Antarctic ice sheet. J. Geophys. Res., v. 102, 27,603–27,621.

Theriault, R.J., St-Onge, M.R., and Scott, D.J. (2001) Nd isotopic and geochemical signature of the Paleoproterozoic Trans-Hudson orogen, southern Baffin Island, Canada: implications for the evolution of eastern Laurentia. Precamb. Res., v. 108, 113–138.

Thomas, D.J., Zachos, J.C., Bralower, T.J., Thomas, E., and Bohaty, S. (2002) Warming the fuel for the fire: evidence for the thermal dissociation of methane hydrate during the Paleocene–Eocene thermal maximum. Geology, v. 30, 1067–1070.

Thomas, L. (2002) Coal geology. Chichester, UK, John Wiley and Sons.

Thomas, R.J., Eglington, R.N., Bowring, S.A., Retief, E.A., and Walraven, F. (1993) New isotope data from a Neoproterozoic porphyritic granitoid–charnockite suite from Natal, South Africa. Precamb. Res., v. 62, 83–101.

Thomas, W.A. (1977) Evolution of Appalachian–Ouachita salients and recesses from reentrants and promontories in the continental margin. Amer. J. Sci., v. 277, 1233–1278.

Thomas, W.A., and Astini, R.A. (2002) Ordovician collision of the Argentine Precordillera with Gondwana, independent of Laurentian Taconic orogeny. Tectonophysics, v. 345, 131–152.

Timmons, J.M., Karlstrom, K.E., Dehler, C.M., Geissman, J.W., and Heizler, M.T (2001) Proterozoic multistage (ca. 1.1 and 0.8 Ga) extension recorded in the Grand Canyon Supergroup and establishment of northwest- and north-trending tectonic grains in the southwestern United States. Geol. Soc. Amer. Bull., v. 113, 163–180.

Torsvik, T.H., and Van der Voo, R. (2002) Refining Gondwana and Pangea: estimates of Phanerozoic non-dipole (octupole) fields. Geophys. J. Internat., v. 151, 771–794.

Torsvik, T.H., Carter, L.M., Ashwal, L.D., Bhushan, S.K., Pandit, M.K., and Jamtveit, B. (2001) Rodinia refined or obscured; palaeomagnetism of the Malani igneous suite (NW India). Precamb. Res., v. 108, 319–333.

Torsvik, T.H., Van der Voo, R., and Redfield, T.F. (2002) Relative hotspot motions versus true polar wander. Earth Planet. Sci. Lett., v. 202, 185–200.

Toteu, S.F., Van Schmus, W.R., Penaye, J., and Michard, A. (2001) New U–Pb and Sm–Nd data from north-central Cameroon and its bearing on the pre-Pan African history of central Africa. Precamb. Res., v. 108, 45–73.

Touret, J., and Dietvorst, P. (1983) Fluid inclusions in high-grade anatectic metamorphites. J. Geol. Soc. London, v. 140, 635–649.

Trettin, H.P. (1991) The Proterozoic to Late Silurian record of Pearya. In Geology of the Innuitian Orogen and Arctic Platform of Canada and Greenland (H.P. Trettin, Ed.). Geology of Canada, no. 3, Geol. Surv. Canada, 241–259.

Trompette, R. (1997) Neoproterozoic (~600 Ma) aggregation of western Gondwana: a tentative scenario. Precamb. Res., v. 82, 101–112.

Tucker, R.D., Krogh, T.E., and Raheim, A. (1990) Petrologic evolution and age-province boundaries in the central part of the Western Gneiss region, Norway: results of U–Pb dating of accessory minerals from Trondheimsfjord to Geiranger. In Mid-Proterozoic Laurentia–Baltica (C.F. Gower, T. Rivers, and A.B. Ryan, Eds.). Geol. Assoc. Canada Spec. Paper 38, 149–174.

Tull, J.F. (2002) Southeastern margin of the middle Paleozoic shelf, southwesternmost Appalachians: regional stability bracketed by Acadian and Alleghanian tectonism. Geol. Soc. Amer. Bull., v. 114, 643–655.

Unrug, R. (1992) The supercontinent cycle and Gondwana assembly: component cratons and the timing of suturing events. J. Geodynamics, v. 16, 215–246.

Vail, J. (1985) Pan-African (late Precambrian) tectonic terrains and the reconstruction of the Arabian–Nubian shield. Geology, v. 13, 839–842.

Valley, J.W. (2001) Stable isotope thermometry at high temperatures. Min. Soc. Amer. Rev. Mineralogy, v. 43, 365–402.

Van der Voo, R. (1993) Paleomagnetism of the Atlantic, Tethys and Iapetus Oceans. Cambridge, Cambridge Univ. Press.

Van Kranendonk, M.J., and Collins, W.J. (1998) Timing and tectonic significance of late Archaean, sinistral strike-slip deformation in the central Pilbara structural corridor, Pilbara Craton, Western Australia. In The tectonics and metallogenic evolution of the Pilbara Craton (M.E. Barley and S.E. Loader, Eds.). Precamb. Res., v. 88, 207–231.

Van Schmus, W. R., Bickford, M.E., and 23 others (1993) Transcontinental Proterozoic provinces. In Precambrian: conterminous U.S. (J.C. Reed, Jr., M.E. Bickford, R.S. Houston, P.K. Link, D.W. Rankin, P.K. Sims, and W.R. Van Schmus, Eds.). The geology of North America, v. C-2, Geol. Soc. Amer., 171–334.

Van Schmus, W.R., Bickford, M.E., and Turek, A. (1996) Proterozoic geology of the east–central Midcontinent basement. In Basement and basins of eastern North America (B.A. van der Pluijm and P.A. Catacosinos, Eds.). Geol. Soc. Amer. Spec. Paper 308, 7–32.

Vandenberg, A.H.M. (1999) Timing of orogenic events in the Lachlan Orogen. Australian J. Earth Sci., v. 46, 691–701.

Vanderhaeghe, O., Ledru, P., Thieblemont, D., Egal, E., Cocherie, A., Tegyey, M., and Milesi, J.-P. (1998) Contrasting mechanism of crustal growth; geodynamic evolution of the Paleoproterozoic granite–greenstone belts of French Guiana. Precamb. Res., v. 92, 165–194.

Veevers, J.J. (1995) Emergent, long-lived Gondwanaland vs. submergent, short-lived Laurasia; supercontinental and Pan-African heat imparts long-term buoyancy by mafic underplating. Geology, v. 23, 1131–1134.

Veevers, J.J. (2001) Atlas of billion-year earth history of Australia and neighbours in Gondwanaland. Sydney, Gemoc Press.

Veevers, J.J. and Tewari, R.C. (1995) Gondwana master basin of peninsular India between Tethys and the interior of the Gondwanaland province of Pangea. Geol. Soc. Amer. Memoir 187.

Vernikovsky, V.A., and Vernikovskaya, A.E. (2001) Central Taimyr accretionary belt (Arctic Asia): Meso–Neoproterozoic tectonic evolution and Rodinia breakup. Precamb. Res., v. 110, 127–141.

Vervoort, J.D., Patchett, P.J., Gehrels, G.E., and Nutman, A.P. (1996) Constraints on early earth differentiation from hafnium and neodymium isotopes. Nature, v. 379, 624–627.

Viele, G.W. (1989) The Ouachita orogenic belt. In The Appalachian–Ouachita orogen in the United States (R.D. Hatcher, Jr., W.A. Thomas, and G.W. Viele, Eds.). The Geology of North America, v. F-2, Boulder, Geol. Soc. Amer., 555–561.

Vincent, S.J., and Allen, M.B. (2001) Sedimentary record of Mesozoic intracontinental deformation in the eastern Junggar Basin, northwestern China; response to orogeny at the Asian margin. In Paleozoic and Mesozoic tectonic evolution of central Asia; from continental assembly to intracontinental deformation (M.S. Hendrix and G.A. Davis, Eds.). Geol. Soc. Amer. Memoir 194, 341–360.

Vine, F.J., and Matthews, D.H. (1963) Magnetic anomalies over ocean ridges. Nature, v. 199, 947–949.

Vinyu, M.L., Hanson, R.E., Martin, M.W., Bowring, S.A., Jelsma, H.K., Krol, M.A., and Dirks, P.H.G.M. (1999) U–Pb and $^{40}Ar/^{39}Ar$ geochronological constraints on the tectonic evolution of the easternmost part of the Zambezi orogenic belt, Northeast Zimbabwe. Precamb. Res., v. 98, 67–82.

von Huene, R., and Scholl, D.W. (1991) Observations at convergent margins concerning sediment subduction, subduction erosion, and the growth of continental crust. Rev. Geophys., v. 29, 279–316.

Wadati, K. (1940) Deep-focus earthquakes in Japan and its vicinity. Proc. Pacific Sci. Congress, v. 1, 139–147, Berkeley, University of California Press.

Waggoner, B. (1999) Biogeographic analysis of the Ediacara biota: a conflict with paleotectonic reconstructions. Paleobiology, v. 25, 440–458.

Walcott, R.I. (1998) Modes of oblique compression; late Cenozoic tectonics of the South Island of New Zealand. Rev. Geophys., v. 36, 1–26.

Wang, X., Metcalfe, I., Jian, P., He, L., and Wang, C. (2000) The Jinshajiang–Ailaoshan suture zone, China: tectonostratigraphy, age and evolution. J. Asian Earth Sci., v. 18, 675–690.

Ward, P.L. (1995) Subduction cycles under western North America during the Mesozoic and Cenozoic eras. In Jurassic magmatism and tectonics of the North American cordillera (D.M. Miller and C. Busby, Eds.). Geol. Soc. Amer. Spec. Paper 299, 1–46.

Wardle, R.J., Gower, C.F., James, D.T., St-Onge, M.R., Scott, D.J., and Garde, A.A. (2002) Proterozoic evolution of the northeastern Canadian Shield. Canadian J. Earth Sci., v. 39, 895 pp.

Wareham, C.D., Pankhurst, R.J., Thomas, R.J., Storey, B.C., Grantham, G.H., Jacobs, J., and Eglington, B.M. (1998) Pb, Nd, and Sr isotope mapping of Grenville-age crustal provinces in Rodinia. J. Geol., v. 106, 647–649.

Warme, J.E., Morgan, M., and Kuehner, H.-C. (2002) Impact-generated carbonate accretionary lapilli in the Late Devonian Alamo Breccia. In Catastrophic events and mass extinctions: impacts and beyond (C. Koeberl and K.G. MacLeod, Eds.). Geol. Soc. Amer. Spec. Paper 356, 489–504.

Warren, J. (1999) Evaporites: their evolution and economics. Oxford, Blackwell.

Wasteneys, H.A., Clark, A.H., Farrar, E., and Langridge, R.J. (1995) Grenvillean granulite-facies metamorphism in the Arequipa massif, Peru. Earth Planet. Sci. Lett., v. 132, 63–73.

Wawrzyniec, T.F., Geissman, J.W., Melker, M.D., and Hubbard, M. (2002) Dextral shear along the eastern margin of the Colorado Plateau; a kinematic link between Laramide contraction and Rio Grande rifting (ca. 75–13 Ma). J. Geol., v. 110, 305–324.

Weber, B., and Koehler, H. (1999) Sm–Nd, Rb–Sr and U–Pb geochronology of a Grenville terrane in southern Mexico; origin and geologic history of the Guichicovi Complex. Precamb. Res., v. 96, 245–262.

Wegener, A. (1912) Die Entstehung der Kontinente. Geologische Rundschau, v. 3, 276–292.

Weil, A., van der Voo, R., Niocaill, C.M., and Meert, J.G. (1998) The Proterozoic supercontinent Rodinia: paleomagnetically derived reconstruction for 1100 to 800 Ma. Earth Planet Sci. Lett., v. 154, 13–24.

West, H.B., Garcia, M.O., Gerlach, D.C., and Romano, J. (1992) Geochemistry of tholeiites from Lanai, Hawaii. Contrib. Mineral. Petrol., v. 112, 520–542.

White, D.A., Roeder, D.H., Nelson, T.H., and Crowell, J.C. (1970) Subduction. Geol. Soc. Amer. Bull., v. 81, 3431–3432.

White, D.J., Forsyth, D.A., Asudeh, I., Carr, S.D., Wu, H., Easton, R.M., and Mereu, R.F. (2000) A seismic-based cross-section of the Grenville orogen in southern Ontario and western Quebec. Canadian J. Earth Sci., v. 37, 183–192.

White, R.W., Clarke, G.L., and Nelson, D.R. (1999) SHRIMP U–Pb zircon dating of Grenville-age events in the western part of the Musgrave block, central Australia. J. Metam. Geol., v. 17, 465–481.

Whitehouse, M.J., Kalsbeek, F., and Nutman, A.P. (1998) Crustal growth and crustal recycling in the Nagssugtoqidian orogen of West Greenland: constraints from isotope systematics and U–Pb zircon geochronology. In Isotopes and crustal evolution; special volume to honour Stephen Moorbath (M.J. Whitehouse and C.R.L. Friend, Eds.). Precamb. Res., v. 91, 365–381.

Whittington, H.B. (1985) The Burgess Shale. Geol. Surv. Canada and Yale Univ. Press.

Wiebe, R.A. (1992) Proterozoic anorthosite complexes. In Proterozoic crustal evolution (K.C. Condie, Ed.). Developments in Precambrian geology 10. Amsterdam, Elsevier, 215–216.

Wignall, P.B., and Twitchett, R.J. (2002) Extent, duration, and nature of the Permian–Triassic superanoxic event. In Catastrophic events and mass extinctions: impacts and beyond (C. Koeberl and K.G. MacLeod, Eds.). Geol. Soc. Amer. Spec. Paper 356, 395–413.

Wilde, S.A. (1999) Evolution of the western margin of Australia during the Rodinian and Gondwanan supercontinent cycles. Gondwana Res., v. 2, 481–499.

Wilde, S.A., Valley, J.W., Peck, W.H., and Graham, C.M. (2001) Evidence from detrital zircons for the existence of continental crust and oceans on the earth 4.4 Gyr ago. Nature, v. 409, 175–181.

Wilde, S.A., Zhao, G., and Sun, M. (2002) Development of the North China Craton during the Late Archaean and its final amalgamation at 1.8 Ga; some speculations on its position within a global Palaeoproterozoic supercontinent. Gondwana Res., v. 5, 85–94.

Williams, G.E. (1993) History of Earth's obliquity. Earth-Sci. Rev., v. 34, 1–45.

Williams, H., Hoffman, P.F., Lewry, J.F., Monger, J.W.H., and Rivers, T. (1991) Anatomy of North America: thematic portrayals of the continent. Tectonophysics v. 187, 117–134.

Willis, B. (1944) Continental drift—ein Marchen. Amer. J. Sci., v. 242, 509–513.

Wilson, J.T. (1965) A new class of faults and their bearing on continental drift. Nature, v. 207, 343–347.

Wilson, T.J., Grunow, A.M., and Hanson, R.E. (1997) Gondwana assembly: the view from southern Africa and East Gondwana. J. Geodynamics, v. 23, 263–286.

Wilson, T.J., Hanson, R.E., and Wardlaw, M.S. (1993) Late Proterozoic evolution of the Zambezi Belt, Zambia; implications for regional Pan-African tectonics and shear displacements in Gondwana, In Assembly, evolution and dispersal; Proceedings of the Gondwana Eight Symposium (R.H. Findlay, R. Unrug, M.R. Banks, and J.J. Veevers, Eds.). Internat. Gondwana Symposium 8, 69–82. Rotterdam, A.A. Balkema.

Windley, B.F., and Razakamanana, T. (1996) The Madagascar–India connection in a Gondwana framework. Osaka, Gondwana Res. Group Mem. 3, Field Sci. Pub., 25–37.

Wingate, M.T.D., Pisarevsky, S.A., and Evans, D.A.D. (2002) Rodinia connections between Australia and Laurentia: no SWEAT, no AUSWUS? Terra Nova, v. 14, 121–128.

Winther, T.K. (1996) An experimentally based model for the origin of tonalitic and trondhjemitic melts. Chem. Geol., v. 127, 43–59.

Woerner, G. (1999) Lithospheric dynamics and mantle sources of alkaline magmatism of the Cenozoic West Antarctic rift system. In Lithosphere dynamics and environmental change of the Cenozoic West Antarctic rift system (F.M. van der Wateren and S.A.P.L. Cloetingh, Eds.). Global and Planetary Change, v. 23, 61–77.

Wyman, D.A., and Kerrich, R. (2002) Formation of Archean continental lithospheric roots: the role of mantle plumes. Geology, v. 30, 543–546.

Wyman, D.A., Kerrich, R., and Polat, A. (2002) Assembly of Archean cratonic mantle lithosphere and crust: plume–arc interaction in the Abitibi–Wawa subduction–accretion complex. Precamb. Res., v. 115, 11–36.

Yang, Z., and Besse, J. (2001) New Mesozoic apparent polar wander path for South China: tectonic consequences. J. Geophys. Res., v. 106, 8493–8520.

Yibas, B., Reimold, W.U., and Anhaeusser, C.R. (2001) The geodynamic evolution of the Neoproterozoic of southern Ethiopia. Gondwana Res., v. 4, 835–836.

Yin, A., Rumelhart, P.E., Butler, R., Cowgill, E., Harrison, T.M., Foster, D.A., Ingersoll, R.V., and Zhang, Q. (2002) Tectonic history of the Altyn Tagh fault system in northern Tibet inferred from Cenozoic sedimentation. Geol. Soc. Amer. Bull., v. 114, 1257–1295.

Yoshida, M. (1995) Assembly of East Gondwanaland during the Mesoproterozoic and its rejuvenation during the Pan-African period, In India and Antarctica during the Precambrian (M. Yoshida and M. Santosh, Eds.). Geol. Soc. India Memoir 34, 25–45.

Young, G.M. (2002) Stratigraphic and tectonic settings of Proterozoic glacigenic rocks and banded iron-formations: relevance to the snowball earth debate. J. African Earth Sci., v. 35, 451–466.

Young, G.M., von Brunn, V., Gold, D.J.C., and Minter, W.E.L. (1998) Earth's oldest reported glaciation: physical and chemical evidence from the Mozaan Group (~2.9 Ga) of South Africa. J. Geol., v. 106, 523–538.

Yumul, G.P., Jr., Imalanta, C.B., Faustino, D.V., and de Jesus, J.V. (1998) Translation and docking of an arc terrane: geological and geochemical evidence from the southern Zambales ophiolite complex, Philippines. Tectonophysics, v. 293, 255–272.

Zartman, R.E., and Doe, B.R. (1981) Plumbotectonics—the model. Tectonophysics, v. 75, 135–162.

Zeitler, P.K., and 10 others (2001) Erosion, Himalayan geodynamics, and the geomorphology of metamorphism. GSA Today, v. 11, no. 11, 4–9.

Zhao, G., Cawood, P.A., Wilde, S.A., and Sun, M. (2002a) Review of global 2.1–1.8 Ga collisional orogens and accreted cratons: a pre-Rodinia supercontinent? Earth-Sci. Rev., v. 59, 125–162.

Zhao, G., Cawood, P.A., Wilde, S.A., Sun, M., and Lu, L. (2000) Metamorphism of basement rocks in the central zone of the North China craton: implications for Paleoproterozoic tectonic evolution. Precamb. Res., v. 103, 55–88.

Zhao, G., Sun, M., and Wilde, S.A. (2002b) Did South America and West Africa marry and divorce or was it a long-lasting relationship? Gondwana Res., v. 5, 591–596.

Zhuravlev, A.Y., and Riding, R. (Eds.) (2001) The ecology of the Cambrian radiation. New York, Columbia Univ. Press.

# Author Index

Abbott, D.H. 33
Abdelsalam, M.G. 214, 237
Acharyya, S.K. 232
Adiyaman, O. 149
Ahall, K.-I. 112, 230
Aichroth, B. 80, 81
Alavi, M. 74, 75
Aleinikov, A.L. 96, 150
Algeo, T.J. 130
Alkmin, F.F. 121
Allen, M.B. 151
Almeida, F.F.M de 224
Alonso-Zarza, A.M. 173
Al-Saleh, A.M. 214–215
Alvarez, L.W. 186
Alvarez, W. 186
Ampferer, O. 19
Amri, I. 37
Anderson, A.L. 97, 112
Anderson, D.L. 158
Anderson, J.B. 136, 157
Anderson, J.L. 41, 112
Andersson, J. 223
Anhaeusser, C.R. 211
Appel, P. 119, 236
Arima, M. 232
Ashwal, L.D. 242
Aspler, L.B. 91, 112
Astini, R.A.103, 122
Attoh, K. 228, 237
Atwater, T. 151
Austrheim, H. 58
Axen, G. Jr. 150

Bada, J.L. 178
Ballance, P.F. 138, 140
Barley, M.E. 44, 93
Barron, C.N. 212, 213
Barth, A.P. 229
Bartlett, J.M. 240
Beck, R.A. 174
Behrendt, J.C. 157
Bekker, A. 169
Bell, K. 62
Benioff, H. 8, 9, 10, 19

Berhorst, V. 238
Berry, E.W. 6
Berry, R. 47, 102
Besse, J. 126, 133, 149
Bettencourt, J.S. 112, 230, 231
Bhattacharya, S. 232
Bhushan, S.K. 97
Bickford, M.E. 91, 97, 112
Bingen, B. 223
Bird, J.M. 20
Biswal, T.K. 225
Black, L.P. 225, 226, 237
Blenkinsop, J. 62
Blewett, R.S. 93
Bogdanova, S. 229
Boger, S.D. 120, 226, 241
Boher, M. 88
Bohlen, S.R. 33
Bolt, B.A. 192
Bond, G.C. 103
Borg, G. 119
Bosworth, W. 141
Botelho, N.F. 234
Botta, G. 178
Boullier, A.-M. 237
Bowring, S.A. 47, 64, 92, 182, 228
Boyle, A.P. 214, 215
Brandon, A.D. 74
Brandon, M.T. 74
Brasier, M.D. 178, 179, 180
Braun, J. 124
Brewer, T.S. 228
Brito Neves, B.B. de 96, 97, 103, 106, 121, 213, 230
Brocks, J.J. 180
Brookfield, M.E. 102
Brown, D. 66, 67
Brueckner, H.K. 237
Buchan, K.L. 98
Bullen, M.E. 87
Burchfiel, B.C. 77, 124
Burollet, P.F. 146
Burrett, C. 47, 102
Burton, K.W. 63
Butterfield, N.J. 181

Byerly, G.R. 180

Cailteux, J. 239
Calver, C.R. 179
Card, K.D. 53
Carignan, J. 62
Carlson, R.W. 64
Carr, S.D. 221, 222
Casey, M. 144, 154
Castle, J.W. 126
CD-ROM Working Group 130
Chacko, T. 37, 91
Chadwick, B. 52
Chamberlain, R.T. 6
Chappell, B.W. 129
Chaudhary, A.K. 232
Chaudhuri, A.K. 96, 97, 105
Chen, J. 226
Chiarenzelli, J.R. 91, 92, 112
Claesson, S. 229
Clark, D.J. 226
Clark, P.O. 159, 161
Cloos, M. 22
Close, D. 124
Cocks, L.R.M. 129, 170, 171
Coleman, D.S. 229
Coleman, R.G. 16
Collerson, K.D. 44
Collins, A.S. 119, 242
Collins, W.J. 41, 232
Collinson, J.W. 123
Comas, M.C. 141
Condie, K.C. 13, 33, 47, 48, 60, 72, 90, 94, 98, 99, 101, 105, 229, 230
Conway Morris, S. 182, 183
Cook, F.A. 228
Corfu, F. 54
Correa-Gomes, L.C. 98
Cosca, M.A. 223, 232
Cottrell, R.D. 28
Courtillot, V. 98, 133
Cox, R. 44
Crawford, A. 123
Crock, J.G. 87
Cronin, T.M. 159, 161, 164
Crowley, T.J. 171
Cullers, R.L. 112
Cunningham, W.D. 151

da Silva Filho, A.F. 234
Dabbagh, M.E. 214
Dallmeyer, R.D. 126
Dalziel, I.W.D. 101, 103, 124, 157
Das, S. 97
Dasgupta, P. 232
Davidson, A. 221, 222
Davidson, J.P. 72
Davies, H.L 35
Davis, G.A. 125
Davis, W.J. 33, 55

Dawson, G.C. 232
Dawson, J.B. 218, 219
De, S.K. 91
de Kemp, E.A. 54
De Lima, E.S. 234
de Wit, M.J. 241
Deblond, A. 226
Debon, F. 77
Decker, K. 149
Dehler, C.M. 169
Delvaux, D. 151
Des Marais, D. J. 178, 180
Dewey, J.F. 20
Dickin, A.P. 210, 221
Dietvorst, P. 57
Dietz, R.S. 10
Dinter, D.A. 141
D'Lemos, R.S. 68
Dobmeier, C. 232
Doe, B.R. 42
Donaldson, J.A. 54
Donnelly, T.W. 37, 141
Dostal, J. 237
Downes, H. 35
Drury, S.A. 241
DuBray, E.A. 218
Dueker, K. 130
Duerr, S.B. 239

Ebinger, C.J. 144, 154
Egan, S.S. 150
Einsele, G. 42
Emery, K.O. 135
Emslie, R.F. 97, 112
Encarnacion, J.P. 122
Engel, M.H. 177
Ernesto, M. 144
Ernst, R.E. 98
Ernst, W.G. 20
Erwin, D.H. 184, 185
Evans, D.A.D. 169, 170
Eyles, N. 164, 165, 169, 170

Faill, R.T. 223
Fan, J. 221, 222
Fanning, C.M. 236
Faure, G. 210
Fedo, C.M. 42, 43
Fedonkin, M.A. 180
Fernandez-Alonso, M. 111
Ferreira, V.P. 234
Ferry, J.M. 197
Fershtater, G.B. 66
Feybesse, J.L. 230, 234
Fielding, E.J. 42
Fitton, J.G. 144
Fitzsimons, I.C.W. 110, 120
Flagler, P.A. 39
Floden, T. 112
Foster, D.A. 124

Fountain, D.M. 32
Fowler, A.D. 55
Frank, W. 234
Frantz, J.C. 234
Freeman, K.H. 178
Freitas Suita, M.T. de 237
Frey, M. 147
Friberg, M. 66
Friedman, R. 222
Frimmel, H.E. 234
Frost, C.D. 37
Fullagar, P.D. 229

Gale, A.S. 173
Gamble, J.A. 35
Gandhi, S.S. 228
Gao, S. 226
Garfunkel, Z. 141
Gaudette, H. 212
Gehling, J.G. 181
Geraldes, M.C. 224
Ghosh, S. 232
Ghuma, M.A. 237
Gibbs, A.K. 212, 213
Glazner, A.F. 74
Golynsky, A. 225
Goodge, J.W. 226, 236
Gorbatschev, R. 229
Gower, C.F. 92, 222, 229
Grantz, A. 136
Gray, D.R. 184
Gray, J. 124, 183
Greenberg, J.K. 37, 41, 214
Gregory, J.W. 154
Grey, K. 182
Griffin, W.L 63
Groenewald, P.B. 225
Grotzinger, J.P. 92
Grow, J.A. 19
Gruau, G. 47
Grunow, A.M. 122
Gueguen, E. 141
Guimares, I. deP. 234
Guiraud, R. 141
Guo, A. 221
Gurnis, M. 94
Gutenberg, B. 7

Haapala, I. 112
Hagadorn, J.W. 179
Hall, R. 156
Hallam, A. 184, 185, 186, 187
Hamilton, W. 140. 149
Hammer, W. 19
Han, T.-M. 180
Hanmer, S. 222
Hansen, E.C. 61
Hansen, V.L. 22
Hanson, G.N. 40
Hanson, R.E. 119, 239

Haq, B.U. 145
Hargraves, R.B. 48
Harmon, R.S. 37
Harris, L.B. 225, 226, 232
Harris, N.B.W. 240
Harrison, T.M. 77
Hartmann, L.A. 46, 88
Hartz, E.B. 103
Hashimoto, M. 70
Hasselbo, S.P. 185
Hatcher, R.D., Jr. 126
Hauck, M.L. 76
Hawkesworth, C.J. 35, 98
Hayes, J.M. 178
Heaman, L.M. 33, 98, 112
Hendrix, M.S. 125
Henry, P. 54
Hess, H.H. 3, 10
Heubeck, C. 151
Hill, K.C. 156
Hodges, K.V. 76
Hoffman, P.F. 101, 194, 158, 164
Holmes, A. 6, 9
Holser, W.T. 178
Holtta, P. 35
Holzer, L. 232
Homewood, P. 150
Hooper, P.R. 35
House, M.R. 184
Housh, T. 47
Humphreys, E.D. 229
Hunt, P.A. 97, 112
Hurich, C.A. 87
Hynes, A. 221, 222

Ionov, D. 151
Irwin, W.P. 16
Isozaki, Y. 185

Jackson, S.L. 54
Jacobs, J. 119, 225
Jacobshagen, V. 122
Jahn, B.-M. 226
Januszczak, N. 166, 167
Jaupart, C. 63
Javaux, E.J. 180
Jayananda, M. 93
Joachimski, M.M. 184
Johansson, L. 223
Johnson, M.R.W. 42
Johnson, P.R. 214
Johnson, S.P. 110, 232
Johnston, S.T. 135
Jolivet, M. 151
Joly, J. 6, 9
Jordan, T.H. 81
Juhlin, C. 223
Jung, S. 239

Kalsbeek, F. 87, 228

Kampunzu, A.B. 239
Kano, T. 70
Karlstrom, K.E. 102, 104, 229
Karson, J.A. 16
Katili, J.A. 140, 149
Kay, S.M. 37
Kearey, P. 13
Kelly, N.M. 232
Kempton, P.D. 35, 37
Kennedy, W.Q. 115
Kent, R.W. 144
Keppie, J.D. 223
Kerr, A. 229
Kerrich, R. 54, 57
Ketchum, J.W.F. 221
Kirschvink, J.L. 164
Kirstein, L.A. 144
Kissling, E. 24, 147
Kloppenberg, A. 93
Knauth, P.L. 168
Knoll, A.H. 185
Koeberl, C. 185
Koehler, H. 223
Kogbe, C.A. 146
Kopf, C.F. 228
Kriegsman, L.M. 242
Kristofferson, Y. 135
Krogh, T.E. 33
Kroner, A. 211, 224, 236, 242
Kuester, D. 119
Kuhn, W.R. 161
Kukla, P.A. 239
Kyle, P.R. 99

Laird, M.G. 122
Lamb, S. 72, 73
Lambiase, J.J. 97, 154, 155
Lan, C.Y. 149
Landenberger, B. 41
Landing, E. 182, 183
Landoll, J.D. 218
Larochelle, A. 11
Larson, R.L. 99, 145
Larue, D.K. 37
Laubscher, H. 150
Lawlor, P.J. 223
LeCheminant, A.N. 98, 112
Ledru, P. 89, 90
Lee, C.-T. 64
LeGrand, H.E. 3, 5
Leitch, E.C. 123
Leite, J.A.D. 234
Lenat, J.-F. 144
Lewry, J.F. 91, 92
Li, Z.-X. 102, 125, 126, 148, 226
Lidiak, E.G. 37
Liegeois, J.-P. 119, 237
Lindblom, S. 217
Lindh, A. 230
Lindsay, J.F. 179

Link, P.K. 109
Litherland, M. 224
Liu, J. 151
Lopez, R. 223
Louden, K.E. 221, 222
Lowe, D.R. 168, 211
Lowrie, W. 168, 211
Lu, S. 112, 230
Lucas, S.B. 91, 92
Ludden, J.N. 57, 221, 222
Luepke, J.J. 109
Lyons, T.W. 109

Macambira, M.J.B. 224
Macdonald, R. 112, 154
Macko, S.A. 177
MacLeod, K.G. 185
Maksimov, Ye.M. 151
Mallard, L.D. 69
Manhica, A.D.S.T. 235
Mareschal, J.C. 63
Markwick, A.J.W. 35
Marsh, J.S. 144
Martignole, J. 222
Martin, D.M. 169
Martin, H. 39
Marvin, U.B. 5
Master, S. 56
Matthews, D.H. 11
Mauldin, L.C. 50
Mazumder, R. 93
McDermott, F. 239
McElhinny, M.W. 200, 203
McFadden, P.L. 200, 203
McGhee, G.R., Jr. 184, 187
McGuire, A.V. 214
McKerrow, W.S. 129
McLennan, S.M. 41
McMenamin, D.L.S. 101
McMenamin, M.A.S. 101
McNamara, A.K. 69
Mechie, J. 80
Meen, J.K. 37, 52, 240
Meert, J.G. 101, 104, 110, 111, 116, 129, 169, 170, 171, 202, 226
Meisel, T. 47
Meng, Q.-R. 149
Menon, R.D. 241, 242
Menzies, M. 144, 154
Meredith, D.J. 150
Mereu, R.F. 221, 222
Metcalfe, I. 126, 149, 171
Meyer, M.T. 91
Meyerhoff, A.A. 4
Meyerhoff, H.A. 4
Meyers, J.B. 155
Mezger, K. 33, 232
Milesi, J.P. 230
Milisenda, C.C. 242
Millar, I.L. 157

Miller, B.V. 68
Miller, J.S. 241
Miller, R. McG. 239
Miyashiro, A. 71, 139
Modie, B.N. 168
Moeller, A. 236
Moghazi, A.-K. M. 214
Moldowan, J.M. 178
Moorbath, S. 47
Moores, E.M. 17, 101, 103
Morgan, J. 186
Morgan, W.J. 27
Morley, C.K. 149
Morley, L.W. 11
Morogan, V. 217
Morozov, I.B. 79, 80, 81
Morozova, E.A. 79, 80, 81
Morrison, J. 41
Moser, D.E. 33, 212
Moyen, J.F. 51
Moyes, A.B. 225
Mueller, P.A. 228
Muenker, C. 123
Muhongo, S. 119, 235, 236
Muir, R.J. 137
Mukherjee, A. 97
Mukhopadhyay, D. 93
Munyanwiya, H. 239
Myers, J.S. 112, 225

Nagler, Th.F. 47
Nagy, R.M. 146
Nance, R.D. 68
Naqvi, S.M. 33, 42, 44, 96, 105
Narbonne, G.M. 181
Neal, C.R. 72
Nedelec, A. 242
Nelson, D.R. 93, 232
Nelson, K.D. 42
Neves, S.P. 234
Nicollet, C. 242
Nikishin, A.M. 96
Nisbet, E.G. 56, 179
Nixon, G.T. 35
Noffke, N. 180
Norton, I.O. 133
Nutman, A.P. 40, 44, 228
Nyblade, A.A. 63

Olive, V. 62, 219
Oliveira, L.P. 98
Oliver, G.J.H. 110, 232
O'Reilly, S.Y. 63
Oreskes, N. 5, 13
Orrell, S.E. 91
Ortega-Gutierrez, F. 223

Palfy, J. 185
Pandit, M.K. 97
Paquette, J.L. 241

Park, J.K. 98
Parrish, J.T. 173
Pearson, D.G. 47, 63
Pedrosa-Soares, A.C. 87, 88, 234
Peel, J.S. 182
Pelechaty, S.M. 90
Peresson, H. 149
Persson, P.-O. 230
Peters, K.E. 178
Petford, N. 72
Petit, J.R. 174
Petrov, G.A. 66
Peucat, J.J. 37, 51, 52, 93, 110, 240
Pichamuthu, C.S. 60
Pilger, R.H., Jr. 141
Pimentel, M.M. 237, 238
Pinna, P. 235
Piper, J.D.A. 101, 102, 104, 105, 112, 113, 169
Pisarevsky, S.A. 120
Polat, A. 54
Pollack, H.N. 63
Porada, H. 238
Poudjom Domani, Y.H. 63
Poujol, M. 211
Powell, C.McA. 120
Prave, A.R. 239
Preiss, W.V. 236
Price, R.A. 90, 102
Prodehl, C. 80, 81
Puchkov, V.N. 66, 67
Puchtel, I.S. 33
Puura, V. 112

Qiu, Y.M.M. 226

Ragland, P.C. 35
Raith, J.G. 124
Ramo, O.T. 112
Rankin, D.W. 223
Rapela, C.W. 37
Rast, N. 68
Raymo, M.E. 174
Rayner, R.J. 171
Raza, A. 156
Razakamanana, T. 241, 242
Reichow, M.K. 150, 185
Reinemund, J.A. 140, 149
Ren, J. 151
Restrepo-Pace, P.A. 72
Retallack, G.J. 173
Richter, C.F. 7
Rickers, K. 225
Riding, R. 182
Ritsema, J. 156
Rivers, T. 221, 222, 229
Robertson, A.H.F. 141
Rogers, J.J.W. 13, 35, 37, 41, 50, 69, 88, 89, 90, 92, 93, 97, 102, 105, 108, 109, 110, 117, 179, 182, 214, 217, 237
Rollinson, H.R. 232

Rosen, O.M. 90, 101, 230
Rosing, M.T. 43
Ross, G.M. 109, 228
Rougvie, J.R. 223
Roy, A. 232
Roy, A.B. 97
Rudkevich, M.Ya. 151
Rudnick, R.L. 35, 63
Ruiz, J. 224
Rumvegeri, B.T. 226
Runcorn, S.K. 10
Runnegar, B. 180

Sacchi, R. 224, 236
Sadowski, G.R. 230
Sanchez-Zavala, J.L. 223
Sandberg, C.A. 184
Santos, J.O.S. 224
Santosh, M. 57, 58, 59, 97, 105, 108, 109, 110, 197, 224, 232, 240, 241, 242
Satish-Kumar, M. 58
Sato, K. 46, 47
Schandelmeier, H. 119, 237
Schidlowski, M. 179
Schilling, J.G. 35
Schmid, S.M. 24, 147
Schmitz, M.D. 64
Scholl, D.W. 42
Schopf, J.W. 180
Schrag, D.P. 164
Schumacher, M.E. 150
Schwartz, M.O. 149
Sclater, J.G. 133
Scotese, C.R. 129
Scrimgeour, I. 124
Sears, J.W. 90, 102, 109
Seilacher, A. 181
Selverstone, J. 153
Sengor, A.M.C. 125, 149
Shackleton, R.M. 119
Shaw, C.A. 225, 232
Shaw, D.M. 32, 35
Shaw, R. 124
Sheridan, R.E. 19
Shiraishi, K. 241
Shore, M. 55
Sidder, G.B. 212, 213
Siga, O., Jr. 46, 47
Simmat, R. 232
Simons, F.J. 81
Skehan, J.W. 68
Sleep, N.H. 30, 154, 179
Smith, A.L. 37
Smith, D. 153
Smith, D.R. 223
Smith, I.E.M. 35
Smith, P.L. 185
Smithson, S.B. 80, 81
Solodilov, L.N. 80, 81
Spadea, P. 66, 67

Spakman, W. 156
Spray, J.G. 37
Stampfli, G.M. 149
Stanistreet, I.G. 239
Starmer, I.C. 223
Stein, H.J. 87
Steinmann, G. 15
Stern, R.A. 40
Stern, R.J. 214, 235
Stewart, A.J. 165
Stewart, I.C.F. 214
Stewart, J.H. 224
Stock, J. 151
Storey, B.C. 98, 99, 122
Stott, C.M. 53
Streepey, M.M. 222
Strieder, A.J. 237
Stump, E. 122
Suayah, I.B. 35
Suess, E. 3, 4, 115
Sun, Y. 149
Sykes, L.R. 11

Tait, J. 125
Tarduno, J.A. 28
Tassinari, C.C.G. 224
Taylor, R.N. 37
Taylor, S.R. 35, 41
Tegtmeyer, A. 211
Teklay, M. 235
ten Brink, U.S. 123
Tewari, R.C. 127, 130
Theriault, R.J. 91
Theunissen, K. 111
Thomas, D.J. 187
Thomas, L. 165, 172
Thomas, R.J. 224
Thomas, W.A. 103, 105, 122
Thompson, H.D. 68
Thorkelson, D.J. 135
Timmons, J.M. 109
Torsvik, T.H. 28, 103, 120, 129, 199
Toteu, S.F. 234
Touret, J. 57
Trettin, H.P. 77
Trompette, R. 121
Tsunogae, T. 58, 59, 197
Tucker, R.D. 230
Tull, J.F. 126
Twitchett, R.J. 185

Uchupi, E. 135
Unrug, R. 88, 116
Upton, B.G.C. 112

Vail, J. 214, 215
Valley, J.W. 197
Van der Voo, R. 170, 199, 203
van Heijst, H.-J. 156
Van Kranendonk, M.J. 232

Van Schmus, W. R. 112, 229
Vandenberg, A.H.M. 123, 124
Vanderhaeghe, O. 213
Veevers, J.J. 123, 127, 130
Vernikovskaya, A.E. 124
Vernikovsky, V.A. 124
Vervoort, J.D. 47
Viele, G.W. 77
Vincent, S.J. 151
Vine, F.J. 11
Vinyu, M.L. 239
von Huene, R. 42

Wada, H. 58, 241
Wadati, K. 8, 9, 10
Waggoner, B. 181, 182
Walcott, R.I. 157
Wang, X. 149
Ward, P.L. 151
Wardle, R.J. 228
Wareham, C.D. 225
Warme, J.E. 184
Warren, J. 165, 172
Wasteneys, H.A. 72
Wawrzyniec, T.F. 153
Weber, B. 223
Wegener, A. 3, 7
Weil, A. 104
West, H.B. 35
White, R.W. 226

Whitehouse, M.J. 228, 232
Whittington, H.B. 183
Wiebe, R.A. 97
Wignall, P.B. 184, 185, 186, 187
Wilde, S.A. 43, 121, 225, 230
Wilkinson, B.H. 130
Williams, G.E. 164
Williams, H. 90
Willis, B. 6
Wilson, J.T. 23
Wilson, T.J. 239
Windley, B.F. 30, 119, 241, 242
Wingate, M.T.D. 103
Winther, T.K. 37, 40
Woerner, G. 157
Wyman, D.A. 54

Yang, Z. 126, 149
Yibas, B. 235
Yin, A. 151
Yoshida, M. 119, 236, 241
Yoshikura, S. 57
Young, G.M. 168, 169
Yumul, G.P., Jr. 140

Zartman, R.E. 42
Zeitler, P.K. 42
Zhao, G. 105, 108, 110, 230
Zhuravlev, A.Y. 182

# Subject Index

$\delta^{13}C$ 178–179
$\delta^{18}O$ 168
$\varepsilon_{Hf}$ 207
$\varepsilon_{Nd}$ 206
$\varepsilon_{Nd}$ and $\varepsilon_{Sr}$ relationship 207
$\varepsilon_{Os}$ 208
$\varepsilon_{Sr}$ 205
Abiotic 178
Abitibi belt 53–54
Abu Khrug ring complex 60, 218
Acadian orogeny 125
Acasta Gneiss 47
Accreting margins 14–15
Accretion of earth 31
Accretionary wedge 21, 70
Achankovil Shear Zone 240
Acraman impact 182
Active margins 13
Adamastor Ocean 106
Adelaidian trough 236
Adiabatic path 195
Adirondack highlands 222
Adriatic Sea 140
Aegean Sea 140
Afghanistan 133, 148–149
Afif terrane 215
Africa 154–156
African hotspots 154–156
African rifts 154–156
Akitkan orogen 107, 230
Akiyoshi belt 70–71
Al-Amar suture 215
Albany belt 102, 225–226, 231–232
Albany–Fraser belt 107, 225, 231–232
Albedo 163
Alboran basin 140
Aldan craton 89, 230
Aleutian Island arc 138
Alleghanian orogeny 126
Alno Island 60, 217
Alps 23–25, 147–148
Altay Range 80, 126, 150–151
Altyn Tagh fault 150–151
Amasia 158
Amazonia 106, 117, 230–231, 235
AMCG complexes 97–98, 111–112

Amerasian basin 136
Amphibolite facies 196
Anabar–Angara craton 89
Anatolia 133, 148–149
Ancient Gneiss Complex 212
Andes 73–74, 153–154
Andesite line 138
Angara belt 107, 230
Annamia 129
Anorogenic magmatism 60
Antarctic Peninsula 157
Antarctic (Southern) Ocean 136–137
Antarctica 157
Antler orogeny 125
Apennines 140
Appalachian coal basin 172
Appalachian Mountains 102
Appalachian rifts 106
Appalachian salients 106
Apparent polar wandering 201–203
Apparent polar wandering curves (APWs) 9, 201–203
Apulia 140, 147, 148
Aqaba–Dead Sea transform 148–149
Ar Rayn terrane 215
Arabia 133, 148–149
Aracuai belt 234–235
Arafura Sea 156–157
Araguaia belt 237–238
Aravalli orogen 107
Arc–trench gap 20–21, 70, 136
Archaea 179
Archean and Proterozoic comparison 55–56
Archean climate 168
Archean gray gneiss 39
Arctic Ocean 135
Arctic Ocean ridge 136
Arctica 90–91
Arequipa block 106
Arequipa massif 72
Armorican massif 69
Armorican terrane 69
Arsikere Granite 51, 53, 60, 217
Asia (growth) 125–126
Aspidella 181
Assam syntaxis 149

## Subject Index

Asthenosphere 8
Atlantic Ocean 135
Atlantic-type margins 10
Atlantica 88–90
Atlas Mountains 140
Atmosphere (composition) 160–161
Atmosphere (density) 162
A-type subduction 19
Aulacogen 27
AUSMEX connection 103
Australia 156–157, 231–232
Austroalpine terrane 69
AUSWUS connection 102
Avalonian terranes 68–70, 116
Axial shear zone (Madagascar) 240–242

Backarc basin 21, 139–140
Bahama–Guinea fracture zone 135
Balkanides 140
Baltic shield 229
Baltic/Ukrainian craton 89
Baltica 92
Baluchistan 148–149
Banda arc 151–152
Banded iron formation 45
Barberton Mountain region 43, 211–212
Barents shelf 136
Basalt plateaus 143–144
Basaltic achondrite best initial ratio (BABI) 205
Basin and Range 17, 152–153
Bastar craton (see Bhandara craton)
Belt–Purcell group 109
Ben Ghnema batholith 117, 235–237
Benioff zones 8–9
Bhandara craton 89
Bioturbation 177
Bismarck spreading center 139
Black body radiation 160
Black Sea 150–151
Blue-green algae (see Cyanobacteria)
Blue Ridge Mountains 223
Blueschist facies 22, 29–30, 196
Bohai Gulf 150–151
Bohemian terrane 69
Bolide impacts 176, 188
Borborema area 117, 234–235
Boundary clay (Cretaceous–Tertiary) 187
Brasilia belt 237–238
Brasiliano oceans 106
Brasiliano orogeny 115, 128, 232–234
Brazilian craton 89
British coal basins 172
Brito-Arctic basalt 143–144
B-type subduction 19
Bulk modulus (incompressibility) 190
Bundelkhand craton 89
Bunger Hills 102, 107, 226, 231–232
Burgess Shale 183
Burin Peninsula 183
Burrowing 182

Cadomian terranes 68–70, 116, 126
Calcareous microplankton 185
Calcareous nanoplankton 185
Caledonide belt 81
Cameroon line 155
Canada basin 136
Canning evaporite basin 172
Cantarito rhyolites 36–37
Cape Breton terrane 69
Capricorn orogen 107, 231–232
Carbon isotopic system 178
Carbonatite 62, 217–219
Cardamom massif 36–37, 240
Caribbean Sea 141
Carolina terrane 69
Carpathian Mountains 140
Cascade Mountains 162
Caspian Sea 150–151
Celebes spreading center 139
Cenozoic climate 173
Central African fold belt 234–235
Central Arabia 117, 235
Central Indian Tectonic Zone 107, 231–232
$CH_4$ 160–161
Charlie Gibbs fracture zone 135
Charniodiscus 181
Cheshire evaporite basin 172
Chewore ophiolite 110
Chichibu belt 70–71
Chicxulub impact 185
Chondrites 204
Chondritic uniform reservoir (CHUR) 204
Cimmerian terranes 116, 148–149
Clarno Formation basalt 34–35
Climate controls 159–163
Climate history 168–174
Closepet Granite 51
$CO_2$ 160–161, 217
Coal 160, 165, 172
Coastal Batholith of Peru 72
Coastal plutonic suite of Canada 151–152
Cocos plate 138
Collision 66–68
Colorado plateau 152–153
Columbia assembly 104–110
Columbia dispersal 110–112
Composite Arc Belt 221–222
Concordia 208–210
Confined orogens 87–88
Conformable lead 208–210
Congo basin 120
Congo craton 89–117
Congo/Kasai craton 89–117
Conrad discontinuity 31–32
Continental crust (accretion) 33–42
Continental crust (destruction) 42–46
Continental crust (origin) 31–33
Continental crust (volume) 46–48
Continental drift 3, 5–7
Continental movements 147–149

Continental shelves 7
Continental-margin arcs 72–74
Continental-margin magmatic suites 72–74
Continentality 162
Contractionism 4
Convection 195
Convection cells 159–160
Convection currents 9
Coppermine homocline 228
Coppermine River Basalts 98, 228
Coral Sea 139
Core 7, 192–193
Coriolis effect 160
Cratonic stabilization 55–56
Cratons 31, 50
Cratons (history) 50–56
Cretaceous flooding 145–146
Cretaceous–Tertiary boundary 185–186
Croll–Milankovitch cycles (see Milankovitch cycles)
Crust 7–8
Curie temperatures 201
Cyanobacteria 179

Dahomeyide orogen 117, 235–237
Damaran belt 117, 225, 235, 238–239
Darling belt 102, 236
Deccan basalt 28, 143–144
Delamerian orogeny 123
Delamination 82–83
Delaware evaporite basin 172
Delhi orogen 107
Depths to the Moho and their changes through time 82–83
Deserts 160–162
Destructive (subducting) margins 19–23
Detachment faults 19–20
Devonian extinction 183
Dharwar block 113, 240
Dharwars 52
Diagenesis 44
Diamonds 29–30
Dickinsonia costata 181
Dinarides 140
Disconformable lead 208–210
Discordia 208–210
Dom Feliciano belt 117, 234–235
Dronning Maud Land 224–225
Dropstone 165, 167
Dzungger (Junggar) basin 126, 150–151

East African orogen 235–236
East African rift system 18, 27, 154–155
East Antarctica 231–232
East European (Russian) platform 66–68, 126, 229
East Gondwana 117
Easterlies 160
Eastern Asia 230
Eastern Australia 89

Eastern Dharwar craton 89
Eastern Ghats 102, 107, 225, 230–232, 240
Eburnian orogen 107, 230–231
Eccentricity 161
Eclogite 22
Eclogite facies 196
Ediacaran 181–182
Elastic material 190
Elburz Mountains 148–149
Elk Point evaporite Basin 172
Entendeka volcanic suite 143
Epeiric seas 145
Epidote-amphibolite facies 196
Equatoria (Romanche) zone 135
Equilibration pressure 197–198
Equilibration temperature 196–198
Equilibrium adiabat 195
Equilibrium thermal gradient 195
Esmond–Dedham terrane 69
Ethiopian flood basalts 143
Ethiopian–Yemen basalt plateau 27, 143–144
Eukaryote evolution 180
Eurasia collision zone 147–149
Eurasia extension 150–151
Eurasian basin 136
European coal basins 172
European Geotraverse 79–81
Eustatic changes in sealevel 144–146
Evaporites 160, 165, 172
Evolution of organisms 179–187
Exterior thrust belt 222

Failed arm 27
Falkland/Malvinas–Agulhas fracture zone 135
Farallon plate 138
Felsic granulite 36–37
Felsic rocks 31
Felsic rocks (generation) 39–41
Fenitization 217
Fiji plateau 139
Fixism 4
Fluid inclusions 58–59
Fluids 56–60, 217–220
Flysch 23
Fore-arc basin 20–21
Foredeep trench 20–21
Foreland basin 22–23
Fort Simpson batholithic suite 228
Fossil ornamentation 166
Fossils and climate 165–167
Foxe orogen 107, 227–228
Fracture zones 12, 26
Franklin Mountains 223
Fraser orogen 225–226, 231–232
Frasnian–Famennian extinction 184
Frontenac–Adirondack belt 221–222

Galapagos spreading ridge 138
Gangdese batholith 76–77
Gardar province 111–112

Gariep belt 117, 234–235
Gariep–Dom Feliciano ocean 121
Gawler craton 89
Geochron 209
Geographic poles 199
Geomagnetic poles 199
Glacial periods 169–170
Glacial periods (origin) 163–164
Gnowangerup dikes 111–112
Godavari rift 97–109
Goias massif 238
Gondwana 114, 116
Gondwana glaciers 115, 171
Gondwana Land 3, 115
Gorda plate 138
Granite–rhyolite terrane (North America) 97, 110–111
Granite–rhyolite terranes 97, 110–111
Granulite facies 32–33, 196
Gravitational fractionation 31
Great Bear intrusive suite 109
Great Falls tectonic zone 107, 227–228
Great Rift Valley (see East African rift system)
Greater (High) Himalayas 76–77
Greenhouse climate in Cretaceous 173
Greenhouse gases 160
Greenschist facies 196
Greenstone belts 53–55
Grenville age 101–112, 118, 221–222
Grenville belt of eastern Canada 221–223
Grenville front 221–222
Grenville rocks in North America 223
Grunehogna craton 143
Guapore shield 89
Guiana (Guyana) craton 89, 212–214
Gulf Coast evaporite basin 172
Gulf of California (Sea of Cortez) 152–153
Gulf of Mexico 141–142
Gyres 162–163

Halekote trondhjemite 52
Half graben 17
Havre spreading center 139
Hawaii–Emperor seamount chain 29, 138
Hawaiian hotspot 138
Hearne craton 89
Heat flow 63, 194–195
Heat production 9, 194
Hellenides 140
Hercynian orogeny (see Variscan orogeny)
HF 217–218
Hida belt 70–71
Highland/Southwestern Complex (HSWC) 240, 243
Himalaya 76–77, 148–149
Hindu Kush 148–149
$H_2O$ 217
Hoggar massif 117, 236–237
Hotspot tracks 28–29

Iapetus Ocean 128
Iapetus rift margin 104–106
Iberian peninsula 140
Iberian terrane 69
Iceland 28
Iceland rhyolite (liparite) 36–37
Ichnofossils 177
Iforas belt 236–237
Illinois coal basin 172
Imnaha basalt 34–35
Incipient charnockite 57–58
Incompatible elements 31
India 133, 148–149, 231–232, 235
Indian Ocean 132–133
Indian rift valleys 96–97
Indo-Burman ranges 76–77
Indochina 133
Indonesian subduction zone 134
Indus–Tsangpo suture 76–77
Innuitian orogeny 77–78
Inside (eastern) margin of Laurasia 124
Intercratonic orogens 85–87
Interior magmatic belt 102, 222–223
Internal orogens 90–91
Intra-Alpine terrane 69
Intracontinental rifts 17–19
Intracratonic orogens 85–87
Intracrustal melting 41
Intraoceanic island arcs 20–21, 70–72
Intraoceanic subduction 19
Ionian Sea 140
Iran 133, 148–149
Iridium 186
Ironstone (see Banded iron formation)
I–S line 124
Island arcs 20–22, 70–72
Isobaric cooling (IBC) 197
Isochores 197
Isostasy 4–5
Isothermal decompression (ITD) 197
Isua 43–44
I-type granites 124
Izanagi plate 138
Izu–Bonin arc 139

Japan arc 70–72, 139
Japan (Sea of) 139
Juan de Fuca plate 138
Junggar (Dzunggar) basin 126, 150–151
Jura Mountains 81
Juvenile crust 48

Kaap Valley 211
Kaapvaal craton 43, 89, 113, 210
Kabbaldurga, India 61
Kalahari block 106, 117, 235
Kaoko belt 117, 234–235
Kapuskasing zone 32, 34–35, 53
Karakum 126

Karoo rift basins 155
Kazakhstan block 66–68, 89, 126
Kenoran 55
Kenorland 91–92
Keratophyre 36–37
Kerguelen plateau 137
Kerogen 178
Ketilidian orogen 107, 227, 229
Kibaran orogeny 102, 117, 226
Kimberley craton 89
Kodaikanal massif 240
Kohistan island arc 76–77
Kola–Karelia belt 107, 229
Komatiite 29–30, 54–55
Konigsbergian–Gothian magmatic suite 107
K–T boundary (*see* Cretaceous–Tertiary boundary)
Kufra basin 236–237
Kula plate 138
Kunlun Range 126, 148–149
Kurile arc 139
Kurosegawa terrane 70–71
Kuunga orogeny 116

Labradorian batholithic suite 107, 227, 229
Lachlan orogen 123–124
Lake Baikal 150–151
Lake Nyos 155
Lake Victoria 27
Lanai basalt 34–35
Land bridges 4
Land plants 183
Lapse rate 161
Laramide orogeny 153
Large blocks that collide with each other 66–68
Large igneous provinces 98
Large-ion-lithophile elements (LILE) 31
Late-Devonian extinction 184
Late-Eocene faunal change 186–187
Late Triassic–Early Jurassic extinction 185
Lau spreading center 139
Laurasia 89, 124–130
Laurentia 104
Lesser Antilles 21
Lesser Himalaya 76–77
Lewisian 89
Lhasa block 133
LILE-rich rocks (origin) 41–42
Limpopo belt 107, 231–232
Linear rift systems 17
Lithosphere 8
Llano area 102, 223
Lomonosov Ridge 136
London platform 69
Longonot volcano 18
Low-velocity zone (LVZ) 8, 80–81
Lower crust 31–32, 79–80
Lufilian belt 117, 235, 238–239
Lu–Hf isotopic system 207
Luis Alves Ocean 106

Lurio (Lurio–Namama) belt 102, 107, 224, 231–232, 236
Lutzow–Holm complex 240–241

MacKenzie dikes 98, 111–112
Madagascar 117, 235
Mafic dyke swarms 98
Mafic granulite 34–35
Mafic rocks 32–33
Mafic underplating 82–83
Magnetic anomalies 201
Magnetic dating 201
Magnetic declination 200
Magnetic dipole 199
Magnetic field 199
Magnetic inclination 200
Magnetic polarity 200
Magnetic poles 9, 199
Magnetic quiet zone 99
Magnetic remanence 200
Magnetic reversals 200–201
Magnetic stripes 11–12, 201
Magnetic time scale 200
Magnitogorsk arc 66–68
Mahanadi–Lambert lineament 109
Main Boundary Thrust 76–77
Main Central Thrust 76–77
Main Uralian fault 66–68
Makkovikian orogen 107, 227, 229
Makran accretionary prism 148–149
Malani igneous suite 97
Malopolska terrane 69
Manila trench 139
Mantle 192–193
Mantle discontinuities 7–8
Mantle solidus 29–30
Mantle wedge 20–21
Mariana island arc 139
Mariana spreading center 139
MARID vein 219
Massif Central 69
Matsoku kimberlite 219–220
Maudheim province 102, 117, 225
Mauritanide belt 126
Mawson continent 110
Maximum packing in supercontinents 85
Mazatzal orogen 107, 227, 229
Median Tectonic Line 70–71
Mediterranean Sea 140–141
Melanesian arc 139
Mélange 22–23
Mendocino Fracture Zone 152–123
Mesoproterozoic 56
Mesoproterozoic climate 169
Mesozoic climate 173
Metallic meteorite 131
Metamorphic facies 196
Metasedimentary (paragneiss) belts 53–55
Metasomatism of SCLM 61–64
Mid-Atlantic Ridge 135

Mid-continent rift system 106
Middle Cambrian black shale fauna 183
Mid-ocean ridge basalt, 34–35
Mid-ocean ridges 9
Milankovitch cycles 161
Minnesota River gneiss terrane 53–54
Mino–Tanaba terrane 70–71
Mirnvy region 226
Mohorovicic discontinuity (MOHO) 7, 31–32, 82–83
Molasse 23
Molecular fossils 177
Mongolian orogenic belt 126
Moodies Formation 212
Moraines 165
Moscow evaporite Basin 172
Mountain roots 7
Mozambique belt 224, 235–236
Mozambique ocean 128
Multicellular organisms 181
Murzuk basin 236–237
Musgrave belt 102
Mwembeshi shear zone 239

Nabitah suture 215
Nagercoil massif 240
Nagssugtoqidian orogen 107, 227–228
Nain province 89
Namama belt (see Lurio–Namama belt)
Napier complex 89, 117, 225, 235, 240–241
Nappes 23
Narakay volcanic suite 228
Natal belt 102
Natrocarbonatite (see Sodium carbonate lavas)
Nazca plate 138
Nena 92
Neoarchean supercontinent 112–113
Neoproterozoic 56
Neoproterozoic climate 169–170
Neotethys 137
New Brunswick terrane 69
New England orogen 124
New Guinea 156–157
New Hebrides spreading center 139
New Quebec orogen 107, 227–228
New Zealand 156–157
Newfoundland–Azores–Gibraltar fracture zone 135
Ngorongoro 218
Nile–Uweinat craton (see West Nile craton)
Nimrod orogeny 226
North Algerian basin 140
North America 151–153
North Anatolian fault zone 148–149
North China (Sino-Korean) craton 89, 126
North Sea 150
Nova Scotia terrane 69
Nubian–Arabian craton (shield) 89, 214–216, 235–237

Oaxaquia 102, 223–224
Obduction 16
Obliquity 161–163
Ocean currents 162–163
Oceanic crust 72
Oceanic plateaus 72
Okhotsk Sea 139
Ol Doinyo Lengai 60, 218–219
Onshore wind 172
Ophiolite conundrum 17
Ophiolites 15–17
Orbital effects 161
Ordovician extinction 183
Ordovician glaciation 183
Organic activity (preservation) 176–178
Organic compounds produced abiotically 176–178
Organic compounds produced by organisms 176–178
Orogenic belts (types) 85–87
Ouachita Mountains 77–78
Outer (western) margin of Gondwana 122
Outer (western) margin of Laurasia 123–124
Outwash plains 165
Oxygen isotopes 167–168

Pachemel (Pachelma) belt 107, 229
Pacific Ocean 137–140
Pacific-type margins 10
Paired metamorphic belts 70–71
Palaeopangaea 102, 104
Palau–Yap arc 139
Paleocene–Eocene faunal change 186–187
Paleocene–Eocene Thermal Maximum (PETM) 186–187
Paleoproterozoic 56
Paleoproterozoic climate 168–169
Paleoproterozoic glaciation 168–169
Paleotethys 133
Paleozoic climate 170–171
Paleozoic–Mesozoic boundary 184–185
Paleozoic North America 77–78
Paleozoic orogeny 122
Palghat–Cauvery shear zones 240
Palni massif 240
Pamir syntaxis 149
Pampia block 106
Pan-African orogeny 114–117, 234–242
Panamparema block 106
Pangea 114, 122–130, 184
Pangea accretion 122–126
Pangea dispersal 95, 131, 140–144
Pangea (movements of blocks to form) 127
Pannonian basin 140
Panthalassa 131
Paradox evaporite basin 172
Paragneiss belts 53, 55
Paraguay belt 237–238
Parana basalt 28, 238

Parana basin 143–144, 238
Parece Vela spreading center 139
Parnaiba block 106
Parnaibo evaporite basin 172
Parvancorina 181
Passive-margin sediments 74–75
Passive margins 13, 18
Past climates (methods for inferring) 164–167
PDB 168, 178
Pearya 77–78
Peninsular Range batholith 151–152
Penokean orogen 89, 107, 227–228
Peri-Franciscan Ocean 106
Permanentism 4
Permian–Triassic extinction 184–185
Permo-Carboniferous glaciation 170–171
Petrologic Moho 82–83
Pharusian ocean 106, 117, 235–237
Philippine arc 139
Phoenix plate 138
Pilbara craton 89, 113
Plagiogranite 36–37
Plate margins 13–14
Plates 13–14
Pleistocene glaciation 174
Plume head 28
Plume toe 28
Plumes 27–28, 33, 94–95
Polar fronts 160
Pontides 140–148
Post-orogenic granite 36–37, 55–56
Post-stabilization history 56–62
Precambrian–Cambrian transition 182–183
Precession 161
Precordillera of Argentina 122
Prince Charles Hills 117, 226, 235
Prograde metamorphism 197
Prokaryotes 179–180
Proterozoic 55–56
Provencal basin 140
$p$–$T$–$t$ path 197–198
Pull-apart basins 26
Pyrenees 140–148

Qaidam basin 150–151
Qiangtang block 133
Qinling–Dabei orogen 148–149
QUARTZ seismic profile 79–80

Racklan orogen 108–109, 227
Radiation 161
Radioactivity 7
Rae province 89
Rain forests 162
Rain shadows 161
Ranotsara shear zone 240
Rapakivi granites
Rayner belt 102, 107, 225, 231–232, 240–241
Rb–Sr isochron 204–205

Rb–Sr isotopic system 204–205
Recycling crustal growth rate 48
Red River–Ailongshan fault 148–149
Red Sea 27
Reduced heat flow 194
Reefs 160, 164–165
Reindeer zone 91–92
Releasing bend 26
Re–Os isotopic system 207–208
Restraining bend 26
Retrograde metamorphism 197
Reunion Island 144
Reykjanes spreading center 15
Rhine graben 150
Ridge crests 11
Rift valleys 96–97
Rigidity modulus 190
Rinkian orogen 107, 227–228
Rio de la Plata craton 89, 106, 117, 235
Rio Grande rift 152–153
Rio Negro–Juruena belt 107, 230–231
Riphean basins 96
Rocky Mountains 153
Rodinia assembly 101–104
Rodinia dispersal 104–106
Rokelide–Goianide Ocean 106
Rokelide orogen 126
Rondonian belt 107, 110–111, 230–231
Roraima Group 213
Ross orogen 122, 124
r–r–r triple junctions 27
Russia coal basins 172
Ryoke–Abukuma belt 71
Ryukyu arc 139

Saharan Metacraton 237
Salina (Michigan) evaporite basin 172
San Andreas fault 23, 25
Sanbagawa belt 70–71
Sao Francisco/Congo-Kasai craton 235
Sao Francisco craton 89
Sao Luis craton 106
Sayan Range 126
Scotia arc 137
Seafloor spreading 10–11
Secondary (S) waves 170
Seismic discontinuity 191–192
Seismic Moho 82–83
Seismic refraction 191–193
Semail ophiolite 36–37
Sevier orogenic belt 153
Seychelles–Mascarene ridge 134
Shadow zone 192–193
Sheeted dikes 16
Shields 31
Shikoku spreading center 139
Sial 31
Sialic rocks 32
Siberian basalts 150, 184

Siberian plate 126
Sibumasu terrane 133
Silsilah complex 218
Sima 31
Singhbhum craton 89
Sino-Korean craton (*see* North China craton)
Siwaliks 76–77
Skeletal organisms 182
Slab avalanches 94
Slave craton 89
Small continental blocks that accrete to continental margins 68–70
Small shelly fossils 182
Sm–Nd isotopic system 205–207
Snell's law 191
Snowball Earth 164
Snowbird zone 228
Sodium carbonate lavas 216
Solar irradiance (luminosity) 161
Solomon arc 139
Songimvalo block 211
Songliao microcontinent 126
Sonoma orogeny 125
Sor Rondane Mountains 240–241
Sorong fault 156–157
South America 153–154
South China 133, 148–149
South China Sea 139
South China (Yangtze) craton 89, 226
South Fiji spreading center 139
South Tibetan Detachment 76
Southeast Indian Ocean ridge 134
Southern Africa 231–232
Southern (Antarctic) Ocean 136–137
Southwest China flysch basin 126, 148
Southwest Indian Ocean ridge 134
Southwest Pacific island arcs 138–140
Southwest Scandinavian gneiss complex 110
Species diversity 166
Spreading rates 15
Spriggina 181
$(^{87}Sr/^{86}Sr)_i$ 204–205
Sri Lanka 240
St. Francois Mountains 97
Statherian rifts 96
Steinmann trinity 15
Steynsdorp terrane 211
Stony meteorites 31
Stromatolites 180
S-type granites 124
Subcontinental lithospheric mantle (SCLM) 32, 61–63, 80–81
Subduction beneath continental margins 22–23
Subduction under arriving terranes 77–78
Subduction zone basalt 34–35
Subduction zones 8, 10
Sunsas belt 102, 224
Supercontinent assembly 85–93
Supercontinent cycle 157–158
Supercontinent dispersal 95–99

Supercontinents 3
Superior craton (province) 53–55, 89
Superplumes 98–99
Suwanee terrane 69
Svecofennian orogen 107
Sveconorwegian orogen 102, 223
Sverdrup evaporite basin 172
SWEAT (Southwest North America East Antarctica) connection 101

$T_{CHUR}$ 46, 206
$T_{DM}$ 46, 206
$T_{RD}$ 208
Taconic orogeny 125
Taimyr belt 124
Taltson magmatic suite 227–228
Taltson–Thelon orogen 89, 227–228
Tanakura line 70–71
Tanzania craton 89, 113, 117
Tarim basin 150–151
Tarim block 89, 126
Tasman orogen 123
Tasman Sea 139
Tauride Range 148–149
Tectonic control of glaciation 164
Tectonic crustal growth rate 48
Tectosphere 81
Tethys 76–77, 131
Thelon orogen 227–228
Thermal conductivity 194
Thermal gradients 28–30, 82–83, 194–197
Thermal infrared 160–161
Tibesti massif 117, 236–237
Tibet 148–149
Tienshan Range 126, 150–151
Tillite 165
Tocantins region 117, 237–238
Tonga arc 139
Torngat orogen 107, 227–228
Tornquist Ocean 69
Torres–Walvis Ridge 135
Trace fossils 177
Trade winds 160
Traditional crustal growth rate 48
Trans-African lineament 146
Transamazonian event 213
Transamazonian orogen 107, 230–231
Transantarctic Mountains 102, 226, 236
Transfer fault 17
Transform faults 11–12
Transform margins 23–27
Trans-Hudson orogenic belt 53, 91, 227–238
Transition zone in southern India 51–53, 60–61
Trans-North China orogen 107, 230
Trans-Saharan orogen 236–237
Triassic–Jurassic extinction 185
Tribrachidium 181
Triple junction 27
Tristan da Cunha 155

## Subject Index

Tropic of Cancer 163
Tropic of Capricorn 163
True polar wandering 9
TTG (tonalite–trondhjemite–granite) 36–37, 39–40
Turbulent mantle 29
Tyrrhenian Sea 140

Uatuma Group 213
Uinta rift 109
Ukraine coal basins 172
Ultramafic rocks 31
Umudhua terrane 211
Ungava/Cape Smith orogen 107, 227–228
Unkar Group 109
Upper crust 31–32
Ur 93
Ural Range 66, 68, 79–80, 126
Urd suture 215
Ust-Urt terrane 126
U–Th–Pb isotopic system 208–210

Variscan orogeny 81, 125
Vascular plants 183
Vestfold Hills 89
Vijayan Complex 240, 243
Vitiaz arc 139
Volcanism (effect on evolution) 187–188
Volhyn belt 229

Wadati–Benioff zones (see Benioff zones)
Wajid Formation 214–216
Wanni Complex 240, 243
Wathaman (Wathaman–Chipewyan) batholith 89–90
Wernecke Group 228
West Africa 106, 155, 235

West African craton 89, 117, 236–237
West Antarctica 157
West Congo belt 117
West Gondwana 117
West Nile craton 89, 117, 235–237
West Philippine spreading center 139
West Siberian Basin 79–80, 145–146, 150–151
West Texas 102
Westerlies 160
Western Australia 235
Western Dharwar craton 50–53
Western Gneiss region 230
Western interior coal basin 172
Western Interior Seaway 145–148
Windmill Islands 102, 107, 226, 231–232
Woodlark spreading center 139
Wopmay (Bear) province 91, 107, 227–228
Wyoming craton 89

Yangtze craton (see South China craton)
Yarlung–Tsangpo suture (see Indus–Tsangpo suture)
Yarrol orogen 124
Yavapai orogen 107, 227–229
Yellowstone hotspot 89
Yilgarn craton 89
Younger Granite (Egypt) 214–216

Zagros Mountains 74–75, 148–149
Zagros simply folded belt 74–75
Zagros suture 74–75
Zambezi orogenic belt 110, 117, 235, 238–239
Zechstein evaporite basin 172
Zimbabwe craton 89
Zircon dating 208–209
Zircons 43, 46
Zircon–$T_{DM}$ crustal growth rate 48

www.ingramcontent.com/pod-product-compliance
Ingram Content Group UK Ltd.
Pitfield, Milton Keynes, MK11 3LW, UK
UKHW062306230426
12049UKWH00005B/124